Lecture Notes in Physics

For information about Vols. 1–151, please contact your bookseller or Springer-Verlag.

Vol. 152: Physics of Narrow Gap Semiconductors. Proceedings, 1981. Edited by E. Gornik, H. Heinrich and L. Palmetshofer. XIII, 485 pages. 1982.

Vol. 153: Mathematical Problems in Theoretical Physics. Proceedings, 1981. Edited by R. Schrader, R. Seiler, and D.A. Uhlenbrock. XII, 429 pages. 1982.

Vol. 154: Macroscopic Properties of Disordered Media. Proceedings, 1981. Edited by R. Burridge, S. Childress, and G. Papanicolaou. VII, 307 pages. 1982.

Vol. 155: Quantum Optics. Proceedings, 1981. Edited by C.A. Engelbrecht. VIII, 329 pages. 1982.

Vol. 156: Resonances in Heavy Ion Reactions. Proceedings, 1981. Edited by K.A. Eberhard. XII, 448 pages. 1982.

Vol. 157: P. Niyogi, Integral Equation Method in Transonic Flow. XI, 189 pages. 1982.

Vol. 158: Dynamics of Nuclear Fission and Related Collective Phenomena. Proceedings, 1981. Edited by P. David, T. Mayer-Kuckuk, and A. van der Woude. X, 462 pages. 1982.

Vol. 159: E. Seiler, Gauge Theories as a Problem of Constructive Quantum Field Theory and Statistical Mechanics. V, 192 pages. 1982.

Vol. 160: Unified Theories of Elementary Particles. Critical Assessment and Prospects. Proceedings, 1981. Edited by P. Breitenlohner and H.P. Dürr. VI, 217 pages. 1982.

Vol. 161: Interacting Bosons in Nuclei. Proceedings, 1981. Edited by J.S. Dehesa, J.M.G. Gomez, and J. Ros. V, 209 pages. 1982.

Vol. 162: Relativistic Action at a Distance: Classical and Quantum Aspects. Proceedings, 1981. Edited by J. Llosa. X, 263 pages. 1982.

Vol. 163: J.S. Darrozes, C. Francois, Mécanique des Fluides Incompressibles. XIX, 459 pages. 1982.

Vol. 164: Stability of Thermodynamic Systems. Proceedings, 1981. Edited by J. Casas-Vázquez and G. Lebon. VII, 321 pages. 1982.

Vol. 165: N. Mukunda, H. van Dam, L.C. Biedenharn, Relativistic Models of Extended Hadrons Obeying a Mass-Spin Trajectory Constraint. Edited by A. Böhm and J.D. Dollard. VI, 163 pages. 1982.

Vol. 166: Computer Simulation of Solids. Edited by C.R.A. Catlow and W.C. Mackrodt. XII, 320 pages. 1982.

Vol. 167: G. Fieck, Symmetry of Polycentric Systems. VI, 137 pages. 1982.

Vol. 168: Heavy-Ion Collisions. Proceedings, 1982. Edited by G. Madurga and M. Lozano. VI, 429 pages. 1982.

Vol. 169: K. Sundermeyer, Constrained Dynamics. IV, 318 pages. 1982.

Vol. 170: Eighth International Conference on Numerical Methods in Fluid Dynamics. Proceedings, 1982. Edited by E. Krause. X, 569 pages. 1982.

Vol. 171: Time-Dependent Hartree-Fock and Beyond. Proceedings, 1982. Edited by K. Goeke and P.-G. Reinhard. VIII, 426 pages. 1982.

Vol. 172: Ionic Liquids, Molten Salts and Polyelectrolytes. Proceedings, 1982. Edited by K.-H. Bennemann, F. Brouers, and D. Quitmann. VII, 253 pages. 1982.

Vol. 173: Stochastic Processes in Quantum Theory and Statistical Physics. Proceedings, 1981. Edited by S. Albeverio, Ph. Combe, and M. Sirugue-Collin. VIII, 337 pages. 1982.

Vol. 174: A. Kadić, D.G.B. Edelen, A Gauge Theory of Dislocations and Disclinations. VII, 290 pages. 1983.

Vol. 175: Defect Complexes in Semiconductor Structures. Proceedings, 1982. Edited by J. Giber, F. Beleznay, J.C. Szép, and J. László. VI, 308 pages. 1983.

Vol. 176: Gauge Theory and Gravitation. Proceedings, 1982. Edited by K. Kikkawa, N. Nakanishi, and H. Nariai. X, 316 pages. 1983.

Vol. 177: Application of High Magnetic Fields in Semiconductor Physics. Proceedings, 1982. Edited by G. Landwehr. XII, 552 pages. 1983.

Vol. 178: Detectors in Heavy-Ion Reactions. Proceedings, 1982. Edited by W. von Oertzen. VIII, 258 pages. 1983.

Vol. 179: Dynamical Systems and Chaos. Proceedings, 1982. Edited by L. Garrido. XIV, 298 pages. 1983.

Vol. 180: Group Theoretical Methods in Physics. Proceedings, 1982. Edited by M. Serdaroğlu and E. İnönü. XI, 569 pages. 1983.

Vol. 181: Gauge Theories of the Eighties. Proceedings, 1982. Edited by R. Raitio and J. Lindfors. V, 644 pages. 1983.

Vol. 182: Laser Physics. Proceedings, 1983. Edited by J.D. Harvey and D.F. Walls. V, 263 pages. 1983.

Vol. 183: J.D. Gunton, M. Droz, Introduction to the Theory of Metastable and Unstable States. VI, 140 pages. 1983.

Vol. 184: Stochastic Processes – Formalism and Applications. Proceedings, 1982. Edited by G.S. Agarwal and S. Dattagupta. VI, 324 pages. 1983.

Vol. 185: H.N. Shirer, R. Wells, Mathematical Structure of the Singularities at the Transitions between Steady States in Hydrodynamic Systems. XI, 276 pages. 1983.

Vol. 186: Critical Phenomena. Proceedings, 1982. Edited by F.J.W. Hahne. VII, 353 pages. 1983.

Vol. 187: Density Functional Theory. Edited by J. Keller and J.L. Gázquez. V, 301 pages. 1983.

Vol. 188: A.P. Balachandran, G. Marmo, B.-S. Skagerstam, A. Stern, Gauge Symmetries and Fibre Bundles. IV, 140 pages. 1983.

Vol. 189: Nonlinear Phenomena. Proceedings, 1982. Edited by K.B. Wolf. XII, 453 pages. 1983.

Vol. 190: K. Kraus, States, Effects, and Operations. Edited by A. Böhm, J.W. Dollard and W.H. Wootters. IX, 151 pages. 1983.

Vol. 191: Photon Photon Collisions. Proceedings, 1983. Edited by Ch. Berger. V, 417 pages. 1983.

Vol. 192: Heidelberg Colloquium on Spin Glasses. Proceedings, 1983. Edited by J.L. van Hemmen and I. Morgenstern. VII, 356 pages. 1983.

Vol. 193: Cool Stars, Stellar Systems, and the Sun. Proceedings, 1983. Edited by S.L. Balliunas and L. Hartmann. VII, 364 pages. 1984.

Vol. 194: P. Pascual, R. Tarrach, QCD: Renormalization for the Practitioner. V, 277 pages. 1984.

Lecture Notes in Physics

Edited by H. Araki, Kyoto, J. Ehlers, München, K. Hepp, Zürich
R. Kippenhahn, München, H. A. Weidenmüller, Heidelberg
and J. Zittartz, Köln
Managing Editor: W. Beiglböck

231

Hadrons and Heavy Ions

Proceedings of the Summer School
Held at the University of Cape Town
January 16 – 27, 1984

Edited by W. D. Heiss

Springer-Verlag
Berlin Heidelberg GmbH

Editor

W. D. Heiss
University of the Witwatersrand, Department of Physics
1 Jan Smuts Avenue, Johannesburg, 2001 South Africa

ISBN 978-3-540-15653-6   ISBN 978-3-540-39563-8 (eBook)
DOI 10.1007/978-3-540-39563-8

2153/3140-543210

PREFACE

Heavy ion physics has in the recent past grown far beyond the scope of traditional nuclear physics. In fact, the extreme conditions that may prevail during the collision are expected to shed light on fundamental aspects of the constituents of matter. It appears that the two branches of physics - nuclear and elementary particle physics - merge again after having developed separately over several decades.

The Organizing Committee of the Course 'Hadrons and Heavy Ions' has tried to highlight this recent development so as to underline this fascinating 'interdisciplinary' field. These proceedings provide a very readable account of the course presented by six outstanding lecturers who gave coverage to the mean field approach to HI collisions, exotic nuclear shapes, overcritical fields, hadronic degrees of freedom, the QCD approach to nuclear interaction and the quark gluon plasma. While it is natural for a fairly young scientific activity to be controversial in certain aspects, the lecturers performed their task in a lucid fashion, thus stimulating numerous discussions during and after the sessions.

The beautiful surroundings, the campus of the University of Cape Town itself and the facilities, could only help to foster an intimate atmosphere in which lecturers and audience mingled to the benefit of all. The Organizing Committee gratefully acknowledges the support given by the University.

In common with previous Advanced Courses in Theoretical Physics, this Course was generously sponsored by the Council for Scientific and Industrial Research (CSIR) of South Africa. The Organization of Theoretical Physicists is indebted for the financial and organizational support made available by the CSIR.

We owe a last word of thanks to the editors of 'Lecture Notes in Physics' for their cooperation over the appearance of these notes.

Johannesburg, South Africa   W.D. Heiss
April 1985

LECTURERS

W. Greiner, University of Frankfurt

J.H. Hamilton, Vanderbilt University

S.E. Koonin, Caltech

J. Rafelski, University of Cape Town

W. Weise, University of Regensburg

L. Wilets, University of Washington

ORGANIZING COMMITTEE

G. Delic, University of the Witwatersrand, Johannesburg.

C.A. Engelbrecht, University of Stellenbosch.

H. Fiedeldey, University of South Africa, Pretoria.

F.J.W. Hahne, University of Stellenbosch.

W.D. Heiss (Chairman), University of the Witwatersrand, Johannesburg.

Ms A. Schnetler (Secretary), CSIR, Pretoria.

S.M. Perez, University of Cape Town.

O.A. van der Westhuysen, CSIR, Pretoria.

TABLE OF CONTENTS

S.E. Koonin, H.B. Geyer

Mean-Field Approximations in Heavy-Ion Collisions 1
    Introduction 3
    TDHF - formal development 6
    Application of TDHF to HI collisions 14
    Relation and extension of TDHF to other formalisms 35
    Statistical dissipation models 51

J.H. Hamilton

New Vistas of the Shapes and Structures of Nuclei
Far off Stability 67

W. Greiner, J. Reinhardt, U. Heinz, B. Müller, U. Müller
Th. de Reus, P. Schlüter, M. Seiwert, G. Suff

Quantum Electrodynamics of Strong Fields 95
    Supercritical fields: general overview 97
    On the vacuum in field theories 116
    The vacuum of quantum electrodynamics 118
    The charged vacuum in supercritical fields
    (single particle aspects) 121
    Second quantization of the Dirac field: the vacuum state 131
    The decay of the vacuum state 140
    Vacuum charge and vacuum energy 148
    Solution of the point charge problem 157
    Superheavy quasimolecules 164
    Positrons from heavy ion collisions 169
    Could the line structure be of trivial origin? 188
    Conversion processes in single atoms and in
    super-critical compound systems 194
    On the existence of giant nuclei
    and giant nuclear molecules 203
    Extensions of the semi-classical approach. The interplay
    between reaction-dynamics and positron spectroscopy 214
    Summary and outlook 220
    References 224

W. Greiner, P.O. Hess

On the Structure of Giant Nuclear Molecules 227
    Introduction 229
    Definition of variables 230
    Coordinate symmetries 231
    The model Hamilton function and its quantization 233
    Solution of the Schrödinger equation 234
    Application to the molecular system ^{238}U-^{238}U 236
    Conclusion 237

W. Greiner, D. Vasak, B. Müller

Pion Bremsstrahlung in Subthreshold Heavy Ion Collisions 241

W. Weise

Pions and Other Hadronic Degrees of Freedom in Nuclei 251
    Introduction 253
    The nucleon-nucleon interaction 254
    Pion-nucleon coupling in relativistic quark models 259
    Virtual pions in nuclei 269
    Pion-nucleon scattering 276
    Pions in nuclear matter 279
    Nuclear spin-isospin response 288
    Pion-nucleus scattering and related processes 293
    Gamow-Teller and magnetic isovector transistions 300
    Hyperons in nuclei 310

L. Wilets

Quark Models of Hadronic Interactions 317
    Modelling quantum chromodynamics 319
    The mean field approximation 324
    Quantum alternatives to the mean field 329
    Small amplitude oscillations 333
    Large amplitude motion and collisions 336
    Recoil and projection 344
    Gluons and color 350
    Transition from nuclear matter to the quark plasma 354
    Summary and prospects 357
    Acknowledgments, References 358

J. Rafelski, M. Danos

Nuclear Matter Under Extreme Conditions 361

    Introduction 363

    The world of quarks and gluons 368

    From quark bag to quark-gluon plasma 378

    Strangeness in the quark-gluon plasma 398

    Thermodynamics of the interacting hadronic gas 420

    Formation and cooling of a baryon rich quark-gluon plasma

    in nuclear collisions 428

    Summary 447

    Acknowledgements 449

    References 450

List of Participants 457

MEAN-FIELD APPROXIMATIONS IN

HEAVY-ION COLLISIONS

Steven E. Koonin
W.K. Kellogg Radiation Laboratory
California Institute of Technology
Pasadena, California 91125, U.S.A.

Manuscript compiled from lecture notes

by

H.B. Geyer
Institute for Theoretical Nuclear Physics
University of Stellenbosch
Stellenbosch, South Africa

Introduction

Confronted by the task of describing low-energy heavy-ion collisions, one might *a priori* take the point of view that since nuclei are complicated strongly interacting many-fermion systems for which no 'first principles description' (from quantum chromodynamics (say)) exists, nuclear structure already constitutes a very difficult enterprise. In addition to this the collision of two such complicated objects is a very non-equilibrium process, seemingly rendering any description well-nigh impossible.

While this may at first sound very disconserting some hope still remains. Nuclear spectroscopy studies of the past reveal that a hierarchy of degrees of freedom govern the behaviour of nuclei in equilibrium and near-equilibrium situations and good models exist for the description of nuclei and nuclear properties under such circumstances.

These properties might broadly be classified as being either indicative of collective, single-particle or statistical behaviour. The first group recognizes the cooperative movement of nucleons for which one has the phenomenological liquid drop picture or the more microscopic random phase approximation (RPA). Single particle properties, due to valence nucleons moving 'around' an inert core, are well accounted for by the shell model or more refined Nilsson model and these pictures are very well supported by microscopic Hartree-Fock (HF) and Brückner-Hartree-Fock theories. Finally the properties of higher lying nuclear states can be accounted for in terms of statistical Hauser-Feshbach theory or random matrix models.

A natural question now arises about the theoretical description of the collision between nuclei, namely whether the above pictures, valid for near-equilibrium nuclear structure aspects, can be extended to bear on non-equilibrium situations such as heavy-ion (HI) collisions.

The advantages of such a line of attack consist of:

(1) A firm grounding in known phenomenology, i.e. the description of HI collisions

is not treated in isolation form the rest of nuclear physics;

(2) one can work with established many-body techniques and technology;

(3) the proposed extension probes the limits of the various equilibrium pictures;

(4) although indirectly, the success of models for HI collisions might eventually be traced back to the nucleon-nucleon (NN) interaction.

Returning to nuclear structure one finds that the central paradigm is the existence of independent nucleons. The motivation for attempting a dynamical extension of this independent particle picture can then be found in the relatively small excitation energy and long mean-free path for nucleons near the Fermi surface.

This situation is shown schematically in fig. 1a where a single excited nucleon has an excitation energy E^*/A above a sea filled to the Fermi energy ε_F. Due to the residual NN-interaction the 1-particle configuration decays to the 2-particle−1-hole configuration in fig. 1b.

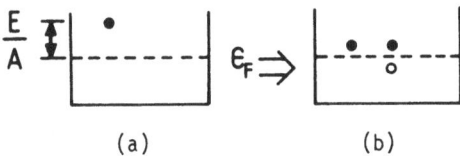

(a)   (b)

Fig. 1.   Schematic illustration of the "two-body" damping
           of a single-particle configuration into a two-
           particle-one-hole state.

Assuming an isotropic differential NN cross-section ($\frac{d\sigma}{d\Omega} = \frac{\sigma_0}{4\pi} \cong 20$ mb/sr, typically for low energies) the decay rate may simply be estimated from a calculation of classical collisions, taking into account Pauli blocking effects. For high energies ($E^*/A \gg \varepsilon_F$) the mean free path of a nucleon is $\lambda = 1/\rho\sigma^0 = 0.23$ fm at $\rho = 0.17$ fm^{-3}, corresponding to $k_F = 1.4$ fm^{-1}, $\varepsilon_F = 41$ MeV). At low energies ($E^*/A \ll \varepsilon_F$) Pauli blocking largely limits the number of possible states, thereby increasing the mean-free path to $\lambda = \frac{4}{3}(\frac{\varepsilon_F}{E^*/A})^2/\rho\sigma_0 \cong 530$ fm/$(E^*/A)^2$ with a corresponding lifetime $\tau = \lambda/v_F = 5.9 \times 10^{-21}$ sec/$(E^*/A)^2$, E^*/A in MeV. Despite contributions due to e.g. thermal excitations of the Fermi sea, a typical 'low energy' heavy ion has $E^*/A \sim 1$ MeV implying a lifetime of some 10^{-21} sec which is comparable to the collision time.

This estimate gives one a flavour of the background against which it seems natural to extend the description to time-dependent situations where one has a physical picture of nucleons moving freely in a common time-dependent potential generated by themselves.

Next, time-dependent Hartree-Fock (TDHF) theory is presented as a mathematical formalism presenting just such a picture. The degree of success with which it can describe experimental data is then investigated while possible improvements are finally considered.

Complementary to this microscopic approach, the second major part of the lectures deals with statistical theories where relevant macroscopic degrees of freedom and equations governing them are considered.

A recent review that covers in more detail some topics in common with the present lecture notes is that by Negele |31| which also contains a very useful bibliography.

1. TDHF - formal development

The system to be described consists of A nucleons interacting through a non-singular, galilean invariant two-body interaction V. In the usual second-quantized notation the many-body hamiltonian is

$$H = \sum_{\alpha\beta} t_{\alpha\beta} \, a_\alpha^\dagger \, a_\beta + \sum_{\alpha\beta\gamma\delta} V_{\alpha\beta\gamma\delta} \, a_\alpha^\dagger \, a_\beta^\dagger \, a_\delta \, a_\gamma \quad . \tag{1.1}$$

The greek labels refer to a complete set of single-particle basis states, $\{\phi_\alpha\}$, where α describes spatial, spin and isospin coordinates of a nucleon. The creation and annihilation operators ($a^\dagger$ and a) satisfy fermion anticommutation rules

$$\{a_\alpha^\dagger, \, a_\beta^\dagger\} = \{a_\alpha, \, a_\beta\} = 0$$

$$\{a_\alpha^\dagger, \, a_\beta\} = \delta_{\alpha\beta}, \tag{1.2}$$

while the one- and two-body matrix elements t and V, both Hermitian, are given by

$$t_{\alpha\beta} = \int \phi_\alpha^*(\tau) \, t(\tau) \, \phi_\beta(\tau) \, d\tau = t_{\beta\alpha}^* \tag{1.3}$$

and

$$V_{\alpha\beta\gamma\delta} = \int \phi_\alpha^*(\tau) \, \phi_\beta^*(\tau') \, V(\tau, \, \tau') \, \phi_\gamma(\tau) \, \phi_\delta(\tau') \, d\tau d\tau'$$

$$= V_{\beta\alpha\delta\gamma} = V_{\gamma\delta\alpha\beta}^* \quad . \tag{1.4}$$

Integration over the symbolic variable τ includes sums over the spin and isospin parts of the indices.

1.1 Derivation of TDHF by truncation

The exact many-body wave function evolving through the Schrödinger equation

$$i\hbar \, \frac{\partial \Psi}{\partial t} = H\Psi \tag{1.5}$$

is of course intractable, but also contains far more information than experimentally accessible. Especially in HI collisions, gross characteristics are being measured (total KE loss, fragment mass, scattering angle, etc.) in the final state. It seems therefore appropriate to deal with a reduced quantal description provided in terms of the time-dependent one-body density matrix

$$\rho_{\alpha\beta}(t) \equiv <\Psi(t)|a_\beta^\dagger a_\alpha|\Psi(t)> = \rho_{\beta\alpha}^*(t) \qquad (1.6)$$

which is hermitian, as indicated, and deals with expectation values of one-body operators.

Indeed, for any one-body operator $\Theta = \sum \Theta_{\alpha\beta} a_\alpha^\dagger a_\beta$ its expectation value is linked to the one-body density matrix through

$$<\Psi(t)|\Theta|\Psi(t)> = \sum_{\alpha\beta} \Theta_{\alpha\beta} \rho_{\beta\alpha} = \text{tr } \Theta\rho \quad . \qquad (1.7)$$

Information about one-body quantities is therefore contained in ρ, including the number of particles, A, for which one has

$$N = \sum a_\alpha^\dagger a_\alpha \Rightarrow N_{\alpha\beta} = \delta_{\alpha\beta}$$

and therefore

$$\text{tr } \rho = A. \qquad (1.8)$$

The time-evolution of ρ can be obtained directly from eqs. (1.5) and (1.2), namely

$$i\hbar \frac{d}{dt} \rho_{\alpha\beta} = <\Psi| [a_\beta^\dagger a_\alpha, H]|\Psi>$$

$$= t_{\alpha\nu} \rho_{\nu\beta} - \rho_{\alpha\nu} t_{\nu\beta} + \tfrac{1}{2}(\rho_{\sigma\alpha\mu\nu}^{(2)} \tilde{V}_{\mu\nu\sigma\beta} - \rho_{\mu\sigma\beta\nu}^{(2)} \tilde{V}_{\alpha\nu\sigma\mu}). \qquad (1.9)$$

Here we are summing over repeated indices and have introduced the anti-symmetric matrix element

$$\tilde{V}_{\alpha\beta\gamma\delta} = V_{\alpha\beta\gamma\delta} - V_{\alpha\beta\delta\gamma} = -\tilde{V}_{\alpha\beta\delta\gamma} = \tilde{V}_{\beta\alpha\gamma\delta} , \qquad (1.10)$$

as well as a two-body density matrix $\rho^{(2)}$, defined by

$$\rho_{\alpha\beta\gamma\delta}^{(2)} = <\Psi|a_\gamma^\dagger a_\delta^\dagger a_\alpha a_\beta|\Psi> = -\rho_{\beta\alpha\gamma\delta}^{(2)} = -\rho_{\alpha\beta\delta\gamma}^{(2)}$$

$$= \rho_{\delta\gamma\beta\alpha}^{(2)*} \quad . \qquad (1.11)$$

The evolution of ρ is therefore dependent on a second dynamical quantity $\rho^{(2)}$, the hierarchy terminating at $\rho^{(A)}$. The complete system of equations involving the coupling of all A density matrices is therefore equivalent to solving the exact

Schrödinger equation.

Our goal of retaining only gross one-body characterization of the final HI state will therefore imply an approximation in the above hierarchy of equations governing all density matrices. The TDHF approximation terminates the hierarchy at $\rho^{(2)}$, assuming

$$\rho^{(2)}_{\alpha\beta\gamma\delta} \approx \rho_{\alpha\delta} \, \rho_{\beta\gamma} - \rho_{\beta\delta} \, \rho_{\alpha\gamma} \, , \qquad (1.12)$$

thus retaining only those two-body correlations prescribed by the Pauli principle while neglecting all 'true' correlations of dynamical origin.

With this assumption about $\rho^{(2)}$, the equation of motion (1.9) now determines ρ completely. In terms of the HF potential

$$W_{\alpha\beta} \equiv \rho_{\delta\gamma} \, \tilde{V}_{\alpha\gamma\delta\beta} \qquad (1.13)$$

eq. (1.9) now becomes

$$i\hbar \frac{d}{dt} \rho_{\alpha\beta} = \left[t, \rho \right]_{\alpha\beta} + \left[W, \rho \right]_{\alpha\beta} \qquad (1.14)$$
$$\equiv \left[h, \rho \right]_{\alpha\beta}$$

where the matrix multiplication given explicitly in eq. (1.9) is still understood and we have introduced the HF single-particle hamiltonian $h = t + W$. Since $W = W(\rho)$ it is clear that the TDHF equations (1.14) are non-linear in ρ.

A more convenient form of the TDHF equations presents itself when one realizes that the assumption (1.12) implies that ρ is a projector, namely $\rho = \rho^2$. (This follows when $\alpha = \delta$ is considered in expressions (1.11) and (1.12) and summed over. Using eq. (1.8) one then has from (1.11) $\rho^{(2)}_{\alpha\beta\gamma\alpha} = \langle \Psi | a^\gamma N a_\beta | \Psi \rangle = (A - 1)\rho_{\beta\gamma}$, while (1.12) yields $\rho^{(2)}_{\alpha\beta\gamma\delta} = A\rho_{\beta\gamma} - \rho_{\beta\alpha}\rho_{\alpha\gamma}$, the combination then implying $\rho^2 = \rho$.) All eigenvalues of ρ are therefore either 0 or 1. Furthermore exactly A of them are 1 since $tr\rho = A$. This implies a spectral representation for ρ:

$$\rho_{\alpha\beta} = \sum_{j=1}^{A} \psi_j(\alpha) \, \psi_j^*(\beta) \qquad , \qquad (1.15)$$

the A single-particle wavefunctions $\psi_j(\alpha)$ spanning the space of unit eigenvalues:

$$\sum_{\beta} \rho_{\alpha\beta} \, \psi_j(\beta) = \psi_j(\alpha) \qquad . \qquad (1.16)$$

Furthermore, these wavefunctions are orthonormal:

$$\sum_{\beta} \psi_j^*(\beta) \, \psi_k(\beta) = \delta_{jk} \quad . \tag{1.17}$$

From the relation (1.15) it follows that under any unitary transformation of the wavefunctions ψ, ρ is invariant as is any one-body observable. Also implied by the same relation is that this representation of ρ is consistent with the choice of a determinantal many-body wavefunction

$$\Psi = (A!)^{-\frac{1}{2}} \det \{\psi_1 \, \psi_2 \, \dots \, \psi_A\}, \tag{1.18}$$

each of the single-particle wavefunctions evolving according to the Schrödinger-like TDHF equation

$$i\hbar \frac{\partial}{\partial t} \psi_j(\alpha) = \sum_{\beta} h_{\alpha\beta} \, \psi_j(\beta). \tag{1.19}$$

In the TDHF approximation the original A-dimensional linear Schrödinger equation (1.5) has therefore been replaced by A coupled, non-linear 1-dimensional equations (1.19).

The structure contained in the TDHF eqs. (1.19) is exemplified by considering a spin and isospin independent two-body interaction which is local and galilean-invariant:

$$V_{r_1 r_2 r_3 r_4} = \delta(\frac{\vec{r}_1 + \vec{r}_2}{2} - \frac{\vec{r}_3 + \vec{r}_4}{2}) \delta \; (\vec{r}_1 - \vec{r}_2 - (\vec{r}_3 - \vec{r}_4)) \; v(\vec{r}_3 - \vec{r}_4).$$

In definition (1.13) this implies a coordinate space HF potential

$$W(\vec{r}_1, \vec{r}_2) = \delta(\vec{r}_1 - \vec{r}_2) \int d\vec{r}_3 \; v(\vec{r}_2 - \vec{r}_3) \; \rho(\vec{r}_3) - v(\vec{r}_1 - r_2) \; \rho(\vec{r}_1, \vec{r}_2)$$

$$\equiv \delta(\vec{r}_1 - \vec{r}_2) \; W_H(\vec{r}_1) - W_F(\vec{r}_1, \vec{r}_2), \tag{1.20}$$

which contains a local Hartree term as well as a non-local Fock term.

The coordinate space density, $\rho(\vec{r})$, is the diagonal part of the one-body density matrix, namely

$$\rho(\vec{r}) = \rho(\vec{r}, \vec{r}) = \sum_{j=1}^{A} |\psi_j(\vec{r})|^2, \tag{1.21}$$

while the TDHF eqs. (1.19) become

$$i\hbar \frac{\partial}{\partial t} \psi_j(\vec{r}) = -\frac{\hbar^2}{2m} \nabla^2 \psi_j(\vec{r}) + W_H(\vec{r}, t) \, \psi_j(\vec{r}) - \int d\vec{r}' \, W_F \; (\vec{r}, \vec{r}', t) \, \psi_j(\vec{r}') \tag{1.22}$$

in coordinate space.

TDHF is thus seen to embody the original idea whereby each single-particle wave function was envisaged to evolve through a Schrödinger-like equation (albeit non-local) in a common time-dependent mean-field generated by the instantaneous configuration of all the other nucleons.

The development thus far emphasises that all non-trivial many-particle correlations are neglected in TDHF where the wavefunction is approximated by a single time-dependent Slater determinant.

A useful class of solutions to the TDHF eqs. (1.19) is the static solutions which contain a harmonic time-dependence through $\psi_j(t) = e^{-i\varepsilon_j t} \psi_j$, ε_j being real single-particle energies. For these solutions the wavefunctions satisfy non-linear eigenvalue equations $h[\phi] = \varepsilon_j \phi_j$, h depending on the ϕ's through the one-body density matrix.

In conjunction with the appropriate effective interactions (to be described shortly), these static solutions yield an accurate description of both bulk and shell aspects of nuclear ground states.

1.2 Variational derivation of TDHF

A derivation of the TDHF equations which illuminates some complementary aspects is inspired by the variational procedure leading to the static HF equations.

Recall that static HF follows from a variational principle, namely that one demands that $\langle\Psi|H|\Psi\rangle$ be stationary with respect to norm-preserving variations of a determinantal wavefunction, i.e.

$$\frac{\delta}{\delta\psi_j^*(r)} \langle\Psi|H|\Psi\rangle = 0 \;\forall\; j \;, \qquad (1.23)$$

Ψ being represented as $\Psi = (A!)^{-\frac{1}{2}}\det(\psi_1...\psi_A)$. In the static HF equations that emerge, the single-particle energies ε_j play the role of Lagrange multipliers ensuring norm preservation, $|\psi_j|^2 = 1$.

Extending this procedure to the time-dependent regime, one considers an 'action' I, a functional of Ψ^* and Ψ, defined as

$$I[\Psi^*, \Psi] = \int dt \; \langle\Psi| \; i\hbar \frac{\partial}{\partial t} - H|\Psi\rangle$$

$$\equiv \int dt \; L[\Psi^*, \Psi], \tag{1.24}$$

i.e. I is the expectation value of a Schrödinger form subsequently written suggestively as the time integral of some Lagrangian density.

Demanding stationarity of I with respect to norm-preserving variations of a general Ψ^* naturally recovers the full Schrödinger equation (1.5). Restricting the class of Ψ's to those parametrized in determinantal form (1.18), however, leads to the TDHF equations (1.19) in the form

$$i\hbar \frac{\partial \psi_j(\alpha)}{\partial t} = \frac{\delta H}{\delta \psi_j^*(\alpha)} \tag{1.25}$$

with $H = \langle\Psi|H|\Psi\rangle$ the 'hamiltonian' functional

$$H = \sum_j \psi_j^*(\alpha) \; t_{\alpha\beta} \; \psi_j(\beta) + \tfrac{1}{2} \sum_{ij} \psi_i(\alpha) \; \psi_j(\beta) \; \tilde{V}_{\alpha\beta\gamma\delta} \; \psi_i^*(\delta) \; \psi_j^*(\gamma)$$

$$= \langle\Psi|t + \tfrac{1}{2}W|\Psi\rangle = \langle\Psi|h - \tfrac{1}{2}W|\Psi\rangle \quad . \tag{1.26}$$

The 'action' (1.24) therefore leads to the 'hamiltonian' equations (1.25), suggesting that TDHF can be viewed as describing a Slater determinant evolving optimally in the sense of minimizing the expectation value of the many-body Schrödinger operator. Furthermore TDHF is equivalent to a set of interacting classical fields associated with the Lagrangian and Hamiltonian functionals introduced above, implying that TDHF is in some sense a classical or at best semi-classical approximation to the full Schrödinger equation.

It should be pointed out that the above derivation of the TDHF-equations differs from the static case in being based on a stationarity principle, in contrast to the minimization involved in the latter case.

Realising that static HF deals with energy minima, one could contemplate small oscillations about these minima - as sketched below this naturally leads to the random phase approximation (RPA) and an interpretation as the simplest TDHF normal modes about a static HF solution.

Consider a Thouless-type wave function constructed from a static solution $|\phi\rangle$ with time-dependent expansion coefficients $c_{ph} = c_{ph}(t)$, i.e.

$$|\Psi(t)\rangle = \exp\{\sum_{ph} c_{ph}(t)\, a_p^\dagger a_h\}|\phi\rangle, \tag{1.27}$$

yielding, to second order in the c's:

$$|\Psi(t)\rangle \sim [1 + \sum_{ph} c_{ph}\, a_p^\dagger a_h - \tfrac{1}{2}\sum_{ph}|c_{ph}|^2 + \tfrac{1}{2}(\sum_{ph} c_{ph}\, a_p^\dagger a_h)^2]|\phi\rangle. \tag{1.28}$$

Demanding

$$\frac{\delta}{\delta c_{ph}^*}\int dt\, \langle\Psi|i\hbar\frac{\partial}{\partial t} - H|\Psi\rangle = 0 \tag{1.29}$$

to second order in the c's, then yields a set of first order linear equations for the c's, namely

$$i\hbar\dot{c}_{ph} = (\varepsilon_p - \varepsilon_h) + \sum_{p'h'} (V_{ph'hp'}\, c_{p'h'} + V_{pp'hh'}\, c_{p'h'}^*) \tag{1.30}$$

where the ε's are single particle energies of the static HF solution.

Taking a harmonic time dependence for the c's,

$$c_{ph} = X_{ph}\, e^{-i\omega t} + Y_{ph}^*\, e^{i\omega t} , \tag{1.31}$$

leads to the usual RPA eigenvalue equation. Here, however, the RPA emerges as a set of classical equations - one finds a set of frequencies governing the normal modes of oscillations about the minima, but nowhere is one forced to associate the mode frequencies with excitations of the system. This connection is to be provided from outside, via path integrals for instance. The above 'ħ-less derivation' of RPA is another signal that TDHF is a semi-classical approximation.

1.3 Conservation laws

Pursuing the analogy with classical mechanics a bit further, one realises that TDHF solutions admit to a number of conservation laws. Apart from the desirability of such laws for a physically plausible theory, they also provide useful checks in numerical computations.

(i) Norm: From eq. (1.19) it follows that each orbital evolves in time through the same hermitian hamiltonian h, implying that the overlap matrix $\langle\psi_j|\psi_i\rangle$ is time-independent. Hence the norm of the TDHF wave function is time-independent, as is the expectation value of the number operator.

(ii) Energy: The time evolution of the expectation value of any operator Θ in the TDHF wave function Ψ is

$$i\hbar \frac{d}{dt} \langle\Psi|\Theta|\Psi\rangle = \sum_j \left[\frac{\delta\langle\Theta\rangle}{\delta\psi_j(\alpha)} \left(i\hbar \frac{\partial\psi_j(\alpha)}{\partial t}\right) + \frac{\delta\langle\Theta\rangle}{\delta\psi_j^*(\alpha)} \left(i\hbar \frac{\partial\psi_j^*(\alpha)}{\partial t}\right) \right] \qquad (1.32)$$

where $\langle\Theta\rangle = \langle\Psi|\Theta|\Psi\rangle$ is viewed as a functional of $\{\psi_j\}$ and $\{\psi_j^*\}$ and Θ is considered not to have any intrinsic time dependence. From eq. (1.25) it then follows that

$$i\hbar \frac{d\langle\Theta\rangle}{dt} = \sum_j \left[\frac{\delta\langle\Theta\rangle}{\delta\psi_j(\alpha)} \frac{\delta H}{\delta\psi_j^*(\alpha)} - \frac{\delta\langle\Theta\rangle}{\delta\psi_j^*(\alpha)} \frac{\delta H}{\delta\psi_j(\alpha)} \right] \qquad (1.33)$$

which is analogous to the classical Poisson bracket of Θ with H. Energy conservation in TDHF follows immediately by putting Θ = H above.

(iii) Momentum: With $\Theta = \vec{P}$, the total momentum operator, in eq. (1.33) it follows easily that $\frac{d\langle\vec{P}\rangle}{dt} = 0$ if $[H, \vec{P}] = 0$, as expected for a two-body interaction which is galilean invariant. Intimately connected with this property is the existence of translating static HF solutions. Boosting a static HF determinant by an exponential involving the centre of mass in its phase, i.e. constructing

$$\Psi = \exp (i \vec{K}\cdot\vec{R})\Phi = \exp (i \vec{k}\cdot\sum \vec{r}_j)\Phi \qquad (1.34)$$

as an initial TDHF solution, results in a solution representing uniform translation in time with velocity $\vec{v} = \hbar\vec{k}/m = \hbar\vec{K}/mA$, where $\vec{R} = \sum\vec{r}_j/A$ is the centre-of-mass coordinate. In this case the time-dependent single-particle wave functions displaying the above property are

$$\psi_j(\vec{r}, t) = \exp\left[-i(\varepsilon_j + \hbar^2 k^2/2m)t/\hbar\right] \exp (i\vec{k}\cdot\vec{r}) \phi_j(\vec{r} - \vec{v}t). \qquad (1.35)$$

(iv) For angular momentum one has $\frac{d}{dt}\langle\vec{J}\rangle = 0$ if $[H, \vec{J}] = 0$ where $\Theta = \vec{J} \equiv$

$$\sum_j\left[(\vec{r}_j \times \vec{p}_j) + \vec{s}_j\right]$$ has been used in eq. (1.33).

2.   Application of TDHF to HI collisions

2.1   Choice of the effective interaction

From static HF theory one knows that the strong repulsion in the bare NN interaction
leads to divergent matrix elements and therefore precludes any naive TDHF calculations based on the bare interaction.

Also, from nuclear structure applications of static HF theory, one knows that a renormalized interaction is required to cure these divergences. This is accomplished
through the independent pair approximation whereby ladder sums of v, namely the
G-matrix, replaces the bare interaction. The complete solution of the resulting
Brückner-Hartree-Fock (BHF) equations constitutes a tremendous technical task, already for time-independent situations and in applications to static properties of
finite nuclei an approximation that has a proven track record is the density-
dependent HF (DDHF) which uses a slightly adjusted nuclear matter G-matrix in a
local density approximation.

In contrast to BHF calculations which give too large binding energies and too small
r.m.s. radii for nuclear ground states, the DDHF approach has remarkable quantitative
success in the description of these and other nuclear properties.

In applications to heavy-ion collisions the non-locality still present in the DDHF
approximation causes some computational difficulties which can be alleviated by the
so-called density matrix expansion. This expansion capitalises on the fact that in
DDHF the energy is a functional of the one-body density matrix $\rho(r, r')$ which can be
parametrized in analogy with the nuclear matter case. There the known analytic
structure of $\rho(r, r')$ shows that the off-diagonal behaviour of $\rho(r, r')$, in terms
of $(r - r')$ is parametrized solely in terms of the density itself, i.e. in terms of
$k_F \sim \rho^{\frac{1}{3}}$. The philosophy behind the density matrix expansion is therefore to use the
same kind of expression for $\rho(r, r')$ appearing in DDHF, but to parametrize the off-
diagonal behaviour of $\rho(r,r')$ in terms of its near diagonal derivatives, i.e.
$\rho(r,r')$ is parametrized in terms of $\rho = \sum_i |\psi_i|^2$ and $\tau = \sum_i |\nabla\psi_i|^2$.

For the DDHF energy functional one then has

$$E = E[\rho, \tau] \qquad (2.1)$$

and variation with respect to the single-particle wave functions ψ_i then leads to a set of local equations for the ψ_i's.

The final solution to these equations for static systems is then practically equivalent to a full DDHF treatment.

In TDHF applications the actual interactions used are of the Skyrme type which take this philosophy of the density matrix one step further, namely the energy is taken as an analytic functional of ρ and τ, involving six parameters that are adjusted to ground state properties of doubly-magic nuclei such as the binding energy, rms radius and single-particle level spacing.

The specific form of the Skyrme interaction is the sum of a 'two-body' and 'three-body' term,

$$V = V^{(2)} + V^{(3)}, \qquad (2.2)$$

where, in terms of the relative momenta $\vec{k}$ and $\vec{k}'$ of two nucleons

$$<\vec{k}|V^{(2)}|\vec{k}'> = t_0 (1 + x_0 P_\sigma) + \tfrac{1}{2}t_2 (k^2 + k'^2) + t_1 \vec{k}\cdot\vec{k}'$$
$$+ i W_0 (\vec{\sigma}_1 + \vec{\sigma}_2)\cdot(\vec{k} \times \vec{k}'). \qquad (2.3)$$

Here t_0, x_0, t_1, t_2 and W_0 are constant, P_σ is the spin exchange operator while $\vec{\sigma}_1$ and $\vec{\sigma}_2$ are nucleon Pauli spin matrices. The t_0 term acts in relative s-waves, t_2 in both s- and d-waves and t_1 in p-waves. W_0 is the spin-orbit strength. The three-body part simulates the density dependence of more fundamental considerations and is taken as

$$V^{(3)} = t_3 \, \delta(\vec{r}_1 - \vec{r}_2) \, \delta(\vec{r}_2 - \vec{r}_3) \qquad (2.4)$$

i.e. a zero-range interaction with strength t_3.

For a spin-saturated system the hamiltonian functional is readily expressed in terms of the density

$$\rho(\vec{r}) = \sum_j |\psi_j(\vec{r})|^2, \qquad (2.5)$$

the current density

$$\vec{J}(r) = \sum Im(\psi_j^* \nabla\psi_j(\vec{r})), \qquad (2.6)$$

and the kinetic energy density

$$\tau(\vec{r}) = \sum_j |\nabla_j(\vec{r})|^2, \qquad (2.7)$$

as

$$E = \int dr\, H(\vec{r}) \qquad (2.8)$$

where

$$H = \frac{\hbar^2}{2m}\tau + \frac{3}{8}t_0\rho^2 + \frac{1}{16}(3t_1 + 5t_2)(\rho\tau - J^2)$$

$$+ \frac{1}{64}(9t_1 - 5t_2)|\nabla\rho|^2 + \frac{1}{16}t_3\rho^3$$

$$+ \text{"spin-orbit"} + \text{"coulomb" terms.} \qquad (2.9)$$

Here isospin symmetry has also been assumed, the lack of which would have introduced terms depending explicitly on neutron and proton densities. In eq. (2.9) the first term is the kinetic energy, the t_0-term gives two-body s-wave contributions, the $(\rho\tau - J^2)$ term gives rise to non-locality effects while the $|\nabla\rho|^2$ term is large in the nuclear surface and can be identified with the surface energy. The t_3 term expresses the density dependence of the effective nucleon-nucleon interaction.

Using the variation (1.25) the Hartree-Fock Hamiltonian h_q for nucleon species $q(q \equiv p(\text{proton}) \text{ or } n(\text{neutron}))$ is obtained schematically as

$$h_q = -\nabla\cdot(\frac{\hbar^2}{2m_q^*(\vec{r})})\nabla + U_q(\vec{r}) + \frac{1}{2i}(\nabla\cdot\vec{I}_q + \vec{I}_q\cdot\nabla)$$

$$+ \text{'spin-orbit'} + \text{'coulomb'} \qquad (2.10)$$

which contains a local HF part

$$U_q(\vec{r}) = t_0\, ((1 + \tfrac{1}{2}x_0)\rho - (x_0 + \tfrac{1}{2})\rho_q)$$

$$+ \tfrac{1}{8}(t_2 - 3t_1)\nabla^2\rho + \tfrac{1}{16}(3t_1 + t_2)\nabla^2\rho_q$$

$$+ \tfrac{1}{4}(t_1 + t_2)\tau + \tfrac{1}{8}(t_1 - t_2)\tau_q + \tfrac{1}{4}t_3(\rho^2 - \rho_q^2), \qquad (2.11)$$

a kinetic energy part involving a density dependent effective mass through

$$\frac{\hbar^2}{2m_q^*(\vec{r})} = \frac{\hbar^2}{2m} + \frac{1}{4}(t_1 + t_2)\rho + \frac{1}{8}(t_2 - t_1)\rho_q, \qquad (2.12)$$

and a current contribution involving

$$\vec{I}_q = -\frac{1}{2}(t_1 + t_2)\vec{J} - \frac{1}{4}(t_2 - t_1)\vec{J}_q. \qquad (2.13)$$

The fact that h has now effectively been reduced to a local operator offers significant computational advantage.

Five linear combinations of the six parameters t_0, t_1, t_2, t_3, x_0 and W_0 are fixed by the nuclear volume, surface, symmetry, coulomb and spin-orbit energies. The remaining linear combination essentially describes the balance between the effective mass and the density dependent t_3-term which produces saturation in nuclear matter.

Calculations with the Skyrme-type interaction constitute one of the most successful microscopic descriptions of nuclear properties throughout the periodic table. Fig. 2.1 is illustrative of results obtained by the Orsay group [1]. Note that this calculation is based on spherical Slater determinants - use of deformed Slater determinants reduces the discrepancies in binding energies to 1- 2 MeV.

Given the success of such static calculations, one is encouraged to adopt these interactions for dynamical situations where E/A is not too large. The results of such calculations are now discussed.

2.2 Numerical methods

Through TDHF the possible description of a heavy-ion collision has been reduced to the solution of the A coupled non-linear equations (1.19). The initial conditions describe the target and projectile boosted towards each other with the appropriate wave-number, i.e. the initial wave functions are of the type

$$\psi_i(o) = \phi_i e^{\pm i\vec{k}\cdot\vec{r}}, \qquad (2.14)$$

with ϕ_i a static solution.

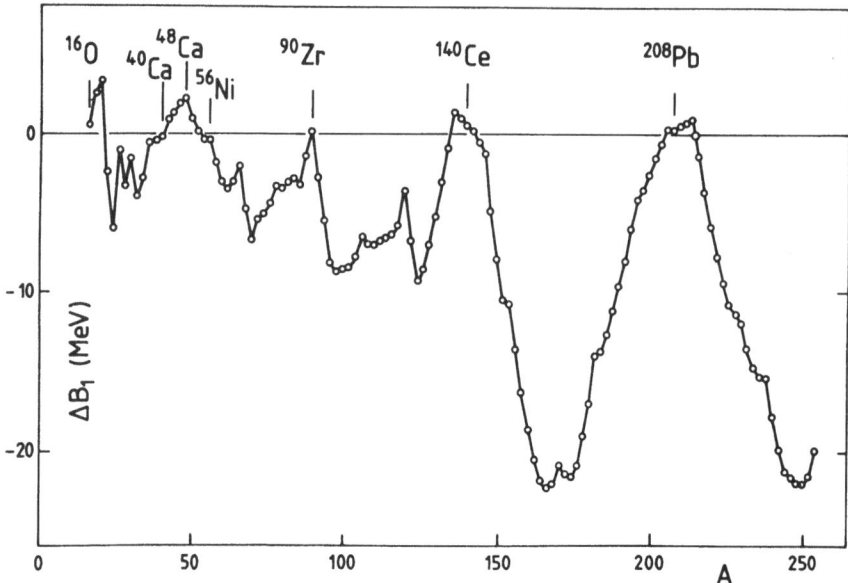

Fig. 2.1 Discrepancies between experimental and calculated nuclear binding energies illustrating the success of Skyrme type interactions throughout the periodic table. (From ref. 1.)

Initial calculations in 1975 |2, 4| were done with only one space dimension taken into consideration, i.e. collisions between slabs of nuclear matter were studied. At present the TDHF equations can reliably be solved in three space dimensions.

It should first be noted that due to reasons concerning the numerical stability of $\nabla^2\rho$, these terms in the Skyrme interaction are replaced by direct Yukawa two-body interaction energies corresponding to a range a = 0,46 fm |3|. Readjusting t_0 and x_0 then leaves all static properties unchanged.

The TDHF equations are solved numerically on a discrete space-time mesh, where a single-particle wave function takes on the value $\psi_{ijk}^{(n)} \equiv \psi(i\Delta x, j\Delta y, k\Delta z, n\Delta t)$ at mesh point $x = i\Delta x$, $y = j\Delta y$, $z = k\Delta z$ and $t = n\Delta t$. On such a mesh the hamiltonian (2.10) is a sparse matrix acting on the spacial indices. As an illustration, for a one-dimensional problem with a local potential U, i.e.

$$H = \frac{-\hbar^2}{2m} \frac{\partial^2}{\partial x^2} + U(x),$$

(2.15)

the simplest discrete approximation is

$$(h\psi)_i = \frac{-\hbar^2}{2m}\left[\frac{\psi_{i+1} - \psi_{i-1} - 2\psi_i}{(\Delta x)^2}\right] + W_i\psi_i \tag{2.16}$$

where $W_i = W(i\Delta x)$ |4|. Here h is a tri-diagonal matrix. Higher order discretiza-
tions are possible and become essential for three-dimensional calculations where a
relatively coarse mesh is used to reduce the number of variables to be integrated
|5, 6|.

Time evolution proceeds from an approximation to the time evolution operator, namely

$$\psi^{(n+1)} = \exp(-ih\Delta t)\psi^{(n)} \cong \left[\sum_{k=0}^{5} \frac{(-ih\Delta t/\hbar)^k}{k!}\right]\psi^{(n)} \tag{2.17}$$

where h is assumed to be time-independent. In practice the time-dependence of h
necessitates the use of a two-point predictor-corrector scheme in conjunction with
eq. (2.17) |4|.

At each time step the Coulomb and Yukawa potentials are obtained from the correspond-
ing Poisson and Helmholtz equations |5, 7|.

Typical three-dimensional calculations involve a mesh size of 30 x 30 x 16 points
(the last dimension being the one normal to the scattering plane) with a mesh spacing
$\Delta x = 0,4 - 1,2$ fm. Time steps $\Delta t = 4-6 \times 10^{-23}$ sec are usual and a heavy-ion collision involves the evolution of about two hundred wave functions for several hundred
time steps.

The boundary condition requires vanishing of the wave functions outside the chosen
'box' - in practice one checks that nothing hits the 'walls' during the collision.

Several useful numerical checks are available to check the validity of solutions
obtained. Firstly one can check time reversibility by evolving the collision back-
wards, although all aspects of the algorithms used are not yet fully reversible.
Secondly the constancy of conserved quantities mentioned in section 1.3 can be
checked while invariance to mesh size serves as a further check.

2.3 Assorted assumptions to reduce computational work

Although full-dimensional TDHF calculations can be performed for lighter systems |5|, it is especially for the heavier systems that various approximations are introduced to reduce computation.

In the axial approximation |8| the system is assumed to remain invariant under rotations about the line joining the mass centres of the colliding ions, facilitating the factorization

$$\psi_k(\vec{r}, t) = \exp(im_k \phi) \, \psi_k(r_\perp, z, t), \qquad (2.18)$$

with $(r$, z and ϕ the usual cylindrical coordinates. Hereby the TDHF problem is re- duced to two spatial dimensions yielding an order of magnitude reduction in computa- tional cost. Ion-ion orbital angular momentum effects are simulated by rotating the symmetry axis about the normal to the scattering plane at angular velocity $\omega = L/I[\rho]$. The moment of inertia actually used depends on whether the ions are separated or have stuck together |8|. It should be noted that in this approximation non-axial shapes are excluded as well as angular momentum transfer between fragments.

Another geometrical assumption which has been used with a great deal of success, is the frozen approximation |9, 10| which capitalizes on the fact that the smooth mean- field shows little tendency to deform the solution in the direction normal to the scattering plane. This property facilitates a factorization of the form

$$\psi_k(\vec{r}, t) = \psi_k(x, y, t) \, \chi_k(z) \qquad (2.19)$$

where z is the cartesian coordinate normal to the scattering plane, while the time-in- dependent functions $\chi_k(z)$ are taken to be oscillator functions matched to the ground states of the colliding ions. While this simplification again introduces an effect- ive two-dimensional spatial problem, it is superior to the axial approximation in allowing angular momentum transfer and triaxial shapes. The remarkable accuracy of the frozen approximation in comparison with a full three-dimensional calculation is illustrated in figure 2.2.

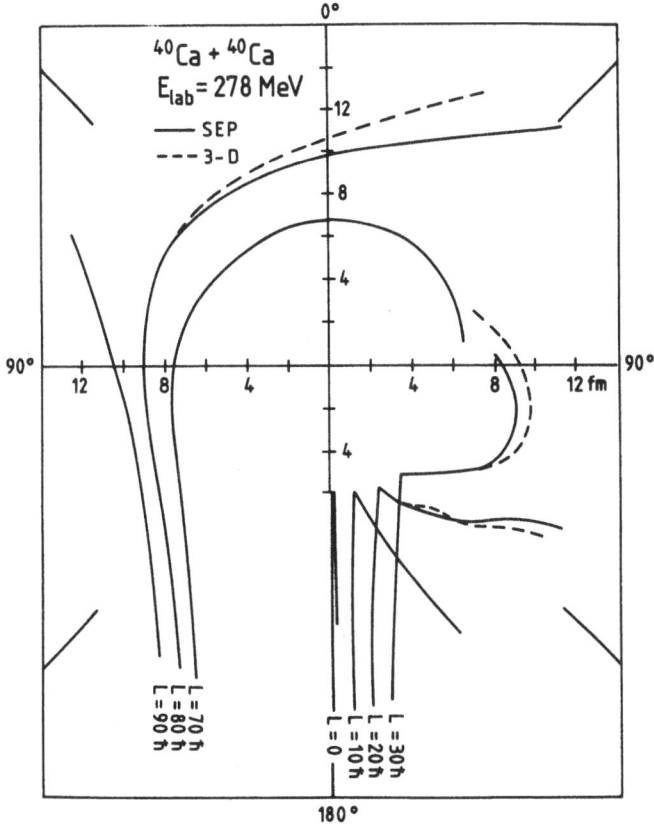

Fig. 2.2 A comparison of fragment separation trajectories $\vec{R}(t)$ calculated in the frozen (separable) approximation and using the full 3-dimensional TDHF formalism. A single trajectory indicates that the two curves are indistinguishable.

Other symmetries and approximations that are introduced when appropriate include isospin symmetry, where $\rho_n = \rho_p$ and an effective charge is introduced, reflection symmetry with $\psi_k(\vec{r}, t) = \pm \psi_k(-\vec{r}, t)$ for symmetric systems and the filling approximation in open shell nuclei where valence nucleons are uniformly distributed over the last shell to ensure spherical solutions. Furthermore the spin-orbit interaction has been neglected in all calculations to date because of its indicated small contribution.

2.4 Heavy-ion phenomenology relevant to TDHF calculations

In terms of impact parameter one distinguishes four regimes for heavy-ion collisions, namely those characteristic of elastic, quasi-elastic, deep inelastic and fusion collisions.

The first case corresponds to very peripheral collisions with trajectories which are essentially slightly perturbed Rutherford trajectories. At slightly smaller impact parameter quasi-elastic scattering takes place where only a few nucleons can be transferred between the two projectiles with possible excitation of discrete states. Even smaller impact parameters lead to deep inelastic scattering with substantial energy loss, equilibration phenomena and in general a large scale collective evolution of the system. At the smallest impact parameters fusion takes place where an equilibrated system is formed which only decays (by splitting or particle emission) long after the collision.

The balance between these possibilities is further determined by factors such as the bombarding energy, the masses and the charges involved.

To date TDHF has only been applied to fusion and deep inelastic scattering, the other two regimes exhibiting mostly quantum mechanical phenomena where a detailed description of the final states is required.

Deep inelastic (DI) scattering is the dominant reaction mode for heavy systems or for light systems at higher energies. The final states in DI collisions are predominantly binary, namely two fragments emerge with masses close to the original target and projectile masses as illustrated in figure 2.3. A second characteristic of DI collisions is strong energy damping with final energies below the coulomb barrier. These low energies indicate that radial motion of the ions has stopped and that they are 'pushed apart' by the coulomb force. The energies being in fact below E_{coul} indicates a high degree of deformation when the ions finally fly apart - figure 2.4 is indicative of these aspects. DI collisions are furthermore observed to be 'fast' or 'direct' processes. Angular distributions of the emerging fragments follow the grazing angle very well as function of energy and in fact peak at this angle as shown in figure 2.5.

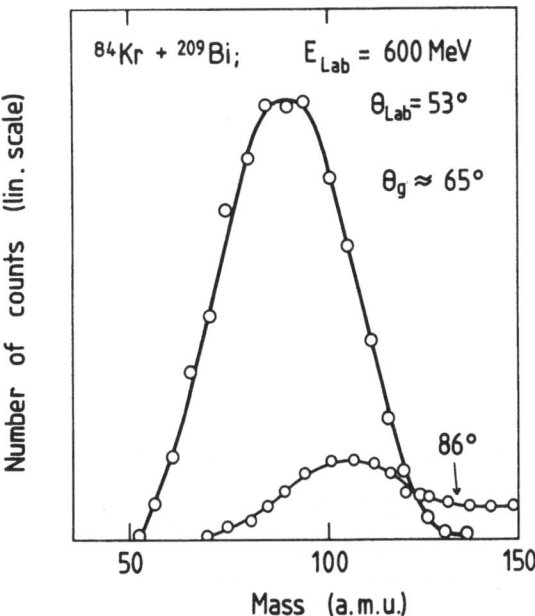

Fig. 2.3   Illustration at different scattering angles of the observation that DI
scattering leads to a final state where the fragment masses correspond
closely to the target and projectile masses

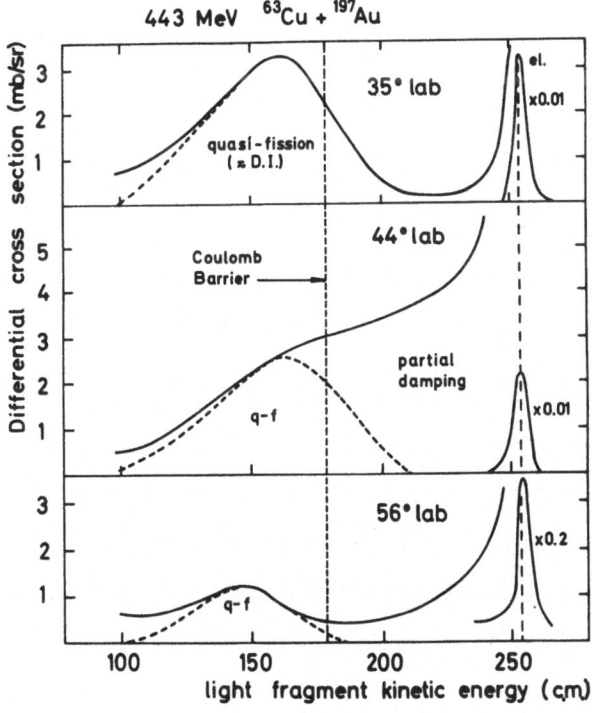

Fig. 2.4

Strong energy damping in DI collision of ^{63}Cu + ^{197}Au at 443
MeV showing final energies below
the Coulomb barrier.

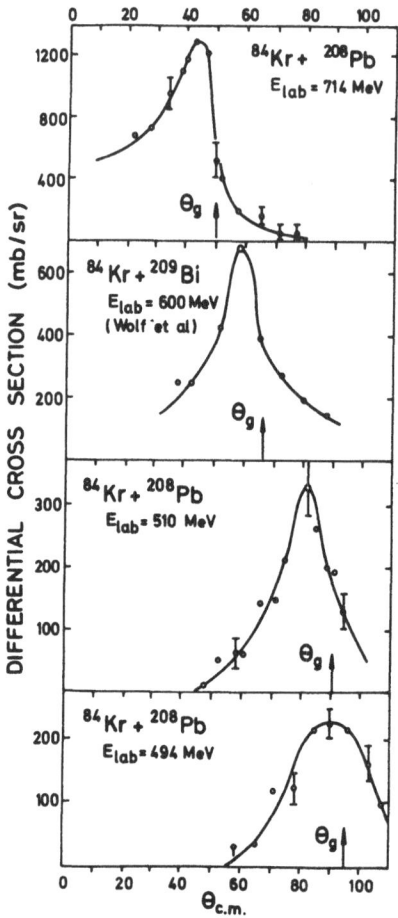

Fig. 2.5 DI reactions are relatively fast and direct processes inferred from the
fact that angular distributions peak near the grazing angle as shown here.
(From ref. 33.)

The process is therefore a direct one taking place on a short time scale during
which the system 'remembers' its initial configuration.

A further useful correlation between observables is expressed on a Wilczynski plot
which shows the differential cross-section as function of the kinetic energy and
scattering angle of the fragments. Figure 2.6 shows this for ^{86}Kr + ^{139}La at
$E_{c.m.}$ = 377 MeV. In this plot one can effectively track the energy loss as function
of angle, the latter serving as a 'clock' in the sense that the energy loss can be
related to the interaction time.

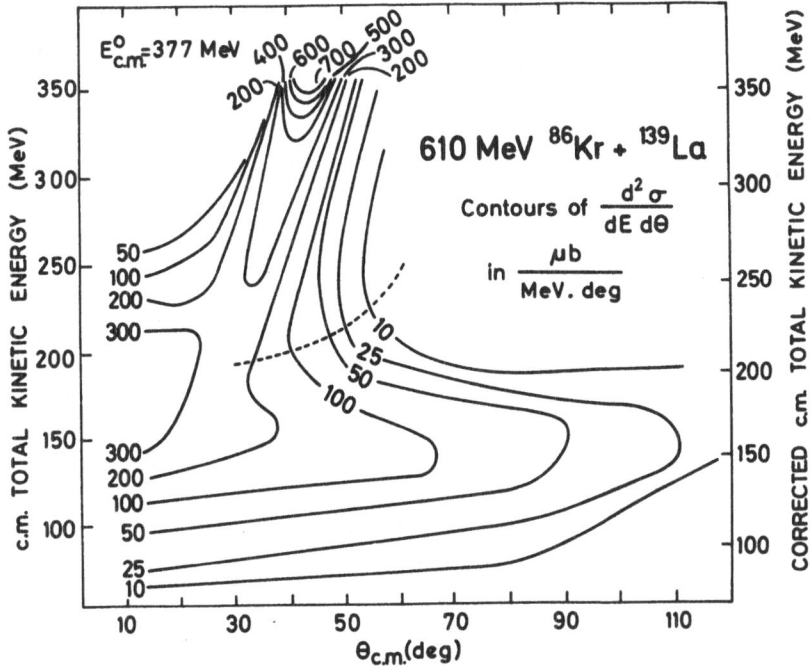

Fig. 2.6 Wilczynski plot |ref. 34| for ^{86}Kr + ^{139}La at $E_{c.m.}$ = 377 MeV

Finally atomic number Z can be correlated with kinetic energy loss as shown in figure 2.7 where elastic scattering peaks around the incident Z and broadens with increasing kinetic energy loss - indicative of a diffusion process.

Conclusions to be drawn from the data considered here are that one encounters equilibration of the initial relative kinetic energy in the large quantum mechanical system of a few hundred nucleons; the scattering angle serves as a 'clock' pertaining to the relative times different collision processes take; a combination of dynamical and statistical effects can either lead to 'fast modes' where the relevant degrees of freedom equilibrate quickly as is the case for the charge to mass ratio, the radial relative motion, as well as the tangential relative motion, or to 'slow modes' such as the mass asymmetry.

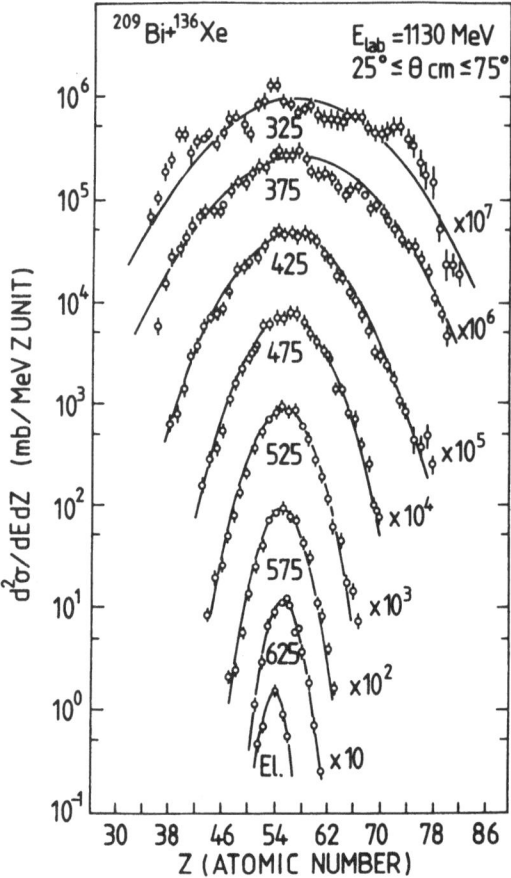

^{209}Bi+^{136}Xe E_{lab} =1130 MeV
25° ≤ θ cm ≤ 75°

Fig. 2.7 Broadening of elastic scattering peaks with increasing kinetic energy
loss - indicative of a diffusion process

These are representative of the main phenomenological features that are expected to
be reproduced in a TDHF calculation. In fact, equilibration features observed in
many HI collisions and alluded to above are considered among the most interesting
phenomena of HI physics and their accurate description constitutes a stringent test
for the mean field approximation and the numerical calculations.

2.5 Extracting observables from the numerical TDHF calculations

Faced with the impossibility of handling a density as function of all the space-time
coordinates of all the HF orbitals, one naturally considers taking moments. One of
the simplest possibilities is to diagonalize the inertia tensor for the mass distribution, locate the principal axes and then define a fragment separation

coordinate between the two centres of mass located on opposite sides of the largest moment of inertia (see figure 2.8).

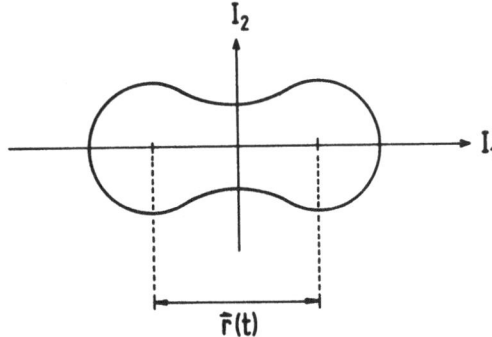

Fig. 2.8 Definition of fragment separation coordinate

Trajectories corresponding to the $\vec{r}(t)$ so defined is shown in figure 2.9 for various impact parameters. As expected the most grazing trajectories show Rutherford behaviour. The more central collisions are 'caught' by the ion-ion potential, leading in some cases to fusion, while for the head-on (small L) collisions the system 'bounces', i.e. comes apart again.

Another observable that is accessible from the calculations is the fusion cross-section. Operationally this quantity is defined by those initial conditions which lead to a long-lived compound system undergoing several rotations. Such behaviour occurs in figure 2.10 while a calculation of $\sigma_{fus}(E)$ from

$$\sigma_{fus}(E) = \pi \lambda^2 \sum_{\ell_<}^{\ell_>} (2\ell + 1)$$

(2.20)

is compared with experiment in figure 2.11 |11|.

Other observables can be obtained by (naive) quantum mechanical interpretation of the TDHF determinant, e.g. 'a fragment mass' is defined through

$$A_R = <N_R>$$

(2.21)

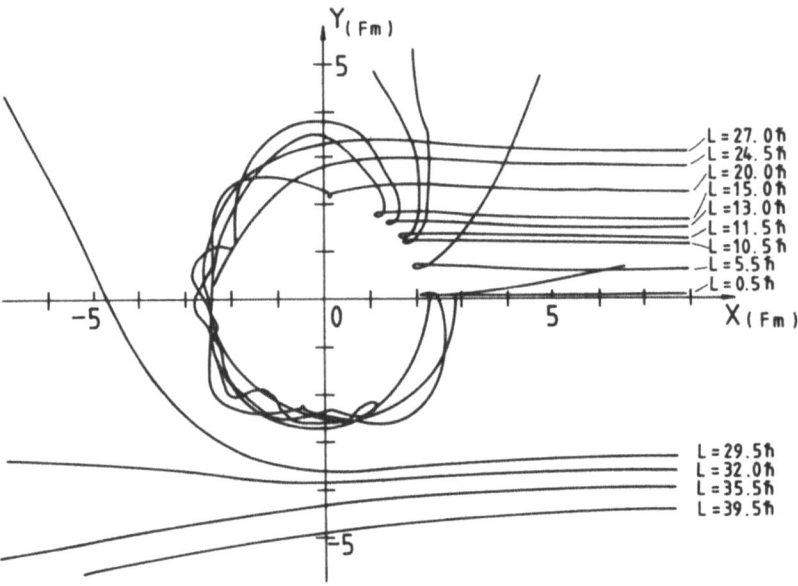

Fig. 2.9 Trajectories of the fragment separation vector $\vec{r}(t)$ for ^{16}O + ^{16}O at E_{Lab} = 105 MeV. The curves are labelled by the initial orbital angular momentum. (From ref. 5)

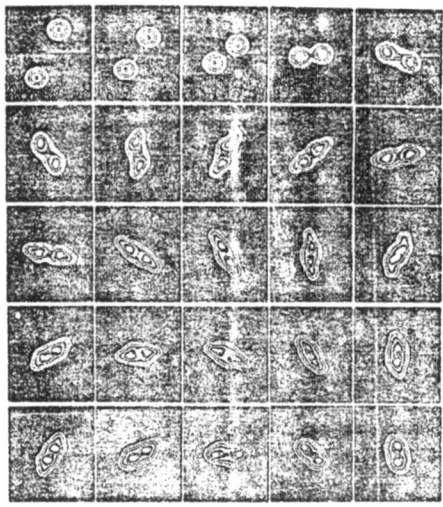

Fig. 2.10 Contour lines of the density integrated over the coordinate normal to the scattering plane for an ^{16}O + ^{16}O collision at E_{Lab} = 192 MeV and incident angular momentum L = 42 $\hbar$. The time interval between pictures is 10^{-22} sec. (From ref. 5).

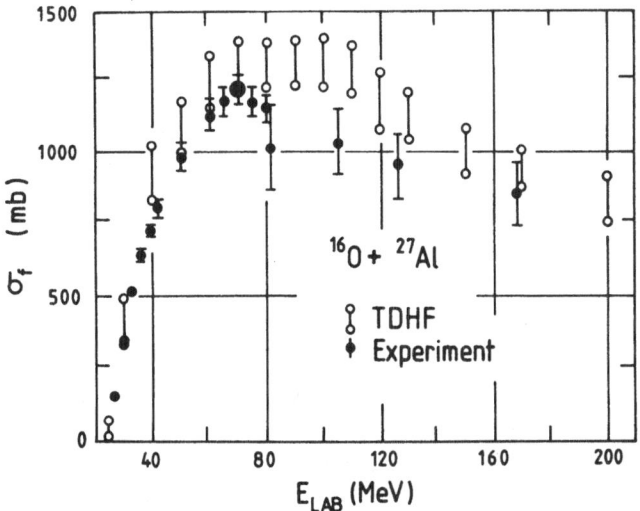

Fig. 2.11 Comparison between experimental and TDHF calculated fusion cross-sections
for ^{16}O + $^{27}A\ell$ at various laboratory energies

with N_R the number operator in terms of the second quantized nucleon field
operators,

$$N_R = \int_{z \geq 0} d^3r \; \psi^\dagger(\vec{r}) \; \psi(\vec{r}). \tag{2.22}$$

Here R is referring to the 'right hand' fragment while the expectation value is
taken with respect to the TDHF determinant.

The mean-square dispersion in A_R,

$$\Delta A_R^2 = \langle N_R^2 \rangle - \langle N_R \rangle^2$$

$$= tr \; (\rho_R - \rho_R^2) \quad (\rho_R(r, r') = \Theta(z) \; \rho(r, r') \; \Theta(z')) \tag{2.23}$$

can then be related to Γ_A, the full width at half maximum, through

$$\Gamma_A^2 = 8 \; \ell n \; 2 \; \Delta A_R^2 \tag{2.24}$$

which follows from an assumed gaussian dispersion.

Note that A_R involves the expectation value of a one-body operator, while ΔA_R^2 involves the expectation value of a two-body operator. As pointed out earlier one
therefore expects the one-body TDHF theory to do rather poor in the latter case.
Extentions to remedy this deficiency are considered in section 3.4.

Returning to actual contact with experiment one finds that the fusion cross-section shows sensitivity to the interaction used as shown for ^{40}Ca + ^{40}Ca below. In all cases the barrier is overestimated and the predicted behaviour is mainly correlated with the surface energy coefficient occurring in a particular interaction.

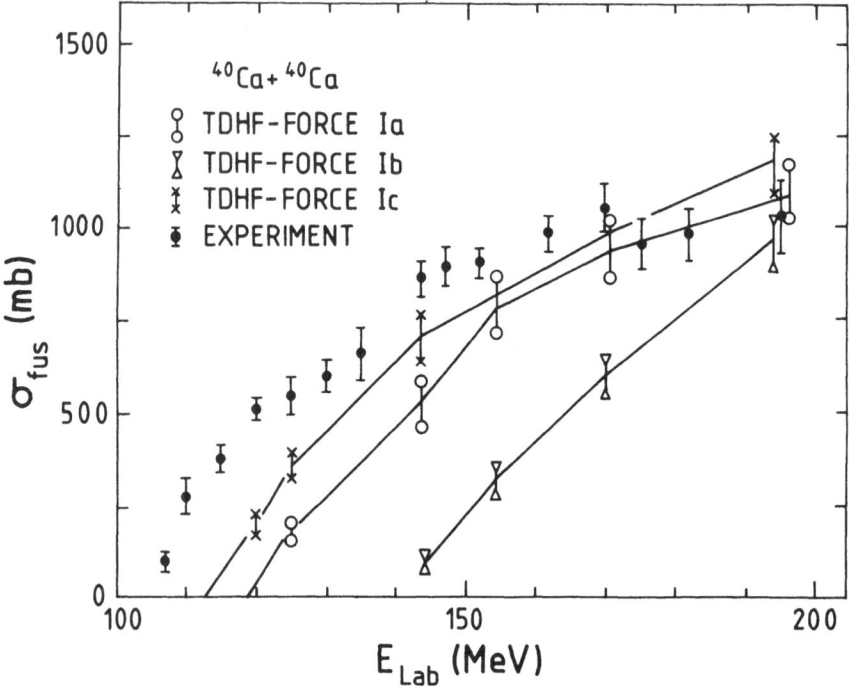

Fig. 2.12 TDHF fusion cross-sections calculated with different effective interactions show consistent overestimation of the barrier. (See ref. 6 for
further discussion.)

In TDHF fusion can be regarded as sufficient randomization of single particle orbitals of the two colliding ions as their separate potential wells merge into a common potential well. Since this can happen only through the mean field there is little tendency towards transverse deformation as particles pass through the con-necting "neck" with the result that the heavy-ion system remains very prolate.

In light systems with the energy slightly above the interaction barrier energy, i.e. $E \gtrsim E_B$, geometry dictates fusion for all $\ell < \ell_>(E)$ where $\ell_>(E)$ is the critical max-imum angular momentum for fusion as indicated in eq. (2.20). In the energy range $E \gg E_B$ TDHF predicts fusion for $\ell_< < \ell < \ell_>$ with $\ell_< \neq 0$. This behaviour, reminiscent of hydrodynamics, is as yet unconfirmed experimentally due to the technical difficulties involved.

The existence of a dynamical, energy-dependent lower angular momentum limit to fusion in TDHF can be traced to the long mean-free path assumption. Figure 2.13 shows a non-fusion head-on $^{16}O + ^{16}O$ collision at 2 MeV per nucleon centre-of-mass energy ($E_{\ell ab}$ = 128 MeV). Initially each ^{16}O nucleus is described by four spatial orbitals 1s and $1p_m$ (m = 0, ±1) bound self-consistently and resulting in a spherical density distribution at t = 0. By t = 0.15 sufficient overlap between the two nuclei has taken place to reduce the mean-field potential barrier between them, allowing orbitals to 'flow' from one nucleus to the other. The $1p_{m=0}$ orbitals has the highest linear momentum along the symmetry axis, thus crossing the compound

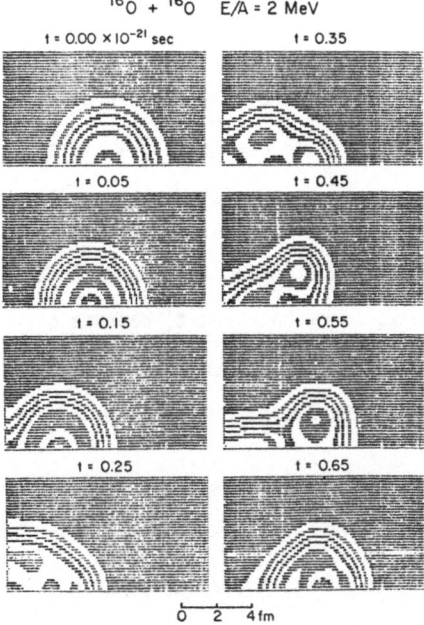

Fig. 2.13 Density contour maps for a head-on $^{16}O + ^{16}O$ collision at E/A = 2 MeV (shown in centre-of-mass frame). The density for $Z \geq o$ only is shown because of rotational symmetry around the horizontal and reflection symmetry around the vertical axis. (Ref. 7).

system first, striking and deforming the opposite potential wall (t = 0.35). In this frame the torroidal $1p_{m = \pm 1}$ orbitals can be seen trailing behind. By t = 0.45 the $1p_{m = 0}$ orbital has 'bounced' off the wall moving now left through the still right moving $1p_{m = \pm 1}$ orbitals reaching its original side by t = 0.55, scission taking place at t = 0.65. Note that a fair amount of orbital 'trapping' or particle exchange is observed in such a collision.

In heavier systems fewer calculations have been performed but do point in general to a great sensitivity to the interaction as well as the geometry, i.e. whether the calculation is performed in two or three dimensions. Figure 2.14 shows a calculation for head-on collisions of ^{86}Kr + ^{139}La (the interaction time τ_{int} is defined as the time for which ρ_{min} (along the symmetry axis) > $\frac{1}{2}\rho_{nuclear\ matter}$). Figure 2.15 illustrates the sensitivity of the same system to various initial conditions, i.e. bombarding energies. In general fusion takes place at intermediate energies where good agreement with experiment is found. At low energies the non-fusion is traced to reflection of orbitals with little energy loss while systems at higher energies do not fuse because of transparency associated with the long mean-free path.

In the regime of deep inelastic scattering (DIS) one finds that for light systems peripheral collisions tend to 'come back apart' because of rotational instability connected with a large L-value while for central collisions the tendency is associated with vibrational instability connected with transparency.

For heavy systems undergoing DIS one finds relatively unremarkable behaviour. One interesting aspect concerns a high degree of 'orbital exchange' without any significant mass exchange. In a recent |12| systematic study of ^{136}Xe + ^{209}Bi TDHF shows an unexpected Z/A drift as function of bombarding energy, namely a tendency towards symmetric breakup at energies of 940 MeV and 1130 MeV and away from symmetry at 1422 MeV. This is in agreement with experiment (although the specific reason is difficult to trace), but quite distinct from diffusion models which predicts a tendency towards symmetric breakup for all cases.

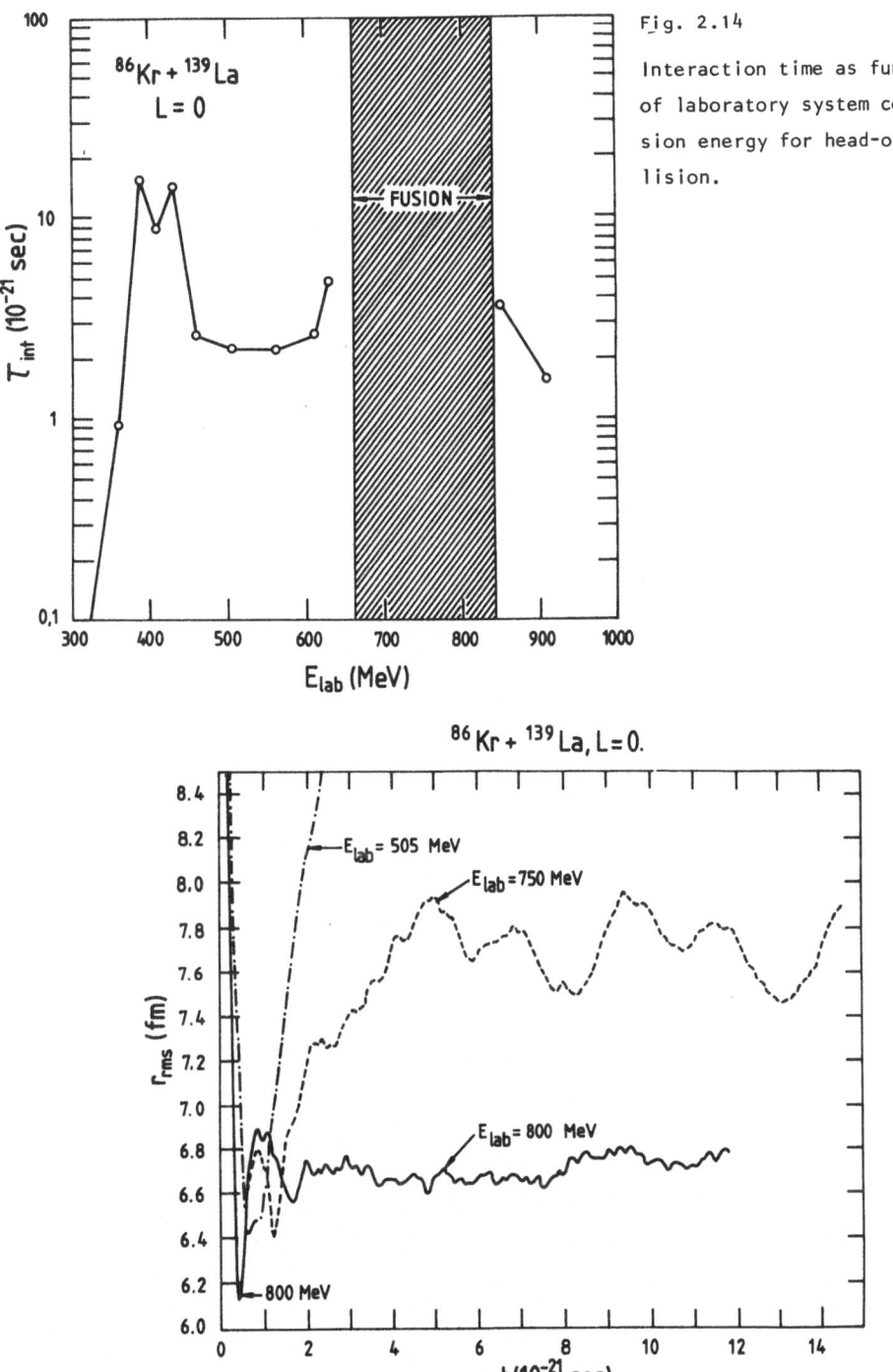

Fig. 2.14

Interaction time as function of laboratory system collision energy for head-on collision.

Fig. 2.15 Possibility for eventual fusion (expressed here in terms of rms radius as function of interaction time) shows great sensitivity to the initial conditions. (See ref. 32)

In summary the application of TDHF to DIS shows good agreement for $d^2\sigma/d\Theta dE$, predicts small mass and charge transfers ΔZ and ΔA, also in good agreement with experiment while Γ_A is much too small as already alluded to.

A final interesting feature of realistic TDHF-calculations concerns the often predicted appearance of prompt, energetic low-density jets of matter |13| as shown in figure 2.16, for $^{12}C + ^{197}Au$ at relatively high energy, namely $E_{Lab}/A = 30$ MeV. These jets are emitted on a time scale comparable to the transit time of a projectile nucleon crossing the compound system ($\ll$ collision time) with densities $\rho \sim .01\rho_{nucl.\ matter}$. Furthermore almost all of these nucleons originate from the projectile, i.e. behave like free fermi gas particles escaping from a potential well.

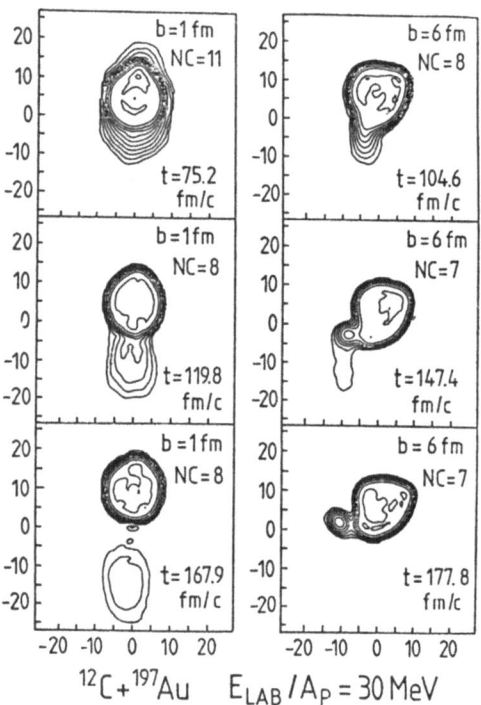

$^{12}C + ^{197}Au$ $E_{LAB}/A_P = 30$ MeV

Fig. 2.16   Appearance of energetic low density jets of matter at relatively high energies predicted by realistic TDHF calculations

3. Relation and extension of TDHF to other formalisms

3.1 TDHF and fluid dynamics

In view of existing fluid-like considerations in the description of nuclear collective motion and given the various conservation laws TDHF admits, it is tempting to cast the TDHF-equations in a similar fluid dynamical framework, reducing dynamical variables to only density, current and perhaps temperature fields.

This reduction is most economically accomplished through the Wigner distribution function

$$f(\vec{r}, \vec{k}, t) = \int d^3s \, \exp(-i\vec{k}\cdot\vec{s}) \, \rho(\vec{r} + \tfrac{1}{2}\vec{s}, \vec{r} - \tfrac{1}{2}\vec{s}, t) \tag{3.1}$$

defined in terms of the coordinate-space one-body density matrix $\rho(\vec{r}, \vec{r}', t)$ encountered in eq. (1.6). This function has the convenient property that the density, current density and kinetic energy density are all expressed as expected classically:

$$
\left\{
\begin{array}{c}
\rho(\vec{r}, t) \\
\vec{j}(\vec{r}, t) \\
\tau(r, t)
\end{array}
\right\}
= (2\pi)^{-3} \int d^3k
\left\{
\begin{array}{c}
1 \\
\vec{k} \\
\vec{k}^2
\end{array}
\right\}
f(\vec{r}, \vec{k}, t) \quad . \tag{3.2}
$$

In many-body calculations the fact that f is not necessarily positive definite is not a serious problem since enough wave functions contribute to f so that it is almost always positive whenever it is 'large'.
The evolution of f follows directly from the Wigner transform of the TDHF equation (1.14):

$$\frac{\partial f}{\partial t} + \frac{\hbar\vec{k}}{m}\cdot\frac{\partial f}{\partial \vec{r}} + \frac{2}{\hbar} \sin\left(\frac{\hbar}{2} \partial_r^{(1)}\cdot\partial_k^{(2)}\right) W^{(1)} f^{(2)} = 0. \tag{3.3}$$

Here W denotes the Wigner transform of the TDHF potential (1.13),

$$W(\vec{r}, \vec{k}, t) = \int d^3s \, \exp(-i\vec{k}\cdot\vec{s}) \, W(\vec{r} + \tfrac{1}{2}\vec{s}, \vec{r} - \tfrac{1}{2}\vec{s}, t), \tag{3.4}$$

and the labels (1) and (2) indicate how the gradient operators act on the functions W and f.

Eq. (3.3) is a linear equation in f and can be interpreted as a quantum Vlasov equation, the analogy becoming exact in the classical limit where W is taken to be local

and where only the first term of the sin-expansion survives for $\hbar \to 0$. In this limit
TDHF emerges as a system of classical independent particles moving in their common,
time-dependent mean field.

In 2 and 3 dimensions f is a time-dependent function of respectively 4 and 6 variables
and properties of f are difficult to trace. Schematic one-dimensional TDHF calcula-
tions have, however, been carried out |14| for the collision of two slabs of nuclear
matter (one of finite thickness and the other semi-infinite). The results are shown
in figure 3.1 and the corresponding f, shown as contour plots, appear in figure 3.2.
The solid line represents the total density, while the dashed line only takes into
account orbitals of the finite projectile. One observes penetration of the target
slab with some density perturbation.

The contour plots of the Wigner function f reveals a 'gliding' of projectile over
target, resulting from the relatively long lifetime of particles near the Fermi
surface - a consequence of the Pauli principle.

Momentum moments of the quantum Vlasov equation (3.3) result in local conservation
laws analogous to those of hydrodynamics and one can thus obtain a formal hydrodynamic reduction of TDHF via the Wigner transform. One finds, for example, for a
local TDHF potential $W \neq W(\vec{k})$ from the zeroth and first moments the conservation of

matter and momentum:

$$\frac{\partial \rho}{\partial t} + \vec{\nabla} \cdot \vec{J} = 0; \tag{3.5}$$

$$\left(\frac{\partial}{\partial t} + \vec{U} \cdot \vec{\nabla}\right)\vec{U} = -(\rho^{-1} \vec{\nabla}\rho + \vec{\nabla}W). \tag{3.6}$$

$\vec{U} = \vec{J}\rho^{-1}$ is the velocity field and the isotropic pressure p is related to the kinetic
energy density τ by $p = \frac{1}{3}(\tau - J^2 \rho^{-1})$.

Classical hydrodynamics imposes a truncation on the moment hierarchy (which is not
closed in itself) through an equation of state, e.g. $P = P(\rho)$. Such a truncation is,
however, intimately connected with the idea of local equilibrium, the very state of
affairs not encountered in TDHF due to the long mean-free path assumption.

A proper TDHF based rheology is therefore expected to be very different from ordinary hydrodynamics. Furthermore a rheology for one-body dynamics is complicated by the intrinsic spatial non-locality of TDHF and various attempts have been made at incorporating two-body collisions at a phenomenological level. This is briefly discussed in the next section.

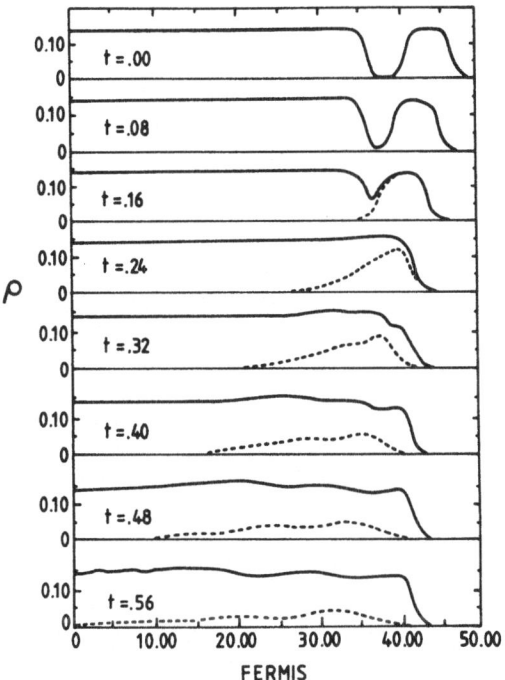

Fig. 3.1   Time-dependent density distributions for a slab-slab collision at 2 MeV/nucleon.   The time unit is 10^{-21} sec.

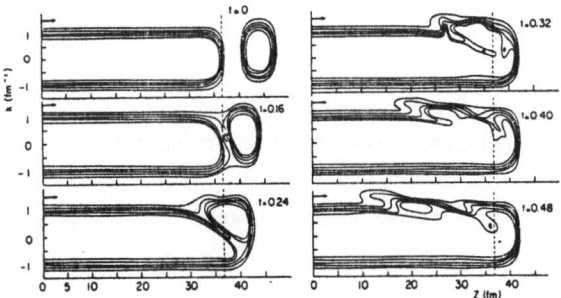

Fig. 3.2   Contour plots for the Wigner function $f(k_z = k, z, t)$ for the collision shown in fig. 3.1. The dashed vertical line marks the initial edge of the semi-infinite slab, while the horizontal arrow indicates the fastest nucleons in the initial projectile distribution

3.2   Incorporation of two-body collisions

One attempt |15, 16| at incorporating two-body collisions uses a truncation of the Green's function hierarchy based on arguments about rapid time variation in $g^{(2)}$. The resulting equations involving only $g^{(1)}$ lead to the following structure:

$$\rho(\vec{r}, \vec{r}', t) = \sum_\lambda n_\lambda(t)\, \psi_\lambda(r)\, \psi_\lambda^*(r') \tag{3.7}$$

where ψ_λ evolves in a HF-like way,

$$i\,\frac{\partial\psi_\lambda}{\partial t} = h[\rho]\psi_\lambda, \tag{3.8}$$

while n evolves as

$$\frac{dn_\lambda}{dt} = \frac{\pi}{\hbar} \sum_{\mu\nu\sigma} \delta(\varepsilon_\lambda + \varepsilon_\mu - \varepsilon_\nu - \varepsilon_\sigma) \left[(1 - n_\lambda)(1 - n_\mu)n_\nu\, n_\sigma \right.$$
$$\left. - n_\lambda\, n_\mu(1 - n_\nu)(1 - n_\sigma)\right] |{<}12|\nu'|34{>}|^2, \tag{3.9}$$

where

$$\varepsilon_\lambda(t) = {<}\psi_\lambda(t)|h|\psi_\lambda(t){>}. \tag{3.10}$$

Formally n therefore evolves via a 'collision integral'.
Problems with this briefly outlined program include:

(i)   one has to introduce some smearing of the discrete levels in order to utilize $\delta(\varepsilon)$;

(ii)   the choice of the residual interaction ν' might introduce double counting problems connected to an already chosen V_{eff} in TDHF;

(iii)   conservation laws are not any more guaranteed and 'static' solutions are found to evolve in time;

(iv)   a number of additional free parameters have to be introduced.

A more recent attempt |17| which shows promise starts from the phenomenological equation

$$\frac{d\rho}{dt} = -i[h, \rho] + \tau^{-1}(\rho_0 - \rho) \tag{3.11}$$

where τ is a relaxation time (available from microscopic nucleon-nucleon scattering calculations) while ρ_0 is the uniform fermi-gas density matrix with occupation numbers adjusted to have locally the same density, current and kinetic energy density as ρ. This guarantees for example that tr ρ is preserved since eq. (3.11) implies

$$\frac{d}{dt} \text{tr } \rho = -i \text{ tr } [h, \rho] + \tau^{-1} \text{ tr } (\rho_0 - \rho) \tag{3.12}$$

which is zero because of the construction of ρ_0 and the fact that tr $[h, \rho]$ = 0.

While schematic one-dimensional calculations based on eq. (3.11) are promising |17|, realistic implementation is complicated by the fact that not only the evolution of the orbitals $\psi_\lambda(t)$ must be tracked, but in fact the full density matrix $\rho(\vec{r}, \vec{r}', t)$.

3.3 Time-dependent mean-field methods

3.3.1 The Hubbard-Stratonovich transformation

The representation thus far indicates that TDHF is a useful zeroth-order description which is, however, burdened by the following deficiencies:

(i)   Its relation to the exact Schrödinger equation is obscure and, moreover, the non-linearity of the equations precludes any attempt at extracting an S-matrix;

(ii) as it stand there is no way in which systematic corrections can be generated, even in principle;

(iii) the interpretation of the wavefunction is unclear and has been addressed only in the most naive terms.

In what follows the Hubbard-Stratonovich representation |18, 19| is shown to address these problems and provide a link between exact many-body evolution and one-body evolution which can be used to show that TDHF emerges as a member of a particular class of approximations to a more general theory.

Starting from a second quantized fermion hamiltonian

$$H = \frac{\hbar^2}{2m} \int dx \ \nabla\psi^\dagger \cdot \nabla\psi + \tfrac{1}{2}\int dx \ dx' \ \psi^\dagger(x)\psi^\dagger \ (x') \ v(x - x') \ \psi(x')\psi(x) \tag{3.13}$$

where $v(x - x')$ is a local two-body interaction, the anti-commuting property of the field operators $\psi(x)$, $\psi^\dagger(x)$ leads to the equivalent from

$$H = \int dx \left[\frac{h^2}{2m} (\nabla\psi^\dagger \cdot \nabla\psi) - \tfrac{1}{2} v(0) \rho(x) \right]$$

$$+ \tfrac{1}{2} \int dx\, dx'\, \rho(x)\, v(x - x')\, \rho(x')$$

$$\equiv K + \tfrac{1}{2}(\rho, v\rho) \tag{3.14}$$

We have introduced the one-body density operator $\rho(x) = \psi^\dagger(x)\,\psi(x)$ and a convenient 'inner product' notation. Note that $\rho(x)$ is hermitian and that ρ's at different x's commute.

What one is ultimately seeking is an appropriate approximation to the time-evolution operator $\exp(-iHT)$. Consider therefore N successive evolutions over a short time $\Delta t = T/N$ for which

$$\exp(-iHT) = \{\exp(-iH\Delta t)\}^N. \tag{3.15}$$

To lowest order in Δt one has for each step

$$\exp(-iH\Delta t) = \exp\{-i(K + \tfrac{1}{2}(\rho, v\rho))\Delta t\}$$
$$\approx \exp(-iK\Delta t)\, \exp(-i/2(\rho, v\rho)\Delta t). \tag{3.16}$$

Further manipulation is based on the gaussian integral identity

$$\int_{-\infty}^{\infty} \exp(-\pi x^2 - 2\sqrt{\pi}\, ax)dx = \exp(a^2). \tag{3.17}$$

In the present context this identity is used in the form

$$\exp\left(-\tfrac{i}{2} v\rho^2 t\right) = \left(\frac{vt}{2\pi i}\right)^{\tfrac{1}{2}} \int_{-\infty}^{\infty} d\sigma\, \exp\left(\tfrac{i}{2} \sigma^2 vt\right) \exp(-i\sigma\rho vt), \tag{3.18}$$

where σ is an auxilliary one-body field. This identity shows that a two-body evolution can thus be regarded as a 'weighted superposition' of one-body evolutions. Substitution of the identity (3.18) into approximation (3.16) yields [18]

$$\exp(-iH\Delta t) \approx (\det v)^{\tfrac{1}{2}} \int D\,[\sigma(x)]\, \exp\left(\tfrac{i}{2}(\sigma, v\sigma)\Delta t\right) \exp(-ih_\sigma \Delta t) \tag{3.19}$$

where v is viewed as a matrix in x, x' and the measure D of the functional is defined in terms of differentials $d\sigma_q$ ($\sigma_q \equiv$ q-th Fourier component of σ). h_σ is a hamiltonian-like operator,

$$h_\sigma = K + (\sigma, v\sigma). \tag{3.20}$$

Utilizing expression (3.19) N times, each term with a field σ_n, gives for the full

evolution operator

$$\exp(-iHT) = (\det v)^{N/2} \int D[\sigma_1] \cdots D[\sigma_N] \exp \{\frac{i}{2} \sum_{j=1}^{N} (\sigma_j \, v\sigma_j)\Delta t$$

$$\times \exp (-ih_{\sigma_N} \Delta t) \cdots \exp (-ih_{\sigma_1} \Delta t). \tag{3.21}$$

Care should be exercised in the ordering of the operators $\exp (-ih\sigma_j \Delta t)$ which do

not commute in general. In the continuous limit one replaces $\sigma_j(x) \to \sigma(x, t)$ and by

realizing that the product of exponentials is just the evolution operator $U(T)$ for

the time-dependent one-body hamiltonian $h_\sigma = h_\sigma(t) = K + (\sigma(t), v\rho)$, the final re-

sult

$$\exp (-iHT) = \int D[\sigma(x, t)] \exp (\frac{i}{2} \int_0^T (\sigma(t), v\sigma(t))dt) \, U_\sigma(T) \tag{3.22}$$

is obtained.

Equation (3.22) expresses the many-body evolution operator as a functional integral

over all real fields, $\sigma(x, t)$. Each field configuration in turn contributes a one-

body evolution operator U_σ, weighted by the 'gaussian' factor $\exp (\frac{i}{2} \int_0^T (\sigma, v\sigma)dt)$ and

evolving from

$$i \frac{dU_\sigma}{dt} = h_\sigma U_\sigma. \tag{3.23}$$

Before actually applying this expression, the following remarks are in order. An

analogous expression can be obtained in the more general case of a non-local potential

v |20, 21|. Here the field $\sigma(x, t)$ is replaced by a non-local quantity $\sigma_{ij}(t) =$

$\sigma_{ji}(t)$ where the indices refer to any orthonormal single-particle basis. The Hubbard-

Stratonovich representation has other incarnations, depending on the actual choice of

grouping the field operators. For example, a pairing type problem is appropriately

addressed by the introduction of a pairing field where $\eta(x, x')$ and $\eta^\dagger(x, x') =$

$\psi^\dagger(x) \psi^\dagger(x')$ is used instead of $\rho(x)$ |18, 20, 22, 23|.

A semi-classical or mean-field approximation to the Hubbard-Stratanovich representa-

tion is now considered. The matrix element of eq. (3.22) between arbitrary initial

and final states is

$$\langle f|e^{-iHT}|i\rangle = \int D[\sigma] \exp(iS[\sigma])|\langle f|U_\sigma|i\rangle| \tag{3.24}$$

with

$$S[\sigma] = \tfrac{1}{2} \int_0^T dt \ (\sigma, \ v\sigma) + \text{Im} \ \ell n \ <f|U_\sigma(T)|i>. \tag{3.25}$$

U_σ being a one-body operator, S is most easily evaluated when $|i>$ and $|f>$ are determinants, although the expressions hold for general states. Only the phase of $<f|U_\sigma|i>$ appears in the 'action' S; inclusion of the modulus as well would give complex stationary fields.

The analogy with a Feynman path integral suggests one considers a stationary phase approximation. Let $\bar{\sigma}(x, t)$ therefore be the stationary field and imagine a small variation $\delta\sigma(x, t)$ around σ at some time t. Since

$$i \ \dot{U}_\sigma(t) = h_\sigma(t) \ U_\sigma(t) \tag{3.26}$$

one has

$$\delta U_\sigma(T) = -iU_{\bar{\sigma}}(T, \ t) \ \delta h_\sigma(t) \ U_{\bar{\sigma}}(t)$$

$$= -U_\sigma(T, \ t) \ (\delta\sigma, \ v\rho) \ U_{\bar{\sigma}}(t, \ o). \tag{3.27}$$

The stationary constraint is then

$$\delta S = 0 = (\delta\sigma, \ v\bar{\sigma}(t)) - \text{Re} \left[\frac{<f(t)|U_{\bar{\sigma}}(T, \ t) \ (\delta\sigma, \ v\rho)U_{\bar{\sigma}}(t, \ o)|i(t)>}{<f(t)|i(t)>} \right] \tag{3.28}$$

or, for $\bar{\sigma}$ itself,

$$\bar{\sigma}(x, \ t) = \text{Re} \left[\frac{<f(t)|\rho(x)|i(t)>}{<f(t)|i(t)>} \right] \tag{3.29}$$

where the 'matrix' $v(x, \ x')$ is assumed invertible. The states $|i(t)>$ and $|f(t)>$ evolve forward and backward under $U_{\bar{\sigma}}$, namely

$$|i(t)> = U_{\bar{\sigma}}(t, \ o)|i>$$

$$<f(t)| = <f|U_{\bar{\sigma}}(T, \ t) \tag{3.30}$$

rendering the denominator in eq. (3.29) time-independent.

At this stage the following remarks are in order. The full evolution operator has now been approximated through the stationary field by a one-body evolution operator. The equation (3.29) determining $\bar{\sigma}$ is however highly non-linear since $\bar{\sigma}$ is determined in terms of wave functions which evolve under a hamiltonian already containing $\bar{\sigma}$. This is reminiscent of TDHF but different in that the evolution is dependent on the choice of the final state. (Replacing $<f(t)|$ by the adjoint of $|i(t)>$ TDHF is formally recovered.)

The stage is now set for systematic corrections to the present approximation by actually performing the gaussian and higher order integrals corresponding to fluctuations about $\bar{\sigma}$. Diagrammatic expansions can thus be generated but prove largely unilluminating and difficult to utilize. These expansions do however raise the question of what the 'small parameter' is in terms of which the expansion about $\bar{\sigma}$ is being made. $\hbar$ cannot be a candidate, since the kinetic part involves $\hbar$ in the denominator. In simple cases the number of particles or rather the 'collectivity' seems appropriate but the general question has not been answered.

Finally, the classical character of the stationary-phase approximation is exemplified by the existence of a conserved energy E, for example, where

$$E = \mathrm{Re}\left[\frac{<f(t)|h(t)|i(t)>}{<f(t)|i(t)>}\right] - \tfrac{1}{2}\,(\sigma(t),\,v\sigma(t))$$

and $^{dE}/_{dt} = 0$ follows from the equations of motion for $|i(t)>$ and $<f(t)|$. This form of E is clearly reminiscent of the usual HF energy.

3.3.2 Time-dependent problems and mean-field approximation for the S-matrix

Rather than considering the full scattering problem, we first consider a simpler time-dependent problem with many-body hamiltonian

$$H = H_0 + V(t) \quad (\lim_{|t| \to \infty} V(t) = 0) \tag{3.31}$$

where H_0 includes one- and two-body interactions while $V(t)$ is a time-dependent one-body potential. While this problem is illustrative of the mean-field approximation for the S-matrix, it is much simpler in having eliminated all the spatial aspects involved in a real scattering process.

Given an initial state $|i>$, we now seek the properties of

$$|i(t)> = U(t,\,o)|i> \tag{3.32}$$

especially for $t \to \infty$.

The mean-field formalism is now first illustrated for the case where we consider the expectation value of some operator A in the final state $|i(t)>$.

In the Hubbard-Stratonovich representation we have

$<A> \equiv <i(t)|A|i(t)>, \quad t \to \infty$

$= <i|U(o, T) A U (T, o)|i>, \quad T \to \infty$

$= \int D[\sigma_i] \; D[\sigma_f] \exp \left(i \int_0^T \{(\sigma_i, v\sigma_i) - (\sigma_f, v\sigma_f)\} dt \right) <i|U_{\sigma_f} (o, T) A \; U_{\sigma_i} (T, o)|i>.$

$\hspace{11cm} (3.33)$

Using the stationary phase approximation discussed in the previous section $<U_{\sigma_f} A \; U_{\sigma_i}>$ is evaluated by introducing an auxilliary field ϕ in the hamiltonian through the addition of a term $\phi_i(t) A$ (or $\phi_f(t) A$) and demanding

$$\frac{\delta}{\delta\phi(T)} <i|U_{\sigma_f}^{(\phi_f)} \; U_{\sigma_i}^{(\phi_i)} |i>|_{\phi = o} = 0. \hspace{3cm} (3.34)$$

For the case of a one-body operator A one finds a stationary solution at $\bar{\sigma} = \sigma_f = \sigma_i \; |24|$ giving

$\sigma_i = <i(t)|\rho|i(t)>, \hspace{6cm} (3.35)$

$<A> = <i(T)|A|i(T)>, \quad T \to \infty \hspace{4cm} (3.36)$

namely one has recovered TDHF. This shows that the optimum evolution of an initial state for a one-body operator at the mean-field level is indeed given by TDHF.

For the case of a two-body operator A one immediately runs into non-trivial self-consistency conditions on the mean field $\bar{\sigma}$ which complicates matters tremendously.

Retaining the hamiltonian (3.31) we now seek to calculate the amplitude for excitation from an initial state $|\beta>$ to a final state $|\beta'>$ and therefore consider the S-matrix element

$S_{\beta'\beta} = \lim_{t \to \infty} <\beta'|U_o(o, t) \; U(t, -t) \; U_o(-t, o)|\beta> \hspace{3cm} (3.37a)$

$= \lim_{t \to \infty} \exp i(E_\beta + E_{\beta'})t) \; <\beta'|U(t, -t)|\beta>. \hspace{2cm} (3.37b)$

Here $U_o(U)$ refers to evolution under $H_o(H)$ and the second line follows when the channel states $|\beta\rangle$ and $|\beta'\rangle$ are eigenstates of H_o.

Direct resource to the mean-field approximation for the latter equation is now unsatisfactory for the following reasons. The eigenstates $|\beta\rangle$ and $|\beta'\rangle$ of H_o are already the solutions to a complicated many-body problem and must of necessity be approximated, rendering in general invalid the step from eq. (3.37a) to eq. (3.37b). Furthermore, even with exact channel states, $\langle\beta'|U(t, -t)|\beta\rangle$ is not time-independent due to the non-linearity of the mean-field approximation. This implies that the resulting S is not asymptotically constant.

Direct application of the mean-field approximation to eq. (3.37a) is therefore much more profitable, avoiding the above difficulties (e.g. S is time-independent). This involves the introduction of three σ-fields σ_i, σ and σ_f, one for each U, corresponding to the preparation, interaction and analysis stages of the initial state.

For the S-matrix we then have

$$S_{\beta' \beta} = \lim_{t \to \infty} \int D[\sigma_f]\, D[\sigma]\, D[\sigma_i]\, \exp\left(\tfrac{1}{2}\, \phi(\sigma, v\sigma)dt\right)$$

$$\times \langle\beta'|U_{\sigma_f} (o, t)\, U_\sigma(t, -t) U_{\sigma_i} (-t, o)|\beta\rangle \qquad (3.38)$$

with U_σ corresponding to $h_\sigma = K + (\sigma, v\sigma) + (V, \rho)$ and U_{σ_i}, σ_f corresponding to h_{σ_i}, $\sigma_f = K + (\sigma_{i,f}, v\rho)$. This is schematically shown in figure 3.3. The stationary condition is summarized by the statement that at any time τ around the loop the field σ is the mixed expectation value of the density operator between $|\beta\rangle$ evolved clockwise to τ and $\langle\beta'|$ evolved counter-clockwise to the same point. This implies, for example

$$\sigma(x, t) = \mathrm{Re}\left[\frac{\langle\beta'|U_{\sigma_f} (o,t)U_\sigma(t,\tau)\rho(x)U_\sigma(\tau, -t)\, U_{\sigma_i}(-t, o)|\beta\rangle}{\langle\beta'|U_{\sigma_f}(o, t)U_\sigma(t, -t)U_{\sigma_i} (-t,o)|\rho\rangle}\right]. \qquad (3.39)$$

The S-matrix is obtained by evaluating the integrand in expression (3.38) at the stationary field (summing all the contributions if more than one such field exist).

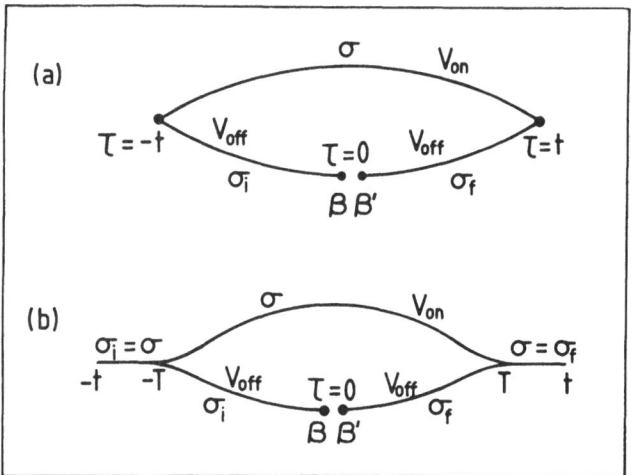

Fig.3.3(a) Schematic illustration of the 'loop' involved in the interaction picture
mean field approximation to the S-matrix element $S_{\beta'\beta}$

(b) Collapse of the loop ends when $V(\tau) = 0$ for $|\tau| > T$. (From ref. 25)

The procedure for approximating S outlined above has the appealing result that a
self-cancelling takes place which renders S time-independent, that is independent of
the end-point time t (as long as this is large enough so that $V(t) = 0$). Furthermore this S can be shown to be time-reversal invariant if H has this property. How-
ever, S is no longer unitary, since each different pair of states $|\beta\rangle$ and $|\beta'\rangle$
generate a different stationary field. In applications thus far S is, however, nearly
unitary.

As an application $|25|$ we consider the Lipkin-Meshkov-Glick model $|26|$ modified by a
time-dependent external potential, namely for H we take the form (3.31) with

$$H_0 = \frac{\varepsilon}{2} \sum_{n=1}^{N} \sum_{s=\pm 1} s\, a_{ns}^+ a_{ns} - \frac{V}{2} \sum_{n,n'=1}^{N} \sum_{s=\pm 1} a_{n-s}^+ a_{n'-s}^+ a_{n's} a_{ns} \qquad (3.40)$$

which describes N fermions in two levels separated by energy ε and interacting in
such a way that pairs of particles are transferred from one level to the other, as
indicated. H_0 can be expressed in terms of quasi-spin operators which satisfy an
SU(2)-algebra, namely

$$H_0 = \varepsilon J_z - \tfrac{1}{2}V(J_+^2 + J_-^2) = \varepsilon J_z - V(J_x^2 - J_y^2) \qquad (3.41)$$

where

$$J_z = \tfrac{1}{2} \sum_n (a_{n1}^+ a_{n1} - a_{n-1}^+ a_{n-1}) \qquad (3.42a)$$

$$J_+ = \sum_n a_{n1}^+ a_{n-1} \; ; \quad J_- = J_+^\dagger \quad . \qquad (3.42b)$$

The operator $J^2 = J_x^2 + J_y^2 + J_z^2$ commutes with H, so that states appear in multi-
plets characterized by $J(J+1)$, J taking on the values $\tfrac{1}{2}N$, $\tfrac{1}{2}N-1$ ... Only the
highest $J = \tfrac{1}{2}N$ states need be considered (H has the same form for every J), since
lower J states correspond to the maximum multiplet for smaller values of N. H can
conveniently be diagonalized numerically in a $|JM\rangle$ basis.

The time-dependent V(t) added to the above H_o is taken as

$$V(t) = \vec{f}(t) \cdot \vec{J} \qquad (3.43)$$

where typically a gaussian form is adopted for f. This problem is exactly solvable
by expanding the time-dependent wave function in the basis of numerically determined
eigenstates of H and integrating the resulting N + 1 coupled channels equations.
The evolution operator U(t) acquires the form of a rotation operator in quasi-spin
space, namely

$$U(t) = \exp(-2i\vec{\alpha}(t) \cdot \vec{J}) \qquad (3.44)$$

The mean-field $\vec{\sigma}$, entering through terms $(\sigma, v\rho) = -2V(\sigma_x J_x - \sigma_y J_y)$ is then de-
termined in terms of $\vec{\alpha}$ by eq. (3.39). Ref. |25| presents a detailed discussion.
In figure 3.4 we show results involving a comparison of (1) exact excitation amplitudes, (2) time-dependent mean-field amplitudes obtained by a self-consistent
solution of the loop equations and (3) amplitudes obtained from interaction picture
TDHF where U_σ is replaced by U_{TDHF} taking into account only time-local self-
consistency.

One observes satisfactory agreement with improvement as N increases. Calculation of
$|S_{oo}|$ as function of interaction strength $\chi = NV/\varepsilon$ also shows good agreement over a
wide range of strengths |25|.

A more realistic application has recently been made to atomic p + He scattering |27|

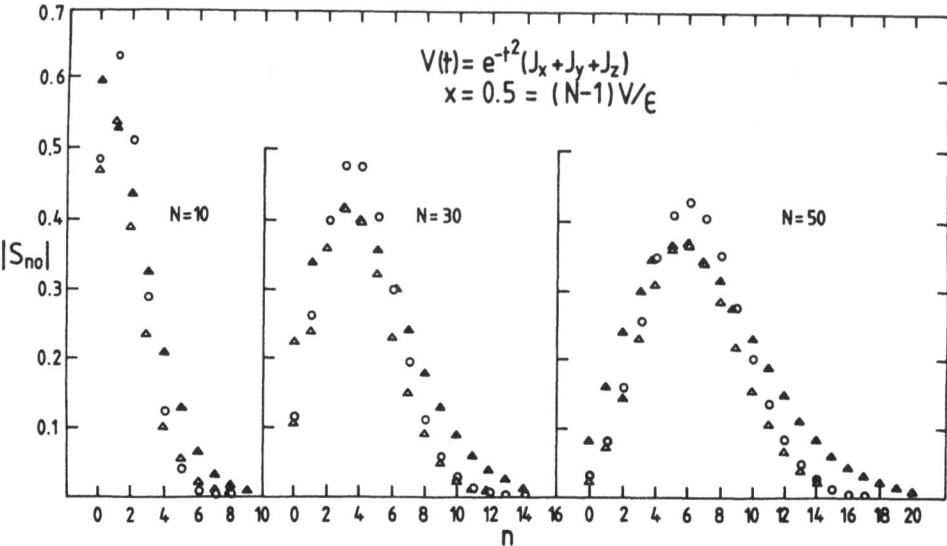

Fig. 3.4   Excitation amplitudes $|S_{no}|$ for the forced Lipkin model. o - exact results;
Δ - mean field approximation; ▲ - interaction picture TDHF. (From ref. 25)

which compared very favourably with experiment and other calculations tailored to
the atomic problem.

3.4 A variational principle of Balian and Veneroni

We consider here the problem of an improved treatment of the interpretation of the
TDHF 'wave function' which has so far been implemented in strict analogy with the
usual quantum mechanical wave function.

While this analogy has led to satisfactory results for one-body observables with
$<A> = tr\rho A$, it is for two- and many-body observables that improvement is called for.
(We recall the unsatisfactory results obtained e.g. in eq. (2.23) where
$\Delta A^2 = <A^2> - <A>^2 = tr\,A\rho A\,(\rho - 1)$ was calculated.

Balian and Veneroni |28| have attacked this problem form the following more general
viewpoint. Consider the density matrix D(t) of a many-body problem and let A be a
general many-body observable. For a given hamiltonian H we are interested in
calculating the final state (final time t_1) expectation value of A, namely

$$<A>_f = Tr\, A\, D(t_1) \quad , \tag{3.45}$$

D(t) evolving according to

$$\frac{dD}{dt} = i[D, H] \tag{3.46}$$

implying

$$D(t_1) = \exp(-iH(t_1 - t_0))D(t_0) \exp(iH(t_1 - t_0)) \tag{3.47}$$

where $D(t_0)$ is some prescribed initial value.

Equivalently one could use the Heisenberg picture where the observable A carries the time-dependence while D is time constant. Here one has $<A>_f = \text{Tr } A(t_0, t_1)D$, $A(t_0, t_1) = \exp(iH(t_1 - t_0)) A \exp(-iH(t_1 - t_0))$, A evolving as d/dt_0 $(A(t_0, t_1)) = i[H, A]$ with 'initial' condition $A(t_1, t_1) = A$ at the final time t_1.

We therefore have

$$<A>_f = \text{Tr } A D(t_1) = \text{Tr } A(t_0)D = \text{Tr } A(t) D(t) \tag{3.48}$$

independent of t corresponding to D evolving forward in time from time t_0, A evolving backward from time t_1, or a combination where both evolve toward some intermediate time t.

While this contains nothing new, one now proceeds to construct a functional

$$I(t_0, t_1) = \text{Tr } D(t_1) A(t_1) - \int_{t_0}^{t_1} dt(\text{Tr } A \frac{dD}{dt} + i \text{ Tr } A [H, D]) \tag{3.49}$$

where unrestricted variation with respect to $D(t)$ and $A(t)$ $(D(t_0) = D$ and $A(t_1) = A$ fixed) recovers the exact Schrödinger equation while $I = <A>_f$.

Approximations in this scheme now clearly correspond to restrictions on the variational space. From the equations

$$\text{Tr } \delta A(\frac{dD}{dt} + i[H, D]) = \text{Tr } \delta D(\frac{dA}{dt} + i[H, A]) = 0, \tag{3.50}$$

valid for general A and D, such a restriction now leads to a set of coupled equations in the parameters defining $A(t)$ and $D(t)$. For $<A>_f$ the restricted variation has the consequence that it now depends on both the initial state and the particular observable, implying that the approximate D is no longer an ordinary density matrix but rather a calculational tool for obtaining the result of an observation, given an initial state. TDHF can be recovered from the described formalism by restricting D to the form $D = \exp(-Q(t))$, $Q(t)$ and the observable $A(t)$ taken to be general one-body operators.

$$Q(t) = m(t) + \sum_{\alpha\beta} M_{\alpha\beta}(t) \, a_\alpha^+ \, a_\beta, \qquad (3.51a)$$

$$A(t) = b(t) + \sum_{\gamma\delta} B_{\gamma\delta}(t) a_\gamma^+ \, a_\delta. \qquad (3.51b)$$

For a two-body observable one finds, in contrast to the above situation where TDHF is recovered, a complicated set of equations with no transparent interpretation and specifically no obvious connection with TDHF. However, for the dispersion of a one-body observable Q (which involves two-body operators as in eq. (2.23)) one can get around this difficulty by introducing

$$A(\lambda) = \exp(\lambda Q) = \exp\{\lambda \, (b + \sum B_{\alpha\beta} \, a_\alpha^+ \, a_\beta)\} \qquad (3.52)$$

where λ is a small parameter and b and B known for given Q. The evaluation of the functional I is facilitated by the above choice of A in that it involves now an algebra of quadratic exponentials. This results in final time expectation values

$$<Q>_f = \frac{TrQD}{TrD} = \frac{d}{d\lambda} \left(\ell n \; Tr \; A(\lambda) \; D(t_1) \right)|_{\lambda = 0} \qquad (3.53)$$

and

$$<\Delta Q^2>_f = \frac{d^2}{d\lambda^2} \left(\ell n \; Tr \; A(\lambda) \; D(t_1) \right)_{\lambda = 0}. \qquad (3.54)$$

The important point that emerges from an analysis of these results is that a pre-scription for the calculation of $<Q>_f$ can be obtained |28| which utilizes existing TDHF-technology, i.e. while one succeeds in generalizing naive TDHF, the generalisa-tion can be accommodated within the existing formalism. The resulting prescription consists of

(i) Doing TDHF from initial time t_0 to final time t_1, starting with an initial TDHF wavefunction $\psi(t_0)$ (associated density matrix ρ) and ending with $\psi(t_1)$;

(ii) constructing at t_1 a new wavefunction $\psi'(t_1)$ as $\psi'(t_1) = \exp(i\epsilon Q)\psi(t_1)$, namely a wave function obtained through a unitary transformation generated by Q;

(iii) doing TDHF backward from $\psi'(t_1)$ to $\psi'(t_0)$ (associated density matrix σ) for $\langle\Delta Q^2\rangle_f$ one then has the result

$$\langle\Delta Q^2\rangle_f = \lim_{\varepsilon \to 0} \frac{1}{2\varepsilon} \, tr \, (\rho - \sigma)^2. \tag{3.55}$$

Apart from the fact that the prescription follows from a rigorous formulation while retaining existing TDHF technology, it has the additional advantage of treating centre of mass motion correctly.   In a very recent application to mass dispersion in $^{16}O + ^{16}O$ head-on collisions   (the operator Q being the number operator A_R in one half-plane) encouraging results show an increase in the width $\langle\Delta A_R^2\rangle$ of a factor two as compared to conventional TDHF widths which are always too small.

4. Statistical dissipation models

4.1 Motivation and background

The reason why the above general class of models, including branches like transport models, linear response models and friction models, have any bearing on the description of heavy ion collisions can be traced to the fact that these collisions deposit large amounts of energy in the emerging fragments. At excitation energies of several hundred MeV nuclear properties show mainly statistical behaviour as a result of the high level density (thousands/MeV) and the associated average properties of levels.

On the other hand the initial state of the heavy ion collision is not statistical - all of the energy is organized into coherent relative motion of the target and projectile. In fact, many features of heavy ion collision data as summarized on a Wilczynski plot, suggest the existence of a relaxation process necessarily responsible for the transformation of "coherent" kineric energy into statistical heat of the fragments, namely observables such as for example fragment charge, mass, angular momentum, etc. show a gradual evolution from some initial (coherent) value to a broad distribution of values. This is of course commensurate with the idea of equilibration and a picture of cold nuclei colliding, forming a hot lump of nuclear

matter and eventually fissioning. Further support for thinking about heavy ion collisions in statistical terms comes from the fact that much of the experimental data
is inclusive with a grossly characterized final state and only a few observables
actually probed. These include mostly the energy E, charge Z, mass A, scattering
angle Θ and, less often, angular momentum L. All of these considerations indicate
that there is no need for a detailed following of all the system coordinates (as in
TDHF, for example). Rather, one hopes to identify a few "relevant" degrees of freedom, the dynamics of which is to be treated in detail, while the reamining coordinates are to be lumped in a heat bath, the influence of which is treated in some
average way. This is shown schematically in figure 4.1 where it is indicated that
one is mainly concerned with energy transfer from the relevant to irrelevant degrees
of freedom.

This proposed separation of variables raises the following questions, namely:

(i) What constitute the relevant and irrelevant degrees of freedom?

(ii) How is the latter group to be modelled and coupled to the other coordinates
and can they actually be treated statistically?

(ii) Which equations determine the relevant degrees of freedom and should they be
of classical or quantum nature?

(iv) How are the transport coefficients in these equations to be determined in
terms of known nuclear properties?

As far as the separation is concerned, the choice of relevant degrees of freedom is
largely determined by what is observed, namely E, Z, A, Θ and L already mentioned,
together with nuclear shape characterization. While some of the other questions are
addressed in what follows, we observe here the importance of relating calculations
in this framework to the rest of nuclear phenomenology in order to evaluate the
validity of the whole approach.

We now consider an estimation of the number of degrees of freedom involved in the
description of a nucleus at excitation energy E*, for which we return to the Fermi

gas paradigm of section 1. First one notes that this number is not equivalent to the level density at E* - the extreme situation of one (classical) degree of freedom corresponds in fact to a very high level density!

At E* = 0 we have a degenerate system with moe and more particle-hole (p-h) pairs appearing with increasing E*. Furthermore an average (optimal) number of p-h pairs $\bar{n}$ is associated with a given E* - for fewer ph pairs than $\bar{n}$ phase space is not used efficiently while larger than $\bar{n}$ values result in too little energy for each pair.

The density ρ_n of n p-h pairs, and hence $\bar{n}$, can be estimated in the following way due to Ericson.   Consider the expression

$$\rho_n(E^*) = (n!)^{-2} \int_0^\infty g \, d\varepsilon_{p_1} \cdots \int_0^\infty g \, d\varepsilon_{p_n} \cdot \int_0^\infty g \, d\varepsilon_{h_1} \cdots \int_0^\infty g \, d\varepsilon_{h_n}$$

$$x \, \delta(\textstyle\sum \varepsilon_{p_i} + \sum \varepsilon_{h_i} - E^*) \tag{4.1}$$

where g is the single particle level density at the Fermi surface, empirically known to be given by $g \approx (A/13)$ MeV^{-1}. Performing the integral results in

$$\rho_n(E^*) = g \frac{(gE^*)^{2n-1}}{(n!)^2 (2n-1)!} . \tag{4.2}$$

The total level density ρ is given by

$$\rho = \sum_{n=1}^\infty \rho_n \sim \exp(2\sqrt{aE^*}), \quad (a = \frac{\pi^2}{6} g) \tag{4.3}$$

from which the peak value $\bar{n}$ is found as that n for which ρ is stationary. Using Stirling's approximation for n!(n! $\cong \sqrt{2\pi n} \, n^n e^{-n}$) this gives

$$\bar{n} \cong (\frac{gE^*}{2})^{\frac{1}{2}}. \tag{4.4}$$

For typical values A ~ 200, E* ~ 100 MeV one finds $\bar{n}$ ~ 25 - 30, which is roughly the number of coordinates involved. This shows that while the statistical assumption is probably tolerable, one is nowhere near the solid state regime with ~ 10^{23} degrees of freedom.

We now consider the statistical assumptions in heavy ion collisions from the point of view of different time scales involved. The collective time scale over which a

typical collision takes place (i.e. the time scale provided by collective dynamics)

can be extracted from TDHF (and other analyses) and is found to be τ_{coll} ~ several

x 10^{-21} sec.

Several microscopic time scales can be identified. It is clear that the recurrence

time (Poincaré time) ~ $(\rho(\varepsilon^*))^{-1}$ is extremely long as compared to τ_{coll} and has no

bearing on the present considerations. The next longest time is the decay time for

a p-h excitation, shown in section 1 to be ~ 10^{-21} sec, which is still not short

enough to account for statistical equilibration of the irrelevant degrees of free-

dom in a time comparable to τ_{coll}.

The shortest microscopic time scale is provided by the transit time τ_{sp} for a (pro-

jectile) particle across the (target) nucleus. For a Fermi velocity of

80 fm/10^{-21} sec and typical nuclear radii one finds τ_{sp} ~ 10^{-23} sec, suggesting that

during a heavy ion collision nucleons near the Fermi surface undergo many collisions

with the walls of the generated "potential". In classical terms the changing

geometry of the potential "walls" from which nucleons "bounce" is therefore the

major randomizing agent underlying the statistical picture.

From the above point of view nuclei are therefore one-body systems dominated by the

long mean free paths of nucleons, the coordinates of which constitute the micro

degrees of freedom. The relevant micro processes are transfer   through the "neck"

of the potential (provided by the colliding nuclei) and collisions with the changing

walls. The relevant macro degrees of freedom considered here include nuclear shape,

average nucleon number in either half plane (in terms of the original locations of

the colliding nuclei) and average neutron/proton number radio.

Next we consider the equations determining the macro variables and the calculation

of the various transport coefficients involved.

4.2 Transport equations

Pursuing the idea of describing heavy-ion, and more specific, deep inelastic collisions in terms of the coupling between a few collective degrees of freedom and a large number of intrinsic coordinates, we have arrived at the situation shown schematically in figure 4.1.

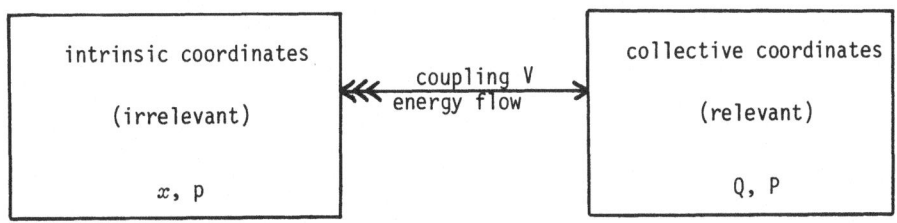

Fig. 4.1

Just from the point of view of the phase space involved, the dominant energy and momentum flow is a dissipative one from the few collective coordinates (nuclear shape, Z, A, L, E and Θ of the fragments) to the many intrinsic coordinates (location of individual nucleons in the (di-) nuclear shape). The actual number of intrinsic co-ordinates is, however, only of the order ≤ 40 (as shown in the previous section) indicating that these coordinates can be subject to rather large statistical fluctua-tions resulting in reverse energy and momentum flow to the collective coordinates. Consequently one has the situation that the statistical fluctuations of the intrinsic coordinates result in a non-deterministic evolution of the collective coordinates, leading to a description of collective coordinates in terms of distributions of values rather than specific values.

The above picture is further evidenced by the fact that the masses associated with the collective coordinates are not very large on a nuclear scale (5 ~ 10 nuclear masses) and can therefore be significantly affected by fluctuations of the intrinsic coordinates. An often drawn analogy is that with Brownian motion where continuous micro-collisions slow down a heavy test particle (dissipation), while there is also the possibility of this particle executing a random walk if initially at rest.

One of the most fundamental implementations of the above scheme is embodied in the Fokker-Planck equation. Here the collective coordinates are introduced in a probabil-istic sense from the outset, namely one introduces a classical distribution function f(Q, P, t) which gives the probability of finding the collective coordinates at (Q, P) in phase space at time t. From the initial conditions $Q = Q_0$ and $P = P_0$ at t = 0 it clearly follows that

$$f(Q, P, t = 0) = \delta(Q - Q_0) \, \delta(P - P_0). \tag{4.5}$$

Coupling of the collective coordinates to the heat bath represented by the intrinsic coordinates now leads on the most microscopic level to a Langevin equation, namely the equation of motion for a collective coordinate under the fluctuating force exerted by the heat bath. Assumptions about the time scale of the fluctuations can further simplify this equation to the Fokker-Planck equation which summarizes the behaviour of a collective coordinate in terms of just two coefficients, namely a friction or dissipation coefficient γ and a momentum diffusion coefficient D_p. In the Fokker-Planck equation

$$\frac{\partial f}{\partial t} + \frac{P}{M} \frac{\partial f}{\partial Q} - \frac{\partial V}{\partial Q} \frac{\partial f}{\partial P} = \frac{\partial}{\partial P} (\gamma P f) + \frac{1}{2} \frac{\partial^2}{\partial P^2} (D_p f) \tag{4.6}$$

the ℓ.h.s. is recognized as Vlasov piece encountered already in eq. (3.3) in another context) in which V represents a conservative potential, while the r.h.s. gives the heat bath contribution in terms of γ and D_p, both related to autocorrelation functions of the fluctuating force.

The effect of the r.h.s. terms is to distort the classical phase space trajectory associated with the Vlasov piece into a whole set of trajectories realized within some probabilistic envelope.

The interpretation of the coefficients γ and D_p become evident by considering moments of f and the time evolution of the associated expectation value. Hence one has e.g.

$$<Q> = \int dP \, dQ \, Q \, f(P, Q, t) \tag{4.7a}$$

$$<P> = \int dP \, dQ \, P \, f(P, Q, t) \tag{4.7b}$$

and for the time evolution of <Q>

$$\frac{d}{dt} <Q> = \int dP \, dQ \, Q \, \frac{\partial f}{\partial t}$$

$$= \int dP \, dQ \, Q \cdot (-\frac{P}{M} \frac{\partial f}{\partial Q})$$

$$= \int dP \; dQ \; (\tfrac{P}{M})f = \tfrac{1}{M}<P>, \tag{4.8}$$

where the second line uses the Fokker-Planck equation (4.6) and the fact that the non-surviving terms all contain a total derivative of P which vanishes upon integration because $f \xrightarrow[\pm\infty]{} 0$. In a similar way one can show

$$\frac{d}{dt} <P> = -<\frac{\partial V}{\partial Q}> - \gamma<P> \tag{4.9}$$

and

$$\frac{d}{dt} <\Delta P^2> = \frac{d}{dt} (<P^2> - <P>^2)$$

$$= D_P - 2\gamma<\Delta P^2>. \tag{4.10}$$

The solution

$$<\Delta P^2> = \frac{D_P}{2\gamma} (1 - e^{-2\gamma t}) \tag{4.11}$$

facilitates a connection between the friction and momentum diffusion coefficients γ and D_P. When thermal equilibrium is reached by the collective coordinate coupled to the heat bath, equipartition relates $<\Delta P^2>$ to the temperature T by

$$<\Delta P^2>_{t\to\infty} \cdot \frac{1}{2M} = \tfrac{1}{2}T,$$

which, when combined with eq. (4.11), yields the Einstein relation

$$D_P = 2\gamma M T, \tag{4.12}$$

a special case of the more general fluctuation-dissipation result.

It is interesting to note that the Fokker-Planck equation does not explicitly contain a term of the form $\tfrac{1}{2}(\partial^2/\partial Q^2)D_Q f$, since the original microscopic description only involves energy and momentum coupling of the test particle, but no direct coupling to its position coordinate. Nevertheless one does find a spread of Q in time, namely

$$\frac{d}{dt} <\Delta Q>^2 = \frac{2T}{\gamma M} - \frac{2}{M} (<\Delta Q^2> \frac{\partial^2 V}{\partial Q^2})_{<Q>} \tag{4.13}$$

where V is assumed to be smooth. From this result it is then possible to extract a position diffusion coefficient D_Q given by

$$D_Q = \frac{2T}{\gamma M}. \tag{4.14}$$

This coefficient plays a role in a simplified Fokker-Planck equation, often involved in the analysis of data, namely

$$\frac{\partial f}{\partial t} = -\frac{\partial}{\partial Q}(vf) + \frac{1}{2}\frac{\partial^2}{\partial Q^2}(D_Q f), \tag{4.15}$$

which essentially corresponds to an overdamped approximation. The equation describes a random walk in terms of a drift coefficient v and can be solved for constant v and D_Q. The result is a spreading and drifting Gaussian

$$f(Q, t) = (2\pi D_Q t)^{-\frac{1}{2}} \exp\left(-(Q - vt)^2/2D_Q t\right) . \tag{4.16}$$

Note that in this case one finds no Einstein relation, i.e. v and D_Q are unrelated.

The Fokker-Planck equation is well suited for the description of smoothly changing collective variables. For situations with rapid variation (as also encountered in heavy ion collisions, e.g. in the form of changes in fragment charge) one needs a more general description as offered by what is generally known as a master equation, which is just a statement about probability conservation.

One imagines being able to group the levels of a quantum mechanical system in such a way that the levels in a specific group have more or less similar macroscopic properties. Probability conservation is expressed by

$$\frac{d}{dt}P_s(t) = -\sum_m W_{s \to m} P_s(t) + \sum_m W_{m \to s} P_m(t) \tag{4.17}$$

where

$s(m)$ represents a specific group of levels, $W_{s \to m}$ is the total rate for a level in group s to evolve into any level in group m, averaged over the levels in s, while P_s is the probability to find the system in level s.

The master equation (4.17) can be cast into the Fokker-Planck form (4.15) whereby the drift and diffusion coefficients v and D_Q can be related to the underlying transition probabilities. In order to do this, the level index s is treated as a continuous variable suggesting the substitution

$$W_{s \to m} \to W_{sm} \rho(m)$$

where $W_{sm} = W_{ms}$ is the square of some matrix element and $\rho(m)$ the level density for group m. Under this assumption the master equation takes the form

$$\frac{d}{dt} P_s(t) = \int dm \, W_{sm} \, (P_m \, \rho(s) - P_s \, \rho(m)) \qquad (4.18)$$

from which one recognizes an equilibrium solution

$$P_s^{(o)} = \rho(s) / \sum_m \rho(m) \qquad (4.19)$$

corresponding to a uniform population of phase space.

The Fokker-Planck form (4.15) is obtained if one assumes the functional form $W_{sm} = W(\frac{1}{2}(s + m), s - m)$, where W is a slowly varying function in $\frac{1}{2}(s + m)$ while sharply peaked and symmetric near s=m. A standard moment expansion of the integrand in eq. (4.18) around m = s then yields

$$\frac{\partial}{\partial t} P_s(t) = -\frac{\partial}{\partial s} \{c_1(s) \, P_s(t)\} + \frac{\partial^2}{\partial s^2} \{c_2(s) \, P_s(t)\} \qquad (4.20)$$

$$c_2(s) = \rho(s)\mu_2(s); \quad c_1(s) = \frac{1}{\rho(s)} \frac{d}{ds} (\rho(s) \, c_2(s))$$

$$\mu_2(s) = \frac{1}{2} \int_{-\infty}^{\infty} W(s, \xi)\xi^2 \, d\xi \qquad . \qquad (4.21)$$

From a comparison of eqs. (4.15) and (4.20) the coefficients γ and D_Q can now be related to microscopic quantities in an obvious way.

As it stands, the master equation is not *a priori* applicable to heavy-ion collisions for which some of the inherent assumptions are manifestly violated. In the first place W is assumed time-independent while the collective motion in a heavy-ion collision "switches" the evolution for, say, charge transfer on and off during the collision. While this is easily remedied by giving W time-dependence, a more serious problem concerns the Markovian assumption, namely that given $P = P(t_0)$ at some initial time the subsequent evolution is completely determined, i.e. the equation is first order in time. The violation of this assumption is traced to the long

mean free path of nucleons participating in the heavy-ion collision and the fact
that their motion at a given time subsequently reflects a complete time history or
memory of previous motion.

These considerations suggest a generalization of the form

$$\frac{d}{dt} P_s(t) = - \sum_m \int_{-\infty}^{\infty} d\tau \; K_{s \to m} \; (t, \tau) P_s(t + \tau) + \sum_m \int_{-\infty}^{\infty} d\tau \; K_{m \to s} \; (t, \tau) P_m(t + \tau),$$

$$(4.22)$$

where K acts as a memory kernel. If K is short-ranged in τ, one can put $P_s(t + \tau) \approx P_s$
and eq. (4.15) is then recovered with W appearing as a τ-integral of K.

As an illustration of how one arrives at an equation of the form (4.22) in a microscopic situation, we use a formulation of the intrinsic degrees of freedom, original-
ly due to Hofmann and Siemens |29|. Consider a nucleus where the collective variables
are identified with the nuclear shape.

Fig. 4.2

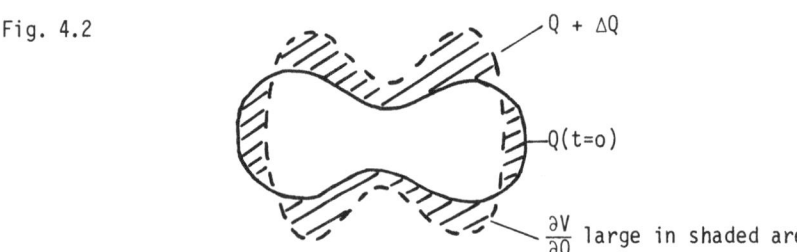

$Q + \Delta Q$

$Q(t=o)$

$\frac{\partial V}{\partial Q}$ large in shaded area

With the shape prescribed at t = 0 (say Q(t = 0) = 0) we now consider the response
of the intrinsic system to a change in Q and the subsequent change in the potential
V. Having a linear response picture in mind, the hamiltonian is therefore taken as

$$H = T + V(x, Q(t))$$

$$\cong H_0(x, p) + \frac{\partial V}{\partial Q} Q(t) \qquad (4.23)$$

where $\frac{\partial V}{\partial Q}$ is surfaced peaked, i.e. large in the shaded areas indicated above and
$H_0(x, p) = \sum p^2/2m + V(x, Q = 0)$. (The initial choice for Q implies V(t = 0) = 0.)

From an initial value ρ_0 at t = 0, the density matrix ρ of the intrinsic system
evolves as

$$i\dot{\rho} = [H, \rho] = [H_0, \rho] + [V, \rho]. \qquad (4.24)$$

Introducing the interaction picture quantities $\hat{\rho}$ and $\hat{V}(\hat{\rho} \equiv \exp(iH_o t)\rho \exp(-iH_o t)$

and similarly for $\hat{V}$ this equation is solved by the Dyson series

$$\hat{\rho} = \rho_0 - i\int_0^t [\hat{V}(t'), \rho_0]dt' + \frac{(-i)^2}{2!} \int_0^t dt' \int_0^{t'} dt'' [\hat{V}(t'), [\hat{V}(t), \rho_0]] + \cdots \quad (4.25)$$

The energy change of the intrinsic system is given by

$$\frac{dE}{dt} = \frac{d}{dt} \text{tr} (H\rho) = \text{tr} \frac{dH}{dt} \rho \quad (4.26)$$

where we have used

$$\text{tr } H \frac{d\rho}{dt} = -i \text{ tr } H [H, \rho] = i\text{tr } [H, H]\rho = 0. \quad (4.27)$$

Therefore, retaining the first order term in eq. (4.25), we have

$$\frac{dE}{dt} = \text{tr} \frac{d\hat{V}}{dt} \hat{\rho}$$

$$= \dot{Q} \text{ tr } \frac{\partial V}{\partial Q} \rho_0 - i\int_0^t dt' \text{ tr } \left[\frac{\partial \hat{V}}{\partial t}(t), \hat{V}(t')\right]\rho_0 , \quad (4.28)$$

where the first term has a "force x velocity" structure expected from a conservative

potential. Interpretation of the second term in $\frac{dE}{dt}$ is facilitated by a Taylor ex-

pansion $Q(t) = \dot{Q}t + \frac{1}{2}\ddot{Q}t^2 + \cdots$ in the functional form assumed for V in eq. (4.23).

Up to this order the combination $V \frac{\partial V}{\partial t}$ therefore contains terms with contributions

$(\frac{\partial V}{\partial Q})^2 \dot{Q}^2$ and $(\frac{\partial V}{\partial Q})^2 \dot{Q}\ddot{Q}$, the first of which can be interpreted as a dissipation term,

while the second is the time derivative of a kinetic energy, the coefficient of which

can be identified with a mass. Evaluating the integrals under the assumption of a

short relaxation time then gives the following expressions for the dissipation co-

efficient γ and the mass M:

$$\gamma = \sum_{mn} \rho_n |\langle n| \frac{\partial V}{\partial Q} |m\rangle|^2 \delta'(E_n - E_m) \quad (4.29a)$$

$$M = \sum_{mn} \rho_n |\langle n| \frac{\partial V}{\partial Q} |m\rangle|^2 P(E_n E_m)^{-3} \quad (4.29b)$$

where the indices enumerate eigenstates of H_o, ρ_n is the n-th element of the un-

perturbed density matrix, δ' denotes a derivative of the delta function and P in-

dicates that the principal value is to be taken.

Note that the second expression is analogous to that obtained from cranking, while the first implies, through the delta function, that no dissipation occurs for a finite system with discrete energy levels. In order to introduce dissipation in such systems would therefore require some averaging over levels and the introduction of an appropriate width.

In terms of a response function $\chi(t, t')$ the general structure of $\frac{dE}{dt}$ is

$$\frac{dE}{dt} = \dot{Q} \ tr \ \rho_0 \ \frac{\partial V}{\partial Q} + \int_0^t dt' \ \dot{Q}(t)\chi(t - t')Q(t') \qquad (4.30)$$

with

$$\chi(t - t') = 2\Theta(t - t') \sum_{mn} \rho_n |<n| \frac{\partial V}{\partial Q} |m>|^2 \ \sin \ (E_n - E_m)(t - t'). \qquad (4.31)$$

While the detailed structure of χ depends on the specific choice of H, the general behavious is indicated schematically below in figure 4.3.

Fig. 4.3

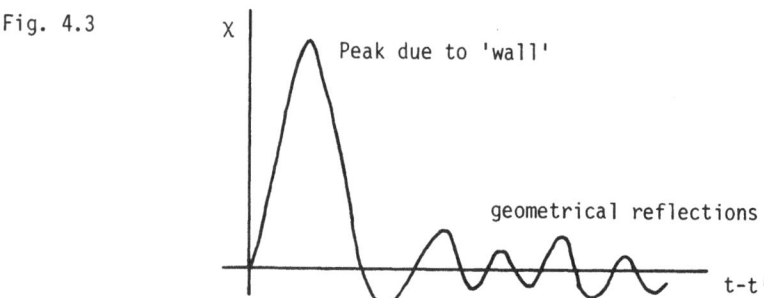

The first large peak corresponds to the response of all the particles as they initially interact with the wall of the potential and semi-classical considerations show that its occurrence is largely independent of the shape of the potential. The irregular oscillations that follow are concerned with motion throughout the potential and are largely determined by geometry of that potential.

As already indicated, in a local time expansion approximation for Q(t, t'), it is now possible to relate the quantities γ and M to microscopy through the response function χ as shown in expression (4.29). Unfortunately, both γ and M diverge when these expressions are calculated for a realistic potential in a single particle framework - a result of the long time tails associated with the single particle states and subsequent "ringing". These divergences can be remedied by introducing

an additional randomizing agent in the system, namely a finite single particle life-

time, which is most naturally accomplished by using an optical model potential.

An approach which retains the spirit of the foregoing discussion in a simplifying

phenomenological framework is the one-body dissipation picture of Swiatecki, Randrup

and collaborators |30|. In its simplest form one derives the "wall formula" for dis-

sipation by considering the elastic collision of a nucleon (initial velocity $\vec{v}$) with

an infinitely massive confining wall (area A and initial velocity $\vec{u}$). Taking into

account the collisions of all the Fermi gas nucleons, one finds from standard statis-

tical mechanics the rate at which energy is transferred

$$\frac{dE}{dt} = \tfrac{1}{4} \rho \, \bar{v} \, u^2 \, A$$

$$\rightarrow \tfrac{1}{4} \rho \, \bar{v} \int d^2 S \, u^2(S) \tag{4.32}$$

where ρ is the mass density (taken to be nuclear matter density), $\bar{v}$ is the mean nu-

cleon speed and the second line is a generalization to an arbitrary surface S. At

this stage it is already apparent that some further generalization is required since

$dE/dt \neq 0$ for uniform translation ($\vec{u}$ a constant) - a problem which can be traced to

the fact that the long mean free path is ignored and only a first collision between

nucleon and wall is taken into account. The remedy takes the form

$$\frac{dE}{dt} = \tfrac{1}{4}\rho v \left| \int d^2 S \, u^2(S) + \int dS \, dS' \, u(S)\gamma(S, \, S')u'(S) \right. \tag{4.33}$$

(suggested by a linear response analysis) in which $\gamma(S, S')$ is a dissipation kernel

connecting different parts of the wall surface and determined mainly by geometry.

This expression can e.g. be shown to have proper translational invariance and to

account for the "ringing" of the response function.

Another common idea which specializes one-body dissipation ideas to the case of a di-

nuclear shape, is known as transfer-induced transport. In its simplest form two

spherical nuclei A and B overlap, the connecting window area $\Delta\sigma$ being geometrically

determined in terms of the separation distance. Assuming independent nucleon move-

ment in each nucleus, one recognizes three contributions to the force exerted on

"container" A, namely collisions of nucleons in A with the walls of A, momentum transfer by nucleons of B entering A and vice versa. The force $\vec{F}$ on A and hence momentum
change can then be obtained from statistical dynamics expressions and is given by

$$\vec{F}_A = \tfrac{1}{4}\rho\bar{v}\ \Delta\sigma\ (2\vec{u}_{||} + \vec{u}_{\perp}) \tag{4.34}$$

where the notation of eq. (3.32) is used and $\vec{u}_{||}$ ($\vec{u}_{\perp}$) is the component of relative
average velocity $\vec{u}_A - \vec{u}_B$ parallel (perpendicular) to the normal on $\Delta\sigma$. Similar ideas
can be applied in the case of mass and charge transfer and the torque.

While some conceptual problems remain - in the final case the classical picture of
particles moving through the neck is problematic as the nucleon wavelength can be comparable to the neck diameter - the phenomenological statistical dissipation models
outlined above do have a track record of quite good agreement with experimental data.
Their combination of geometrical ideas with statistical mechanics results for semi-
infinite nuclear matter lead to a picture with hardly any adjustable parameters and
subject to relatively simple numerical analysis. The main drawback remains the large
dependence on assumptions that serve as input.

REFERENCES

1. M. Beiner, H. Flocard, N. Van Giai and P. Quentin,   Nucl. Phys. A238 (1975) 29

2. S. E. Koonin,   Ph.D. Thesis, Massachusetts Institute of Technology (1975) Un-published

3. P. Hoodboy and J.W. Negele,   Nucl. Phys. A288 (1977) 23

4. P. Bonche, S.E. Koonin and J.W. Negele,   Phys. Rev. C13 (1976) 1226

5. H. Flocard, S.E. Koonin and M.S. Weiss,   Phys. Rev. C17 (1978) 1682

6. P. Bonche, B. Grammaticos and S.E. Koonin,   Phys. Rev. C17 (1978) 1700

7. S.E. Koonin et al., Phys. Rev C15 (1977) 1359

8. S.E. Koonin, Phys. Lett. 61B (1967) 227

9. K.R. Sandhya-Devi and M.R. Strayer,   J. Phys. G4 (1978) L97; Phys. Lett. 77B (1978) 135

10. S.E. Koonin, et al., Phys. Lett. 77B (1978) 13

11. K.R. Sandhya-Devi, A.K. Dhar and M.R. Strayer,   Phys. Rev. C23 (1981) 2062

13. H.S. Köhler,   Acta Phys. Rol. B13 (1982) 107

14. H.S. Köhler and H. Flocard,   Nucl. Phys. A323 (1979) 189

15. H. Orland and R. Schaeffer,   Z. Phys. A290 (1978) 191

16. C.Y. Wong and H.H.K. Tang,   Phys. Rev. Lett. 40 (1977) 1070

17. H.S. Köhler,   Nucl. Phys. A378 (1982) 159

18. J. Hubbard,   Phys. Lett. 3 (1959) 77

19. R.L. Stratonovich,   Sov. Phys. Doklady 2 (1958) 416

20. S. Levit, J.W. Negele and Z. Paltiel,   Phys. Rev. C21 (1980) 1603

21. H. Reinhardt,   Nucl. Phys. A331 (1979) 353; A346 (1980) 1

22. H. Kleinert,   Phys. Lett. 69B (1977) 9

23. H. Kleinert and H. Reinhardt,   Nucl. Phys. (1979) 331

25. Y. Alhassid and S.E. Koonin,   Phys. Rev. C23 (1981) 1590

26. H. Lipkin, N. Meshkov and A.J. Glick,   Nucl. Phys. 62 (1965) 188

27. K.R.S. Devi and S.E. Koonin,   Phys. Rev. Lett. 47 (1981) 27

28. R. Balian and M. Veneroni,   Phys. Rev. Lett. 47 (1981) 1353

29. H. Hofmann and P.J. Siemens,   Nucl. Phys. A257 (1976) 165

30. J. Randrup and W.J. Swiatecki,   Ann. Phys. 125 (1980) 193

31. J. Negele,   Rev. Mod. Phys. $\underline{54}$ (1982) 913

32. K.T.R. Davies, K.R. Sandhya-Devi and M.R. Strayer.   Phys. Rev. C$\underline{24}$ (1982) 2576

33. R. Vandenbosch   et al., Phys. Rev. C$\underline{17}$ (1978) 1672

34. J. Wilczynski,   Phys. Lett. $\underline{47}$B (1973) 484

NEW VISTAS OF THE SHAPES AND STRUCTURES OF NUCLEI FAR OFF STABILITY

J.H. Hamilton
Vanderbilt University
Physics Department
Nashville, TN 37235, U.S.A.

INTRODUCTION

One of the important motivations for the study of all aspects of nuclear structures is that the nucleus is our major, if not unique, testing ground for an important intermediate realm of quantum many-body physics; namely, systems with numbers of particles sufficiently small as to not be treatable by statistical methods yet more than a single or few particle systems. Much remains to be done, both experimentally and theoretically, to understand the greater diversity of nuclear motions and shapes being observed today and the interplay of collective and single particle motions. Prior to the last decade most of our knowledge came from stable or near stable nuclei. In the last ten years, the development of new accelerators and new experimental techniques has produced an explosion in our knowledge of the level structures of nuclei far from the valley of beta stability. Such studies have had major impacts on our understandings of the shapes and structures of nuclei, transforming and extending many of our older ideas of nuclear shapes and structures. Beams of heavy ions which have become available in the last decade have made possible many of the important discoveries.

In my first two lectures, some of the highlights of recent discoveries which have changed our understandings will be presented. These include: the coexistence of spherical and deformed shapes in the same nucleus in contrast to nuclei having either permanent spherical shapes or permanent well-deformed shapes (how shape coexistence changes with N, Z and spin I will be described); an unexpected new region with large deformation is established around $N = Z = 38$; the importance of having the shape driving forces for both protons and neutrons reinforcing each other to produce large deformation or spherical shapes; and ground state octupole deformation. Here we will draw in part on a forthcoming review of Hamilton, Hansen and Zganjar[1] on nuclei far from stability as well as the original articles. A new method of analysis to test the applicability of the rotation energy formulae to various bands in deformed and transitional nuclei will be considered along with various consequences of this method including how to extract the aligned angular momentum in rotational aligned bands. High moments in the nuclear shape such as large negative hexadecapole moments, β_4 and β_6 moments, will be shown to have important consequences in our theoretical understandings and in future experimental research. Some other recent reviews in this area include Hansen[2] and Hamilton[3,4,5].

SHAPE COEXISTENCE

Baranger and Sorenson[6] have summarized our understandings of a decade ago on the shapes and structure of nuclei which were grouped into three categories: A) spherical, B) "hard" deformed, and C) "soft" transitional nuclei. They noted "spherical and hard deformed nuclei maintain their shapes through time," that is, both these classes of nuclei have "permanent" shapes[6]. The regions where these shapes were found could be seen in Fig. 1 from their paper. Spherical nuclei occur when N and/or Z are near the magic numbers for closed shells and the hard deformed nuclei in regions well removed from the closed shells. The then known regions of deformed nuclei were in the rare earths and actinides (see Fig. 1). A new region of deformation was predicted far off stability where both $50 \leq N, Z \leq 82$ (region 1 of Fig. 1), and this region has been clearly identified now. The space between regions A and B were occupied by group C, the "soft" transitional nuclei which had no well-defined shape but could vary through different shapes with small deformation, $\beta \lesssim 0.15$. A nucleus had one of these shapes, and the general features of the low-lying energy levels of a nucleus were set by its

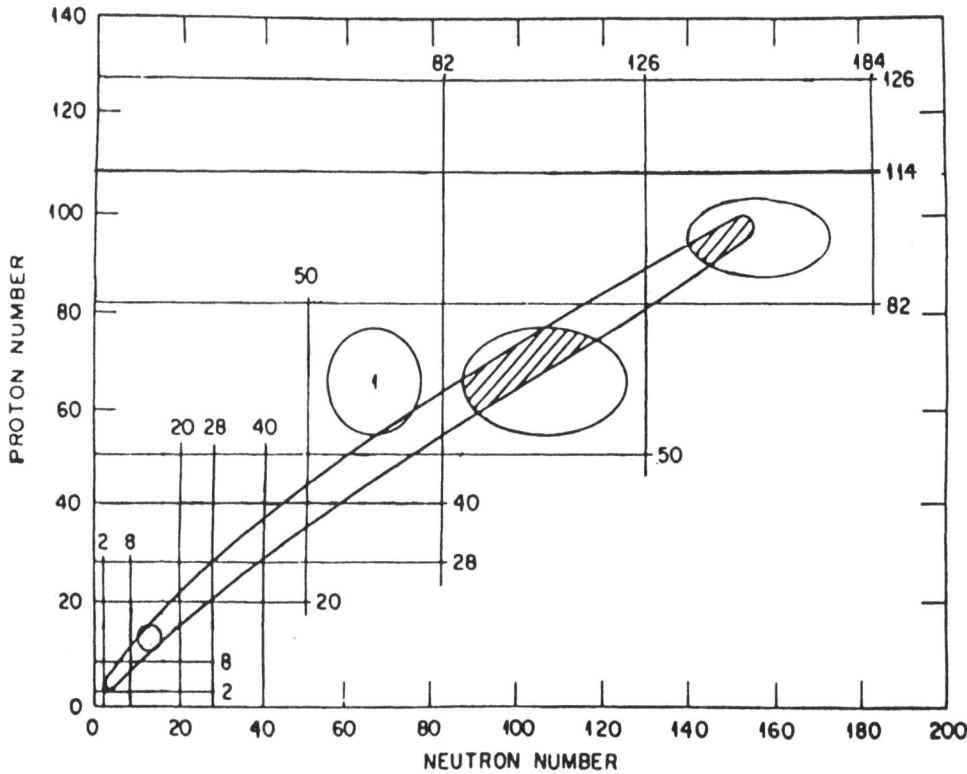

FIGURE 1. A 1969 chart from Baranger and Sorenson[1] of the nuclides as functions of N and Z with the nuclei in the valley of beta stability inside the long ellipse. Closed shell magic numbers are shown by lines and the known deformed nuclear regions are enclosed by smaller ellipses or circles along with a predicted new deformed region, Circle 1.

fixed (A,B) or soft (C) shape. Now let us see how our views have changed by looking at nuclei far off stability.

There had been isolated references to other possibilities than these three shapes. Shape isomers were predicted by Hill and Wheeler[7]. Morinaga[8] introduced both spherical and deformed shapes to explain the high energy, first few levels of double magic ^{16}O. Greiner[9] introduced the idea of shape coexistence to explain the low lying levels in ^{110}Cd. Still in 1966, Soloviev[10] and in 1974 Kumar[11] urged experimentalists to search for nuclei with excited levels with deformation quite different from the ground states. However, these isolated predictions of shape coexistence were not considered in our general understanding of nuclear shapes and level structures.

In simultaneous and separate studies, the full coexistence of bands of levels built on quite different shapes were found in ^{72}Se and $^{184,186,188}Hg$ (Hamilton[12,13]) to firmly establish the longstanding theoretical predictions. These discoveries broke down the picture of every nucleus having only one permanent shape (either A, B, or C). Now nuclear shape coexistence is seen in many regions of the periodic table including even stable nuclei with a closed proton shell like ^{116}Sn with Z = 50 (for more examples, see Hamilton[4,13] and the 1981 Helsingør Conference Proceedings). The shape of a nucleus now is seen as a dynamic variable, and strong competition can occur between the different forces which separately drive the nucleus toward prolate, oblate, triaxial and spherical shapes as N changes for a given Z or vice-versa. There is similar competition as a function of the nuclear spin.

For many nuclei the categories A, B, and C do characterize their shapes and structure up to relatively high energy, for example, deformed rare earth and actinide nuclei with $I^\pi \lesssim 10^+$. However, theoretical and experimental studies of nuclei at high angular momentum have indicated new structures, shape changes, and shape coexistence at high spins for these nuclei, too. In a review of shape coexistence, Hamilton[4] noted that nuclei which are far from stability because of their high rotational motion provide another full range of examples of coexisting structures. Some examples will be presented later.

The light mass mercury isotopes far off stability with A = 184-188 provide a classic example of the coexistence of bands of levels built on well-deformed and near-spherical shapes as illustrated in Fig. 2 (see Hamilton[13] for a more complete discussion

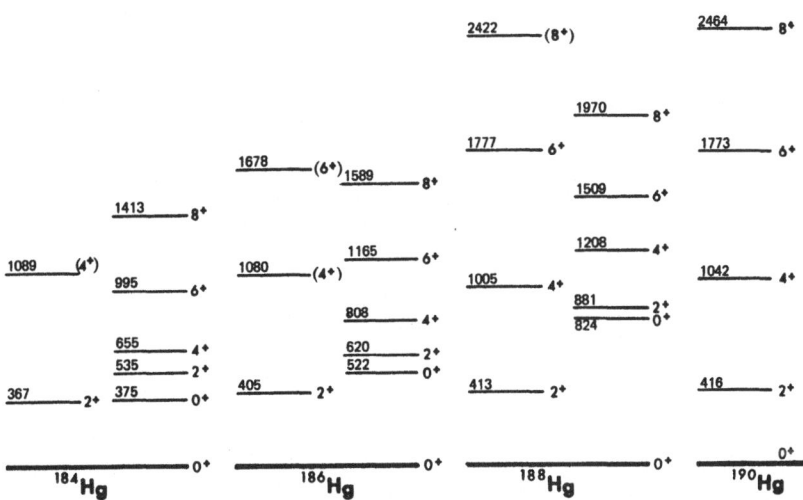

FIGURE 2. Energy levels in $^{184-190}$Hg showing the bands built on the coexisting near-spherical ground states and deformed 0_2^+ states[4,13].

and references). Bands characteristic of both shapes cross and coexist with high purity of shape above and below the mixed states where the two bands cross. This discovery was surprising since Hg nuclei are only two protons away from the Z = 82 closed proton shell at lead where near-spherical structures were expected to dominate. These data can be understood in terms of two minima in the potential energy surfaces, one near-spherical and one with large deformation. However, such a well-developed barrier between these two minima as indicated by the data was a further surprise. Studies[14] of the neighboring odd A nuclei have revealed the high j, $h_{9/2}$, orbital from above the Z = 82 closed shell gap formed by the $h_{11/2}-h_{9/2}$ split drops rapidly with energy as N decreases to far off stability as shown in Fig. 3. The structure of the deformed bands can be understood microscopically in terms of the promotion of a pair of particles to the $h_{9/2}$ orbital.

The energy of the 0_2^+ band head is seen to drop rapidly from 825 keV in ^{188}Hg to 372 keV in ^{184}Hg (see Fig. 2). This drop could indicate that ^{182}Hg or ^{180}Hg could be deformed in its ground state despite their nearness to the 82 closed proton shell. Duvall and Barrett[15] have applied the Interacting Boson Model to these coexisting chapes by adding and mixing IBA configurations characteristic of near-spherical and

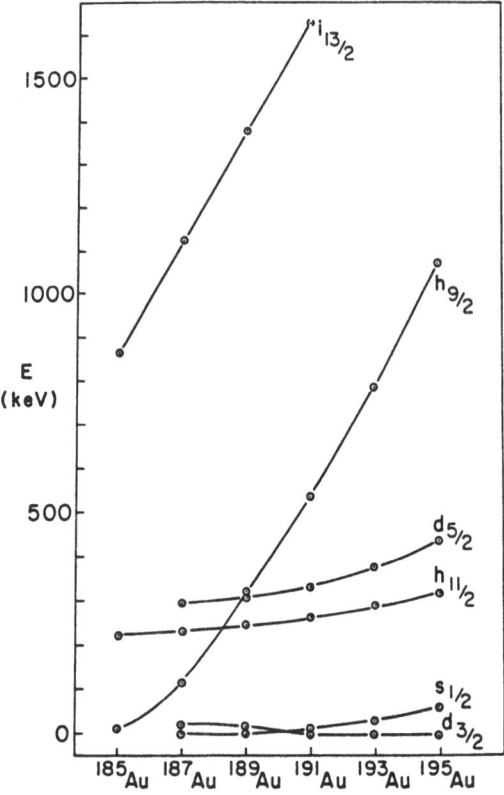

FIGURE 3. Systematics of various single particle levels in $^{185-195}$Au including the $h_{9/2}$ level which drops sharply to become the ground state of ^{185}Au (Ref. 14).

well-deformed shapes. Barfield et al.[16] predict from an IBA analysis of $^{184-190}$Hg that the deformed band should rise 70-100 keV in ^{182}Hg compared to the energies of the corresponding band in ^{184}Hg. More recent potential energy surface calculations in both the modified oscillator and Woods-Saxon potentials show a minimum in the energy of the well-deformed band head around A = 183-184 (Bengtsson et al.[17]).

To test these ideas and predictions, we have studied the reactions ^{156}Gd(^{32}S,4n)^{184}Hg and ^{154}Gd(^{32}S,4n)^{182}Hg at the Holifield Heavy Ion Research Facility. The energy levels in ^{182}Hg were identified for the first time (Ma et al.[18]) as shown in Fig. 4. The ground state of ^{182}Hg is established as near spherical like ^{184}Hg and the heavier isotopes. The near-spherical band is crossed at 2$^+$ by a well-deformed band. The energies of the 0$^+$, 2$^+$ and 4$^+$ members of the deformed band were extracted from the rotational energy formulae as applied to the 6$^+$ to 12$^+$ levels. The experimental energies of these levels undoubtedly are perturbed to some extent by mixing of the two bands as indicated by the small difference in the 4$^+$ energies. Because of a possible small energy shift, it is impossible to say whether the 0$_2^+$ band head is really below or just above the 2$_1^+$ level.

In Fig. 5 the energies of the levels in ^{182}Hg are compared with the heavier Hg isotopes. One sees that the deformed band energies established in ^{182}Hg all drop compared

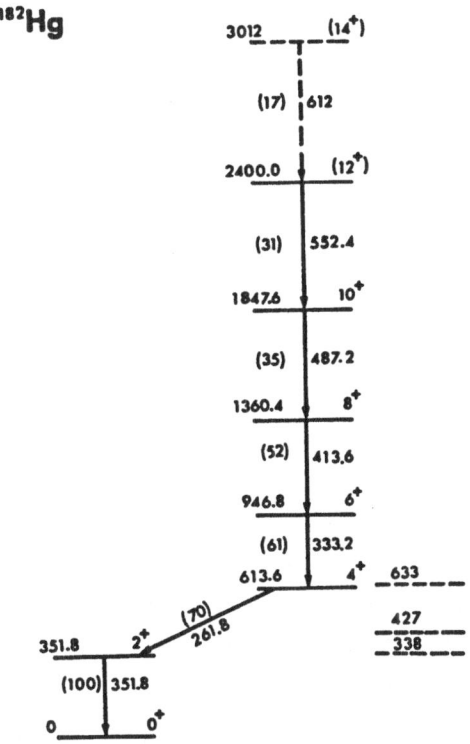

FIGURE 4. Energy levels identified for the first time in [182]Hg from the reaction [154]Gd([32]S,4n)[182]Hg (Ref. 18).

to [184]Hg and the heavier isotopes. This drop is in contrast to the 70 to 100 keV rise predicted for the deformed band energies in [182]Hg based on the Interacting Boson Model by Barfield et al.[16], as well as the recent potential energy surface calculations of Bengtsson et al.[17] These [182]Hg results provide important new data for extending our theoretical understanding of the competition of the near-spherical and well-deformed shapes in these far from stability mercury isotopes near the Z = 82 closed shell. An important feature of these coexisting structures in the light mercury isotopes is that they sensitively probe the single particle spectrum as a function of deformation far from stability. Such information is important in providing the basis for calculations into heavier, unknown regions such as the superheavy nuclei.

These nuclei provide further important tests of models which describe how two quasi-particles can couple to a core and rotation align their angular momenta with the core (see Stephens[19] for a description of the rotation alignment model). In these light mercury nuclei, one has the new possibility that the pairs of quasiparticles can couple to the quite different cores. Theoretical calculations indicate that the near-spherical shape is oblate and the well-deformed shape is prolate, although these oblate and prolate associations are not established experimentally. In any case, one has the opportunity for the first time to probe in the same nucleus as a function of spin how the energies differ for the bands formed from the couplings of different pairs of quasiparticles to the two cores with these quite different deformations.

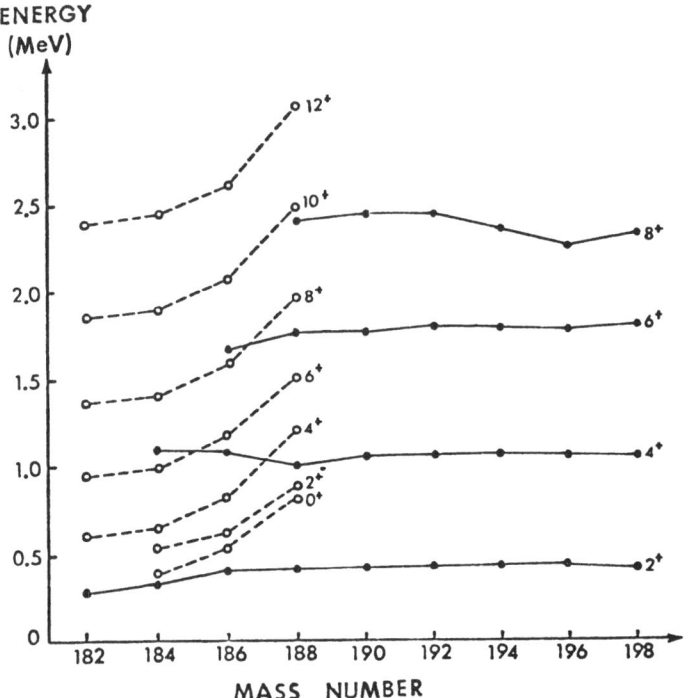

ENERGY
(MeV)

MASS NUMBER

FIGURE 5. Systematics of energy levels in $^{182-198}$Hg. Of particular interest are the newly identified deformed band levels in ^{182}Hg which are connected by dashed lines. The 4^+-12^+ levels in ^{182}Hg continue to drop in energy compared to the heavier isotopes in contrast to recent IBA[16] and potential energy surface calculations[17].

Recently ^{188}Hg (Ref. 20), ^{186}Hg (Ref. 21) and ^{184}Hg (Ref. 22) have been studied to spins of the order of 22^+ to 24^+ in heavy ion reactions. Indeed, a variety of band crossings are observed as seen in Fig. 6. We are currently seeking to understand the nature of the multiple bands observed in these nuclei. It is already clear that the promotion of two $h_{9/2}$ particles to give rise to the deformed structure of the bands built on the excited 0_2^+ states has a blocking effect on the allowed rotation aligned structures.

Simultaneously with its discovery in the light mercury isotopes shape coexistence was found in ^{72}Se (Hamilton et al.[23]). The striking similarity of the moments of inertia in ^{184}Hg and ^{72}Se (Fig. 7) which start out rising vertically as in a spherical vibrator and then bending horizontally as in a deformed rotor illustrate the close parallel between the two different structures and their crossing at spins 2^+ to 4^+ in these two nuclei which are so different in mass. Strongly deformed structures were not expected in the A = 70 region because when both N and Z are less than 50, N and Z are never very far from a magic number, especially when 40 is considered magic as shown in Fig. 1 from 1969. Nuclei in the A = 70-80 region were considered more-or-less as non-descript near-spherical vibrators and so received little attention theoretically or even experimentally before that work[23]. Even with it[23], the recent discovery of a new region of very strong ground state deformation around N = Z = 38 (Hamilton et al.[24], Piercey et al.[25]) was quite unexpected. These discoveries along with the richness of collective and single particle motions discovered[26] and the rapidity at which the different band structures can change with the addition of only two protons or two neutrons[26] have made the A = 64-84 region a very important new testing ground for nuclear models.

The energy levels of ^{76}Kr (Fig. 8, Ref. 25) illustrate the variety of collective and

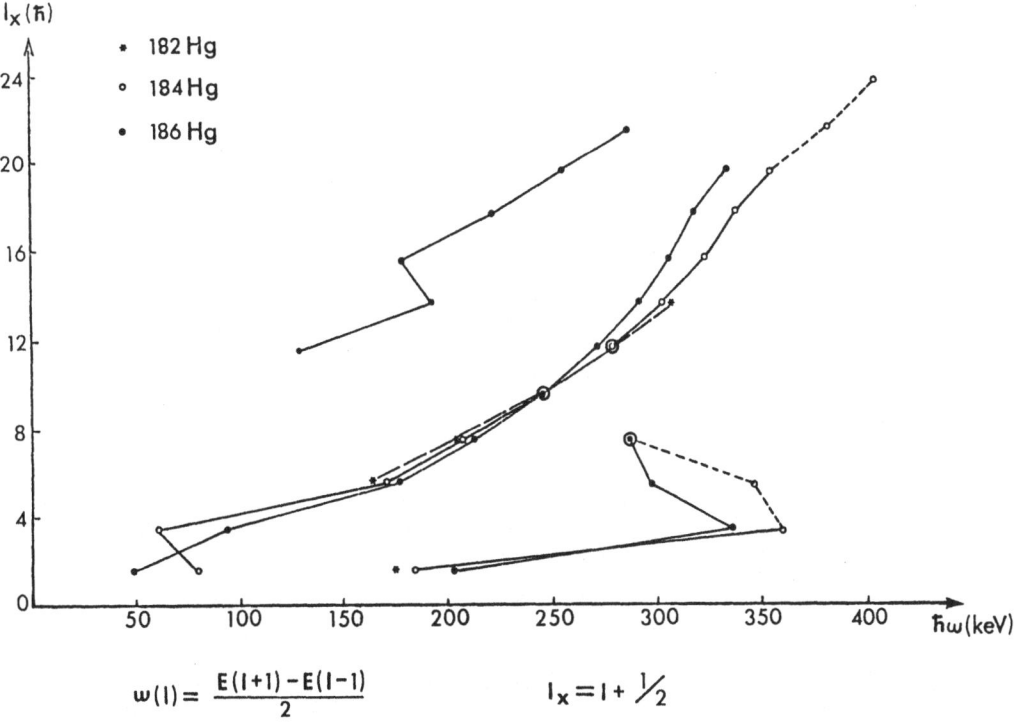

$$w(I) = \frac{E(I+1) - E(I-1)}{2} \qquad I_x = I + \tfrac{1}{2}$$

FIGURE 6. Rotational frequency as a function of spin for the near-spherical and deformed bands in ^{184,186}Hg (Refs. 21, 22).

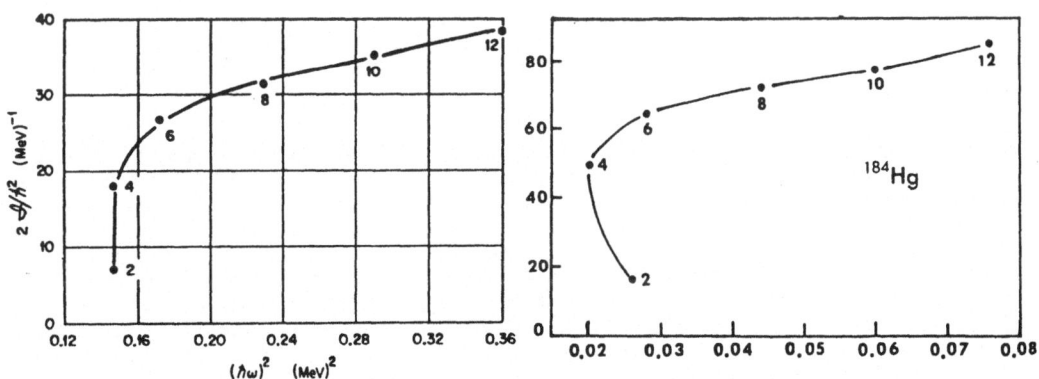

FIGURE 7. The moments of inertia of the yrast cascades of ^{72}Se and ^{184}Hg are shown.

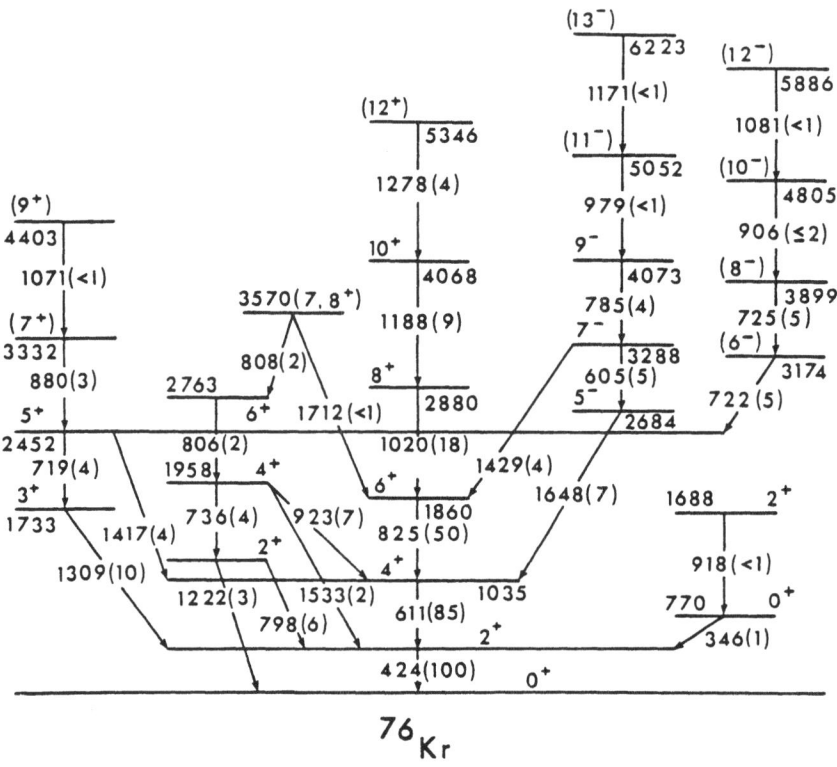

FIGURE 8. Energy levels in ^{76}Kr established from in-beam studies[25]. The levels are dominated by highly collective bands.

single particle motions now observed in this region: coexisting bands built on states with quite different deformation, γ-type vibrational bands with properties very much like those of such bands in well-deformed nuclei, even-parity bands built on a pair of high j$(g_{9/2})^2$ orbitals rotation aligned with the core, and odd- and even-spin nega-tive parity rotation aligned bands with only one particle in a g$_{9/2}$ orbital.

Our recent studies of ^{74,76}Kr led to the discovery[24,25] of strong ground-state defor-mation when both N and Z are at or near 38 and illuminated the origin of the shape co-existence in this region. Simultaneously and independently, the calculations of Möller and Nix[27] of the ground state masses and shapes for over 4000 nuclei indicated that nuclei in the region around N = Z = 38 should have the strongest ground state deformation of any nuclei with β ≈ 0.35-0.4. Likewise, the recent analysis of the levels of $^{74-80}$Kr in the collective potential energy surface approach of the Frankfurt group (Seiwert et al.[28]) yielded deformations of the order of 0.4 for ^{74,76}Kr. Subse-quent studies in the light Sr nuclei (Barclay et al.[29,30], Lister et al.[31]) have con-firmed the discovery of large deformation around N = Z = 38. The conjunction of to-tally independent experimental evidence[24,25] and theoretical prediction[27] of a new region of strong ground-state deformation was quite unique.

Let us give a brief overview. The levels in ^{74}Kr and ^{76}Kr (Fig. 8) are dominated by rotational band structures[25] which have surprisingly large B(E2)'s, for example, 78-120 single particle units in ^{76}Kr in going from the 2$^+$-0$^+$ to 10$^+$-8$^+$ transition[25,32] For comparison, in ^{68,72}Ge the B(E2)$_{exp}$/B(E2)$_{sp}$ for the 2-0 and 4-2 transitions are the order of 10-20 spu. In contrast to ^{72}Se, UNISOR studies of the ^{76}Rb decay (Piercey et al.[33]) show the levels which feed the anomalously low energy 0_2^+ level [a(2^+)-2^+ - 0_2^+ 882-917 keV cascade] have a more typical, near-spherical vibrational spacing.

Summarizing the results, the ground states of ^{74,76}Kr are seen to be well deformed and the 0_2^+ level and states built on it in ^{76}Kr associated with a near-spherical shape[24,25] in contrast to the reverse situation in ^{72,74}Se where the ground states are near spherical and the 0_2^+ deformed. In ^{74,76}Kr the 0_1^+ and 0_2^+ states interact and push each other apart to enlarge the 2_1^+ - 0_1^+ energies. These enlarged energies at first masked the ground state deformation. Similar interactions may have led to large $2_1^+ \rightarrow 0_1^+$ energies and so masked ground state deformations in other regions. Indeed, such an interaction has been used by Wood[34] to explain quite similar behavior in the light Pt isotopes where the high spin states show characteristics of much larger deformation than indicated by the 2_1-0_1 energies.

More recent studies of the levels in ^{80}Sr (Barclay et al.[29], Lister et al.[31]) and ^{78}Sr (Barclay et al.[30], Lister et al.[31]) confirm the large deformation. The systematics of the 2_1^+ - 0_1^+ energies, which in the absence of interaction as noted above, are an indication of nuclear deformation, for Sr nuclei from ^{80}Sr near the N = 50 spherical gap to $^{78}_{38}$Sr$_{40}$ are shown in Fig. 9. There the rapid drop in the 2_1^+ - 0_1^+ energy clearly signifies the rapid onset of large deformation, presumably going to a maximum deformation at N = Z = 38 in ^{76}Sr.

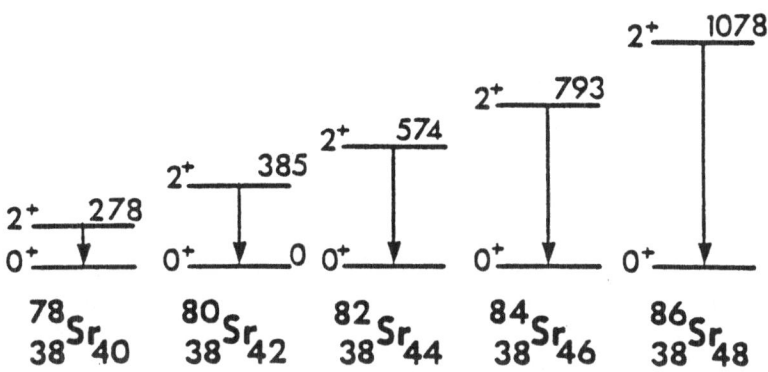

FIGURE 9. The 2^+ energies in $^{78-86}$Sr illustrate the sudden onset of large deformation.

In a unique conjunction of theory and experiment, this new region of strong deformation first discovered in ^{74,76}Kr was predicted by the calculations of Möller and Nix[27] who calculated the nuclear masses and ground state shapes for 4023 nuclei from ^{16}O to ^{279}112 with a Yukawa-Plus-Exponential Macroscopic Model and a folded Yukawa single-particle potential. Their calculations[27] predict that nuclei with both N and Z at or near 38 should be among the most strongly deformed ones in nature, with $\beta \sim 0.4$.

Potential energy surfaces have recently been calculated for $^{76-80}$Kr in a Frankfurt-Vanderbilt collaboration[28]. The ^{76}Kr surface is shown in Fig. 10. Two minima are seen, one at large deformation and one near spherical. There is mixing, but the ground state wave function is centered in the deformed minimum with large β and the 0_2^+ wave function in the near spherical one. The low lying levels and their B(E2)'s are nicely reproduced by the fits. For ^{78,80}Kr the surfaces show the nuclei are soft to γ deformation so the coexistence quickly disappears with increase in N.

Shape coexistence is now seen in many regions of the periodic table as described in the review of Hamilton[4]. Here we consider a few other examples. Shape coexistence in the heavy Zr isotopes was considered earlier by Sheline et al.[35] and Ramayya and

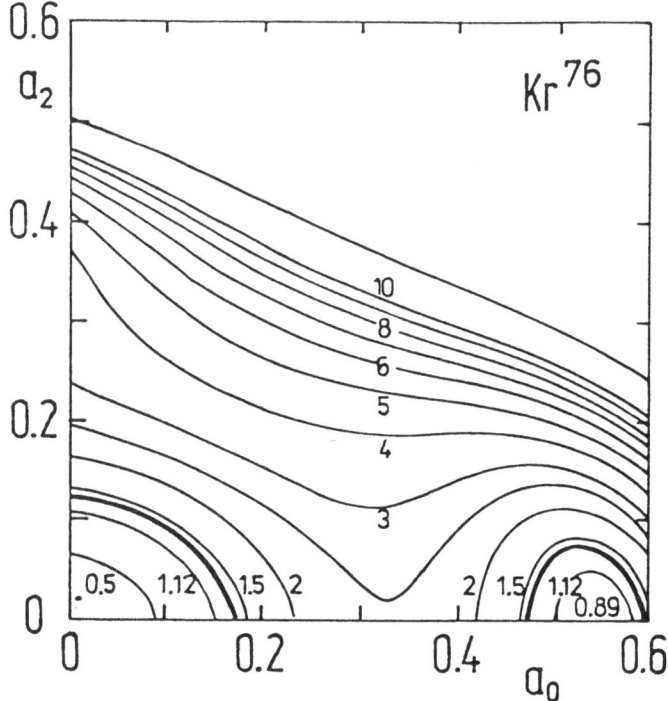

FIGURE 10. Potential energy surface for ^{76}Kr (Ref. 28). Note two minima are at large deformation and one at zero deformation, Note $a_0 \sim \beta$.

Hamilton[36] to account for the spherical ground states and very low energy 0_2^+ states in ^{96,98}Zr and ground state deformation in ^{100}Zr (Creifetz et al.[37]). However, Ramayya and Hamilton[36] already had noted that the work of Flynn et al.[38] indicated the low energy 0_2^+ state in ^{96}Zr is not a deformed band head. Subsequent studies of the properties of the 0_2^+ state in ^{98}Zr likewise did not support a coexistence picture (Sistemich et al.[39]). There is a very sudden onset of large deformation at N = 60 in this region as discussed in the next section. The lowest energies known for any 0_2^+ state in even-even nuclei are at the remarkably low energies of 215.5 (Fig. 11) and 331.3 keV in ^{98}Sr and ^{100}Zr, respectively (Schussler et al.[40], Kahn et al.[41]). These two unusual 0_2^+ states are interpreted[40,41] as built on near spherical shapes which coexist with the deformed ground states of ^{98}Sr and ^{100}Zr. In ^{98}Sr (Fig. 11) a high energy 2^+ state, characteristic of the vibrational 2^+ state of a near-spherical nucleus, is seen on top of the 0_2^+ state to support shape coexistence.

At Helsingnør, Ragnarsson[42] noted that their calculations (Bengtsson et al.[43]) indicate that in ^{100}Zr with N = 60 there are two coexisting close lying 0^+ states and that for ^{98,102}Zr with N = 58 and 62 one expects coexisting 0^+ states with spacings of 1.0-1.5 MeV. Recent studies of ^{98}Zr have found evidence for a 0_3^+ state at 1436 keV that has properties which indicate it has a shape quite different from the ground and 0_2^+, 853 keV level (Kawade et al.[44]). These data are consistent with the expected deformed band levels of Bengtsson et al.[43] and provide support for the shape coexistence model. However, farther away from N = 60 the energy differences between the spherical and deformed shapes rise very rapidly and it becomes difficult to observe any coexistence (Bengtsson et al.[43]). The calculations support the interpretation in ^{96}Zr that the low energy 0_2^+ state is probably not a deformed state. A similar sharp crossing of spherical and deformed shapes with shape coexistence observed in only a narrow range

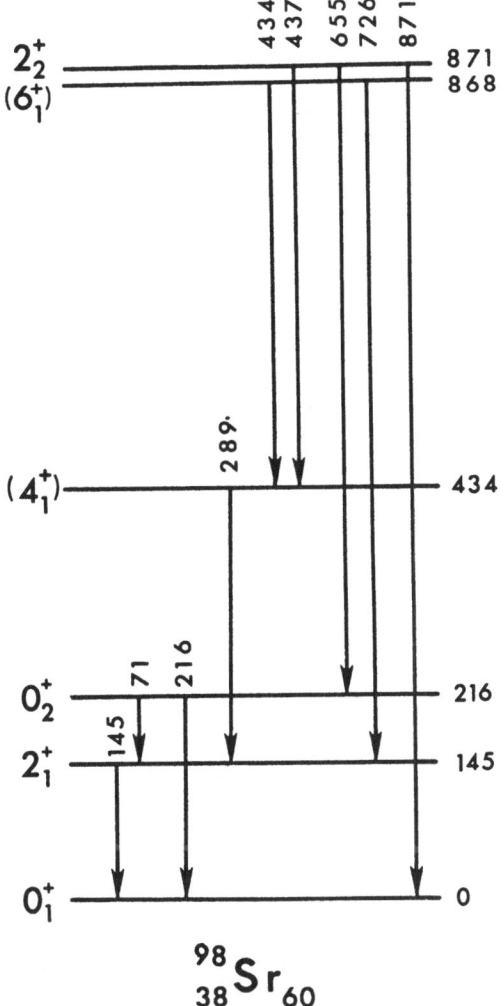

FIGURE 11.   Energy levels in ^{98}Sr. (Ref. 40.)

of neutron numbers around N = 60 in the Sr nuclei also is indicated.

Other important types of shape coexistence are seen in odd A nuclei, some of which are described in more detail in earlier reviews[4,45]. Briefly illustrating, in the odd A Au-Tl nuclei, for example, one finds bands built on hole states which have one type core (for example in Tl spherical Pb cores) and particle states built on quite different cores (eg. in Tl oblate Hg cores). Most fascinating was the discovery of Zganjar et al.[46] in ^{187}Au two $h_{9/2}$ bands built on Pt cores with quite different shapes as evidenced by the large E0 admixtures in the ΔI = 0 transitions between the bands. Here one has a unique opportunity to study particle-core coupling models where the odd particle is coupled to different cores in the same nucleus. Moreover, the odd parti-cle probes the structure of the core itself as, for example, in the ^{187}Au studies the $h_{9/2}$ particle probes the role of the promotion of a pair of $h_{9/2}$ particle across the gap to generate the deformed structures as first emphasized by Wood[34].

It was likewise unexpected to find in the closed proton shell tin nuclei, midway between the double closed shell $^{100}_{50}Sn_{50}$ and $^{132}_{50}Sn_{80}$, deformed, excited bands coexisting with the spherical ground states, as first reported by Ramayya and Hamilton[36] in $^{116}_{50}Sn_{66}$ and now well established there by Bron et al.[47]. Similar shape coexistence has been predicted in closed proton shell lead nuclei far off stability in, for example, ^{188}Pb (Frauendorf and Pashkevic[48]).

Another important new field of shape changes with regions of coexisting and competing structure involves the nucleus as its spin increases. Even the best examples of nuclei with large prolate deformation are expected to make a transition to oblate shapes where the angular momentum and shape are generated by the aligned motions of all the single particles around spin 48 in rare earth nuclei with subsequent further changes at still higher spins to a super deformed prolate shape just prior to fission, see for example Bohr and Mottelson[49]. These changes at very high spins have been the focus of much research. Changes in the moment of inertia associated with the alignment of one, two, and three pairs of particles have been established, for example, in ^{158}Er by groups in Berkeley, Heidelberg-GSI, Vanderbilt-ORNL and Daresbury with some evidence for the breaking off of rotational motion above spin 48, for example by Yamada et al.[50] A proper treatment of this field will require a full paper in itself. Here we only want to emphasize the fact that shape changes and shape coexistence of one type or another are found as functions of spin in most, if not all, nuclei. A final illustration of the competition of different shapes at even relatively low spins is seen in the transitional nuclei

To illustrate, the "classic" region in the rare earths where a sudden change from near-spherical to well-deformed shapes was found to occur at N = 88 now is seen to exhibit competing shapes at relatively low spins with the near-spherical vibrational structures changing to deformed rotational structures. The results of the similar competition of different forces to drive a nucleus toward different shapes and deformation as a function of spin, I, is seen in Fig. 12 (Bengtsson et al.[43]). Also the competition changes as a function of N and Z for a given I.

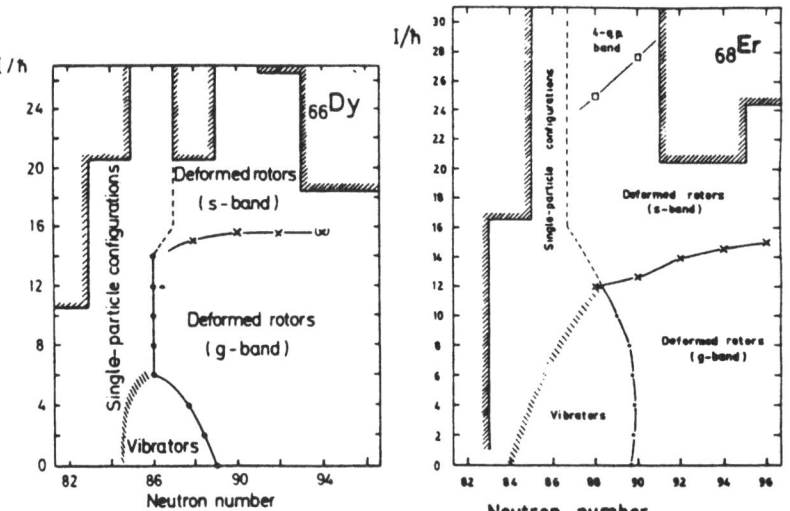

FIGURE 12. Phase diagrams of the regions where different shapes (and models) are found as functions of N and spin I obtained from experimental quantities for even Dy and Er isotopes. The critical N-values of the shape transitions (points) are determined from N(λ) plots while the critical angular momentum, I_c, for the transition to two-quasiparticle (crosses) and four-quasiparticle bands (squares) are extracted from ordinary back-bending plots, I vs. ω. Above the shadowed line, no experimental data are available. In many cases, the transition from one phase to another is not sharp but occurs gradually. (From Ref. 43.)

REINFORCING OF PROTON AND NEUTRON SHELL GAPS ON COMPETING SHAPES

An additional fascinating discovery has come from the recent studies in the regions of the light $_{34}$Se-$_{36}$Kr-$_{38}$Sr and heavy $_{28}$Ni nuclei with N = 38-40 and of Z = 38-40, the heavy Sr-Zr-Mo nuclei around N = 60 as first pointed out by Hamilton[4] and considered in more detailed recently by Hamilton et al.[51] Shape coexistence for nuclei with N $\approx$ 38-40 is related to the number of protons which delicately controls whether a deformed shape or near-spherical shape is lowest in this region. The origin of strong deformation and shape coexistence in this region can be understood by looking at the gaps in the single-particle spectrum that stabilize the nuclear shape.

Strong evidence for a spherical subshell closure for Z = 40 is found when N is 50 because then the neutrons strongly prefer a spherical shape, as first seen in $^{90}_{40}$Zr$_{50}$. Very recently the levels of $^{68}_{28}$Ni$_{40}$ were identified for the first time by Bernas et al.[53] These two nuclei have energy levels characteristic of double closed shell spherical nuclei to support 40 as a closed shell for N and Z. However, as Z or N move away from 28 or 50 the level density for a spherical shape becomes very high and the minimum of the proton deformation energy moves to deformed shapes and similarly for the neutrons which have almost identical single-particle levels. It is the reinforcement of the driving forces for both protons and neutrons toward a spherical shape when N(Z) = 40 by Z(N) = 28 (50) that leads to the strong double magic character of ^{90}Zr and ^{68}Ni.

Away from the Z(N) = 28 and 50 closed shells, Hamilton et al.[24] and Piercey et al.[25] first emphasized the gaps in the Nilsson single particle levels at N and Z $\simeq$ 38 at large prolate deformation, $\beta \approx 0.3$, in addition to the spherical gaps at $\beta \approx 0$ for N = Z = 40. However, the situation is more complex than just large prolate-spherical competition as recently emphasized by Hamilton et al.[51] Bengtsson et al.[52] have presented details of the potential energy surfaces, ground state masses, and shapes calculated for 4023 nuclei from ^{16}O to ^{279}112 with a Yukawa-plus-exponential macroscopic model and a folded-Yukawa single particle potential. Their calculated single particle levels for neutrons (protons are similar) are shown in Fig. 13. These levels differ in two respects from the modified oscillator (Nilsson) levels which were used earlier[24,25]. There are still gaps at N and Z = 38, but at larger prolate deformation $\beta_2 \sim 0.35$ which is in agreement with the values found in ^{74}Kr and ^{100}Sr as well as spherical gaps at 40. However, there are also gaps at N and Z = 36 for oblate shapes with equally large deformation ($\beta_2 \sim -0.35$). In addition, there are gaps for both prolate and oblate shapes at somewhat smaller deformation for Z = 34. Thus, there should be three competing shapes, large oblate, large prolate and near-spherical shapes in this region. Both the oblate and prolate deformed shapes can coexist in a delicate balance with each other and with the near-spherical shape. Large oblate deformation is expected centered around N = Z = 35 and large prolate deformation centered around N = Z = 38. In both cases it is the combination of both N and Z reinforcing each's preference for large deformation that drives the nucleus toward large deformation as first suggested by Hamilton[4].

For N = 38, 40 the protons control the delicate balance of spherical and deformed shapes, so as Z also approaches 38 the deformed minimum drops in energy because of the reinforcement of the neutron push toward deformation by the proton push toward deformation. This drop is in agreement with the experimental evidence. In $^{72,74}_{34}$Se$_{38,40}$ the deformed band becomes yrast at I $\simeq$ 2-4. While in $^{74,76}_{36}$Kr$_{38,40}$ the 36 protons favor deformation even more, and strong ground state deformation is seen with a near-spherical excited band. In $^{78}_{38}$Sr$_{40}$ only the deformed band is seen.

The experimental data on odd A Kr (Refs. 54,55) and Sr (Ref. 31) nuclei indicate that both the $^{74-76}$Kr and $^{78-80}$Sr regions have large prolate deformation. The magnitude of the quadrupole deformation of ^{75}Kr is constrained (Leander[56]) to be $\beta_2 = 0.35 \pm 0.05$ which is in good agreement with the ^{74,76}Kr results and the theoretical calculations. It is a puzzle why the predicted oblate shapes with large deformation have not been seen. However, Bengtsson et al.[52] suggest that this added competition in this region between the oblate and prolate structures with large deformation may show up at high spins in terms of which structure has the lowest energies for the rotation alignment of pairs of particles. Very interesting shape changes already are being seen at

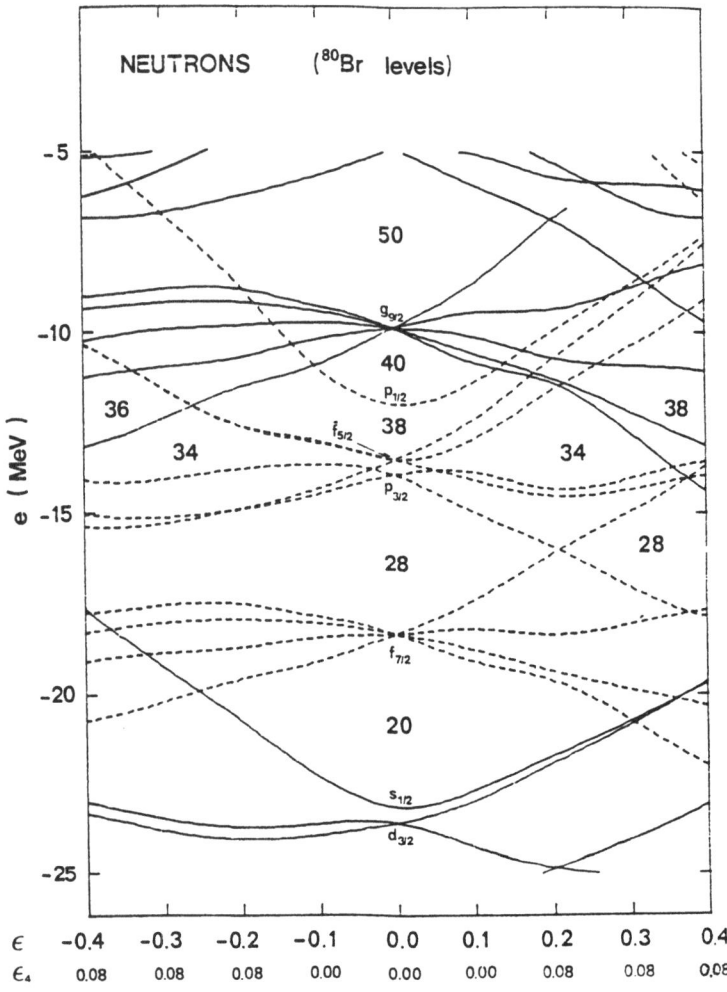

FIGURE 13. The single particle levels calculated with a folded-Yukawa single particle potential[52]. There are gaps at large deformation for both prolate and oblate shapes for 38.

relatively low spins, for example there is a marked shape change when the $g_{9/2}$ band in ^{81}Kr is broken off at $21/2^+$ by a three $g_{9/2}$ particle configuration (Funke et al.[57]) and others are expected at higher spins in both the even-even and odd-A cases.

The heavy Sr, Zr nuclei provide additional data on the competition of the 38 and 40 shell gaps at large and zero deformation and on the importance of the reinforcement of proton and neutron shell gaps at the same deformation on the nuclear shape[4,51]. Arseniev et al.[58] first predicted a new region of deformed neutron rich nuclei with A around 100. The earlier evidence from fusion studies of $^{100}_{40}$Zr$_{60}$ (Creifetz et al.[37]) for a new region of deformation has been confirmed (Kahn et al.[44]). Now there is established a sudden onset of large deformation at N = 60 seen from energy level studies of $^{100}_{40}$Zr$_{60}$ and $^{98}_{38}$Sr$_{60}$ (Wollnik et al.[59], Schussler et al.[40]) as illustrated in Fig. 14

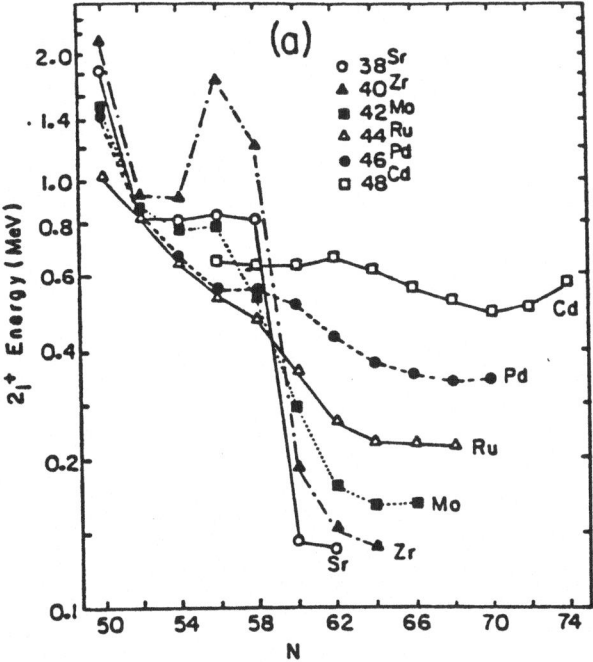

FIGURE 14. The 2^+ energies of heavy Sr to Cd nuclei which show the sudden onset of strong deformation for N = 60 in Sr and Zr nuclei with Z = 38 and 40.

and seen in two neutron separation energies and RMS radii of rubidium nuclei (Epherre et al.[60]). The largest deformation is seen in $^{100}_{38}Sr_{62}$ (Azuma et al.[61]) where surprisingly large, "super" deformation was found. If one scales the 129 keV 2^+_1 energy in ^{100}Sr by $A^{5/3}$ to compare it with ^{240}Pu which has essentially the lowest 2^+_1 energy known, 43 keV, the ^{100}Sr scaled energy for comparison is 30 keV. A similar scaled energy of 28 keV is reported for ^{74}Kr (Piercey et al.[25]) although the exact value there can be larger since it depends on how one extrapolates the high-spin data to low spin. The level structures of odd neighboring A nuclei, first $^{99}_{39}Y_{60}$ (Monnand et al.[62]), and more recently $^{99}_{38}Sr_{61}$, $^{101}_{40}Zr_{61}$, $^{99}_{39}Y_{60}$ and $^{101}_{39}Y_{62}$ (Wohn et al.[63]) have been investigated. In all four cases, rotational bands with large moments of inertia and deformations, $\beta \sim 0.3$-0.4, which are the same magnitudes as those in their even-even neighbors are found.

Azuma et al.[61] related the origin of this new region of very large deformation to the gap at N = 60 in the neutron Nilsson levels at $\beta \sim 0.35$. This region is seen to be more complex than in the earlier calculation[61] with recent calculations[27,52] showing neutron shell gaps at 56($\epsilon \sim 0.2$), 58($\epsilon \approx - 0.2$), and 64($\epsilon \sim 0.35$) in addition to the 60($\epsilon \sim 0.35$). So it is not simply a neutron effect. The light Kr-Sr data discussed above show the importance of N and Z gaps at 38 at very large deformation.

The importance of the proton gap at Z = 38 in deformation in the A = 100 region is seen in that as one goes away from Z = 38 the deformation rapidly decreases for the N = 60 and 62 nuclei, as seen in the rise in the 2^+_1 energies, for example 129.2, 151.9 and 192.2 keV (as seen in Fig. 14) for N = 62, $^{100}_{38}Sr$, $^{102}_{40}Zr$ and $^{104}_{42}Mo$ (Azuma et al.[61], Shizuma et al.[64] and Sistemich et al.[65]) as first pointed out by Hamilton[4] and emphasized more recently by Hamilton et al.[51] Thus, this new region of large deformation undoubtedly also is related to the strong reinforcement of the neutron shell gap at N = 60 at large deformation by the proton gap at large deformation at Z = 38.

Summarizing, the available experimental data and the theoretical calculations show
the importance of the Z = 38 and N = 38 shell gaps at large deformation and emphasize
the importance of the mutual reinforcement of the neutron and proton driving forces
when both N and Z are at or near shell gaps with the same deformation, near zero or
large, to produce spherical or large deformation in nuclei in this region. However,
when N or Z is around 40, a spherical shape is found only when the other particle number is near the strong 28 or 50 closed shells. When both N and Z approach 40 from
either direction, the N = Z = 38 gaps at large prolate deformation dominate. As N
and Z move away from both being at or near 38, the coexistence of both spherical and
deformed shapes plays an important role. Likewise, in the heavy Sr-Zr region the
strongest deformation is found when Z = 38 to reinforce the neutron shell gap at 60.
For a given N, the deformation there rapidly decreases with Z in going away from
Z = 38. The importance of the reinforcement of the proton and neutron driving forces
in determining whether a nucleus has a spherical or a deformed ground state and the
marked switch in a magic number, from 40 being like a strong closed shell to produce
double closed shell $^{90}_{40}Zr_{50}$ and $^{68}_{28}Ni_{40}$ to N and Z both around 38, 40 or Z = 38-40,
N $\geq$ 60 driving a nucleus toward large deformation are both new shell structure phenomena not seen in nuclear near the beta stability line as first noted by Hamilton[4]
and Hamilton et al.[51]

MAGIC NUMBERS FAR OFF STABILITY

One of the important directions in research in nuclei far off stability has been towards the question of how magic far from stability are the magic closed shell numbers
for neutrons and protons obtained from stable or near stable nuclei. Above we have
described the unexpected surprises in the N,Z = 38,40 region. We also noted there
the discovery of another new double closed shell type nucleus $^{68}_{28}Ni_{40}$, which supports
the near stability magic numbers. Considerable effort has been made to probe the regions of $^{132}_{55}Sn_{82}$ and $^{100}_{55}Sn_{50}$ which were expected to be double closed shell nuclei.
The levels of ^{132}Sn are characteristic of a double closed shell structure[66]. As
noted by Blomquist[66] the over 4 MeV gap to the first excited state of ^{132}Sn is strik-
ingly large compared to double magic $^{208}_{82}Pb_{126}$ with its 2.6 MeV gap. No other nucleus
above ^{16}O has a gap so large. Even allowing for the difference in size, ^{132}Sn exhibits
the strongest shell closure of any nucleus from this point of view.

The identification of levels in ^{100}Sn has been more difficult. Nevertheless, studies
now are approaching quite near to ^{100}Sn. Among the closest nuclei where levels have
been established are $^{102}_{48}In_{54}$ (Beraud et al.[67], Treherne et al.[68]) and $^{96}_{46}Pd_{50}$ (Kurcewicz
et al.[69], Piel et al.[70]). The 2^+_1 energy in ^{102}In of 776.8 keV shows a small rise over
those of $^{104-112}In$, 667.9, 632.7, 633.0, 657.7 and 617.4 keV, respectively. This
energy rise is an indication of an approach to a double closed shell but is not as much
as was expected. Roeckl and co-workers at the on-line GSI separator have tried to
approach ^{100}Sn from below in reactions like ^{40}Ca + ^{58}Ni(or + ^{63}Ca). The 2^+_1 energy increase is much more dramatic in $^{96}_{46}Pd_{50}$, 4 protons away from ^{100}Sn, with its 2^+_1 energy
of 1415.5 keV compared to 863.1 keV for $^{98}_{46}Pd_{52}$ (Kurcewicz et al.[69], Piel et al.[70]).
Thus, studies around the two double magic tin nuclei very far off stability support
the magic numbers found near stability.

Another area where one needs to know what are the stabilizing magic numbers for protons and neutrons far from stability is in the quest for the super heavy elements, with
Z in the range of 114 to 126. The synthesis of new super heavy elements with Z $\gtrsim$ 114
has been a major challenge for over a decade as documented in the proceedings of recent conferences[71]. A wide variety of projectiles and targets have been used all the
way up to ^{238}U on ^{248}Cm. The different experimental searches were the subject of two
very recent, major reviews by Flerov and Ter-Akopian[72] and Gäggeler et al.[73] and need
not be repeated here. Summarizing, it is unfortunate that there has been no positive
identification of such super heavy elements to date despite the extensive efforts.
However, major research efforts continue in this area.

Very recently Kumar and Mustafa[74] have suggested that the searches for super heavy
elements have concentrated around the wrong Z. Previous studies have searched in the
region of Z = 110 to 118, centered around 114. Kumar and Mustafa[74] have improved and

extended the previous theory by combining the dynamic deformation model of Kumar[75] with certain aspects of the theory developed by Nix and co-workers[76]. With an improved theory for the six inertial functions of the nuclear deformation and for the potential energy of deformation at large fragment separation, they found a lowering of 36 orders of magnitude in their predicted fission half-life for the nucleus Z = 114, (Z = 298) compared with previous theoretical predictions of Nix and co-workers while that of the nucleus Z = 126 (A = 310) is raised by 29 orders of magnitude. Their total lifetimes including lifetimes for α-decay, β-decay, and e⁻ capture as well as fission, indicate lifetimes in the range of milliseconds or larger around Z = 126. Thus, they suggest that experimental efforts be directed toward this region with reactions such as $^{138}Ba + ^{172}Yb + 401$ MeV $= ^{310}126$ or $^{140}Ce + ^{174}Yb + 418$ MeV $= ^{314}128$. For the sequence starting at $^{314}128$, they expect α-decay energies and halflives of 17.8 MeV, 10^{-15} s, 11.6 MeV, 4 ms and 11.3 MeV, 72 ms, respectively. Their prediction will undoubtedly generate new studies to better explore the region around Z = 126.

NUCLEAR OCTUPOLES

Another interesting development in nuclear structure during the last few years has been the discovery that intrinsic reflection symmetry is spontaneously broken in the ground states of certain heavy nuclei. This possibility was implied by the 'quasimolecular' nuclear model proposed by Bohr[77]. The observation of Stephens et al.[78] of low-lying negative-parity states in doubly even radium nuclei generated considerable interest, but by 1970 a consensus had formed that intrinsic parity mixing through stable octupole shapes does not occur in nuclear ground states. However, better insight has now been achieved into the microscopic origin of intrinsic 'octupole shape' and its consequences for nuclear spectroscopy (Chasman[79], Leander and Sheline[80]), in conjunction with a range of new experiments (Teoh et al.[81], Ahmad et al.[82]). More realistic theoretical calculations do give octupole equilibrium shapes in the Ra-Ac region, and nuclear spectroscopic theories which take into account an adiabatic octupole mode contribute significantly to the understanding of experimental data in the radium, actinium, and light thorium nuclei.

Octupole equilibrium deformations were obtained from theory by using a realistic flat-bottomed radial shape for the single-particle potential (Möller and Nix[27], Leander et al.[83]); all previous investigations had been based on single-particle potentials proportional to r^2, which in the harmonic oscillator limit do not have octupole shell structure at all (Bohr and Mottelson[49]). Furthermore, the surface stiffness to octupole deformation was lowered by taking the finite range of the nuclear force into account.

The spectroscopic signature of reflection asymmetry in molecules is parity doubling, with some small energy splitting (cf. Fig. 15) related to quantum mechanical tunneling. In nuclei, the parity splitting is empirically large relative to the excitation energy of other modes, and this has been used in the past as an argument against octupole deformation. However, recent phenomenological analyses (Zimmermann[84], Rohozinski and Greiner[85]) indicate that the dominant contribution to the observed parity splitting in radium nuclei arises specifically because of the other modes and not because of tunneling. The very close (0.22 keV apart) parity doublet found[82] in the ground state of ^{229}Ps is thus an accident rather than the expected rule.

In order to cope with the empirical parity splitting in the analysis of nuclear spectra, Leander and Sheline[80] devised a core-particle coupling model where the core parity splitting comes in phenomenologically, just like the core moment of inertia which is familiar from the particle-rotor model. (Note the analogy between parity and spin splitting in the right hand part of Fig. 15). The matrix elements of the single-particle parity operator, $\hat{\pi}$, are calculated microscopically and come to play a role analogous to that of the Coriolis matrix elements of $\hat{j}_+$: the diagonal matrix element determines the parity splitting or 'decoupling' from the octupole mode in the strong-coupling limit, and further decoupling may arise because of the off-diagonal terms. Leander and Sheline[80] deomonstrate that different quasi-particle states in the same nucleus can be expected to exhibit the full range of possibilities from strong coupling to weak coupling because of these off-diagonal terms, when the core is

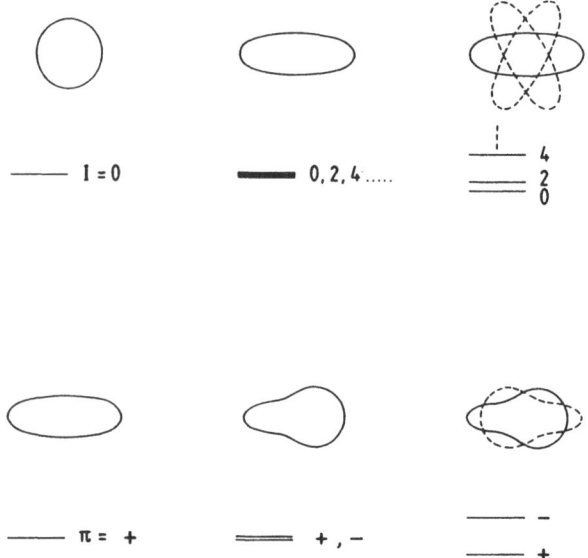

FIGURE 15. Spontaneous symmetry breaking in doubly even nuclei is represented schematically for rotational symmetry (above) and reflection symmetry (below). The figures in the middle are in the intrinsic frame of reference, while those to the right have been projected back into the laboratory frame.

rigidly octupole deformed. They proceed to test the parity-mixed Nilsson wave functions obtained at reflection asymmetric equilibrium shapes by comparing with symmetric wave functions and the available data.

Leander and Sheline[80] find that band-head energies and parity splittings are surprisingly insensitive to the octupole; only in ^{225}Ra was there an orbital so sensitive that the data can distinguish the reflection asymmetric interpretation from an alternative symmetric interpretation. Rotational decoupling factors are very sensitive to octupole deformation, however, which is easy to understand since Y_{30} mixes orbitals with Δj up to 3. Several experimental decoupling factors are available or measurable to test the asymmetric theory. Sheline et al.[86] find in ^{225}Ra evidence for two sets of K = $1/2^{\pm}$ and $3/2^{\pm}$ bands. As expected from the theoretical calculations[80,87], the decoupling parameters for the K = $1/2^{+}$ and $1/2^{-}$ bands have opposite signs and similar magnitude, a = +1.534 and -2.594, respectively. These values are totally inconsistent with any decoupling parameter for a pure Nilsson configuration[86,87] to give strong support for stable octupole deformation. Recently, it also was shown that even at high spins the octupole has a major effect on rotational alignment phenomena (Nazarewicz et al.[88]).

The first detailed spectroscopic calculations of octupole-deformed nuclei were actually made by Chasman[79] using a different approach from that of Leander and Sheline[80]. Chasman[79] employed a large reflection-*symmetric* basis to diagonalize a quadrupole-quadrupole, octupole-octupole and pairing interaction. This method bears the same relation to octupole-deformed nuclei as the spherical shell model does to quadrupole-deformed nuclei (c.f. Fig. 15). It proved to be successful and instructive in the cases where it was applied.

It is difficult to test the calculated stability of the octupole shape except in the shape transitional region. The transition to octupole shapes takes place in theory at A ∿ 229, where octupole shape coexistence was indeed established experimentally (Bemis et al.[89] for ^{229}Th. A calculation treating the octupole as a dynamical mode in the theoretical adiabatic potential was, in fact, able to account for the ^{229}Th data (Leander and Sheline[80]).

Continued exploration of intrinsically parity-mixed nuclei is proceeding in several directions, for example towards other mass regions (Leander et al.[83]), other excitation modes (Rohozinski et al.[90], Iachello and Jackson[91]), and very high spins (Fernandez-Niello et al.[92], Bonin et al.[93], and Nazarewicz et al.[88]). The first high-spin measurements on ^{218}Ra and ^{222}Th have revealed new collective features such as fast in-band E1 transitions. Doubly odd nuclei may also present a worthwhile challenge[80].

HIGHER MOMENTS IN DEFORMED NUCLEI

Although it involves stable nuclei, another very interesting shape phenomena is the large negative hexadecapole deformation in a few prolate rare earth nuclei. Evidence was found earlier for large negative E4 matrix elements in ^{180}Hf (Ronningen et al.[94-95]) and in $^{182-186}$W (Lee et al.[96], and more recently in ^{178}Hf (Soundranayagam et al.[97]). The origins of these very large negative E4 moments is not established. For example, in soft or transitional tungsten isotopes, the proper explanation of these large negative matrix elements may be related to vibrational or other degrees of freedom. This could likewise be true in ^{180}Hf, but it is known to be a good rotor to above spin 12$^+$ (Hamilton et al.[98]) which would suggest there it could be a real static deformation with $\beta_4 \sim -0.22$ in ^{180}Hf as found by Ronningen et al.[95]. To further test the different possibilities we carried out electron scattering off ^{180}Hf at Bates (Maguire et al.[99]). These new data indicate that indeed ^{180}Hf has a very large negative β_4 deformation. These data do not establish whether the tungsten isotopes also have a large static negative β_4 deformation. The nuclear shape of ^{180}Hf with a $\beta_4 \sim -0.2$, as shown in Fig. 16, looks something like peanuts in their hulls. These large negative moments present an interesting challenge.

It turns out that these shapes are not merely interesting curiosities. Rhoades-Brown and Oberacker[100] have shown that nuclei with such large negative hexadecapole deformation have very strongly enhanced fusion cross sections compared to spherical nuclei and factors of 2-5 enhancements over nuclei with only quadrupole deformation. Thus, they may be very useful in making nuclei in the heavy element region beyond element 100. It is interesting that hexadecapole moments are now seen to be important in one of the major problems in physics today -- the decay of the vacuum. As seen in the calculations of the Frankfurt group[101], the hexadecapole moments of the actinides are important in forming the potential pockets which are necessary to give the long sticking times for U on U and U on Cm and so allow the emission of spontaneous positrons to be observed.

Ronningen et al.[102] have looked for the influence of higher moments in the deformation, β_6, in ^{232}Th and ^{234,236,238}U. They not only find evidence for β_6 values but for the first time establish that β_6 goes from a positive value in ^{232}Th, 0.009(2), to a negative value, -0.011(2) in ^{238}U. This change is predicted by the calculations of Nilsson et al.[103] While relatively small, these β_6 deformations are now found to be important in order to obtain agreement between the theoretical and experimental masses in the actinide region[27]. It will be important to extend such measurements of β_6 to higher Z to test the calculations in this region. Such tests will make it possible to do more realistic calculations in the region of Z > 110 also.

SPIN DEPENDENCE OF LEVEL ENERGIES INSIDE COLLECTIVE BANDS

Peker et al.[104] have considered the basic question: what is the actual spin dependence of E_{lev} in the observed collective bands in the various regions of the periodic table? For example, in $K^\pi = 0^+$ collective bands is it a) simple rotational expressed in powers of $[I(I + 1)]$ or, b) rotation-vibration with odd powers of I terms required.

Most rotational models [traditional axial symmetric Bohr-Mottelson model[49], variable moment of inertia model (VMI)[105], triaxial model of Davidov-Fillipov[106] for $\gamma \stackrel{<}{\sim} 15°$,

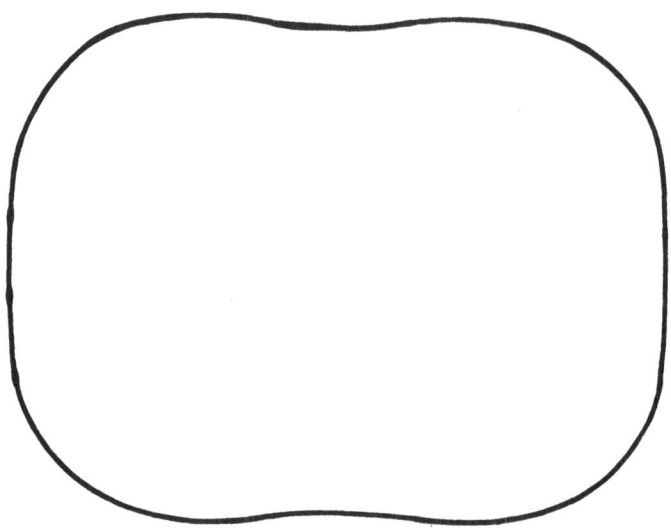

FIGURE 16. Nuclear shape for ^{180}Hf for β_2 and β_4 as shown.

SU(3) limit of IBA-model[107]] describe the spin dependence of $E_{lev}(I, K^{\pi} = 0^+)$ only in $[I(I + 1)]^n$ terms,

$$E_{coll}(I, K^{\pi} = 0^+) = A_1 I(I + 1) + A_2 [I(I + 1)]^2 + . . . \tag{1}$$

which is equivalent to $E_{Rot}(\omega) = \sum_{n=1} \alpha_n \omega^{2n}$. On the other hand, many collective models [anharmonic vibrational models[108] and Boson models that include 0(6) and 0(5) limits of the IBA model[109]] describe the spin dependence of $E_{lev}(I)$ in I^n terms, and so include odd powers of spin.

$$E_{coll}(I) = a_1 I + a_2 I^2 + a_3 I^3 + . . . \tag{2}$$

In recent times, the analysis of the spin dependence of $E_{lev}(I)$ generally have been focused on high-spin states. But with increasing spin the difference between terms $\sim I(I + 1)$ and $\sim I^2$ quickly decreases.

Peker et al.[104] have proposed a very sensitive method for revealing the real spin dependence of $E_{lev}(I)$ inside a collective band. They consider the function $E_\gamma(I \rightarrow I-2) = f(I)$ for intraband levels. Below the band crossing region of higher spins $I \gg K$, it is assumed to be a smooth function of spin. However, contrary to the traditional approach in which the behavior of this function was analyzed in the direction of increasing I, they analyze the behavior of $E_\gamma(I \rightarrow I - 2)$ in the direction of decreasing I. The reason for this choice is that as the spin decreases the significance of higher order terms in Eqs. (1) and (2) quickly decreases, so that more and more one is dealing only with the leading order terms in the spin dependence.

In the case of the function $E_{lev}(I)$ with pure rotational type of spin dependence, $[I(I + 1)]$, we can derive from Eq. (1),

$$E_\gamma(K = 0, I \to I - 2) = A_1\{[I(I + 1) - (I - 2)(I - 1)\} + \ldots \tag{3}$$

By simple substitution, $E_\gamma \equiv 0$ at $I_a = +1/2$. So the curves $E_\gamma(I \to I - 2)$ for all rotational bands with a pure $[I(I + 1)]^n$ (or ω^{2n}) dependence have to intercept the energy axis ($E_\gamma = 0$) at $I_a = +1/2$. Different bands will differ only by different slopes of the curves.

By simple substitution it also can be shown that in all $K \neq 0$ and $K \neq 1/2$ rotational bands the presence in (1) of a signature term does not change the prediction that $I_a^{K \neq 1/2}(E_\gamma = 0) = +1/2$. It means that in any $K \neq 1/2$ collective band any deviation of the observed $I_a(E_\gamma = 0)$ from $+1/2$ definitely indicates the existence of a nonrotational part in the spin dependence of the function $E_{coll}(I)$.

For an anharmonic vibration from Eq. (2) it follows that the leading order effect of anharmonicity in $E_{coll}(I)$ may be described by a term $a_2 I^2$ added to the harmonic vibrator term $a_1 I$

$$E_{coll}(I) = a_1 I + a_2 I^2 \tag{4a}$$

which can be transformed into the formula

$$E_{coll}(I) = aI(I + 1) + bI \tag{4b}$$

which was first proposed by Ejiri et al.[110] In this case $I_a(E_\gamma = 0) = +1/2 - b/2a$, i.e. the linear spin term gives rise to a deviation of I_a from the rotational value $+1/2$. From the experimental observation in certain bands that $E_{coll}(I) > E_{coll}^{GS}(I = 0^+)$ and $E_\gamma(I \to I - 2)$ increases with increasing I, but not so fast as in a true rotational band, it follows that $a > 0$ and $b > 0$ for these bands. It means that for such anharmonic vibrational bands in transitional nuclei ($a > 0$, $b > 0$), one expects that $I_a = +1/2 - b/2A < +1/2$ and often is less than zero. In the limit of a pure harmonic vibrational band with equally spaced levels, E_γ is constant and I_a goes to negative infinity.

Now consider rotational bands with intrinsic moment K and affected by very strong Coriolis coupling with higher bands with $\Delta K = 1$, $[(\hbar^2/2\mathcal{J}) < I,K+1|j_+|>]^2 \Delta E_{K,K+1}(I)]$. Because we want to look at the behavior of the function $E_\gamma(I)$ in the region of $E_\gamma \approx 0$, we may ignore all known (an unknown) higher order effects in comparison with the leading order effects and, therefore, will use a simplified aligned model.

In the first approximation for the energies of the levels in such a rotational band with aligned intrinsic momentum, i_α, rotational momentum R and total momentum $I \approx R + i_\alpha$:

$$E_{rot}(I) = A R(R + 1) \approx A (I - i_\alpha)(I - i_\alpha + 1) = AI(I + 1) + BI \tag{5}$$

with $B = -2A i_\alpha$ and $i_\alpha = -\frac{B}{2A}$. Because $A = \hbar^2/2\mathcal{J} > 0$ and $i_\alpha > 0$ (by definition), the parameter B is negative. The structure of formula (5) is identical with (4b), therefore

$$I_a(E_\gamma = 0) = +1/2 - \frac{B}{2A} . \tag{6}$$

However, because $B/A < 0$ in rotational bands aligned by strong Coriolis coupling, I_a is expected to be larger than $+1/2$. From (5) and (6) it follows that the aligned angular momentum

$$i_\alpha \approx I_a(E_\gamma = 0) - 1/2 . \tag{7}$$

From these considerations it follows that the real character of the spin dependence of $E_{lev}(I)$ inside a band may be found from the experimental function $E_\gamma(I \to I - 2)$. If we assume that inside the band and below the first band crossing this function is smooth, we may extrapolate it in the direction $E_\gamma \to 0$ and find the value of $I_a(E_\gamma = 0)$.

For bands with I_0 (or K) $\lesssim 5/2$, such extrapolation is short enough and therefore is, to a high degree, reliable.

In Fig. 17 are presented examples of such analysis for rare earth and actinide regions[111]. It can be seen that in strongly deformed nuclei with A $\geq$ 152 (and A $\geq$ 230)

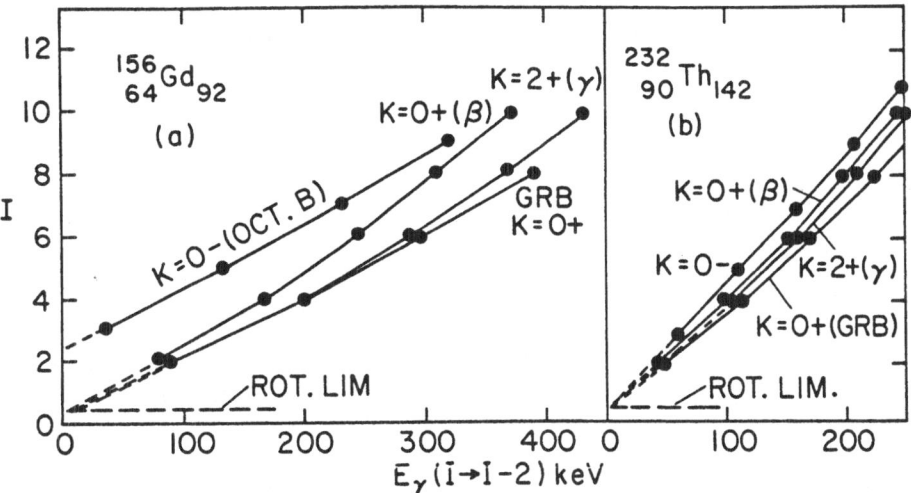

FIGURE 17. Examples of the functions $E_\gamma(I \to I - 2) = f(I)$ for collective bands in strongly deformed nuclei: a) GS-band, γ-band, β-band and aligned octupole vibrational K = 0⁻ in $^{156}_{64}Gd_{92}$ (Ref. 111); b) Collective bands in $^{232}_{90}Th_{142}$ (Ref. 112).

(Figs. 17a,b) the collective bands weakly affected by Coriolis coupling all have $I_a \approx +1/2$ and therefore b $\approx$ 0 (all K = 0⁺ ground state bands, the K $\lesssim 5/2$ bands with small j, large Ω; and β and γ-vibrational bands). These data show that the energies $E_{lev}(I)$ in these bands may be described in terms of purely rotational spin dependence, $I(I + 1)$ [formulas (1), (3)]. On the other hand, the bands in the same nuclei (A $\approx$ 150-180) which are strongly affected by Coriolis coupling (often the octupole vibrational bands (Fig. 17a)), and K $\neq$ 1/2 bands with large j, low Ω; for example, n 5/2⁺[642] - $i_{13/2}$ bands in odd A nuclei (Fig. 18), as well as aligned superbands responsible for backbending that can be treated in terms of a Coriolis alignment model, all have $I_a > +1/2$ with aligned intrinsic momenta $i_a = I_a - 1/2 = B/2A$.

At the same time in all transitional, weakly deformed nuclei the curves for the functions $E_\gamma(I \to I - 2)$ for the ground state bands intercept the $E_\gamma = 0$ spin axis at $I_a < +1/2$ and very often at <0 as illustrated in Fig. 19. Thus, the real function $E_{coll}(I)$ for these bands includes a nonrotational linear spin terms, i.e. it has an anharmonic vibrational spin dependence $\sim \sum_m a_m I^m$ rather than a rotational spin dependence, $\sum_n A_n[I(I + 1)]^n$. It can be seen that in each group as E(2⁺) decreases to approach the value for a strongly deformed nucleus the value of I_a increases toward $I_a(rot) = +1/2$.

It is interesting to note that in K = 0⁺ ground state bands of N = 90 nuclei (Fig. 19b), the function $E_{lev}(I)$ according to the observed deviations from $I_a(rot) = 1/2^+$ (Fig. 19b) generally includes a linear spin term $\sim bI$, and the magnitude of this term generally increases in going away from the Z = 60-64 region. In particular, we note the description of $E_{lev}(I)$ in the yrast bands observed in $^{160}Yb_{90}$ and $^{158}Er_{90}$ has to include a significant linear spin term bI with b/a > I. In extensive theoretical analysis of yrast bands in these nuclei, this linear spin term was completely ignored. This may strongly affect the results of analysis (for examples see Ref. 114). Other transitional regions exhibit similar behavior.

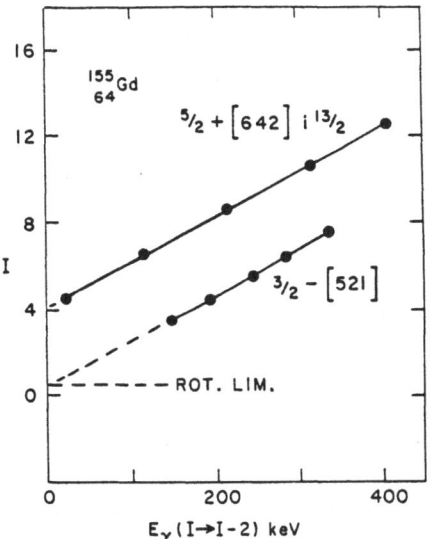

FIGURE 18. Functions $E_\gamma(I \to I - 2) = f(I)$ for nonaligned ($3/2^-[521]$) and strongly a-ligned ($5/2^+[642]i_{13/2}$) bands in deformed odd-A nuclei $^{155}_{64}Gd_{91}$ (Ref. 113)

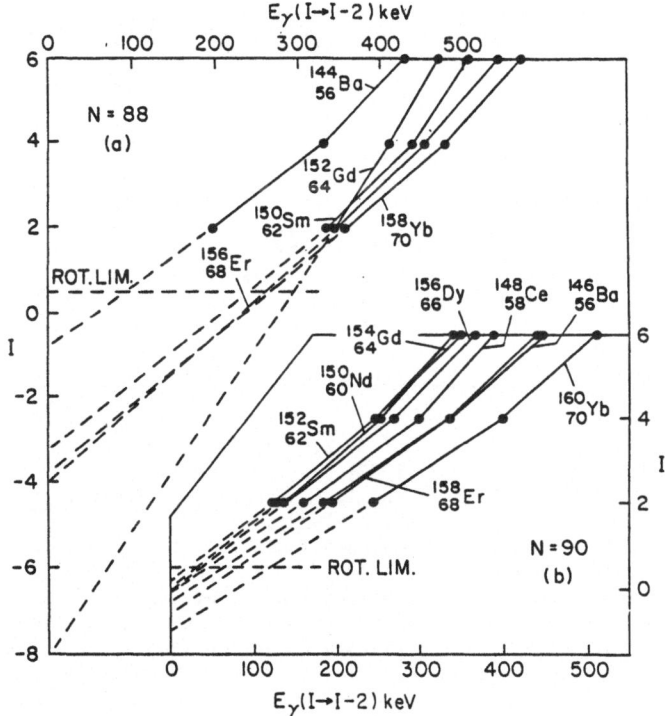

FIGURE 19. Functions $E_\gamma(I \to I - 2) = f(I)$: a) for N = 88 transitional nuclei (Ref. 114); b) for N = 90 transitional nuclei (Ref. 113).

This work was supported by the Director, Office of Energy Research, Division of Nuclear Physics of the Office of High Energy and Nuclear Physics of the U.S. Department of Energy under Contract DE-AS05-76ER05034 with Vanderbilt University. I would like to thank my many co-workers for valuable discussions and for permission to use our results.

REFERENCES

1. J. H. Hamilton, P. G. Hansen, and E. F. Zganjar, in Reports on Progress in Physics (to be published).
2. P. G. Hansen, Ann. Rev. Nucl. Part. Sci. 29, 69 (1979).
3. J. H. Hamilton, in Proc. Extreme States in Nuclear Systems, H. Prade and S. Tesch, eds., (Dresden, GDR, 1980) ZfK-430, Vol. I, p. 33.
4. J. H. Hamilton, in Int. Symp. on Nuclear Collectivity, (Dogashima, 1981), A. Arima and T. Marumori, eds., Inst. Nuclear Studies Tokyo, p. 87.
5. J. H. Hamilton, in Heavy Ion Collisions, R. Bock, ed., (North-Holland, Amsterdam, 1982), Vol. III, p. 575.
6. M. Baranger and R. A. Sorenson, Scientific American (1969), p. 58.
7. D. L. Hill and J. A. Wheeler, Phys. Rev. 89, 1102 (1953).
8. H. Morinaga, Phys. Rev. 101, 254 (1956).
9. W. Greiner, in Int. Conf. on Neutron Capture (Jülich, Germany), (1965).
10. V. G. Soloviev, Phys. Lett. 21, 311 (1966).
11. K. Kumar, in The Electromagnetic Interaction in Nuclear Spectroscopy, W. D. Hamilton, ed., (North-Holland, Amsterdam, 1975), p. 119.
12. J. H. Hamilton, in Proc. Int. Conf. on Selected Topics in Nuclear Structure, V. G. Soloviev et al., eds., (Dubna, USSR, 1976), Vol. II, p. 303.
13. J. H. Hamilton, Nukleonika 24, 561 (1979).
14. E. F. Zganjar, in Future Directions in Studies of Nuclei Far From Stability, J. H. Hamilton et al., eds., (North-Holland, Amsterdam, 1980), p. 49.
15. P. D. Duval and B. R. Barrett, Phys. Lett. 100B, 223 (1981); Nucl. Phys. A376, 213 (1982).
16. A. F. Barfield, B. R. Barrett, K. A. Sage and P. D. Duval, Z. Phys. 311, 205 (1983).
17. R. Bengtsson, G. Leander, W. Nazarewicz and I. Ragnarsson, 1983 (to be published).
18. W.-C. Ma, A. V. Ramayya, J.H. Hamilton, S. J. Robinson, M. E. Barclay, K. Zhao, J. D. Cole, E. F. Zganjar, and E. H. Spejewski, submitted to Phys. Lett.
19. F. S. Stephens, Rev. Mod. Phys. 47, 43 (1975).
20. K. Hardt et al., Z. Phys. A 312, 251 (1983).
21. R. V. F. Janssens et al., Phys. Lett. 131B, 35 (1983).
22. A. V. Ramayya, W.-C. Ma, J. H. Hamilton, M. E. Barclay, K. Zhao, S. Robinson, J. D. Cole, E. F. Zganjar, and E. H. Spejewski, Bull. Am. Phys. Soc. 28, 989 (1983).
23. J. H. Hamilton et al., Phys. Rev. Lett. 32, 239 (1974).
24. J. H. Hamilton et al., in Proc. Fourth Int. Conf. on Nuclei Far From Stability, CERN 81-09 (1981), p. 391.
25. R. B. Piercey et al., Phys. Rev. Letts. 47, 1514 (1981); R. B. Piercey, A. V. Ramayya, J. H. Hamilton, X-J. Sun, Z. Z. Zhao, R. L. Robinson, H.J. Kim and J. C. Wells, Phys. Rev. C25, 1914 (1982).
26. J. H. Hamilton, R. L. Robinson, and A. V. Ramayya, in Nuclear Interactions, B. A. Robson, ed. (Springer Verlag, Berlin, 1979), p. 253.
27. P. Möller and J. R. Nix, At. Nucl. Data Tables 26, 156 (1981); Am. Chem. Soc. 183rd Nat. Meeting, Abs. Nucl. 16 (1982).
28. M. Seiwert, A. V. Ramayya and J. Maruhn, Phys. Rev. 29, 284 (1984).
29. M. Barclay et al., Fall 1981, private communication; and reported in Ref. 4.
30. M. Barclay et al., Spring 1982, private communication; and reported in Ref. 45.
31. J. C. Lister, B. J. Varley, H. G. Price, and J. W. Olness, Phys. Rev. Letts. 49, 308 (1982).
32. G. Winter, J. Döring, L. Funke, P. Kemnitz, and E. Will, in Int. Symp. on Dynamics of Nucl. Collective Motion, Contributed Papers (Mt. Fuji, Japan, 1982), p. 21.
33. R. B. Piercey et al., 1980 UNISOR private communication; and reported in Ref. 4.
34. J. L. Wood, in IV Conf. on Nuclei Far From Stability, CERN 81-09 (1981), p. 612.
35. R. K. Sheline, I. Ragnarsson, and S. G. Nilsson, Phys. Lett. 41B, 115 (1972).

36. A. V. Ramayya and J. H. Hamilton, in Gamma Ray Transition Probabilities, S. C. Pancholi and S. L. Gupta, eds. (Univ. of Delhi Press, 1974), p. 125.

37. E. Creifetz, R. C. Jarod, S. G. Thompson and J. B. Wilhelmy, Phys. Rev. Lett. 25, 38 (1970).

38. E. R. Flynn, J. G. Beery and H. G. Blair, Nucl. Phys. A218, 285 (1974).

39. K. Sistemich et al., Z. Phys. A: Atoms and Nuclei 281, 169 (1977).

40. F. Schussler, J. A. Pinston, E. Monnand, A. Moussa, G. Jung, E. Koglin, B. Pfeiffer, R. V. F. Janssens, and J. Van Lkinken, Nucl. Phys. A339, 415 (1980).

41. T. A. Kahn, W. D. Lauppe, K. Sistemich, H. Lawin, G. Sadler and H. A. Selić, Z. Physik A283, 105 (1977).

42. I. Ragnarsson , in IV Int. Conf. on Nuclei Far From Stability, CERN 81-09 (1981), p. 434.

43. R. Bengtsson et al., IV Int. Conf. on Nuclei Far From Stability, CERN 81-09 (1981), p. 509.

44. K. Kawade et al., Z. Phys. A: Atoms and Nuclei 304, 293 (1982).

45. J. H. Hamilton, Heavy Ion Collisions, in Lecture Notes in Physics, G. Madurga and M. Lozano, eds., Vol. 168, pp. 287-300 (1982).

46. E. F. Zganjar et al., Proc. of Int. Conf. on Nuclei Far From Stability, CERN 81-09 (1982), p. 630.

47. J. Bron, W. H. A. Hesselink, A. Van Poelgeest, J. J. A. Zalmstra, M. J. Uitzinger and H. Verheul, Nucl. Phys. A318, 335 (1979).

48. S. Frauendorf and V. V. Pashkevich, Phys. Lett. 55B, 365 (1975).

49. A. Bohr and B. R. Mottelson, Nuclear Structure, Vol. 2 (Benjamin, NY, 1975).

50. H. Yamada et al., Phys. Lett. B128, 33-36 (1983).

51. J. H. Hamilton et al., submitted to J. Phys. G Letters.

52. R. Bengtsson, P. Möller, J. R. Nix and J. Zhang, 1983, submitted to Physica Scripta.

53. M. Bernas, P. L. Dessagne, M. Langevin, J. Payet, F. Pougheon and P. Roussel, Phys. Letts. 113B, 279 (1982).

54. B. D. Kern, et al., Phys. Rev. C28, 2168 (1983).

55. M. Herath-Banda, A. V. Ramayya and G. A. Leander, private communication (1983); and reported earlier A. V. Ramayya and J. Eberth, Heavy Ion Collisions, (Springer Verlag, Berlin, 1982), p. 317.

56. G. A. Leander, 1983, private communication.

57. L. Funke et al., Phys. Letts. 120B, 301 (1983).

58. D. A. Arseniev, A. Sobieczewski and V. G. Soloviev, Nucl. Phys. A139, 269 (1969).

59. H. Wollnik, F. K. Wohn, K. D. Wünsch, and G. Jung, Nucl. Phys. A291, 355 (1977).

60. M. Epherre et al. Proc. IV Int. Conf on Nuclei Far From Stability, P. G. Hansen and O. B. Nielson, eds., (Helsingnør, 1981), p. 62.

61. R. E. Azuma et al., Phys. Lett. 86B, p. 5 (1979).

62. E. Monnand et al., Z. Phys. A: Atoms and Nuclei 306, 183 (1982).

63. F. Wohn, J. C. Hill, R. F. Petry, H. Dejbakhsh, Z. Berant and R. L. Gill, Phys. Rev. Letts. 51, 873 (1983).

64. K. Shizuma et al., Phys. Rev. C27, 2869 (1983).

65. K. Sistemich, W. D. Lauppe, H. Lawin, H. Seyfart and B. D. Kern, Z. Physik A289, 225 (1979).

66. J. Blomqvist, in IV Int. Conf. on Nuclei Far From Stability, (Helsingnør, 1981), CERN 81-09, p. 536.

67. R. Béraud et al., Z. Phys. A: Atoms and Nuclei 299, 279 (1981).

68. J. Tréherne et al., Z. Phys. A: Atoms and Nuclei 309, 135 (1982).

69. W. Kurcewicz et al., Z. Phys. A: Atoms and Nuclei 308, 21 (1982).

70. W. F. Piel, Jr., G. Scharff-Goldhaber, C. J. Lister, and B. J. Varley, 1983, submitted for publication.

71. Proc. Int. Symp. on Synthesis and Properties of New Elements, Dubna, Sept. 1980 (Pure and Appl. Chem. Vol. 53, Pergamon Press, Oxford, 1981); Proc. Int. Symp. on Superheavy Elements, ed. M.A.K. Lodhi (Pergamon Press, New York, 1978).

72. G. N. Flerov and G. M. Ter-Akopian, Rep. Prog. Phys. 46, 817 (1983).

73. H. Gäggeler et al., Proc. XXI Int. Winter Meeting on Nuclear Physics (Bormio, Italy, 1983), p. 366.

74. K. Kumar, Prog. Part. Nucl. Phys. 9, 233 (1983).

75. C. E. Bemis, Jr. and J. R. Nix, Comments Nucl. Part. Phys. 7, 65 (1977); E. O. Fiset and J. R. Nix, Nucl. Phys. A193, 647 (1972); P. Möller and J. R. Nix, Phys.

Rev. Lett. <u>37</u>, 1461 (1976).

76. K. Kumar and M. G. Mustafa, to be published.
77. A. Bohr, Mat. Fys. Medd. Dan. Vid. Selsk. <u>26</u>, No. 14 (1952).
78. F. Stephens, F. Asaro and I. Perlman, Phys. Rev. <u>96</u>, 1568 (1954); <u>100</u>, 1543 (1955).
79. R. R. Chasman, Phys. Rev. Lett. <u>42</u>, 630 (1979).
80. G. A. Leander and R. K. Sheline, Nucl. Phys. <u>A413</u>, 375 (1984).
81. W. Teoh, R. D. Connor and R. H. Betts, Nucl. Phys. <u>A319</u>, 122 (1979).
82. I. Ahmad, J. E. Gindler, R. R. Betts, R. R. Chasman and A. M. Friedman, Phys. Rev. Lett. <u>49</u>, 1758 (1982).
83. G. A. Leander, R. K. Sheline, P. Möller, P. Olanders, I. Ragnarsson and A. J. Sierk, Nucl. Phys. <u>A388</u>, 452 (1982).
84. R. Zimmermann, Phys. Lett. <u>113B</u>, 199 (1982).
85. S. G. Rohozinski and W. Greiner, Phys. Lett. <u>128B</u>, 1 (1983).
86. R. K. Sheline et al., Phys. Lett. <u>133B</u>, 13 (1983).
87. I. Ragnarsson, Phys. Lett. <u>130B</u>, 353 (1983).
88. W. Nazarewicz, P. Olanders, I. Ragnarsson, J. Dudek and G. A. Leander, Lund-Mph-83/12 (Lund Inst. of Technology), (1983).
89. C. E. Bemis, F. K. McGowan, J. L. C. Ford, W. T. Milner, R. L. Robinson, P. H. Stelson and C. W. Reich, to be published.
90. S. G. Rohozinski, M. Gajda and W. Greiner, J. Phys. G8, 787 (1982).
91. F. Iachello and A. D. Jackson, Phys. Lett. <u>108B</u>, 151 (1982).
92. J. Fernandez-Niello, H. Puchta, F. Riess and W. Trautmann, Nucl. Phys. <u>A391</u>, 221 (1982).
93. W. Bonin, M. Dahlinger, S. Glienke, E. Kankeleit, M. Krämer, D. Habs, B. Schwarts and H. Backe, Z. Phys. <u>A310</u>, 249 (1983).
94. R. M. Ronningen, J. H. Hamilton, L. Varnell, J. Lange, A. V. Ramayya, G. Garcia-Bermudez, W. Lourens, L. L. Riedinger, F. K. McGowan, P. H. Stelson, R. L. Robinson and J.L.C. Ford, Jr., Phys. Rev. <u>C16</u>, 2208 (1977).
95. R. M. Ronningen et al., Phys. Rev. Lett. <u>40</u>, 364 (1978).
96. I. Y. Lee, J. X. Saladin, J. Holden, J. O'Brien, C. Baktash, C. Bemis, Jr., P. H. Stelson, F. K. McGowan, W. T. Milner, J.L.C. Ford, Jr., R. L. Robinson and W. Tuttle, Phys. Rev. <u>C12</u>, 1483 (1975).
97. R. Soundranayagam, J. H. Hamilton, A. V. Ramayya and J. Saladin, private comm.
98. J. H. Hamilton et al., Phys. Lett. <u>112B</u>, 327 (1982).
99. C. F. Maguire, J. H. Hamilton, L. Cleemann, R. Soundranayagam, W. Bertozzi, M. Finn, C. Hyde-Wright and R. B. Piercey, private communication (1983).
100. M. J. Rhoades-Brown and V. E. Oberacker, Phys. Rev. Letts. <u>50</u>, 1435 (1983).
101. M. J. Rhoades-Brown, V. E. Oberacker, M. Seiwert and W. Greiner, Z. Phys. A:Atoms and Nuclei <u>310</u>, 287(1983); M. Seiwert, W. Greiner, V. Oberacker and M. J. Rhoades-Brown, Phys. Rev. <u>C29</u>, 477 (1984).
102. R. M. Ronningen et al., Phys. Rev. Letts. <u>47</u>, 635 (1981).
103. S. G. Nilsson, C. F. Tsang, A. Sobiczewski, Z. Szymanski, S. Wycech, C. Gustafson, I. L. Lamm, P. Möller, and B. Nilsson, Nucl. Phys. <u>A131</u>, 1 (1969).
104. L. K. Peker, S. Pearlstein, J. O. Rasmussen and J. H. Hamilton, to be published.
105. M. A. J. Mariscotti, G. Scharff-Goldhaber and B. Buck, Phys. Rev. <u>178</u>, 1864 (1969).
106. A. S. Davydov and G. F. Filippov, Nucl. Phys. <u>8</u>, 237 (1958).
107. A. Arima and F. Iachello, Ann. of Phys. <u>111</u>, 201 (1978).
108. T. K. Das, R. M. Dreizler, and A. Klein, Phys. Rev. <u>C2</u>, 632 (1970).
109. A. Arima and F. Iachello, Ann. of Phys. <u>99</u>, 253 (1976); ibid. <u>123</u>, 468 (1979).
110. H. Ejiri, M. Ishihara, M. Sakai, K. Katori and T. Inamura, J. Phys. Soc. of Japan <u>24</u>, 1189 (1968).
111. J. Konijn, F.W.N. DeBohr, A. van Poelgeest, W.H.A. Hesselink, M.J.A. DeVoigts, H. Verheul and O. Scholten, Nucl. Phys. <u>A352</u>, 191 (1981).
112. Ch. Brianson et al., <u>Symp. on High Spin Phenomena in Nuclei</u>, ANL (1979), p. 477.
113. G. Løvhøiden et al., Nucl. Phys. <u>A148</u>, 657 (1970).
114. M. Sakai and A. C. Rester, Nucl. Data Tables <u>20</u>, 441 (1977).
115. L. L. Riedinger, Physica Scripta <u>24</u>, 312 (1981).

QUANTUM ELECTRODYNAMICS OF STRONG FIELDS

W. Greiner
Institut für Theoretische Physik
J.W. Goethe Universität Frankfurt, West Germany

In cooperation with

J. Reinhardt, U. Heinz, B. Müller, U. Müller,
Th. de Reus, P. Schlüter, M. Seiwert, and G. Suff

I. SUPERCRITICAL FIELDS: GENERAL OVERVIEW

Since the days of the early Greek natural philosophers our view
of the physical world has been dominated by certain paradigms, i.e.
specific pictures, for selected physical entities. Such entities are
space, time and matter as the basis of natural philosophy or, more
specifically, of physics. Therefore, it is no surprise that our
conception of the "vacuum", intimately connected with the picture of
space, time and matter, ranges among the most fundamental issues in the
scientific interpretation of the world.

The picture of the vacuum has undergone perpetual modifications
during the last twenty-five centuries as the available technologies have
changed; often old, abandoned ideas have been resurrected when new
information became accessible. Many aspects of today's conception of the
vacuum date back to the ancient Greek philosophy, but have only recently
been established by modern experiments.

Over the centuries many different conceptions of the vacuum were
developed by scientists, different vacua as carriers for different kinds
of physical phenomena. We mention Newton's absolute space, on which the
hypothesis of the vacuum as an elastic medium, the "ether", is based.
It was developed in the early 19th century when the wave nature of light
had been firmly established - in close analogy to the theory of
elasticity. In Einstein's theory of relativity and gravity the absolute
space and the "ether" were abandoned and replaced by a bundle of inertial
frames.

Quantum mechanics and quantum field theory, finally, laid the
grounds for our present conception of the vacuum. In today's language,
the vacuum consists of a polarizable gas of *virtual* particles, fluctuating randomly. It is found that, in the presence of strong external

fields, the vacuum may even contain "real" particles. The paradigm

of "virtual particles" not only expresses a philosophical notion, but

directly implies observable effects:

1) The occurence of spontaneous radiative emission from
   atoms and nuclei can be attributed to the action of
   the fluctuations of the virtual gas of photons.

2) The virtual particles cause effects of zero-point
   motion as in the Casimir effect. (Two conducting,
   uncharged plates attract each other in a vacuum
   environment with a force varying like the inverse
   fourth power of their separation).

3) The electrostatic polarizability of the virtual
   fluctuations can be measured in the Lamb shift and
   Delbrück scattering.

However, the most fascinating aspect of the vacuum of quantum

field theory, which will be discussed here, is the possibility that

it allows for the creation of real particles in strong, time-independent

external fields. In such a case the normal vacuum state is unstable

and decays into a new vacuum that contains real particles. This, in

itself, is a deep philosophical insight. But it is more than an

academic problem : first, very strong electric fields are available

for laboratory experiments that are presently in progress; second, it

can be shown that the quantum theory of interacting fields may be

constructed from the vacuum-to-vacuum amplitude $W(J)$ of a quantized

field in the presence of an arbitrary external source J. Effects that

occur in strong external fields may, therefore, in some way be carried

over to strongly coupled, interacting fields as they form the basis

of the strong and superstrong interactions. Only recently have extensive

theoretical studies in Frankfurt and − in the early stages − to some

extent also in the Soviet Union led to new insights and full theoretical

clarification of the strong field problem.

The decay of the vacuum

The decay of the vacuum in strong electrostatic fields is a relative recently recognized phenomenon in quantum electrodynamics that can be studied only via low-energy heavy-ion collisions. The original motivation for developing the new concept of a charged vacuum arose in 1965-70 in connection with understanding the atomic structure of superheavy nuclei expected to be produced by the GSI-heavy ion linear accelerator.

The best starting point for discussing this concept is to consider the binding energy of atomic electrons as the charge of a heavy nucleus is increased, as shown in fig. 1. In view of the large mass of the heavy nucleus compared to the electron mass, the external field approximation is quite appropriate. Solving the Dirac equation in the presence of an electric field from a central point charge gives the well-known fine structure formula, first derived by Sommerfeld from the early theory of the atom:

$$E\ (nj) = m_o c^2 \left[1 + \left(\frac{Z\alpha}{n - |K| + \sqrt{K^2 - z^2 \alpha^2}} \right)^2 \right]^{1/2}$$

$n = 1, 2 \ldots\ldots\ldots\ldots\ldots$ = principal quantum number;

$K = \pm 1, \pm 2, \ldots\ldots\ldots\ldots$ = angular momentum quantum number;

$\alpha = e^2/c\hbar \ \ldots\ldots\ldots\ldots$ = fine-structure constant.

Because of the term $|K^2 - z^2 \alpha^2|^{1/2}$, the above equation breaks down at $Z\alpha > |K|$. Thus all states with $j = 1/2$ cease to exist at $Z = 1/\alpha = 137$, as shown in fig. 1: the corresponding wave function becomes non-normalizable at the origin; the K-shell binding energy goes to $-m_o c^2$. Note from fig. 1 that the energy levels move only very slowly away from the upper continuum as Z rises until Z=137 is approached rather closely.

Thus even in the heaviest known element, the binding is only a small
fraction of the rest energy.

The Z=137 'catastrophe' was well-known but it was argued loosely
that it disappears when the finite size of the nucleus is taken into
account. But, in a paper which started the modern development of
quantum electrodynamics of strong fields, Greiner and Pieper showed in
1969 that the problem is not removed but merely postponed and
reappears at Z~173; the exact value of this critical charge Z_{cr}
depends on many assumptions concerning the potential in the vicinity
of the nucleus, in particular the nuclear radius. One can trace any
level E(nj) down to a binding energy of twice the electronic rest
mass if the nuclear charge is further increased. At the corresponding
charge number, which we shall call Z_{cr}, the state reaches the negative-
energy continuum of the Dirac equation ('Dirac sea') which, according
to the hole-theory hypothesis, is totally occupied by electrons. (Note,
the hole theory of Dirac is completely equivalent in its predictions
to field theory). If the strength of the external field is further
increased, the bound state dives into the continuum. The overcritical
state acquires a width and is spread over the continuum. Still, the
electron charge distribution does remain localized. This insight was
gained in a series of papers by Greiner, Müller and Rafelski in the
early seventies.

When Z exceeds 145, $E(1s_{1/2})<0$, i.e. the binding energy exceeds
the rest mass of the electron. Adding the electron therefore dinimishes
the mass of the atom. It would be energetically advantageous for an
electron to be spontaneously created, thereby reducing the total energy.
This is not possible because it would violate the conservation of
charge and lepton number. Similarly, when $Z>Z_{cr}$ a K-shell electron

is bound by more than twice its rest mass, so that it becomes energetically favourable to create an electron-positron pair. Now, however, the spontaneous appearance of such a pair is not forbidden by any conservation law. The electron becomes bound in the $1s_{1/2}$ orbital and the positron escapes.

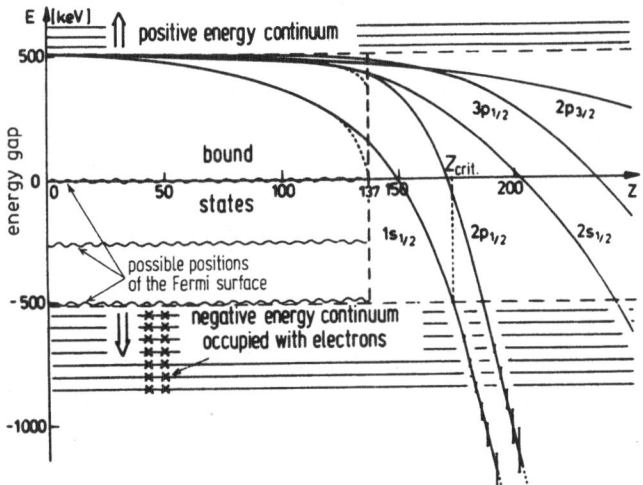

Fig. 1: Lowest bound states of the Dirac-equation for nuclei with charge Z. While the Sommerfeld-eigenenergies (dashed lines) for j=1/2 end at Z=137 the solutions with extended Coulomb potential (full lines) can be traced down to the negative continuum which is reached at critical charge Z_{cr}. The states entering the continuum obtain a spreading width as indicated by the bars (magnified by a factor of 10). If the state was previously unoccupied two positrons will be emitted spontaneously.

We say that *the overcritical vacuum state is charged.* This has the following meaning. As already stated, within the hole theory, which is a lucid model for interpreting the field theoretical (quantum electrodynamical) calculations, the states of negative energy are occupied with electrons. This was postulated by Dirac to avoid the decay of electronic states with emission of an infinite amount of energy. In the undercritical situation we can define a vacuum state

|0> without charges or currents by choosing the Fermi surface (up
to which the levels are occupied) below the lowest bound state: we
set $E_F = -m_o c^2$. The negative-energy continuum states occupied with
electrons represent the model for this vacuum; its infinite charge
is renormalized to zero, and so it is a *neutral* vacuum.   If
now an empty atomic state dives into the negative continuum, it will
be filled spontaneously with an electron from the Dirac sea with
the simultaneous emission of a free positron moving to infinity.
The remaining electron cloud of the supercritical atom is necessarily
negatively charged. Thus, the vacuum becomes *charged*.

An atom with Z>173 and an empty K-shell will spontaneously
shield itself by two K-electrons and emit two positrons of rather
well-defined energy. This two-electron state becomes the stable state,
and it forms in a time scale of about 10^{-20}sec. If the central charge
is further increased to Z=184 (diving point of the $2p_{1/2}$ level), the
vacuum acquires a charge of -4e. With increasing field strength, more
and more electronic bound states join the negative continuum, and
each time the vacuum undergoes a new phase transition and becomes
successively higher charged: the vacuum sparks in overcritical fields.

Clearly, the charged vacuum is a new ground state of space and
matter. The normal, undercritical, electrically neutral vacuum, is
in overcritical fields no more stable: it decays spontaneously into
the new stable but charged vacuum. Thus the standard definition of
vacuum, "a region of space without real particles", is not true in
very strong external fields. It must be replaced by the new and better
definition, the "energetically deepest and stable state that a region
of space can have while being penetrated by certain fields" (see fig.2).

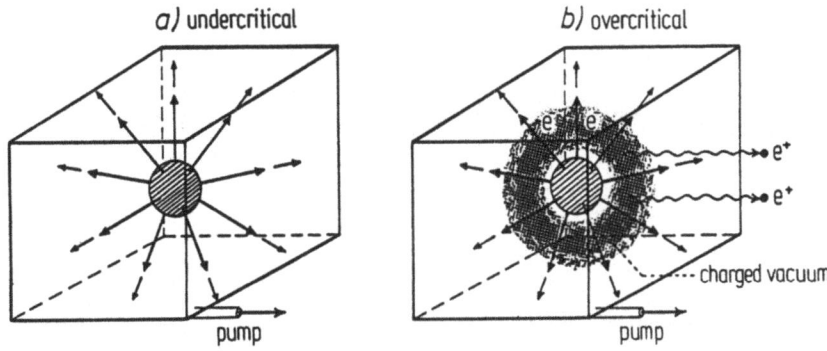

Fig. 2: The space inside box represents the vacuum. The central nucleus
acts as a source of a strong electric field. In the undercritical case (a) the vacuum is empty, i.e. no particles (besides
the central source) are in the box.
In overcritical fields space becomes charged through the emission
of antiparticles. In principle the vacuum is no longer empty
under these conditions. The shaded sphere in the center
represents the giant nucleus, the source of the electric field
(indicated by arrows); the diffuse cloud represents the electrons
of the charged vacuum. If this electron cloud is pumped away,
new positrons (represented by e^+) will be emitted and the
electronic cloud will reappear. The positrons, being in
continuum states, can freely move around and are pumped out
easily. One is again left with the charged vacuum. Hence
under the extreme conditions of supercritical fields the vacuum
is no longer empty; the vacuum is sparking(b).

Superheavy quasimolecules in heavy-ion scattering

Inasmuch as the formation of a superheavy atom of Z>173 is very

unlikely, a new idea is necessary to test these predictions experimen-

tally. That idea, based on the concept of nuclear molecules, was put

forward by Greiner and co-workers in 1969: a *superheavy quasimolecule*

forms temporarily during the slow collision of two heavy ions (the

idea is sketched in fig. 3).

It will be sufficient to form the quasimolecule for a very short instant of time, comparable to the time scale for atomic processes to evolve in a heavy atom, which is typically of the order 10^{-18} - 10^{-20} sec. Consider the case where an U ion is shot at another U ion at an energy corresponding to their Coulomb barrier and the two, moving slowly (compared to the K-shell electron velocity) on Rutherford hyperbolic trajectories, are close to each other (compared to the K-shell electron orbit radius). The two ions can be brought together as close as 16 fm for a time of $\sim 10^{-21}$sec. Then the atomic electrons move in the combined Coulomb potential of the two nuclei, thereby experiencing a field corresponding to their combined charge of 184 (fig. 3). This happens because $v_{ion} \sim c/10$, $v_{el} \sim c$: the ionic velocity is much smaller than the orbital electron velocity, so that there is time for the electronic molecular orbits to be established, i.e. to adjust to the varying distance between the charge centers, while the two ions are in the vicinity of each other.

The condition $v_{ion}/v_{el} \sim 1/10$ is known as adiabaticity. It will not help to make v_{ion} even smaller, so that complete adiabaticity is eventually achieved: for it is a partial breakdown of adiabaticity that makes the inner shells of the quasimolecule ionized, i.e., empty of electrons, which, as we saw earlier, is a necessary prerequisite for the emission of positrons and the accompanying filling of the inner shells with electrons as they dive into the negative continuum.

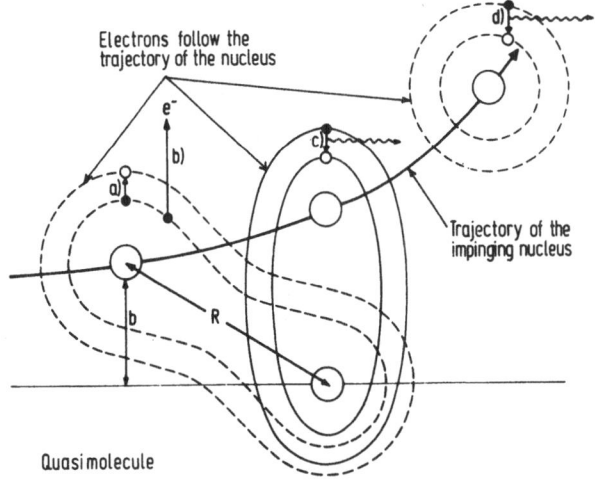

Electrons follow the
trajectory of the nucleus

e⁻

b)

a)

c)

d)

Trajectory of the
impinging nucleus

b

R

Quasimolecule

Fig. 3: The basic concept concerning the formation of quasimolecules
is shown. In the collision of two heavy ions the inner
electrons orbit both nuclei together. The electron orbits
follow the motion of the nuclei. Both nuclei are shown and
their paths are indicated. The distance of closest approach
is related to the impact parameter b. Processes of type
a) (excitations of electrons into higher shells) and of
type b) (excitations of electrons into the upper continuum)
empty the K-shell. Processes c) and d) indicate the molecular
and atomic X-ray transitions, respectively. The molecular
X-rays are emitted from the intermediate quasimolecule,
while the atomic X-rays are emitted from the rearranged atom
after the collision.

When the two U atoms are separated by a large distance, the Z = 184

system is undercritical (i.e., all levels are bound by less than

$2m_oc^2$). It becomes overcritical at small R as the electrons experience

the full combined charge. For the $1s_{1/2}$ level the critical separation

occurs at R_{cr}=35 fm. The diving is very steep as a function of R. The

level energies change rapidly only in the last 150 fm of the approach

to the quasimolecule. This steep diving is important for the production

of K holes (see the schematic Fig. 4).

Dynamical processes in heavy-ion collisions

Several dynamical processes contribute to the ionization
of the inner shells and to the production of positrons in undercritical as well as overcritical systems. This is illustrated in
fig. 4 for a system that becomes overcritical at small distances.
In processes a) and b) one has electron excitation and ionization.
Process c) is the spontaneous filling of a previously produced
vacancy when the level acquires a binding greater than $2m_o c^2$ and
is the decay of the vacuum described earlier. Because of the lack
of full adiabaticity, energy can be drawn from the nuclear motion
to lead to filling of the hole even at distances larger than R_{cr}.
This effect (d,e) may be called an *induced transition*, and its
effect on positron production is twofold: it causes a washed-
out threshold for the spontaneous positron production, and it greatly
enhances the production cross section. f) is the direct pair production
process, which we now proceed to discuss in more detail.

Whereas in ordinary pair production in a Coulomb scattering
process a photon is exchanged between two hadrons only once, now
there are multiple interactions with the joint Coulomb field of both
nuclei. Because of the very strong field, the cross section for the
pair production varies as $(Z_1+Z_2)^{20}$, which means that about 10 (!)
photons are exchanged. This behaviour illustrates the *nonperturbative
character* of this process, which (like the induced decay mechanism)
overwhelms the spontaneous positron production process. The pair
production process f) can be interpreted as the shake-off of the
vacuum polarization (VP) cloud.

It is clear that a K hole is needed for the production of positrons by either the spontaneous or the induced mechanisms. Since
neither the projectile nor the target atom has a K hole to start with,
it has to be produced dynamically via Coulomb excitation or ionization (processes a and b of fig. 4) in the

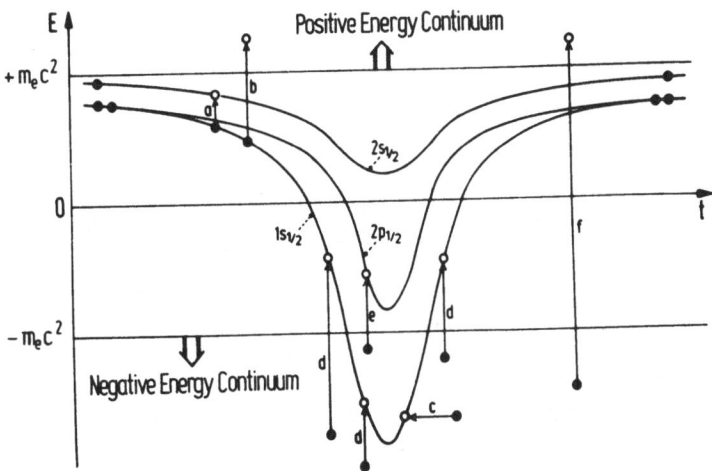

Fig. 4: Dynamical processes connected with positron production in overcritical heavy-ion collisions. The figure shows the inner
electron levels in the quasimolecule as a function of time.
At the deepest point of the 1s level, the colliding nuclei are
at the distance of closest approach.

  a,b: electron excitation and ionization,
    c: spontaneous autoionization of positrons, spontaneous decay
       of the vacuum, "sparking" of the vacuum,

  d,e: induced decay of the vacuum,
    f: direct pair creation.

collision itself. K-hole production occurs whenever the wave-functions

change so rapidly with R that the electrons cannot adjust to the

nuclear motion (breakdown of adiabaticity) and therefore get kicked out

as δ electrons. Because of the rapid change of the wavefunctions at

the onset of diving, vacancy production in the inner shell is

concentrated at small values of R, which is advantageous for the

observation of induced and spontaneous positron emission.

The total K shell vacancy probability in the diving region

for U-U collision at an energy of 1600 MeV is predicted to be about

10%, which is fully confirmed by recent experiments (Greenberg,

Vincent, Bosch, Liesen and others). (This includes both excitation

and ionization, through one-step as well as multistep channels

(see later)). The other 90% of the K electrons adjust to the nuclear

motion, and hence the adiabaticity necessary for the theoretical

treatment is generally valid.

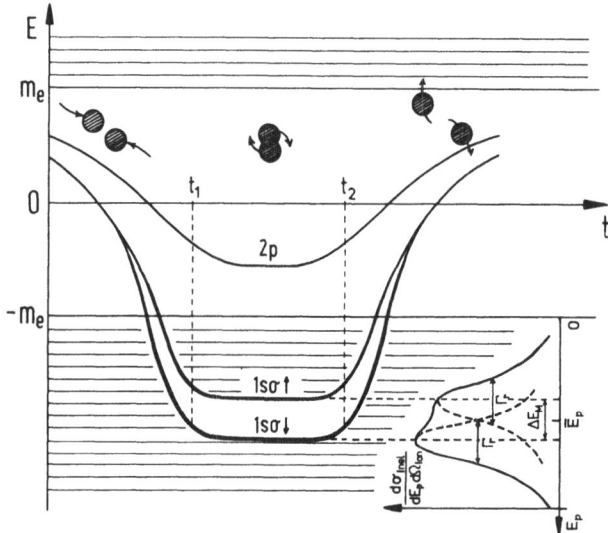

Fig. 5: The innermost shells of the superheavy molecule (atom) as a
function of time. Due to the sticking of the two nuclei, the
superheavy atom lives for the time Δt, thus being able to emit
positrons spontaneously. There are in general two positron
lines because of the Zeeman-splitting due to the strong
magnetic fields from the heavy ion currents.

The energy spectrum for positrons created in an e.g., Uranium-

Uranium collision, consists of three components: the induced, the

direct and the spontaneous one, which add up to a smooth spectrum.

The presence of the spontaneous component leads only to 5-10%

deviations for normal nuclear collisions along Rutherford trajectories.

The question arises: *Is there any way to get a clear qualitative signature for spontaneous positron production, as opposed to detecting it through a quantitative comparison with theory?* Suppose that the two colliding ions, when they come close to each other, stick together for a certain time Δt before separating again. This will in general require the use of bombarding energies slightly above the Coulomb barrier. Then the quasimolecular levels in the over-critical region get stretched out as shown in Fig. 5 which is to be contrasted with Fig. 4. (The splitting in the energy of the 1s - level arises from the Zeeman effect). During the sticking, the energies of the electronic states do not change, and this has two effects: a) the emission of positrons from any given state occurs with a fixed energy; b) the induced production mechanisms do not contribute, whereas the spontaneous production (for overcritical states) continues to contribute.

The longer the sticking, the better is the static approximation. For Δt very long, one sees in the positron spectrum a very sharp line with a width corresponding to the natural lifetime of the resonant positron-emitting state ($\cong 3$ keV for the U-U system).

The observation of such a sharp line will not only indicate the spontaneous decay of the vacuum but also the formation of giant nuclear systems (Z≳180).

Naturally one is also interested in the question of what happens if the two nuclei stick, but for some yet unknown reason the $1s_{1/2}$-level of an overcritical system does not dive, i.e., the neutral vacuum will not decay. Then an oscillatory structure as a function of the positron energy develops, which arises from the delayed interference between

the incoming and the outgoing positron-creation-amplitudes along
the trajectory of the colliding heavy ions. The positron spectrum
will then have an oscillating structure as a function of positron
energy from which the sticking time and even the structure (deformation,
excited states) of the super-heavy nuclear system can be deduced.
In other words, we are dealing here with an *atomic clock* for short-
living exotic nuclei. Because of the non-existence of a spontaneous
amplitude in this case, the spontaneous positron emission line does not
occur.

The search for spontaneous positron emission in heavy-ion
collisions began in 1976 with the first acceleration of uranium beams
at Gesellschaft für Schwerionenforschung (GSI) in Darmstadt, West
Germany. Experiments at this laboratory have utilized three detection
systems, which have pursued complementary aspects of the problem. The
groups are headed by P. Kienle (München), J. Greenberg (Yale) and
D. Schwalm (Heidelberg), and by H. Backe (Mainz) and E. Kankeleit
(Darmstadt). We should note that in connection with these experiments
it was necessary to establish that the conditions for forming quasimolecules could be met for the nuclear velocities required to achieve
internuclear separations sufficiently small to produce overcritical
binding. It was also critically important to demonstrate that the
production probability for $1s\sigma$ vacancies was both large in magnitude
and concentrated at small internuclear separations. There are many
evidences for the formation of quasimolecules in heavy ion collisions
such as δ electrons and molecular-orbital X-rays, inner shell vacancy
production etc. which came from the experimental work of W. Meyerhof
(Stanford) and of J. Greenberg and P. Vincent (Yale). Here we
concentrate on the search for detection of spontaneous positron
production.

One of the first experimental goals in the search for spontaneous positron emission was to determine the rate at which positrons are produced from the atomic processes relative to the rate at which they are produced from nuclear effects such as internal pair conversion of nuclear transitions. The first measurements on the $^{208}Pb-^{208}Pb$ collision system played a particularly important role in this respect and in confirming our theoretical understanding of the dynamic processes of positron production in heavy-ion collisions.

Measurements on Pb-U and U-U collisions have carried these investigations into heavier systems, but under different amd more complex background conditions. To investigate the consequences of this nuclear background in more detail, researchers carried out a systematic investigation of the ratio of positron intensity to γ-ray intensity over a broad range in Z. When Z_u, the combined nuclear charge Z_1+Z_2, exceeds about 160, the total positron production increases in a spectacular way over that expected from nuclear internal pair conversion as it is extrapolated from the positron to γ-ray ratios measured for $Z_u<160$. More precisely, for constant R_{min} and relative velocity, the production of positrons in superheavy collision systems is found to increase as $(Z_1+Z_2)^{20}$. In this striking feature, which seems to have no other analog in nature, the theory of dynamic positron creation in heavy-ion collisions (J. Reinhardt, B. Müller, G. Soff and W. Greiner) again anticipated the experimental results.

The most recent experiments have focussed on studying positron spectra and on extending the investigations to collision systems with higher total nuclear charge. With more comprehensive data, new phenomena have appeared that are connected with the effects being sought. Of special interest are peak-like structures in the positron energy distribution. The most compelling evidence for these comes from

experiments where coincidences between two scattered ions are used
to define clearly events with two-body final states consistent with,
or bordering on, elastic scattering (P. Kienle, Ch. Kozhuharov,
F. Bosch et al., J. Greenberg, D. Schwalm, T. Cowen, J. Schweppe,
P. Vincent et al.). We illustrate these interesting results with an
example.

The uranium-curium collision system, with Z_u = 188, has the
largest combined nuclear charge investigated to date. Fig. 6 shows
positron spectra from uranium-238 and curium-248 colliding at an
energy close to that of the Coulomb barrier.   Particularly striking
in fig. 6 is the well-defined peak centered at an energy of about
320 keV.   The height of this peak above the smoother continuum
is correlated with the choice of two-body final states corresponding
to a selected range of scattering angles for the two heavy ions.

By comparison, if one singles out scattering angles more
forward than those selected in fig. 6 this peak is largely excluded.
One finds a spectrum that mirrors the general shape of the continuum
underlying the peak in Fig. 6.   The continuum distributions are well
represented by the spectra we expect from the dynamically included
processes at the corresponding scattering angles.   As we will see,
it is also significant to find that the measured width of the peak in
Fig. 6 is less than 60 keV.   Moreover, this width is consistent with
the Doppler broadening expected for a positron line spectrum emitted
from a system moving with the velocity of the quasimolecular system.
Therefore, the intrinsic width of the peak is surely less than 60 keV
and, indeed, recent analysis indicates that it is very much smaller
than this value ($\leq$20 keV).

Whatever the source of the peak, it is apparent that we must seek
an explanation outside the scope of the theory based on Rutherford
scattering alone, because this theory of dynamic positron creation does

not allow for narrow peak structures in the positron spectrum.
Deviations from this theory also have been demonstrated for U–U
collisions in other experiments carried out by P. Kienle's group
at GSI. All experiments carried out to date indicate that there is
a new source of positrons – a source that does not originate from the

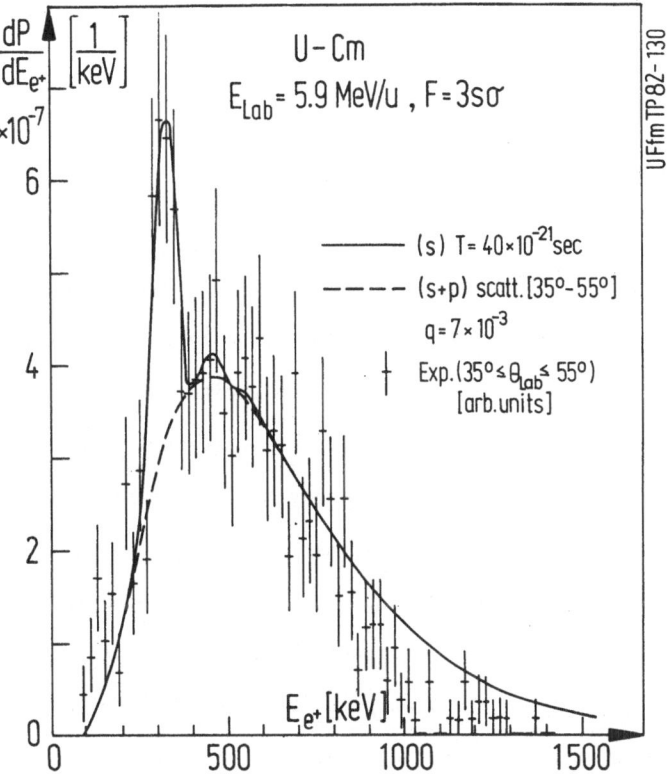

Fig. 6: Positron spectra from 5.8 MeV/amu uranium-curium collisions.
The full curve represents the theory. The line structure can
only be understood, if a rather long-living (~10-19sec) giant
nucleus is formed. These particular experiments were carried
out by Greenberg-Schwalm group, but similar results are
obtained for U+U and U+Th-systems by the group of P. Kienle.
The curve show the theory of Reinhardt, Müller, Soff et al.

known dynamic mechanisms associated in a simple way with the time-
varying electric field produced in Coulomb trajectories.

It has also been excluded to attribute these deviations
from smooth positron spectra to pure trivial nuclear effects.
There are two prominent candidates:

- the internal pair conversion of a nuclear transition
  leading to a positron energy distribution that may be
  peaked,
- the internal pair conversion process followed by the
  capture of the electron into empty atomic orbits, which
  leads to positron line spectra.

Kienle and his coworkers and also Greenberg and Schwalm and their
associates could in fact show that both the X-ray spectra and the
δ-electron spectra measured simultaneously with the positrons are
smooth, i.e. they show no structure. This strongly suggests,
indeed it proves, that the vacuum decay has been observed.

Obviously, a systematic confirmation is required to follow up
on these very suggestive data, but these new developments already raise
the possibility of another important observation. For if, as it is
now proven, the narrow positron peak does indeed represent spontaneous
positron emission, the parent nuclear supercritical charge must exist
for a long time compared to the collision times for scattering below
the Coulomb barrier, as we pointed out earlier.

Therefore, J. Reinhardt, B. Müller, U. Müller and W. Greiner
suggested that the observation of spontaneous positron emission as a sharp
line necessarily implies that, at bombarding energies close to that of the
Coulomb barrier, *metastable giant nuclear composite systems form* with a
rather long lifetime. Widths of 20 keV or less correspond to lifetimes

for the nuclear molecular system longer than about 1000 times the
Rutherford scattering collision time, during which the 1sσ state is
overcritically bound. Indeed, without introducing a time delay it is
difficult to invent any mechanism associated with atomic positron
emission that would explain the narrow peak width found in the U-Cm
spectrum or the positron distribution emitted from the U-U or U-Th
collisions. Such a lifetime could be supplied by the formation of a
rather cold intermediate giant nuclear complex as the nuclei barely
touch in overcoming the Coulomb barrier (see fig. 7).

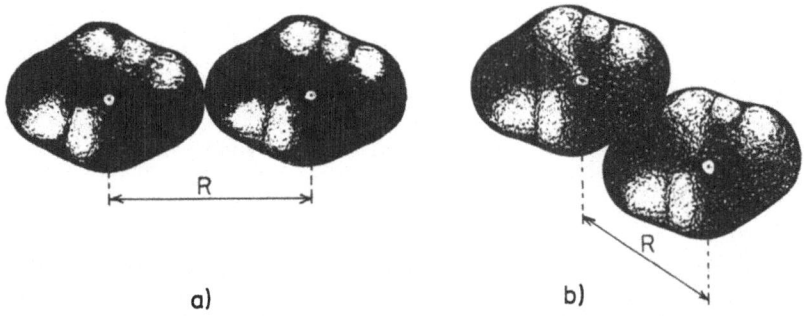

a)   b)

Fig. 7: Two uranium nuclei in typical positions upon contact. Due to
their peculiar shape the touching area in case b) is particularly
large, which leads to a kind of nuclear cohesion; thus
amalgamating the two uraniums to a giant nuclear system. For
a quantitative description of these phenomena we refer to
Mr. Seiwert, T. Pinkston and W. Greiner.

Thus several independent measurements confront us with evidence that
there are peak structures in the positron spectra of collision systems
where the quasiatom can have overcritically bound electrons. We are left
with the task of further identifying unambigouously the sources for the
accompanying (smaller) sidestructures. They are believed to yield
information on the excited states of the giant nuclear system (Raman
satellites).

Of course, identifying the spontaneous emission of positrons,
and thereby obtaining the first observation of the spontaneous decay
of the ground state in a fundamental field theory, is an outstanding
discovery.

II.   ON THE VACUUM IN FIELD THEORIES

During the last decade the understanding of the structure of the
vacuum in field theories has become of general importance. The conjecture,
that the vacuum is not just an empty piece of space, and its philosophical implications have a very long history, dating back to the Greek
philosophers of the Eleatic school and to Aristotle in particular.
Only with the advent of a quantum field theory it became clear that the
vacuum is a physical object, which can be subject of physical experiments.
In modern gauge theories the vacuum necessarily has to have certain
structures; most of them not yet completely understood. Let me only
remind you on some of them:

   a)   In the ϕ^4-theories the vacuum is degenerate, because the $V(\phi)$-

         potential looks as in figure 8 and degenerate groundstates are

         obviously possible.

Figure 8: The groundstate
             (vacuum) in a
             ϕ^4-theory is de-
             generate.

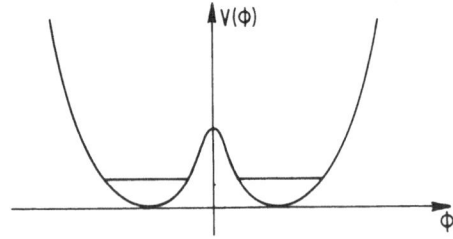

   b)   In some gauge theories the vacuum has to be assumed to contain

         Higgs-fields in order to produce masses for the particles in a

         gauge-invariant way. Such Higgs -fields have not been observed to date. Hence one does not know whether they really exist, but we do not know of any better way to construct gauge invariant, renormalizable theories allowing particles to obtain masses.

c) Particularly in quantum chromodynamics one distinguishes a *true vacuum*, which expells gluo-electric field lines in a similar way as a superconductor repels magnetic field lines. Then, an assembly of real quarks, which are the sources of the gluo-electric field lines, would have to form a bubble (bag) around the true vacuum. In that bubble the true vacuum cannot exist because of the presence of gluo-electric fields and the bubble will be filled with a simple vacuum (sometimes also called per-turbative vacuum). We illustrate these ideas in figure 9.

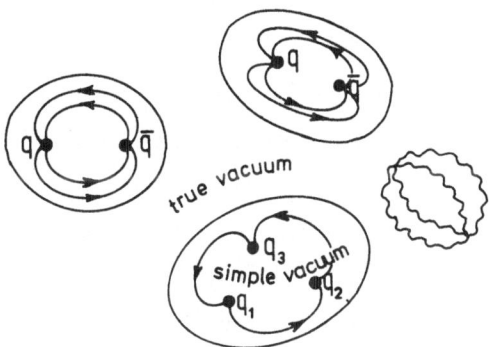

Figure 9: The true vacuum of QCD is liquid-like, with gluon and quark balls of a certain size forming and transforming. Quarks can only exist within the simple vacuum inside; they are caught inside the bubbles.

We shall come back to the structure of the simple vacuum in case of supercritical gluo-fields appearing during the deformation and fission of a bag in one of the next lectures (see D. Vasak, S. Klevansky et. al).

The problem with these vacuum ideas described so far is, that these vacuum models are a priori given. We work with them, but we have so far no tool to change in a relatively easy and controllable way one vacuum structure into another one. Such *"experimenting" with the vacuum* is necessary to substantiate the underlying ideas, or, to say it differently, to convert the models and ideas into true physics.

This is precisely what is achieved with the vacuum of the best field theory we have up to now, i.e. QED.

III. THE VACUUM OF QUANTUM ELECTRODYNAMICS

It is the electron-positron-photon groundstate, which is best described by looking at the single particle spectrum (fig.10) of electrons in an external field. The spectrum devides into a positive and a negative energy spectrum devided by the energy gap of $2m_o c^2$, within which one finds the bound states for attractive external potentials. Let us call the negative energy states ψ_n, the positive energy states ψ_p. Then the field-operator at time t=0 is defined as

$$\hat{\Psi}(\vec{x},t=0) = \sum_p \hat{b}_p \, \psi_p(\vec{x}) + \sum_n \hat{a}^+_n \, \psi_n(\vec{x}) \tag{1}$$

$\hat{b}_p$ are the annihilation operators for electrons and $\hat{a}^+_n$ the creation operators for positrons. This definition guarantees that the energy $\hat{H}_D$ of the electron-positron field is positive definite, i.e.

$$\hat{H}_D = \sum_p E_p \, \hat{b}^+_p \hat{b}_p + \sum_n |E_n| \hat{a}^+_n \hat{a}_n \tag{2}$$

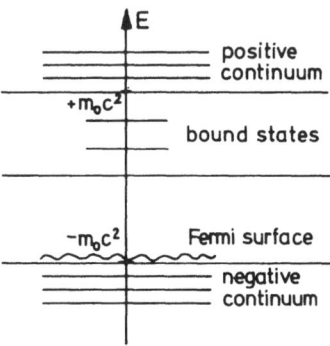

Figure 10: The simple particle spectrum of the Dirac-equation in an attractive external field.

The vacuum $|0>$, characterized by the Fermi energy E_F, i.e. by the $|n>$ with $E_n < E_F$ filled with electrons is best illustrated in hole-theory (Fig.11). A hole in this "Dirac-sea" is then interpreted as a positron. Note, the position of

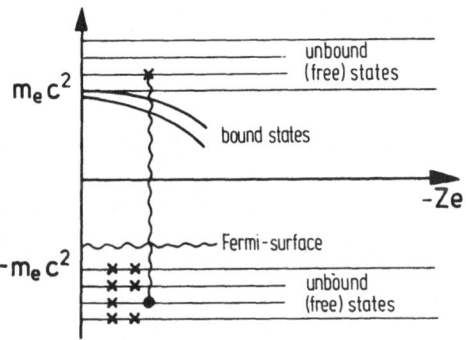

Figure 11: The neutral vacuum of QED can be viewed as the states of negative energy filled with electrons. The infinite charge is renormalized to zero.

the Fermi surface, which divides the single particle states into those which are to be counted as electronic and those which are positronic states is up to nature and can be deduced experimentally by observing e.g. the threshold for e^+e^--pair production. It cannot be choosen at will. This so defined vacuum $|0>$ is neutral:

If one calculates the vacuum expectation value of the charge operator

$$\hat{\rho} = \frac{1}{2} \left[\hat{\Psi}^+ (\vec{x},0), \hat{\Psi}(\vec{x},0) \right]_- \tag{3}$$

which is the zero-component of the current-four vector

$$\hat{j}_\mu = \frac{1}{2} \left[\hat{\bar{\Psi}}, \gamma_\mu \hat{\Psi} \right] \tag{4}$$

one finds

$$<0|\hat{\rho}|0> \equiv \rho_{vac\ pol} = \frac{1}{2} e \left(\sum_n \psi_n^+ \psi_n - \sum_p \psi_p^+ \psi_p \right) . \tag{5}$$

This vanishes for the field-free case because of symmetry (equal number and structure of n- and p-states), but gives the vacuum-polarization charge, $\rho_{vac\ pol}(\vec{x})$ in case an external potential is present. The latter gives rise to a part of the Lamb-shift and is well established. We can say, in weak fields the vacuum is a polarizable medium, characterized under weak external fields by the displacement charge $\rho_{vac\ pol}(x)$, for which

$$\int \rho_{vac\ pol}(\vec{x})\ d^3x = 0 . \tag{6}$$

This *vacuum polarization displacement charge* can be illustrated as in fig. 12. Of particular interest to us here is the *stripping-off of the vacuum polarization charge* in case of the moving ions. This leads to the ejection of the e^+e^- - pairs, which goes like $(Z_1 + Z_2)^{20}$ -power as a function of the colliding charges [1,2]. It has been observed experimentally by Backe, Kankeleit et. a. [3] and by Kienle, Greenberg and associates [4].

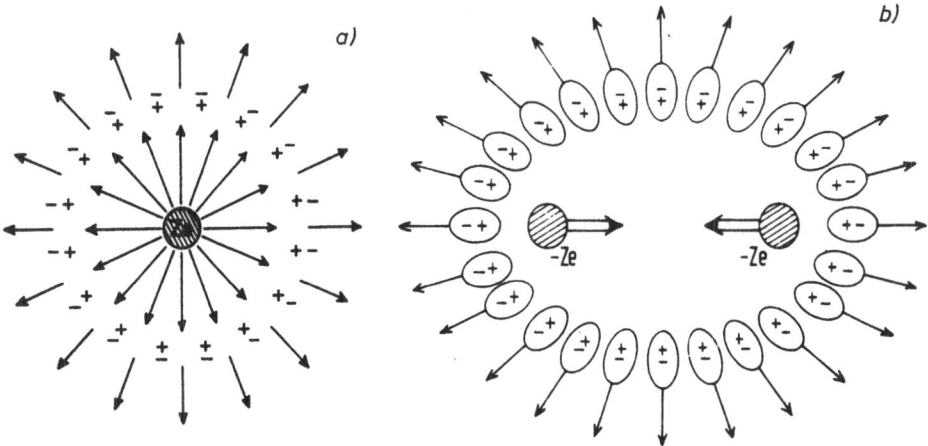

Figure 12: The vacuum polarization charge around a central nucleus (a)
and around two colliding heavy nuclei (b). In the former
case the static s-electrons are shifted in energy somewhat
due to the modification of the Coulomb potential by
$\rho_{vac\ pol}$ (this is part of the Lambshift); in the latter
case the vacuum polarization charge is partially stripped
off because of the motion of the ions.

IV. THE CHARGED VACUUM IN SUPERCRITICAL FIELDS (SINGLE PARTICLE ASPECTS).

In principle, the development of quantum electrodynamics of
strong fields could have proceeded as soon as quantum field theory
had been formulated about 40 years ago. The main obstacle was the fact
that the solution of the Dirac equation (and also Klein Gordon equation)
had not been understood for supercritical potentials. While already
Heisenberg and Euler [5] and also Weisskopf [6] had proposed in the
1930's that strong, external infinitely extended electric fields would
lead to spontaneous pair creation and the subject had been picked up
by several other authors thereafter [7], the modern development which
lead ultimately to a full clarification and, more importantly, to a

new field of physics, by pointing to directions of possible experimental verifications of the theoretical conjectures, started with the work of Pieper and Greiner [8]. The crucial technical step has been the recognition [9,10,11] that the spectrum of the Dirac equation in supercritical fields contains a resonance in the negative energy continuum – continuously connected with the bound particle solutions as the strength of the potential decreases. This discovery was made in the years 1971-72 independently by the Frankfurt group [9,10] and by Zel'dovich and Popov [11] in the Soviet Union. Without this resonance the spectrum is incomplete, and a consistent quantization of the electron field in the supercritical external potential is not possible. Only in a consistent theory that treats the $(Z\alpha)$ effects correctly can further QED effects in order α, which are naturally also present, be considered. However they have been found to be relatively unimportant, and we shall refer to the lectures of M. Gyulassy and P. Mohr at the Lalmoscin Conference for details [13]. The characteristic properties of quantum electrodynamics of strong fields can be derived by considering only effects to order $(Z\alpha)$.

To describe electrons in an external electromagnetic field we use the Dirac equation for spin 1/2 particles:

$$\left[\gamma^{\mu}(P_{\mu}-eA_{\mu}) + m_{o}\right] \Psi(x) = 0, \tag{1}$$

where the four-component vector potential A_{μ} is introduced by minimal coupling. For stationary states in a static electric field, $A_{o} = V(\vec{r})$, the eigenvalue problem is:

$$H_{D} \Psi_{n}(\vec{r}) \equiv (\vec{\alpha}\cdot\vec{p} + \beta m_{o} + V(\vec{r})) \Psi_{n}(\vec{r}) = E_{n}\Psi_{n}(\vec{r}) . \tag{2}$$

In the following we restrict ourselves to spherically symmetric
potentials, $V_o(r)$, where the wavefunction has good angular momentum:

$$\Psi_{\kappa\mu}(\vec{r}) = \begin{pmatrix} g_\kappa(r) \, \chi_\kappa^\mu \\ if_\kappa(r) \, \chi_{-\kappa}^\mu \end{pmatrix} . \tag{3}$$

$g_\kappa(r)$ and $f_\kappa(r)$ are the radial parts of the 'large' and 'small'
components, respectively, and the χ_κ^μ are the spinor spherical harmonics.
The radial equations

$$(\frac{d}{dr} + \frac{\kappa+1}{r}) g_\kappa - (E+m_o-V_o(r)) f_\kappa = 0 \tag{4}$$

$$(\frac{d}{dr} - \frac{\kappa-1}{r}) f_\kappa + (E-m_o-V_o(r)) g_\kappa = 0$$

can be solved analytically for various simple potentials, e.g. for
the point nucleus $V_o(r) = -Z\alpha/r$. The energy eigenvalues are known
as Sommerfeld fine-structure formula:

$$E = m_o \left[1 + \left(\frac{Z\alpha}{n - |\kappa| + \sqrt{\kappa^2 - (Z\alpha)^2}} \right)^2 \right]^{-1/2} , \quad n=1,2,\ldots \tag{5}$$

which exhibits a bifurcation singularity when $(Z\alpha) \rightarrow |\kappa|$:

$$f_{j=\frac{1}{2}}(r), \; g_{j=\frac{1}{2}}(r) \sim r^{\sqrt{1-(Z\alpha)^2}} \xrightarrow[(Z\alpha)>1]{} r^{-i\sqrt{(Z\alpha)^2-1}} \quad .$$

Additional conditions are required to find the selfadjoint extension
of the Hamilton operator H_D.

We therefore turn to the investigation of potentials that are
due to the charge distribution of finite size nuclei:

$$V_o(r) = \begin{cases} -\frac{3}{2} \frac{Z\alpha}{R_n} (1 - \frac{r^2}{3R_n^2}) & : \; o \leq r \leq R_n \\ \\ -\frac{Z\alpha}{r} & : \; R_n < r < \infty \end{cases} \tag{6}$$

Most of the recent calculations have relied on numerical integration techniques both inside and outside the nucleus, although in the special case, Eq. (6), one can proceed further analytically. In order to include also the effect of electron-electron interaction, Hartree-Fock-Slater calculations have been performed. Results for the energy eigenvalues are shown in Fig. 13. The eigenvalues decrease monotonically as the charge increases. None of the eigenvalues or the wave functions exhibit any unusual behaviour at $Z\alpha = 1$. The points at which the levels join the lower continuum are well isolated. The critical value of $Z = Z_{cr}$ is ~ 170, where the 1s level joins the lower continuum. The $2p_{1/2}$ level joins the lower continuum at about $Z = 183$. The exact location of Z_{cr} is an important question for the experimental verification of the theory.

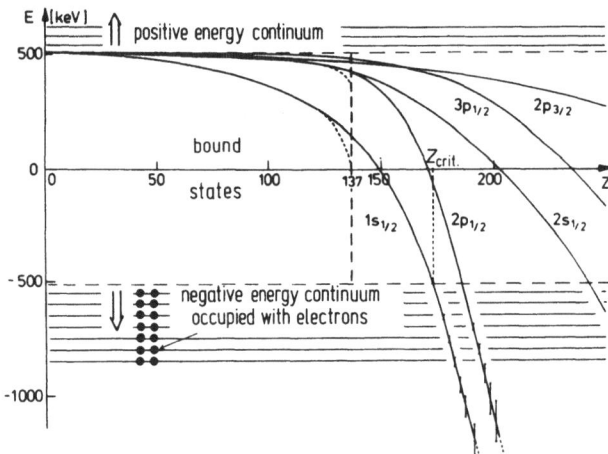

Figure 13: Lowest bound states of the Dirac equation for nuclei with charge Z. While the Sommerfeld fine-structure energies (broken lines) for κ=-1 end at Z≈137, the solutions for extended Coulomb potentials (full curves) can be traced down to the negative-energy continuum which is reached at the critical charge Z_{cr}. The bound states entering the continuum obtain a spreading width,

To describe the supercritical atoms when the nuclear charge exceeds the critical value Z_{cr} and the $1s_{1/2}$ state becomes degenerate with the negative-energy continuum we follow the approximate treatment developed in [9,10] that makes use of Fano's method of treating auto-ionizing states [14]. We search for bound states of the Hamiltonian H_D, Eq. (2) with $V(r;Z) \equiv Z \cdot U(r;Z)$. The quantity $U(r;Z)$ depends on Z only slowly via the $Z-N$-dependent radius of the nuclear charge distribution. Within the range of atomic nuclei considered by us ($170 < Z < 200$) this dependence is very weak and may initially be omitted for convenience:

$$V(r;Z) = Z \cdot U(r) . \tag{7}$$

We now make use of the fact that we know the solution to our problem for $Z = Z_{cr} \simeq 170$ and diagonalize $H(Z = Z_{cr} + Z')$ in the basis of eigenstates given by $H(Z_{cr})$. Let $|\phi\rangle$ be the $1s$ bound state eigenfunction for $Z = Z_{cr}$, i.e.

$$H_D (Z_{cr}) |\phi\rangle = E_o |\phi\rangle \simeq -m_o |\phi\rangle \tag{8}$$

and $|\psi_E\rangle$ be the s-continuum wave functions normalized to energy delta-functions:

$$H_D (Z_{cr}) |\psi_E\rangle = E |\psi_E\rangle , \quad E < -m_o . \tag{9}$$

$|\phi\rangle$ and the $|\psi_E\rangle$ serve as a truncated ("reduced") basis for our diagonalization procedure. In doing so we neglect that small contribution from the higher bound ns states ($n > 1$) which are widely separated from the $1s$ bound state. We will need the matrix elements of $H(Z_{cr} + Z')$ in our truncated basis

$$\langle\phi| H_D (Z_{cr} + Z') |\phi\rangle = E_o + Z'\langle\phi|U|\phi\rangle \equiv E_o + \Delta E_o \tag{10}$$

$$\langle\psi_E|H_D(Z_{cr} + Z')|\phi\rangle = Z'\langle\psi_E|U|\psi\rangle \equiv V_E \tag{11}$$

$$\langle\psi_{E''}|H_D(Z_{cr} + Z')|\psi_{E'}\rangle \equiv E'\delta(E'' - E') + Z'U_{E''E'} . \tag{12}$$

The matrix elements $U_{E''E'}$ describe the rearrangement of the continuum states under the additional potential $U(r)$. For small Z' this effect may be neglected since its influence upon the 1s bound state is of second order. For large Z' the continuum states may be prediagonalized by constructing the solutions of the projected eigenvalue problem

$$\hat{P} \, H_D(Z_{cr}+Z') \, \hat{P} \, |\tilde{\psi}_E\rangle = E|\tilde{\psi}_E\rangle \tag{13}$$

where $\hat{P}=1-|\phi\rangle\langle\phi|$ is the operator that projects out the critical 1s-state. For this continuum the matrix elements $U_{E''E'}$ vanish identically. The projection method which is based on work in nuclear physics [15] has been extensively investigated by J. Reinhardt [16] (see also [13]).

The aim is to find $|\Psi_E\rangle$, a continuum solution to the Dirac equation for $Z>Z_{cr}$. We may expand $|\Psi_E\rangle$ within the space spanned by the truncated basis:

$$|\Psi_E\rangle = a(E)|\phi\rangle + \int_{|E'|>m} b_{E'}(E)|\psi_{E'}\rangle \, dE' \tag{14}$$

We are mainly interested in the effects on the bound state $|\phi\rangle$ finding:

$$|a(E)|^2 = \frac{|V_E|^2}{|E - (E_0+\Delta E_0) - F(E)|^2 + \pi^2|V_E|^4} \tag{15}$$

where $F(E)$ is the principal value integral

$$F(E) = P \int_{|E'|>m} dE' \frac{|V_{E'}|^2}{E - E'} . \tag{16}$$

The quantity $|a(E)|^2$ is the probability that the 1s-electron bound in $|\phi\rangle$ is embedded in $|\Psi_E\rangle$ as the additional charge Z' is "switched on". The quantity $|a(E)|^2$ has an obvious resonance behaviour. If V_E does not depend too strongly on the energy E, we may neglect $F(E)$ with respect to ΔE_0 getting

$$\Gamma = 2\pi|V_{E_0+\Delta E_0}|^2 . \tag{17}$$

Then, indeed, a Breit-Wigner shape is found with the resonance of width Γ peaked around $E_o + \Delta E_o$.

Since we have chosen $E_o \approx -m_o$, $\Delta E_o = -Z'\delta$ describes the energy shift of the bound 1s-state due to the additional charge Z'. The width Γ of the resonance is $\Gamma = Z'^2 \gamma$. Calculations [33] show that except very close to threshold

$$\delta \approx 30 \text{ KeV} , \quad \gamma \approx 0.05 \text{ KeV} . \tag{18}$$

Thus we may explicitly show the Z'-dependence of Eq. (24):

$$|a(E)|^2 = \frac{1}{2\pi} \frac{Z'^2 \gamma}{[(E+m)+Z'\delta]^2 + \frac{1}{4} Z'^4 \gamma^2} , \quad Z' \geq 3 . \tag{19}$$

From Eq. (19) we learn that the bound state $|\phi>$ "dives" into the negative energy continuum for $Z > Z_{cr}$ proportional to $Z' = (Z-Z_{cr})$. At the same time it obtains a width Γ_E within the negative energy continuum that grows like $Z'^2 = (Z-Z_{cr})^2$.

Let us summarize. As the proton number of a nucleus with $Z < Z_{cr}$ is steadily increased, the energy of K-shell electrons is decreased until at $Z = Z_{cr}$ it reaches $-m_o$. During this process the spatial extension of the K-shell electron charge distribution is also shrinking. When Z grows beyond Z_{cr} the bound 1s-state ceases to exist. But this does not mean that the K-shell electron cloud becomes delocalized. Indeed, according to Eq. (15) the bound state $|\phi>$ is shared by the negative energy continuum states in a typical resonance manner over a certain range of energy by Eq. (15). Due to the bound state admixture the negative energy continuum wave-functions become strongly distorted around the nucleus. This additional distortion of the negative energy continuum due to the bound state can be called *real charged vacuum polarization* [17], because it is caused by a real electron state which

joined the "ordinary vacuum states", i.e. the negative energy continuum.
The charge densities induced by all the continuum states superpose to
form an electron cloud of K-shell shape. This electron cloud created
by the collective behaviour of all continuum states contains the charge
of two electrons, since the total probability for finding the 1s-electron
state $|\phi>$ in any of the continuum states is:

$$\int_{-\infty}^{-m_o} dE \, |a(E)|^2 = 1 \, . \tag{20}$$

Thus, the K-electron cloud remains localized in r-space.

We emphasize the surprising fact that it obtains an energy
width Γ. This can be illustrated in the following way: Consider
the Dirac equation with the cut-off Coulomb potential inside a finite
sphere of radius a. Certain boundary conditions on the sphere have
to be fulfilled. In this way the continuum is discretized, see Fig. 14.
Fig. 14a shows the situation at $Z=Z_{cr}$, i.e. before diving. After
diving (Fig. 14b) the 1s-bound state has joined the lower continuum and
is spread over it. One sees that the K-shell electrons still exist,
but are spread out energetically. Therefore a γ-absorption line from
a 1s-2p-transition would acquire an additional width, the spreading

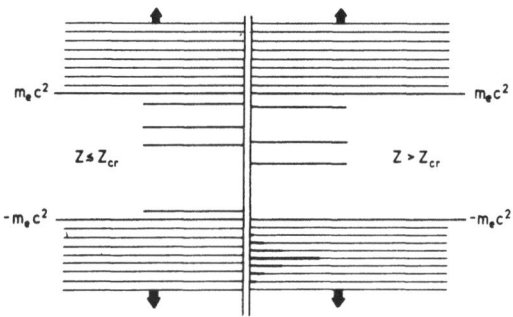

Fig. 14: Spreading of the bound state (solid line) over the negative
energy continuum states (weak lines). Spectrum a) before and b)
after diving.

width. The situation is different when the 1s-bound state is empty while Z is increased beyond Z_{cr}. Then - on grounds of charge conservation - one of the resulting continuum states $|\Psi_E\rangle$ has to be empty, i.e. a positron escapes. The kinetic energy of the escaping positron is not sharp, but has a Breit-Wigner type spectrum given by Eq. (15). Thus the width Γ is also the positron escape width. Of course, the positron-escape-process can be reversed. If positron scattering from nuclei with $Z > Z_{cr}$ were observed, the scattering cross-section would have a resonance at $E = |\Delta E_o|$ with a width Γ. The phase shift of the $s_{1/2}$-positron waves should go through $\pi/2$ at this energy.

Returning now to the discussion of positron spectra we note that the probability per unit time for emission of positrons in the energy interval dE is given by Fermi's "Golden Rule":

$$p(E)dE = \frac{2\pi}{\hbar}|\langle\phi|H|\Psi_E\rangle|^2 \rho(E)dE = \frac{\frac{1}{2\pi}\Gamma_E dE}{|E - (E_o + \Delta E_o)|^2 + \Gamma_E^2/4} \cdot \frac{\Gamma_E}{\hbar} .$$

(21)

This decay must be interpreted (see below) as the *decay of the normal, neutral vacuum into a charged vacuum* (charge 2e for 173<Z<184) in overcritical fields. The normal vacuum state is absolutely stable up to $Z = Z_{cr}$ and becomes unstable to spontaneous decay in supercritical fields. Only the charged vacuum (after two positrons were emitted) is stable in supercritical fields. The vacuum proceeds to become more highly charged as the supercritical fields are further increased. The above results can easily be generalized to several supercritical states embedded in the negative continuum.

The resonance behaviour of a supercritical state can be studied considering the exact continuum solutions [10, 18] of the Dirac equation for the electrostatic potential defined in Eq. (6). A phase shift δ is determined from the ratio of the radial functions at the nuclear surface. The results for $\sin^2(\delta-\delta_o)$ are represented in Fig. 15. The background phase δ_o was calculated using a nucleus with three protons less. The resonance in Fig. 15 is centered at $\varepsilon=-926$ keV and the full width at half maximum is $\Gamma=4.8$ keV. The results for the positions of the $1s_{1/2}$ and $1p_{1/2}$ resonances as functions of the nuclear charge are shown in Fig. 16. It was also shown that one could indeed parametrize the position and width of the resonance as in Eq. (18), for Z not very much greater than Z_{cr}. For values of Z very close to Z_{cr}, it is necessary to include in γ a damping factor which considers that the probability of finding low energy positrons near the nucleus is small when $Z \approx Z_{cr}$.

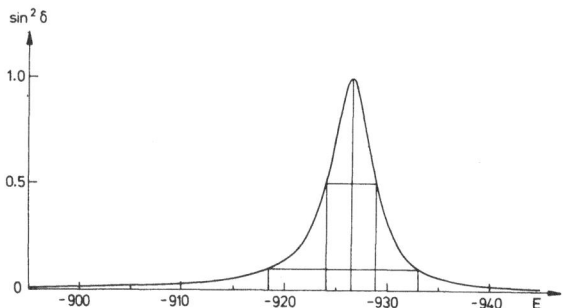

Fig. 15: The energy dependence of $\sin^2(\delta-\delta_o)$ in an overcritical electrostatic potential Z=184.

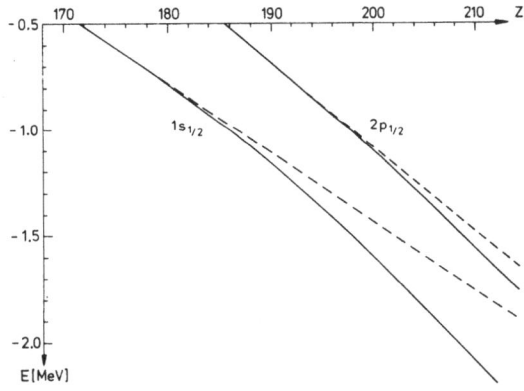

Fig. 16: The positions of the $1s_{1/2}$ and $2p_{1/2}$ resonances as functions of the nuclear charge.

V. SECOND QUANTIZATION OF THE DIRAC FIELD: THE VACUUM STATE

Strictly speaking, the treatment of single-particle orbits in Section IV allows only the conclusion that the single-particle theory breaks down in the presence of a supercritical external field. By physical intuition we are able to guess the nature of the processes causing the break-down, viz. the spontaneous creation of particle-antiparticle pairs. A definite proof for this conjecture, along with the assertion that the supercritical system reaches a new stable configuration, can only be constructed within the framework of a many-particle theory; i.e. relativistic quantum field theory which simultaneously treats particles and antiparticles. In this section we briefly review the second quantization of the Dirac field in the presence of an external potential and discuss the nature of the ground-state [17,19]. The abrupt change of the groundstate (phase transition) for sufficiently strong potentials, accompanied by pair-creation will be the subject of the following section.

In the process of second quantization the Dirac wavefunction $\Psi(\vec{x},t)$ is replaced by an operator-valued distribution $\hat{\Psi}(\vec{x},t)$ that acts in the Fock space of state vectors. The basic meaning of the operator $\hat{\Psi}(\vec{x},t)$ is that it annihilates a particle or creates an antiparticle at time t at the space-point $\vec{x}$. In many cases it is more practical to characterize particles not by position $\vec{x}$ but by a normalizable wavefunction $\Psi_n(\vec{x},t)$. If we have a complete set of such functions, we can divide it into a subset of functions describing particles and one describing antiparticles, which we shall denote symbolically by "n>F" and "n<F", respectively. In the hole picture of Dirac the antiparticle states are considered as filled and therefore the boundary between particle and antiparticle states has the quality of a generalized Fermi surface, from which the symbol "F" is derived. Accordingly we write:

$$\hat{\Psi}(\vec{x},t) = \sum_{n>F} \hat{b}_n \Psi_n(\vec{x},t) + \sum_{n<F} \hat{d}_n^+ \Psi_n(\vec{x},t) . \tag{1}$$

$\hat{b}_n$ annihilates an electron in the single-particle state Ψ_n, $\hat{d}_n^+$ creates a positron in state Ψ_n. In this section we shall restrict ourselves to situations where the external potential is time-independent. Then it is allowed to assume that the functions Ψ_n are stationary states

$$\Psi_n(\vec{x},t) = \phi_n(\vec{x}) \, e^{-iE_n t} \tag{2}$$

where ϕ_n are eigenfunctions of the single particle Hamiltonian

$$H_D = \vec{\alpha} \cdot (\vec{P} - e\vec{A}(\vec{x})) + \beta m_o + V(\vec{x}) . \tag{3}$$

The conjugate operator to $\hat{\Psi}$ is denoted by $\hat{\Psi}^+$, creating an electron (destructing a positron) at $(\vec{x},t)$. By well-known arguments involving Lorentz-invariance and causality the following equal-time anti-commutation relations on the field operators are imposed:

$$[\hat{\Psi}(\vec{x},t),\ \hat{\Psi}(\vec{x}',t)]_+ = [\hat{\Psi}^+(\vec{x},t),\ \hat{\Psi}^+(\vec{x}',t)]_+ = 0$$

$$(4)$$

$$[\hat{\Psi}(\vec{x},t),\ \hat{\Psi}^+(\vec{x}',t)]_+ = \delta(\vec{x}-\vec{x}')\ .$$

The decompositions (1) and (4) lead to the relations

$$[\hat{b}_n,\hat{b}_m]_+ = [\hat{b}_n^+,\hat{b}_m^+]_+ = [\hat{d}_n,\hat{d}_m]_+ = [\hat{d}_n^+,\hat{d}_m^+]_+ = 0$$

$$[\hat{b}_n,\hat{b}_m^+]_+ = [\hat{d}_n,\hat{d}_m^+]_+ = \delta_{nm}\ .$$

$$(5)$$

These equations must be completed by an equation that determines the dynamical evolution of the field operators $\hat{\Psi},\hat{\Psi}^+$ or $\hat{b}_n,\hat{d}_n$. In the absence of two-body interactions it is convenient to work in the Heisenberg picture where the Fock-state-vector is time independent and the dynamics are determined by the operators according to Heisenberg's equations of motion

$$\frac{d\hat{A}}{dt} = i[\hat{H},\hat{A}]$$

$$(6)$$

where $\hat{H}$ is the Hamiltonian of the Dirac field.

Hermiticity of the Hamiltonian is achieved by complete symetrization with respect to the field operator. After a partial integration one finds:

$$\hat{H} = \frac{1}{2}\int d^3x\ [\hat{\Psi}(\vec{x},t),\ H_D\ \hat{\Psi}(\vec{x},t)]_- + \frac{i}{2}\oint d\vec{\Sigma}\cdot[\hat{\Psi}^+,\vec{\alpha}\hat{\Psi}]_-$$

$$(7)$$

For localized states the last surface term in Eq. (7) vanishes, but we shall see that it plays an important role in the transition to supercritical external fields.

In the same way one can construct an operator for the charge-current density:

$$\hat{j}^{\mu}(x) = \frac{e}{2}[\hat{\Psi}^+(x),\gamma^0\gamma^{\mu}\hat{\Psi}(x)]_-\ ,$$

$$(3)$$

and the total charge:

$$\hat{Q} = \frac{e}{2}\int d^3x[\hat{\Psi}^+,\hat{\Psi}]_-\ .$$

$$(9)$$

By explicit calculation it is easy to show that Q is a constant of motion (except for surface effects):

$$\frac{d}{dt} \hat{Q} = i[\hat{H},\hat{Q}] = - \oint d\hat{\Sigma} \cdot \hat{j} . \tag{10}$$

This equation allows us to reqrite Eq. (7) in the following way:

$$\hat{H}_{\ell oc} = \frac{1}{2} \int d^3 x [\hat{\Psi}, H_D \Psi]_- = \hat{H} - \frac{i}{e} \frac{d}{dt} \hat{Q} . \tag{11}$$

Let us for the moment neglect surface effects. In the representation of single-particle states the Hamiltonian and the charge operator take the following form:

$$\hat{H}_{\ell oc} = \sum_{n>F} E_n \hat{b}_n^+ \hat{b}_n + \sum_{n<F} (-E_n) \hat{d}_n^+ \hat{d}_n + E_{vac} \tag{12}$$

and

$$\hat{Q} = e(\sum_{n>F} \hat{b}_n^+ \hat{b}_n - \sum_{n<F} \hat{d}_n^+ \hat{d}_n) + q_{vac} \tag{13}$$

where

$$E_{vac} = -\frac{1}{2n} \sum |E_n| , \quad q_{vac} = -\frac{e}{2}(\sum_{n>F} 1 - \sum_{n<F} 1) . \tag{14}$$

For given external potential, E_{vac} does not depend on the particular state vector of the system and can therefore be discarded renormalizing the zero point of the energy scale. The operator combinations $\hat{N}_n = \hat{b}_n^+ \hat{b}_n$ and $\hat{N}_n = \hat{d}_n^+ \hat{d}_n$ have the properties of number operators counting electrons and positrons, respectively. It is easy to show that they can only take the eigenvalues 0 and 1, as allowed by the Pauli principle. A little consideration tells one that the state of lowest energy, i.e. with the lowest expectation value of $H_{\ell oc}$, is the one that is an eigenstate of eigenvalue zero with respect to all operators $\hat{N}_n$ and $\hat{N}_n$, in combination with the following choice of the Fermi surface F_o which we shall also denote by $E_F = 0$:

$$E_n >o : "n>F_o" , \quad E_n <o : "n<F_o" . \tag{15}$$

In short, we obtain the state of lowest energy by dividing electron
and positron states according to the sign of the energy eigenvalue
and requiring that no particle or anti-particle be present. We
shall call this state the *absolute ground state* or *state of lowest
energy* $|0, F_o>$.

For vanishing external potential the Dirac equation is charge
conjugation invariant, and we have an equal number of states with
$n > F_o$ and $n < F_o$. As a consequence the ground state will have zero
charge: $q_{vac}(F_o) = <0, F_o|\hat{Q}|0, F_n> = 0$. Now consider an external
attractive potential for electrons with a strength parameter λ:

$$V_\lambda(\vec{x}) = \lambda v(\vec{x}) \tag{16}$$

According to the discussion of section IV, for some strength λ_o the
most strongly bound state acquires a binding energy equal to the
rest mass m_o of the electron. For $\lambda > \lambda_o : E(\lambda) < o$ and this level is
counted as a positron state. Therefore it is shifted from the sum
over $n > F_o$ to the sum $n < F_o$.
This changes the balance in the expression for q_{vac}:

$$<0, F_o|\hat{Q}|0, F_o> = q_{vac}(F_o) = eN(\lambda)\theta(\lambda - \lambda_o) \tag{17}$$

where $N(\lambda)$ denotes the number of states with a binding energy exceeding the rest mass m_o. We conclude that beyond a certain strength
of the external potential the lowest energy state of the electron-
positron field carries a non-zero charge (see Fig. 17). We note here
that this state can only be reached if precisely the required number
of electrons is supplied. Interesting as it may be, the lowest energy state is therefore a purely *formal construction* since the charge
operator $\hat{Q}$ is a constant of motion according to Eq. (10) as long as
surface effects can be neglected. When the binding energy of a bound
state is increased beyond m_o, its wavefunction remains localized –
the surface effects vanish.

The situation is fundamentally different when the strength of
the external potential is increased to the point λ_{cr} where one of
the bound states is bound by *twice* the electron rest mass. As discussed in section IV, for $\lambda > \lambda_{cr}$ the bound state becomes imbedded into
the antiparticle scattering states as a resonance state. Due to the
involvement of free scattering states boundary effects can no longer
be excluded and we notice the difference between the local and global

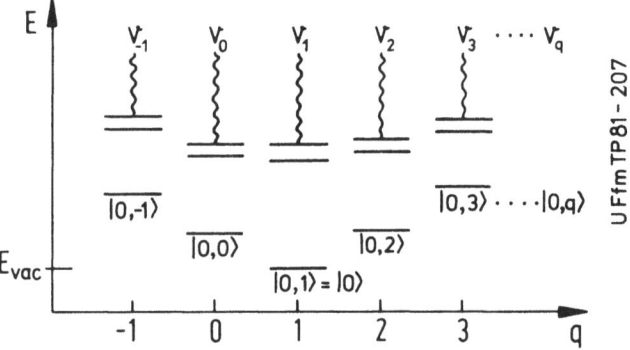

Fig. 17: The groundstates $|o,q\rangle$ for the various charge subspaces
V_q of the total Fock space. The absolute groundstate is $|o,1\rangle$.

state of the system, in particular the possibility to exchange particles with the surroundings develops. According to Eq. (10) the
localized charge of the atomic system needs not to be conserved, if
particles (or antiparticles) cross the boundary and at the same time
the local Hamiltonian $\hat{H}_{loc}$ acquires an imaginary part indicative
of a decay process.

All this means that an atomic system can make a transition from
one charge subspace V_q of the total Fock space to another subspace
$V_{q'}$, by the emission of an antiparticle (or particle). Each subspace
is characterized by a different eigenvalue of the charge operator.
In each sector (subspace) of the Fock space there is a state of lowest
energy, the equilibrium state. It is most easily determined as the

state that minimizes

$$\hat{K} = \hat{H}_{\ell oc} + \mu \hat{Q} \tag{18}$$

where it can be shown that the quantity μ, the chemical potential, must be chosen as $\mu = \dfrac{m_o}{e}$ in order to ensure that pair production is responsible for a transition from one charge sector to another, while one member of the pair is emitted to infinity (see Fig. 18). We thus find the following condition for the equilibrium state:

$$<equil|\hat{H}_{loc} + \frac{m_o}{e}\hat{Q}|equil> = min. \tag{19}$$

By means of Eq. (12,13) the operator $\hat{H}_{loc} + \dfrac{m_o}{e}\hat{Q}$ can be rewritten as:

$$\hat{K} = \sum_{n>F} (E_n + m_o)\hat{b}_n^+\hat{b}_n + \sum_{n<F} (-E_n - m_o)\hat{d}_n^+\hat{d}_n + (E_{vac} + \frac{m_o}{e}q_{vac}) . \tag{20}$$

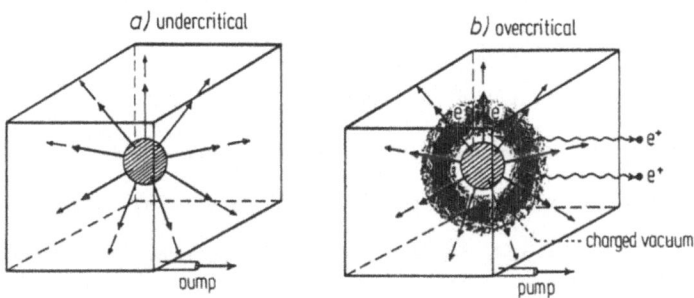

Fig. 18: a) undercritical   b) overcritical

The vacuum in the box is penetrated by the electric field of a central nucleus with charge Z. The nucleus is solely a spectatator furnishing the source of the electric field.

Following the above line of arguments the equilibrium state with the lowest expectation value of $\hat{K}$ is found by requiring

$$\hat{b}_n|equil> = 0 \quad for \quad E_n > -m_o \, (n > F_{-m}) \tag{21}$$

$$\hat{d}_n|equil> = 0 \quad for \quad E_n < -m_o \, (n < F_{-m})$$

i.e. the Fermi energy must be $E_F = -m_o$.

This state is the state of an atomic system subject to a given
external potential in the absence of interference from outside.
In this state, all levels with $E > -m_o$ are particle states and all
levels with $E < -m_o$ are antiparticle states. It is precisely the state
we have called the *charged vacuum state* (for $\lambda > \lambda_{cr}$) in section IV.
We have now shown that a neutral atomic system in weak external
field will develop into the state with $E_F = -m_o$ after the potential
has been increased to arbitrary strength and sufficient time has
elapsed for the equilibrium to be established.

We may go one step further and ask for the charge density contained in the charged vacuum state. The formal expression

$$\langle equil | j^o(x) | equil \rangle = \frac{e}{2} \sum_{n<F} \Psi_E^+(\vec{x}) \Psi_E(\vec{x}) - \frac{e}{2} \sum_{n>F} \Psi_E^+(\vec{x}) \Psi_E(\vec{x}) \qquad (22)$$

needs renormalization, but we can restrict the summation (integration) to a narrow energy interval around the resonance for our
purpose of extracting the real vacuum charge. The results of such
calculations are shown in Fig. 19 in comparison to the K-shell
density of a subcritical atom. Note, in particular, that the charge
density contains a node in the case of the supercritical 2s-state
for Z=255.

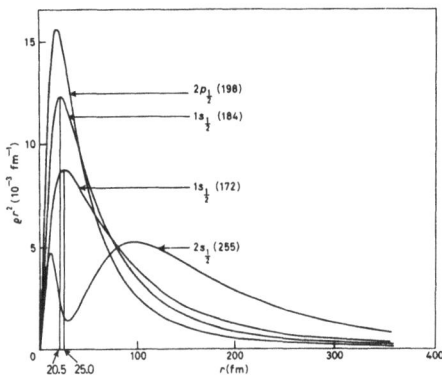

Fig. 19: Vacuum charge
distributions for various
supercritical states in
comparison with the subcritical K-shell at Z=172.

We summarize: In the overcritical case the dived bound state becomes degenerate with the (occupied) negative electron states. Hence *spontaneous e^+e^--pair creation* becomes possible, where an electron from the Dirac sea occupies the bound state, leaving a hole in a continuum state, which escapes as a positron. This is a *fundamentally new process*, which can also be expressed in the following way: The *neutral vacuum of QED becomes* unstable in overcritical electric fields. It decays in about 10^{-20}sec into a *charged vacuum*. The charged vacuum is stable due to the Pauli-principle. It is twice charged, because of the spin-degeneracy two electrons $(\uparrow,\downarrow)$ can occupy the dived shell. After the $2P_{1/2}$-shell dived beyond $Z'_{cr}=183$, the vacuum is four times charged, etc. This change of the vacuum structure is absolutely fundamental; it is *not a perturbative effect*, as all other known QED-effects (vac. polarization, self energy etc.), but a *dramatic phase change*. The *vacuum polarization charge contains now a real component*, not only a displacement charge, and we now have

$$\int < \text{charged vac.} \; |\hat{\rho}| \; \text{charged vac.} > d^3x = 2e, 4e, \text{ etc.}$$

The significance of this vacuum charge is made more transparent in a different way in figure 18. The space in the box is pumped empty by an elementary particle pump. There is only the central nucleus as a source for the electric field left as a spectator. The empty space respresents the neutral vacuum. In the undercritical case a), it is stable. In the overcritical case, however, two positrons are emitted into free states, travelling around in the box and simultaneously two electrons are created in a strongly bound state around the nucleus. The positrons, being in free states, are easily pumped away. The electron cloud represents the charged vacuum. It is the stable groundstate in the overcritical field case. We shall now look closer on the decay itself.

VI. THE DECAY OF THE VACUUM STATE

Three types of experiments involving supercritical external
field can be imagined, as illustrated in Fig. 20:

(i)   a *sub*critical but strong field is made *super*critical for a
       certain finite period of time;

(ii)   a *sub*critical external field is rendered *super*critical and
       made to remain so forever;

(iii) everything is carried out in a *super*critical field.

A prototype of the first kind - the only experiment presently

feasible - is the sub-Coulomb barrier collision of very heavy ions,

such as uranium on uranium, which is discussed in great detail in

ref. 13. The second type of experiment would correspond, for in-

stance, to the creation of a stable supercritical nucleus initially

stripped of electrons by a nuclear fusion process. A typical ex-

periment of the third kind would be resonant positron scattering

on a (entirely hypothetical) supercritical atom and the observation

of the resulting spontaneous pair creation.

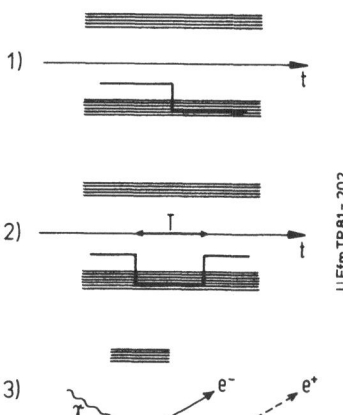

UFtmTP81-202

Fig. 20: The three different
types of experiments with
supercritical states.

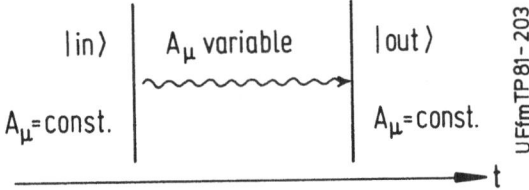

UFmTP81-203

Fig. 21: The in-and out-regions for time-
dependent fields.

In this chapter we want to discuss the three types of experimental

circumstances from the standpoint of quantum field theory.

This will lead us to the conclusion that the instability of the

neutral ground state in supercritical fields, previously asserted

on the basis of several heuristic arguments, is a natural conse-

quence [17] of field theory. To begin with, we have to explain

how the information about observable quantities is extracted from

quantum field theory.

We shall take the point of view that all observations on micros-

copic system undergoing a temporary change are made long before

or long after that change has taken place, at time symbolically

denoted as $t = \pm\infty$ and called the in- and the out-region, respect-

ively. We shall assume that the changes occur on a miscroscopic

time scale and that the system is stationary at $t \to \pm\infty$. We shall

continue to work in the Heisenberg picture, i.e. the Fock space

vectors $|\Omega_i>$ of the system are fixed and all time dependence is

carried by the operators.

We now make use of the two complete sets of solutions of the single-

particle Dirac equation which are eigenstates in the past or in the

future, $\Psi_n^{(\pm)}(\vec{x},t)$, to obtain two different decompositions of the field operator (see Fig. 21)

$$\hat{\Psi}(\vec{x},t) = \sum_{n>F} \hat{b}_n^{(in)} \psi_n^{(+)}(\vec{x},t) + \sum_{n<F} \hat{d}_n^{(in)+} \psi_n^{(+)}(\vec{x},t) \tag{1a}$$

$$\hat{\Psi}(\vec{x},t) = \sum_{n>F'} \hat{b}_n^{(out)} \psi_n^{(-)}(\vec{x},t) + \sum_{n<F'} \hat{d}_n^{(out)+} \psi_n^{(-)}(\vec{x},t). \tag{1b}$$

The two sets F,F' denote the "filled" states (in Dirac's sense) in

the in-and out-regions, respectively, and will be taken as the

vacuum Fermi sets F_{-m}, F'_{-m} in the following. However, since the

formalism remains equally valid for any initial state characterized

by an arbitrary Fermi surface, we shall continue to write F and F'

in this section. The vacuum states before/after the experiment

are then defined by the conditions

$$\hat{b}_n^{(in/out)} |0, \text{ in/out}\rangle = 0 \quad \text{for } n>F/F' \tag{2}$$

$$\hat{d}_n^{(in/out)} |0, \text{ in/out}\rangle = 0 \quad \text{for } n<F/F'$$

When the external field in the in-region differs from that in the

out-region, $F_{-m} \neq F'_{-m}$ and the two vacuum states will not be the

same (although the Fermi energy is the same!). It is important to

realize that the particle operators $\hat{b}_n^{(in)}$, $\hat{d}_n^{(in)}$, (although being

constant in time) differ from the particle operators $\hat{b}_n^{(out)}$,

$\hat{d}_n^{(out)}$ defined in the out-region. The reason for this is that

the $\hat{b}_n^{(in)}$, $\hat{d}_n^{(in)}$ do *not* correspond to physical particles *outside*

of the stationary in-region. Physical particles are defined with

respect to stationary wavefunctions, and the $\psi_n^{(-)}(\vec{x},t)$ are not sta-

tionary in the out-region. Still, it is possible to relate the

operators for in-particles to those for the out-particles by pro-

jecting Eq. (1a) with $\psi_m^{(-)}$ and Eq. (1b) with $\psi_m^{(+)}$. With the single-

particle S-matrix elements

$$S_{mn} = \langle \Psi_m^{(-)} | \Psi_n^{(+)} \rangle \tag{3}$$

one finds that

$$\hat{b}_m^{(out)} = \sum_{n>F} \hat{b}_n^{(in)} + \sum_{n<F} \hat{d}_n^{(in)\dagger} \; S_{mn} \tag{4a}$$

$$\hat{d}_m^{(out)} = \sum_{n>F} \hat{b}_n^{(in)\dagger} + \sum_{n<F} \hat{d}_n^{(in)} \; S_{mn}^* \; . \tag{4b}$$

In an experimental situation a state $|\Omega\rangle$ is prepared in the in-region.
For example, if we were to begin with the vacuum state then
$|\Omega\rangle = |0,in\rangle$; if starting with a single-particle state, we had
$|\Omega\rangle = \hat{b}_k^{(in)\dagger} |0,in\rangle$ etc. The measurements, on the other hand, are
done in the out-region, so the corresponding operators act on the
out-particles. Measuring the number of particles in a given state
$i>F'$ corresponds to taking the expectation value of the operator

$$\hat{N}_i^{(out)} = \hat{b}_i^{(out)\dagger} \hat{b}_i^{(out)} \quad (i>F') \; ,$$

which is easily evaluated with the help of relations (4a,b):

$$N_i = \langle o,in| \hat{N}_i^{(out)} |o,in\rangle = \sum_{n<F} |S_{in}|^2 \; . \tag{5}$$

In the same way, one finds the number of out-antiparticles in a
state $k<F'$ to be given by

$$\bar{N}_k = \langle o,in| \hat{d}_k^{(out)\dagger} d_k^{(out)} |o,in\rangle = \sum_{n>F} |S_{kn}|^2 \; . \tag{6}$$

Since all observables in the out-region such as energy, charge etc.
can be expressed in terms of expectation values of operators expressed in terms of the out-particle operators, every observable
can be calculated from the single-particle amplitudes S_{mn}. This

is a consequence of our neglect of true two-body interactions between Dirac particles as they would arise from the electromagnetic
interactions of electrons with other electrons or positrons. Indeed, it can be shown that Eqs. (4a,b) hold in the Hartree-Fock
approximation to quantum electrodynamics, but become invalid as
correlations are taken into account.

We are now in postion to calculate the effect upon the subcritical vacuum state due to an external potential that becomes
supercritical. We will consider both the cases (ii) and (i):
The sub- and the supercritical potentials we take to be the same
as those discussed in section IV, denoted as V_{cr} and $(V_{cr}+V')$, respectively. We denote here the eigenstates in the subcritical potential by ψ_n, those in the supercritical potential by Ψ_n:

$$\left|-i\vec{\alpha}\cdot\nabla + \beta m_o + V_{cr}\right|\psi_n = \varepsilon_n\psi_n \ , \tag{7}$$

$$\left|-i\vec{\alpha}\cdot\nabla + \beta m_o + V_{cr} + V'\right|\Psi_n = E_n\Psi_n \ . \tag{8}$$

The subcritical and the supercritical bases are connected by a unitary transformation

$$\Psi_n = \sum_m C_{nm}\psi m \tag{9}$$

which was explicitly calculated in section IV. If V' is switched
on at $t = t_o$, we find that the forward-propagating wavefunctions
prior to t_o are given by stationary subcritical functions:

$$\psi_n^{(+)}(\vec{x},t) = \psi_n(\vec{x}) \ e^{-i\varepsilon_n t} \quad (t \leq t_o) \tag{10}$$

whereas the backward propagating functions after t_o are stationary
in the supercritical basis:

$$\psi_m^{(-)}(\vec{x},t) = \Psi_m(\vec{x}) \ e^{-iE_m t} \quad (t \geq t_o) \ . \tag{11}$$

Except for a phase factor, the S-matrix

$$S_{mn} = \langle \psi_m | \psi_n \rangle \, e^{i(E_m - \varepsilon_n)t_o} = C^*_{mn} \, e^{i(E_m - \varepsilon_n)t_o} \tag{12}$$

is given by the complex conjugate of the unitary transformation from the subcritical to the supercritical basis. When we start with the subcritical vacuum state, F_{-m}, containing all states with $\varepsilon_n < -m_o$, we find that the distribution of positrons at $t = \infty$ is given by

$$\bar{N}_E = \sum_{n > F_m} |C_{E,n}|^2 \, , \quad E < -m_o \, . \tag{13}$$

If we reduce our Fock space to the positron continuum and the bound state that becomes supercritical, as we did in section IV, the sum reduces to a single term for the bound state, $n = 0$. In the terminology of section 4, we have $c_{E,0} = a(E)$, and hence:

$$\bar{N}_E = |a(E)|^2 = \frac{\Gamma/2\pi}{(E - E_r)^2 + \Gamma^2/4} \, . \tag{14}$$

Thus in the framework of quantum field theory the neutral vacuum state decays in a supercritical external potential, producing a positron distribution centered around the supercritical quasi-bound state resonance energy E_r.

In order to find out how the decay proceeds with time, we must switch off the additional potential V' after some time T and ask how far the neutral vacuum state has decayed during that period of time. In this situation it is convenient to evaluate S_{mn} at the time of the switch-off, $t = t + T$. For $t < t_o$ and $t > t_o + T$ the functions $\psi_m^{(\pm)}$ propagate in the subcritical potential, whereas they propagate in the supercritical potential for $t_o < t < t_o + T$. This propagation can be described by the homogeneous Green's function for the supercritical potential:

$$S(\vec{x},t;\vec{x}',t') = -i\Sigma_k \Psi_k(\vec{x})\bar{\Psi}_k(\vec{x}')e^{-iE_k(t-t')} \tag{15}$$

After some calculation on finds for the single-particle S-matrix

elements:

$$S_{mn} = \left\langle \psi_m^{(-)}(t_o+T) \middle| \psi_n^{(+)}(t_o+T) \right\rangle = e^{i\varepsilon_m T}\Sigma_k C_{km} C_{kn}^* e^{-iE_k T} \tag{16}$$

If again we restrict the Fock space to the subcritical positron

states and the diving bound state $(n = 0)$, the sum over intermediate

supercritical states becomes an integral over the negative energy

continuum:

$$|S_{mn}|^2 = \left| \int_{-\infty}^{-m_o} dE \; C_{E,m} C_{E,n}^* e^{-iET} \right|^2 \tag{17}$$

In the terminology of the autoionization model in section IV, we

have $C_{E,0} = a(E)$ and $C_{E,\varepsilon}=b_\varepsilon(E)$. Neglecting the continuum-continuum

interaction, the following analytic expressions for these trans-

formation coefficients (in the narrow resonance approximation) are

found

$$a(E) = \frac{V_E^*}{E-E_r + \frac{i}{2}\Gamma} \quad , \quad \Gamma = 2\pi|V_{E_r}|^2 \tag{18}$$

$$b_\varepsilon(E) = \delta(E-\varepsilon) + \frac{a(E)V_\varepsilon}{E-\varepsilon+i\eta}$$

If we start with the subcritical vacuum state F_o, the positron dis-

tribution in the out-region is:

$$\bar{N}_\varepsilon(T) = |S_{\varepsilon,0}|^2 = \left| a^*(\varepsilon)e^{i\varepsilon T} + V_\varepsilon \int_{-\infty}^{-m_o} dE \; \frac{|a(E)|^2}{E-\varepsilon+i\eta} e^{-iET} \right|^2 . \tag{19(}$$

Closing the contour in the lower complex E-plane and neglecting con-

tributions from the finite upper boundary, we find that only the

poles at $E = E_r - \frac{i\Gamma}{2}$ and $E = \varepsilon - i\eta$ contribute to the integral.

The result of the calculation is [17]:

$$\bar{N}_\epsilon(T) = |a(\epsilon)|^2 \cdot |1 - e^{i(\epsilon - E_r + \frac{i}{2}\Gamma)T}|^2 . \tag{20}$$

For times T long compared to the inverse resonance width Γ^{-1}, the positron distribution exponentially approaches that given by Eq. (14), but for small switch-on times T, the distribution is much broader than the resonance width:

$$\bar{N}_\epsilon(T << \Gamma^{-1}) \sim |V_\epsilon|^2 T^2 \left(\frac{\sin\frac{1}{2}T(\epsilon - E_r)}{\frac{1}{2}T(\epsilon - E_r)} \right)^2 . \tag{21}$$

The width in this case is approximately h/T, caused by uncertainty of the energy of the dived bound state due to its short lifetime.

We conclude that the auto-ionization formula for the spectral shape of spontaneously emitted positrons is applicable when the supercritical state lives longer than its natural lifetime $\tau = \Gamma^{-1}$. We note that this condition may not be satisfied in the laboratory tests by means of heavy ion collisions. We can predict already at this point that the spectrum of positrons must be broadened, the width being determined by the time T_{coll}, during which the lowest bound state is supercritical. On the other hand, the probability that the hole in the supercritical bound state has not decayed after a time T, is found to be:

$$1 - N_o(T) = |S_{oo}|^2 = |\int_{-\infty}^{-m_o} dE |a(E)|^2 e^{iET}|^2 = e^{-\Gamma T} . \tag{22}$$

We conclude that the neutral vacuum state decays exponentially in a supercritical potential with a decay time Γ^{-1}. In case (iii) essentially the same result is obtained as in case (i). There may be some difference due to the fact that the preparation of the supercritical vacancy differs in these two cases (for (i) the potential strength is increased, whereas for (iii), e.g., the supercritical vacancy is created by a γ-ray.) That the spectrum of emitted particles (in our case: positrons) may depend on the way of preparation of the initial state, is a phenomenon well-known in quantum mechanics.

VII VACUUM CHARGE AND VACUUM ENERGY

After the meaning and the nature of the vacuum state have been clarified, we are in a position to discuss the total charge and energy associated with it, as given by Eqs. (V,14). These expressions make little sense as they stand, because they involve divergent summations. In order to obtain well-defined expressions one proceeds in a number of steps [20]. First, one demands that in the absence of an external potential the vacuum carries no charge and no energy. After proper subtractions we proceed to the continuum limit and replace the summations by integrations. Finally, we expand in powers of the external potential and apply standard renormalization prescriptions to the lowest order terms. After resumming the series expansion we obtain definite expressions for q_{vac} and E_{vac} that are correct to all orders of the external field and clearly exhibit the phase transition at the critical field strength.

We make the simplifying assumption that there is no external vector potential, $\vec{A} = 0$, and that the electrostatic potential is spherically symmetric, $V(r)$, and of finite range with $V(\infty) = 0$. To be able to count states, we impose a suitable boundary condition on the Dirac wavefunctions at a finite, but large radius R. For good angular momentum waves (IV.3) appropriate boundary conditions are $f(R) = 0$, or $g(R) = 0$. Let us first suppose that the external potential is weak and there is a well-defined gap separating positive energy and negative energy solutions of the Dirac equation as indicated in Fig.22. For every value of the angular momentum quantum numbers κ, μ, the positive energy eigenvalues E_n can then be counted according to growing energy $(0 < E_1 < E_2 < ...)$, while the negative energy eigenvalues E_n can be lined up with decreasing energy $(0 > \bar{E}_1 > \bar{E}_2 > ...)$.

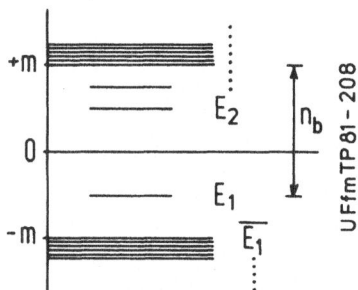

Fig. 22: Numbering of particle and antiparticle states according to energy.

In the case of vanishing external potential, the spectrum is symmetric around $E = 0$, i.e., $E_n = -\bar{E}_n$. Then it is clear that $q_{vac}[V\equiv0] = 0$, but

$$E_{vac}[V\equiv0] = - \sum_{\kappa,\mu} \sum_{n=1}^{\infty} E_n^{\kappa,\mu}[V\equiv0] \tag{1}$$

does not vanish. Since we have to define *some* reference point from which to measure energies, it seems reasonable to take the (infinite) energy of the Dirac vacuum in a vanishing potential as the standard. In this spirit we shall henceforth only be interested in the difference

$$E'_{vac}[V] = E_{vac}[V] - E_{vac}[V\equiv0] . \tag{2}$$

We can then distinguish states for which the energy eigenvalue becomes independent of the boundary condition as R is increased further and further, i.e., the bound states, with $-m_o \leq \lim E_n(R) < +m_o$, and those states that form the continuum in the limit $R\to\infty$. At distances larger than the range of the potential, these states have wavefunctions behaving like $\sin(k_n r + \Delta_n[V])$ with $k_n^2 = E_n^2 - m^2$, and an analogous expression for the antiparticle states. For large R, the boundary condition $f_n(R) = 0$ gives

$$n\pi = k_n R + \Delta_n[V]. \tag{3}$$

The phase shift Δ_n is a functional of the potential. When the boundary at R is made to recede to infinity, $E_n[V]$ and $E_n[0]$ get infinitesimally close, such that

$$E_n[V] - E_n[0] \longrightarrow -\frac{1}{R}\frac{dE}{dk}\Big|_{k_n}(\Delta_n[V] - \Delta_n[0]) . \tag{4}$$

while the separation between the energy eigenvalues approaches zero, the number of states in a given momentum interval dk grows with k according to eq. (3):

$$dn = \frac{R}{\pi}dk + \frac{1}{\pi}\frac{d\Delta}{dk}dk . \tag{5}$$

In the limit $R \to \infty$ the discrete sum over the "continuum" states becomes an integral:

$$\sum_{n=n_b+1}^{\infty} (E_n[V] - E_n[0]) \longrightarrow -\int_0^{\infty} dk\,\frac{dE}{dk}\cdot\frac{1}{\pi}\,\delta_k[V] . \tag{6}$$

Here we have introduced n_b to count the number of positive energy bound states and $\delta[V] = \Delta[V] - \Delta[0]$ for the phase-shift caused by the potential. The full expression for the vacuum energy reads then:

$$E'_{vac}|V| = \sum_{\kappa,\mu}\left\{ \frac{1}{2}\sum_{n=1}^{n_b}(m_o - E_n^{\kappa\mu}[V]) + \frac{1}{2}\sum_{n=1}^{\bar{n}_b}(m_o + \bar{E}_n^{\kappa\mu}[V]) \right.$$

$$\left. + \frac{1}{2\pi}\int_{m_o}^{\infty}dE(\delta_E[V] + \delta_{-E}[V]) \right\} \tag{7}$$

since the bound states correspond to continuum states of infinite-simal kinetic energy in the case of vanishing potential. $\bar{n}_b$ stands for the number of negative energy bound states. For the vacuum charge we obtain in a similar way

$$q_{vac} = -\frac{e}{2} \sum_{\kappa,\mu} \left\{ n_b - \bar{n}_b + \frac{1}{\pi} \int_{m_o}^{\infty} dE \frac{d}{dE} (\delta_E[V] - \delta_{-E}[V]) \right\} . \tag{8}$$

As they stand, eqs. (7,8) still involve divergent expressions. To see this, one has to investigate the high-energy behaviour of the phase shifts $\delta_E^{(\kappa\mu)}$. If the potential $V(r)$ has no singularities, the high-energy limit is exactly described by the WKB approximation which gives the following closed expression for the phase shift:

$$\delta_{WKB} = \int_{r_o[V]}^{\infty} dr \sqrt{(E-V)^2 - \frac{\kappa^2}{r^2} - m_o^2} - \int_{r_o}^{\infty} dr \sqrt{E^2 - \frac{\kappa^2}{r^2} - m_o^2} , \tag{9}$$

where r_o, $r_o[V]$ are the radii where the square roots vanish, respectively (the classical turning points). Taking the limit $E \to \infty$ we find that

$$\lim_{E\to\infty} \delta_E^{(\kappa\mu)} = - \text{sgn } (E) \int_o^{\infty} V (r) dr \tag{10}$$

independent of the angular momentum quantum numbers. As consequence, the sum over κ and μ in eq. (7) diverges:

$$\sum_{\kappa,\mu} \int_m^{\infty} dE \frac{d}{dE} (\delta_E - \delta_{-E}) = \sum_{\kappa,\mu} (-2\int_o^{\infty} V(r)dr - \delta_m^{(\kappa\mu)} + \delta_{-m}^{(\kappa\mu)}) . \tag{11}$$

Since the term linear in V cancels in the phase shift integral occurring in eq. (6), the integral is finite for every angular momentum channel. Still, the quadratic term in V in the sum over κ,μ diverges, as can be seen by a more detailed argument. One easily checks that the V^3 term cancels and that the quartic and all higher terms give finite contributions.

We conclude that, if the phase shift is expanded in powers of the external potential, the divergences arise from the lowest order term that conributes to the vacuum charge and energy, respectively. That the divergences are concentrated in the lowest order terms of a series expansion, is a typical property of a renormalizable field theory. The divergent terms must be absorbed in a renormalization of the electric coupling constant e. If we accept for a moment that this program can be carried through, we can give exact expresions for the vacuum charge and energy. We simply have to subtract the firs order (Born) phase shift $\delta_E^{1\,(\kappa\mu)}$ from the exact phase shift $\delta_E^{(\kappa\mu)}$ for the vacuum charge and the first and second order phase shift $\delta_E^{2\,(\kappa\mu)}$ for the vacuum energy:

$$\delta_E^{\prime\,(\kappa\mu)} = \delta_E^{(\kappa\mu)} - \delta_E^{1\,(\kappa\mu)} \;;\quad \delta_E^{\prime\prime\,(\kappa\mu)} = \delta_E^{\prime\,(\kappa\mu)} - \delta_E^{2\,(\kappa\mu)} \,. \tag{12}$$

If we denote the part of the vacuum charge that is linear in the potential by $q_{vac}^{(1)}$ and the part of the vacuum energy quadratic in V by $E_{vac}^{(2)}$, we find the following expressions

$$q_{vac} = q_{vac}^{(1)} - \frac{e}{2}\,\sum_{\kappa,\mu}\left\{ n_b - \bar{n}_b - \frac{1}{\pi}\delta_m^{\prime} + \frac{1}{\pi}\delta_{-m}^{\prime}\right\} \tag{13}$$

$$E_{vac} = E_{vac}^{(2)} + \frac{1}{2}\,\sum_{\kappa,\mu}\left\{ \sum_{n=1}^{n_b}(m-E_n) + \sum_{n=1}^{\bar{n}_b}(m+\bar{E}_n) + \right.$$

$$\left. + \frac{1}{\pi}\int_m^{\infty}dE\,(\delta_E^{\prime\prime} + \delta_{-E}^{\prime\prime})\,.\right\} \tag{14}$$

We can draw a number of conclusions about the vacuum charge on purely physical grounds. Because of the quantization of electric charge, q_{vac} can take ony values of multiples of e. Since q_{vac} was defined to be zero for vanishing potential, it cannot acquire a nonzero value to any order of perturbation theory for the external potential. To be discontinuous, $q_{vac}[V]$ cannot be of finite order in V. Thus q_{vac} must vanish. However, q_{vac} remains zero even beyond perturbation theory, as long as the potential is not strong enough to mix positive and negative energy states. In this region of "weak" external potentials, Levinson's phase-shift theorem is valid, which states that

$$n_b = \pi\delta_m = \pi\delta'_m \quad , \quad \bar{n}_b = \pi\delta_{-m} = \pi\delta'_{-m} \tag{15}$$

for nonsingular potentials of finite range. The equaliy of $\delta_{\pm m}$ and $\delta'_{\pm m}$ is an expression of the fact that the occurrence of bound states is a nonperturbative phenomenon. Equations (15) immediately yield the result that the vacuum does not carry a charge for weak potentials. In general we have

$$q_{vac} = -\frac{e}{2} \sum_{\kappa,\mu} \left\{ n_b - \frac{1}{\pi}\delta_m - \bar{n}_b + \frac{1}{\pi}\delta_{-m} \right\} . \tag{16}$$

For strong fields q_{vac} acquires a nonzero value in the following way. Let us assume that V is attractive for particles ($V \leq 0$) and that we consider the vacuum state with Fermi energy $E_F = -m$. At the critical strength of the potential, V_{cr}, the lowest bound state crosses the line $E = -m_0$, thereby diminishing n_b by one, and causing a jump of π in the phase shift δ_{-m}. This is illustrated in Fig. 23, where curve 4 corresponds to a potenrial that has just become supercritical. Upon further increase of the potential, the bound state resonance moves through the antiparticle continuum, but does not cause any further change in δ_{-m} until the next bound state dives in.

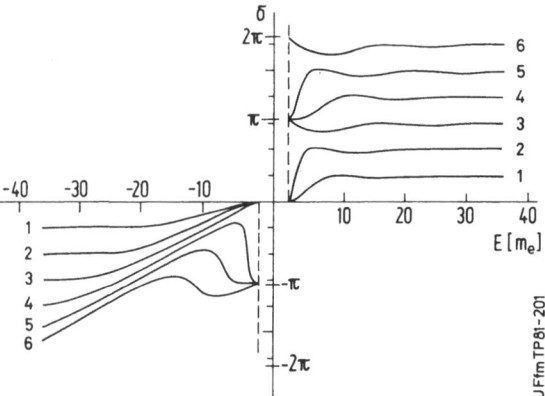

Fig. 23: Phase shifts for square-well potential of increasing depth. The supercritical resonance appears in
curve 4. There are bound states between E=-1,+1 for curves
3 and 6.

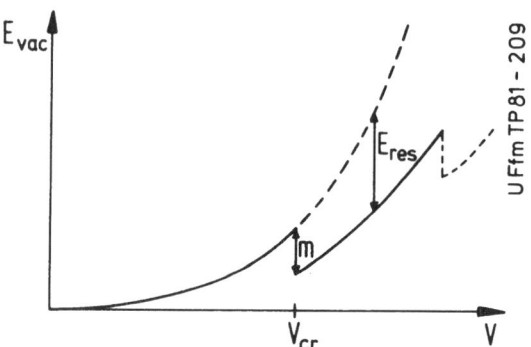

Fig. 24: Energy of the vacuum states as function of potential
strength. The phase transition at V_{cr} leads to sudden lowering
of the energy.

We may use Eq. (14) to find the contribution of the supercri-

tical bound state resonance to the energy of the vacuum state. In

the approximation of the autoionization model, the negative energy

phase shift can be written as

$$\delta_E[V] \simeq \delta_E[V_{cr}] - \arctan \frac{\Gamma}{2(E-E_{res})} , \qquad (17)$$

where E_{res} is the energy of the center of the resonance.   Carrying out the energy integration we find:

$$\int_{m_o}^{\infty} dE \ \delta''_{-E} \ [V] \approx \int_{m_o}^{\infty} dE \ \delta''_{-E} \ [V_{cr}] - (m_o + E_{res})\pi \quad , \tag{18}$$

if the width is much smaller than the diving depth. Keeping in mind that the resonance comes from that bound state which is now missing from the first sum in eq. (14), we obtain the following approximate result for the vacuum energy in a slightly supercritical potential:

$$E_{vac}[V > V_{cr}] = E_{vac}[V_{cr}] + \frac{1}{2} \quad -(m_o - E_{res}) + m + E_{res}$$

$$\tag{19}$$

$$= E_{vac}[V_{cr}] + E_{res} \ .$$

This result corresponds to the picture that the supercritical bound state resonance is occupied by a particle whose energy must contribute to the energy of the charged vacuum an amount E_{res}.
We conclude that the energy of the vacuum state with $E_F = -m_o$ exhibits a discontinuity of size $(-m_o)$ at the critical strength of the external potential. This is illustrated in Fig. 24.

All that is left to do now, is to calculate the lowest-order terms $q_{vac}^{(1)}$ and $E_{vac}^{(2)}$ in Eqs. (12,13). In the diagrammatic perturbation expansion, $q_{vac}^{(1)}$ and $E_{vac}^{(2)}$ are represented by the Feynman diagrams:

By the standard methods of perturbative QED the lowest-order renormalized vacuum current density is found as [39]:

$$\langle 0|\hat{j}^{\mu(1)}_{(x)}|0\rangle_{ren} = \int d^4x' \ \Pi^{(1)}_{ren}(x-x')j^{\mu}_{ext}(x') \quad , \tag{20}$$

where $\Pi_{ren}(p)$ is the first-order polarization function:

$$\Pi^{(1)}_{ren}(x-x') = \int\frac{d^4p}{(2\pi)^4} \ e^{-ip(x-x')}\frac{\alpha}{3\pi} \left[\frac{5}{3} + \frac{4m_0^2}{p^2} - \right.$$

$$\left. - (1+\frac{2m_0^2}{p^2}) \ \sqrt{1-\frac{4m_0^2}{p^2}} \ \ell n \ \frac{\sqrt{1-\frac{4m_0^2}{p^2}}+1}{\sqrt{1-\frac{4m_0^2}{p^2}}-1} \right] \tag{21}$$

As an immediate consequence the total charge induced in the vacuum is zero in lowest-order perturbation theory:

$$q^{(1)}_{vac} = \int d^3x \ \langle 0|\hat{j}^{0(1)}(\vec{x},t|0\rangle_{ren} = 0 \tag{22}$$

if the external charge distribution is static.

The energy of the vacuum state to lowest order is obtained by calculating the electrostatic energy of the induced vacuum charge in the external potential:

$$E^{(2)}_{vac} = \int d^3x \ \langle 0|\hat{j}^{\mu(1)}(x)|0\rangle_{ren} \ A^{ext}_{\mu}(x) \quad . \tag{23}$$

For a static external potential $A^{ext}_0(\vec{x})$ generated by a static charge distribution this reduces to

$$E^{(2)}_{vac} = \int d^3x \ d^3x' \ A^{ext}_0(\vec{x}) \ j^0_{ext}(\vec{x}') \cdot \Pi^{(1)}_{ren}(\vec{x}-\vec{x}') \quad . \tag{24}$$

where $\Pi(\vec{x}-\vec{x}')$ is the three-dimensional analogue of Eq. (21).

VIII SOLUTION OF THE POINT CHARGE PROBLEM [21,22]

In section 2 we circumvented the breakdown of the Sommerfeld formula for $Z\alpha>1$ by introducing a finite size of the potential
source. Still, the academic problem remains what would happen to
a point charge when its strength exceeds the critical value $Z\alpha=1$.
Obviously, we must introduce a reasonable limiting process that
leads to a point source with $Z\alpha>1$. The most straightforward procedure is to introduce a finite radius R and to make R tend to zero.
Another possible procedure is to start from a nonlinear theory
with a limiting electric field strength E_o, such as the Born-Infeld
theory [23] and to consider the limit $E_o \to \infty$. The same answer is
found in this way.

In Fig. 25 we show the energy eigenvalues of several inner
electron states in the potential of a nucleus with Z=150 protons
as a function of the nuclear radius R. It is found that the binding energies of all $j = \frac{1}{2}$ - states start to increase with 1/R for sufficiently small values of R, whereas all higher angular momentum
states readily approach a finite limit. We conclude that all states
with $j = \frac{1}{2}$ have a tendency to become supercritical in the point
charge limit and, as a result, we have to take into account the
screening due to the already supercritical levels when we want
to carry out the limit process. This task is greatly facilitated
by the fact that for all deeply bound ($j = \frac{1}{2}$) states the radial
density has maxima at the same distances from the source (see
Fig.26). The reason for this behaviour is that the Coulomb

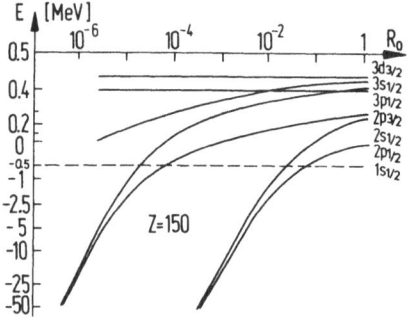

Fig. 25: Single particle
energies of electronic states
in the field of a shrinking
nucleus with charge Z=150 and
radius $R=R_o$ $(2.5Z)^{1/3}$.

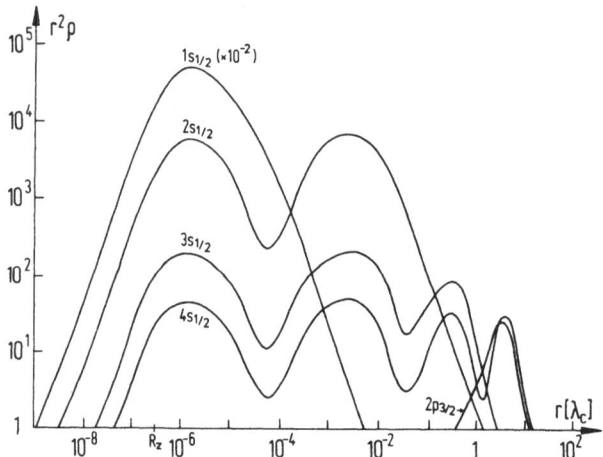

Fig. 26: Single particle densities (in arbitrary units) of some
resonances and bound states for a nucleus with Z=150 and $R_o=10^{-5}$fm.

wavefunctions are Whittaker functions

$$M_{-Z\alpha E/p-1/2,\ \pm i\sqrt{(Z\alpha)^2-1}}\ (2ipr)\ ,\ p^2=E^2-m_o^2$$

which, for Zα>1, pr<<1 and |E|/m>>1 give rise to a radial density distribution

$$pr^2 = N^2 e^{-\pi\gamma} \left[1 - \frac{2}{|\kappa| + (\gamma + Z\alpha)/|\kappa|} \sin\left(2\gamma \ln\frac{r}{R} + \phi\right)\right] \tag{1}$$

independent of E and the sign of κ. The real vacuum polarization density must therefore be also characterized by an undulatory structure with peaks and nodes separated by successive minima and maxima of the sine-function in Eq. (1).

The Thomas-Fermi method, which has been developed to treat the self-consitency of the screening of supercritical potentials [43, see also I], is not directly applicable to the point source problem, because $|\nabla V|/V$ is of the same size as the potential V itself. A better approximation is obtained by writing the Dirac equation (2.4) in the second order form

$$u_{\pm}''(r) + p_{eff}^{(\pm)}(r)^2 u_{\pm}(r) = o \tag{2}$$

with $p_{eff}^{(\pm)}(r)^2 = (E-V)^2 - m_o^2 - \frac{\kappa(\kappa \pm 1)}{r^2} \pm \frac{\kappa}{r} \frac{dV/dr}{E-V\pm m_o} \cdots \tag{3}$

where u_+ stands for the radial functions g and f. One may now solve Eq. (2) in the WKB approximation. The radial variation of p_{eff} is shown in Fig. 27, where it is see that there are two classically allowed regions. The inner region between r_- and r_+ is the region where the supercritical vacuum charge density is located; the outer region beyond r_o contains the normal continuum tail of the wavefunction which does not contribute to the *real* vacuum polarization and will be neglected in our approach.

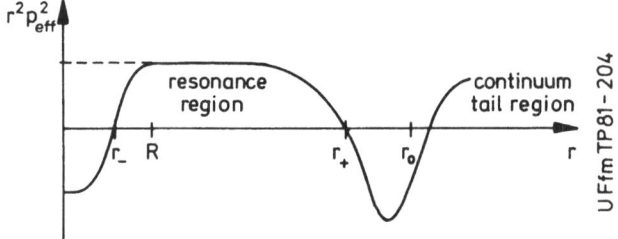

Fig. 27: The quasiclassical momentum for a supercritical nucleus. Dashed line: point charge.

In order to get a workable theory we have to neglect the higher derivatives in Eq. (3). However, this may not be done with the derivatives of the potential itself, but we may write $V(r)=-Z(r)\alpha/r$, with $Z(r) \to Z_N$ at the source, and safely neglect derivatives of $Z(r)$. Obviously, $Z(r)$ is the effectively screened charge. Adding the so-called Langer correction we have

$$q^2_{eff} = p^2_{eff} - 1/4r^2 \tag{4}$$

and we may write down the WKB-solutions:

$$U_\pm(r) = N \left[\frac{E+Z\alpha/r \pm m_0}{q_{(\pm)}} \right]^{1/2} \sin \left[\int_{r-}^{r} q_{(\pm)} dr' + \frac{\pi}{4} \right]. \tag{5}$$

The energies of the quasi-stationary states in the inner classically allowed region (the supercritical bound states) are determined from the Bohr-Sommerfeld condition

$$\int_{r-}^{r+} q_{(+)} dr' = (n + 1/2)\pi. \tag{6}$$

With the abbreviation $\gamma(r)^2 = (Z\alpha)^2 - 1$ we then obtain the following expression for the vacuum charge density:

$$r^2 \rho_{vac} = 4N(r)\ Z(r)\alpha/\gamma(r)\ \sin^2 \left[\int_R^r \gamma(r')dr'/r' + \eta \right] \tag{7}$$

where $N(r_i)^{-1} = \int_{r_i}^{r_{i+1}} dr'\ Z(r')\alpha/\gamma(r')\ \sin^2 \left[\int_R^{r'} \gamma(r'')dr''/r'' + \eta \right]$ \tag{8} is chosen to normalize the total charge contained in any one of the peaks of the vacuum charge distribution between r_i and r_{i+1} (see Fig. 26). For self-consistency the vacuum charge must satisfy the Poisson equation:

$$d^2 Z(r)/dr^2 = 4\pi \, r \, \rho_{vac}. \tag{9}$$

The outer boundary of the vacuum charge distributon is fixed by the Fermi energy $E_F = -m$ giving

$$q_F(r)^2 = -2Z(r)\alpha m_0/r + \gamma(r)^2/r^2. \tag{10}$$

The turning point is r_+^F with $q_F(r_+^F) = 0$. The total screening charge of the vacuum is determined by the condition

$$Z(r_+^F) = Z_N - Q_{vac}, \tag{11}$$

where Z_N is the bare source charge. The solution $Z(r)$ of Eq. (9) is shown in Fig. (28) for $Z_N = 200$ and $R = 10^{-5}$ fm. It is a smooth function with slight wiggles. Since the solution is numerically unstable for very small values of the source radius R, we have chosen to replace it by a step function that jumps by four units whenever the integrated vacuum charge has increased by for electron charges. It is clear that the average slope is the same, but the screening is underestimated by about 2 charge units.

With this recipe it is possible to solve the self-consistent screening equation down to extremely small source radii as illustrated in Fig. 29, where $Z(r_+^F)$ is shown as a function of R for $Z_N = 150$ and $Z_N = 180$. More and more of the source charge is screened with

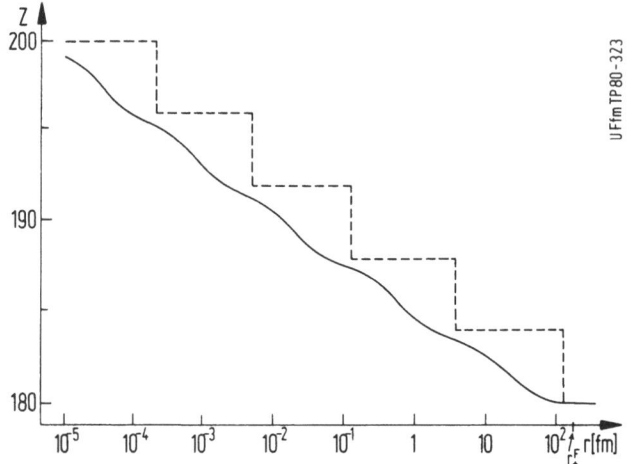

Fig. 28: Screening function Z(r)=-Vr/α for a nucleus with charge Z_N=200 and radius R=10^{-5}fm.——— Solution of the Poisson equation;---approximate Z_{eff}.

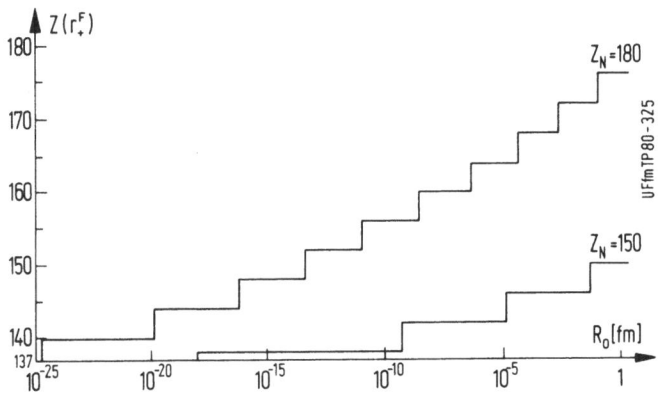

Fig. 29: Effective charge $Z(r_+^F)=Z_N-Q_{vac}$ as function of the radius parameter R_o for two nuclei with charge Z_N=150 and 200.

shrinking R, until the effectively unscreened charge becomes equal to 137. Because we have not taken into account exchange interactions the limit is uncertain by one unit of charge! When the apparent source strength has been reduced to Z=137, no further bound state become supercritical, i.e. the screening no longer increases upon further reduction of the source radius.

We conclude:

(a) Sufficiently point-like charges are screened down to an apparent source strength of Z=137, i.e. Zα=1.

This is a most interesting result, telling us, that *QED does not allow point charges with charge larger than 137*. In other words, the coupling constant of QED (for point particles) can never become larger than 1.

(b) The vacuum charge is arranged in concentric shells ("onion"-layers) around the source. The layers are equidistant in the variable x=γℓn(r/R), as illustrated in the figure 30.

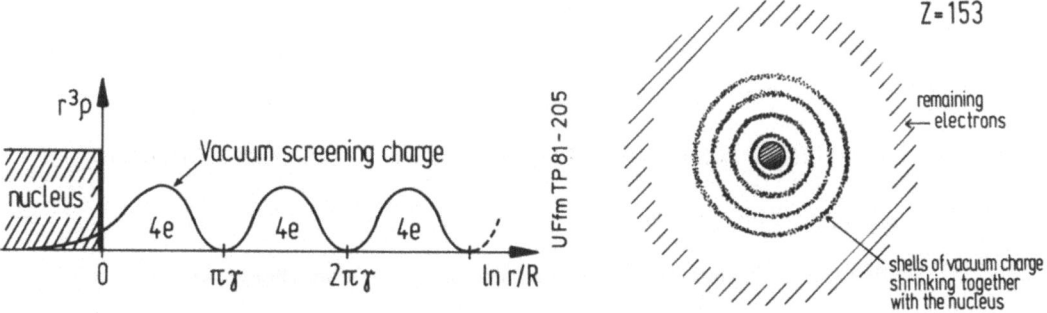

Fig. 30: As the radius R of the central charge shrinks to zero, the vacuum becomes onion-shell-type higher and higher charged, shielding the central charge. In this way point charges higher than 137 are prevented in QED.

(c) The vacuum charge distribution shrinks in scale with the source charge. It also becomes point-like in the limit of a point source.

(d) Virtual vacuum polarization (Uehling potential) only becomes important at $r \sim m^{-1} \exp (3\pi/2\alpha) \sim 10^{-278}$ fm. This is still quite another scale, so we may argue that the external field approximation is still valid for the type of source extension we consider here.

It is tempting to speculate on the relevance of these results for a strongly coupled field theory of elementary particles where, say, α itself becomes large. If taken seriously, they would indicate that in such a theory the vacuum state is very much different from the perturbative free-field vacuum and the interaction between particles is strongly screened by the vacuum rearrangement. We wish, however, to add a word of caution: if α itself is large, radiative corrections cannot be treated perturbatively but must be considered ab initio in the determination of the vacuum state. For a further discussion of this very interesting topic we refer to the Lahnstein-Conference Proceedings [13] - see also D. Vasak et. al.: "Fission of Bags by Spontaneous Quark-Antiquark Pair production in Supercritical Colour Fields"(Lecture 3 and 4 at this Conference).

IX SUPERHEAVY QUASIMOLECULES

Now we come down to earth and ask ourselves whether this can all be experimentally tested; can one make physics out of theoretical ideas? A crucial suggestion was made in 1969 by Greiner [see 24] and also by Gershtein and Zeldovich [25] that during the course of the collision of two heavy ions, superheavy quasimolecules are formed (see fig. 31), due to adiabaticity $(\frac{v_{ion}}{v_{el.}} = \frac{1}{20})$

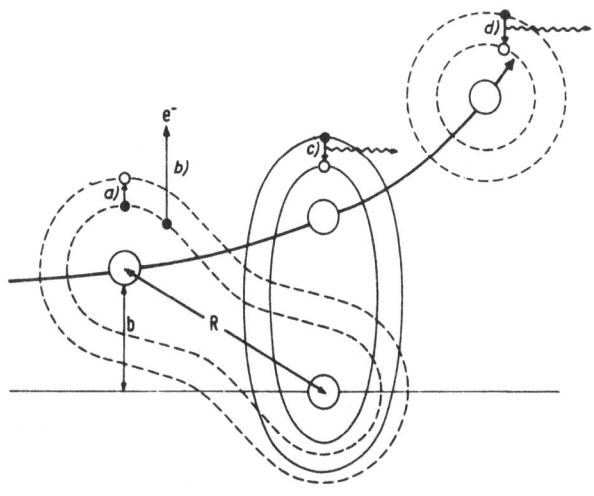

Figure 31: The formation and change of quasimolecules in the course of a heavy ion collision. The various innershell processes are illustrated: ionization (a), quasimolecular x-rays (b), characteristic x-rays (c). Not shown are positron creation processes (for those see fig. 33).

with fast electrons orbiting the colliding, slow nuclei. The proper

quantum-mechanical formulation of the transiently formed superheavy

molecules and the various processes therein required the solution

of the *Two-Center-Dirac-equation*, which was first achieved by Müller

and Greiner [26]. A typical TCD-level diagram is depicted in Fig. 32.

The critical distance, at which the $1s_{1/2}$-level is diving, is about

35 fm. Notice the *strong $\vec{\ell}\cdot\vec{s}$-splitting* of e.g. the $P_{1/2}^-$ and $P_{3/2}^-$

states which is of the order of 600-800 keV!

A time-dependent view of such a level diagram is taken in Fig.

33: t=0 corresponds to the time of closet approach. Due to the time-

dependence and the thereby introduced Fourier-frequencies, a number

of processes happen. They are indicated by vertical arrows. We draw

attention particularly to the K-ionization (process(a)) and to the

dynamical positron creation (process (c), (d), (f)), which represent the

formerly mentioned shake-off of the vacuum polarization cloud (see

fig. 12). The vacuum decay process, indicated by the horizontal arrow

(c) would also go on if all the dynamical processes came to a stand-still, i.e. if a giant nucleus is fused, sitting there forever.

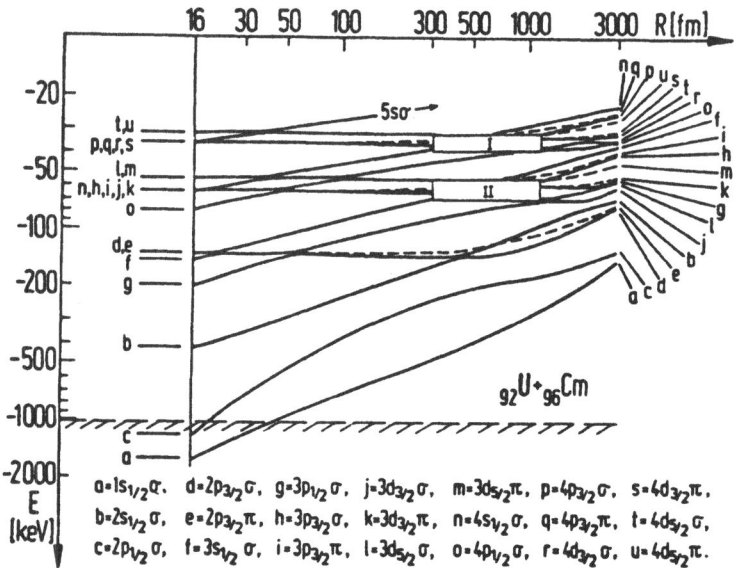

Figure 32: The Two-Center-Dirac -level diagram for U+Cm. The K-shell is diving at a critical distance of about 35 fm. Notice the rather steep diving (double logarithmic scale!) and the enormous fine structure splitting in the giant systems $(Z=Z_1+Z_2=188)$.

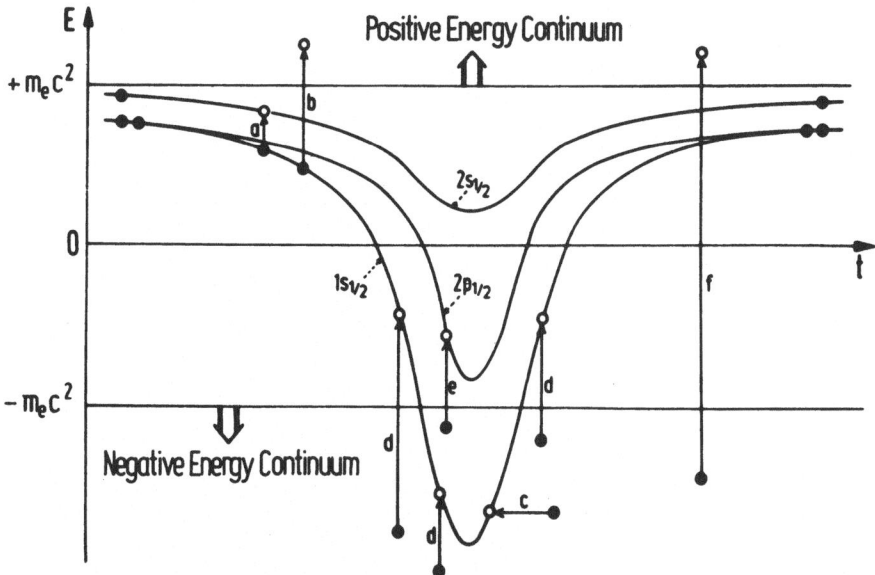

Figure 33: Illustration of the various processes in the time dependent TCD-level-diagram. The vertical arrows indicate dynamical processes connected with the change of the molecular wavefunction during the course of the colision. The horizontal arrow represents the vacuum decay. It always occurs in an overcritical system.

The studies of *quasimolecular x-rays*, in particular the work of Walter Meyerhof [27] and of J. Greenberg and P. Vincent [28] have proven beyond doubt the formation of these intermediate systems in heavy ion collisions. To give some impression of the agreements between theory and experiments in this field I show in fig.34 the impact parameter dependence of the *K-hole-production probability* for the Pb+Cm-system (measurements by Bosch et. al. [29] and fig. 35 a *delta-electron spectrum* for the J+Pb-system (measurements by Koenig et. al. [30]). The excellent agreement with the (parameter free) theory of Soff, Reinhardt et. al. [31] is astounding. The details of the calculations and the measurements can be found in the literature [31,32].

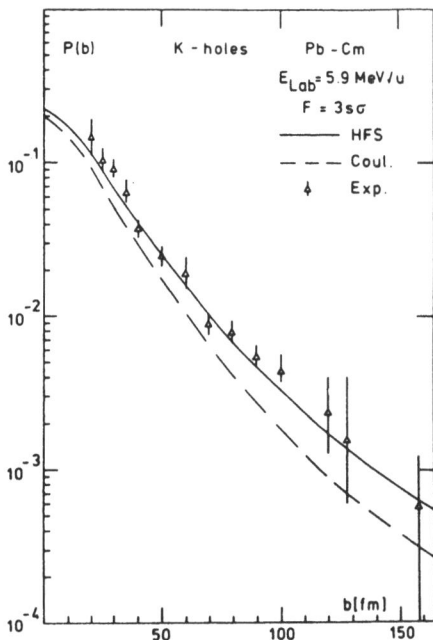

Figure 34: Probability for K-hole creation per Pb-Cm collision at 5.9 MeV/u laboratory energy as function of impact parameter b. The experimental data are from Liesen et. al. [29], the theory is from Th. de Reus, U. Müller, J. Reinhardt, et. al. [31]. No parameters are adjusted. The full curve in the calculation is based on Two-Center-Dirac-Hartree-Fock states containing the electron-electron interaction. The calculations shown by the dashed curve are based on standard TCD-levels (without the electron-electron-interaction).

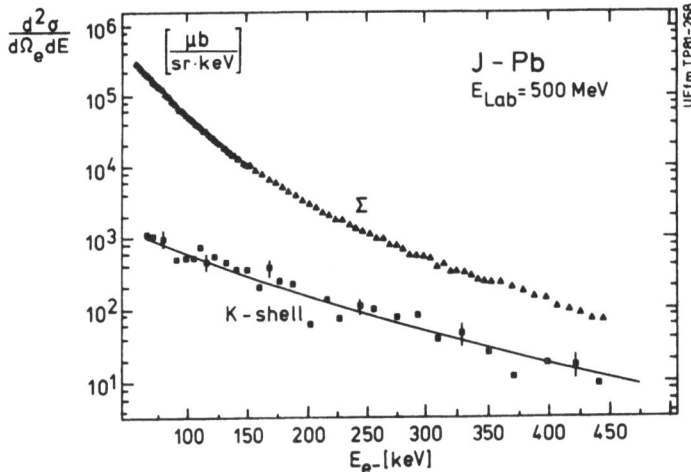

Figure 35: Total measured δ-electron distribution (triangles) and δ-electrons in coincidence with K-vacancies for the J-Pb system versus kinetic energy, Koenig et. al. [30]. The latter data are compared with absolute values of a calculation by de Reus et. al. [31]. No parameters are adjusted.

We remark, that many more details, of the experiments (energy dependence, $(Z_1 + Z_2)$-dependence, etc.) are reproduced by the theory, so that one can rightfully state, that these processes are quantitatively understood.

X. POSITRONS FROM HEAVY ION COLLISIONS

The semi-classical treatment of electron-positron excitation processes is based on the time-dependent two-centre Dirac equation

$$i\partial/\partial t \; \Phi_i \; (\vec{R}(t)) = \hat{H}_{TCD} \; (R(t)) \; \Phi_i \; (R(t)) , \qquad (1)$$

where $\hat{H}_{TCD}$ is the relativistic two-centre Hamiltonian depending on the internuclear separation $\vec{R}(t)$. At non-relativistic bombardig energies it is useful to expand the wavefunction Φ_i into Born-Oppenheimer states ϕ_j given by the instantaneous molecular eigenstates of the Hamiltonian:

$$\Phi_i(\vec{R}(t)) = \sum_j a_{ij}(t) \; \phi_j(\vec{R}(t)) \; \exp \{-i\chi_j(t)\}. \qquad (2)$$

The sum includes an integration over continuum states of positive and negative energy. The phase factors χ_j are conveniently chosen as

$$\chi_j(t) = \int^t dt' <\phi_j(\vec{R}(t'))|\hat{H}_{TCD}(\vec{R}(t'))|\phi_j(\vec{R}(t'))>. \qquad (3)$$

Inserting the expansion of eq. (2) into eq. (1) and projecting with stationary eigenfunctions we obtain a set of *coupled differential equations* for the amplitudes $a_{ij}(t)$

$$\dot{a}_{ij}(t) = -\sum_{k \neq j} a_{ik}(t) \; <\phi_j|\partial/\partial t|\phi_k> \exp \{i(\chi_j - \chi_k)\} , \qquad (4)$$

with the initial condition $a_{ij}(-\infty) = \delta_{ij}$.

After splitting the time derivative operator in terms of a radial and a rotational coupling and neglecting the latter one, the coupled equations (4) may be solved by numerical integration. In the independent-particle approximation excitations of the many-electron system are described by incoherent summation over one-electron transition probabilities.

After the collision the number of particles occupying a state above
the Fermi level, up to which the quasimolecular levels are initially
filled, is

$$N_p = 2 \sum_{k<F} |a_{kp}(\infty)|^2 \quad (p > F) , \qquad (5a)$$

while the number of holes in a state below the Fermi level is

$$N_p = 2 \sum_{k>F} |a_{kp}(\infty)|^2 \quad (p < F) . \qquad (5b)$$

An adequate description of positron production in supercritical
collision systems, where $Z_T + Z_p$ exceeds 173, requires a slight
modification of the formalism set forth here. In a supercritical
system the 1s-state is represented as a resonance in the positron
s-wave continuum and not by a single eigenstate of the Hamiltonian
$\hat{H}_{TCD}$. A formalism that avoids those difficulties and moreover has
heuristic value for the interpretation of the positron creation
process was developed by Reinhardt et al. [32] and later also
discussed by Tomoda and Weidenmüller [33]. The method is based on
the observation that the continuum wavefunction of the supercritical
system at resonance energy $E_p = E_{res}$ is quite similar to the discrete
1s-state in the supercritical case except for an oscillating tail,
small in amplitude but reaching out to infinity. This structure
reflects the occurrence of a tunneling process through the barrier
separating the particle- and antiparticle solutions of the Dirac
equation in a semiclassical picture. Apart from the asymptotic behaviour
the 1s-wavefunction retains many of its properties, e.g., the strong
localization and the radial matrix elements which may be continued
smoothly to the supercritical region if the tail of the wavefunction is
neglected.

This idea can be used to develop a general method to treat resonance scattering. In this context Wang and Shakin [15] introduced a projection formalism for resonances in the nuclear continuum shell model: After having defined a normalizable quasibound wavefunction Φ_R, a new continuum $\tilde{\phi}_{Ep}$ is constructed which spans a subspace orthogonal to Φ_R and replaces the old continuum ϕ_{Ep}. The modified continuum states satisfy the original Dirac equation supplemented by an inhomogeneous term that ensures orthogonality with respect to the resonance wavefunction Φ_R:

$$(\hat{H}_{TCD} - E_p) \mid \tilde{\phi}_{Ep}> = <\phi_R|\hat{H}_{TCD}|\tilde{\phi}_{Ep}> \mid \phi_R> . \qquad (6)$$

If the states Φ_R and $\tilde{\phi}_{Ep}$ are used as part of the basis in eq. (2) the 1s-state Φ_R couples to the new positron continuum by two separate coupling operators

$$\dot{R}<\tilde{\phi}_{Ep}\mid \partial/\partial R\mid \Phi_R> + i/h<\tilde{\phi}_{Ep}|\hat{H}_{TCD}|\Phi_R> . \qquad (7)$$

The second matrix element arises since Φ_R and $\tilde{\phi}_{Ep}$ are not exact eigenstates of the two-centre Hamiltonian $\hat{H}_{TCD}$. It does not depend on the nuclear motion and leads, in the static limit $R(t) = \text{const} < R_{cr}$, to an exponential decay of a hole prepared in Φ_R. The decay width

$$\Gamma = 2\pi \mid <\tilde{\phi}_{Eres}|\hat{H}_{TCD}|\Phi_R>\mid^2 \qquad (8)$$

is identical to the width of the resonance in the unmodified positron continuum.

The formalism thus leads naturally to the emergence of *'induced'* and *'spontaneous' positron creation*, the latter resulting from the presence of an unstable state Φ_R in the expansion basis. In practice, however, this does not result in a marked threshold behaviour at the border of the supercritical region for two reasons. Firstly, both couplings enter via their Fourier transforms depending on the time development of the heavy ion collision. Their contributions have to be added coherently so that in a given collision there is no physical

way to distinguish between them. Secondly, in collisions below the
Coulomb barrier the rapid variation of the quasimolecular potential,
especially in the supercritical region, causes significant contributions
from the dynamical coupling, whereas the period of time for which the
internuclear distance R(t) is less than R_{cr} is usually very short
($\sim 10^{-21}$ sec) as compared with the decay time of the 1s-resonance
($\sim 10^{-19}$ sec).

Therefore, the predicted production rates and energy spectra
of positrons continue smoothly from the subcritical to the supercritical region (see figure 38 below). Qualitative deviations of the
positron production rate in supercritical collision systems are expected
only under favourable conditions: Since the 'spontaneous' and
'dynamical' couplings exhibit a different functional dependence on the
nuclear motion, an increase in collision time can be expected to provide
a clear signature for supecritical collisions. Therefore Rafelski,
Müller and Greiner [34] suggested the study of positron emission in
heavy ion reactions at bombarding energies above the Coulomb barrier,
where the *formation of a di-nuclear system* or of a compound nucleus
would eventually lead to a time delay within the bounds of the critical
distance R_{cr}. During this *sticking time* T the spontaneous decay
of the 1sσ-resonance, by filling dynamically created K-shell holes
under emission of positrons, might be strongly enhanced. This idea
is illustrated in figure 36.

A variety of experiments concerning positron creation have been
performed at the Gesellschaft für Schwerionenforschung (GSI) in
Darmstadt during the past four years. They were the subject of two
contributions (one by the Kienle-group, the other one by the Greenberg-
Schwalm group) to the conference at Florence 1983 to which we refer the
the reader. Here we wish to concentrate on theoretical results and
comparisons with selected experiments concerning non-Coulombic collisions.

First of all some general remarks.

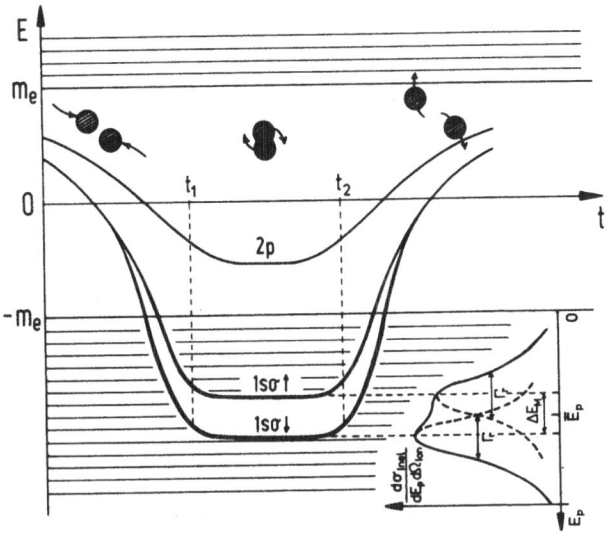

Figure 36: Due to the formation of a giant nuclear system with a certain lifetime T the spontaneous positrons are enhanced. For large T a positron line develops on top of a smooth background stemming from dynamically induced positrons.

We have integrated the modified system of differential equations

(4), (7) in the framework of the monopole approximation including up to

8 bound states and ~17 states in the upper continuum for each angular

momentum channel ($s_{1/2}$ and $p_{1/2}$-waves, i.e. $\kappa = -1$, +1, respectively).

Positron emission rates [32] increase very fast with total nuclear

charge, flattening somewhat for the highest Z-values. If parameterized

by a power law $(Z_T + Z_p)^n$, the exponent takes values of 20 down to 13,

if an initial Fermi-level above $3s\sigma$, $4p_{1/2}\sigma$ is assumed, or even $\cong 29$

for bare nuclei (F=0). This highly nonlinear behaviour clearly expresses

the *non-perturbative nature of the mechanism of positron production* in

such giant systems. Mainly responsible for the enhancement for fully

stripped nuclei is the contribution of the 1s-state which in normal

collisons (F>0) is suppressed by the small K-vacancy probability.

If the K-shell is empty it becomes the dominant final state for pair production due to the strong coupling between the 1s-state and the antiparticle continuum which it approaches and even enters in the supercritical region. In sub-Coulomb barrier collisions $s_{1/2}$ and $p_{1/2}$-waves contribute about equally to the total result.

At this point we must address the major problem in analysing the experimental data. Already for bombarding energies well below the Coulomb barrier $E_c (E/E_c \sim .8)$ the nuclei can be excited by Coulomb excitation, and the emitted photons with energy above 1022 keV can undergo internal pair conversion. Thus one has to measure the γ-spectrum simultaneously and to fold it with the conversion coefficients. Here one has to know - or to assume - the γ-ray multipolarity. Monopole conversion cannot be handled by this method. Up to now, all conclusions on positron production in heavy-ion collisions had to rely on the described procedure for background subtraction.

The first generation of experiments established the dependence of positron excitation rates on the kinematic conditions as well as on the combined charge Z. The Z-dependent increase, which spans an order of a magnitude while $\Delta Z/Z$ is only 12% is well described by theory. Also the shape of the theoretical curves is in good agreement with the experimental data. In the Pb+Pb system and, for smaller distances of closest approach, even in Pb+U collisions the data agree also in absolute values. In the heaviest accessible system U+Cm (Z_u=188) and for larger distances R_{min} the theory has a tendency to overestimate the measured data by up to 25%.

From these data no qualitative signature for the 'diving' of the $1s\sigma$-state in U+U, U+Cm collisions could be extracted, in agreement with theoretical predictions. More sensitive informaton can be obtained by the measurement of energy spectra of positrons detected in coincidence with the scattered ions. Their knowledge is most useful if one wants to find deviations hinting to the positron creation mechanism. Fig 37

shows the earliest published positron spectra from Backe, Kankeleit, et al.

[35] for three collision systems, U+Pd, U+Pb, and U+U, at 5.9 MeV/u

bombarding energy; the ions are detected in an angular window

θ_{lab}=45°±10°. For U+Pd (Z=138) no atomic positrons are expected, the

data can be fully accounted for by nuclear conversion (dashed curve).

Extrapolating this procedure to the U+Pb system (dashed curve) the sum

of background and calculated QED

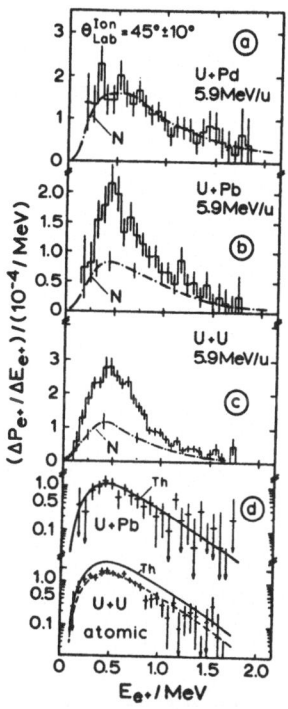

Figure 37: Spectra of emitted positrons in 5.9 MeV/u collisions measured by Backe, Kankeleit, et al. [35] in coincidence with ions scattered in the angular window θ_{lab}=45°±10°. The spectrum in the lightest sysem, U+Pd, is explained by nuclear pair conversion alone (dashed line).
In the U+Pb and U+U systems the sum (full lines) of nuclear and calculated atomic positron production rates is displayed.

positron rates (full curve) is in excellent agreement with the observed

emission spectra.

A more recent total positron production spectrum with heavy ions

deflected into the angular window 25° < θ_{LAB} < 65° has been measured

by the Bohemeyer - Bethge - Greenberg - Schwalm - Vincent - et al.

Group (fig 38). The U+Cm, U+U and U+Pb spectra are shown in fig 38

together with the measured positron background (dashed line). The
theoretical (dynamical) positron spectrum, calculated for Rutherford
trajectories has been added to this background, yielding the full curves
of fig 38. Obviously the agreement with the measurements is quite
satisfactory.

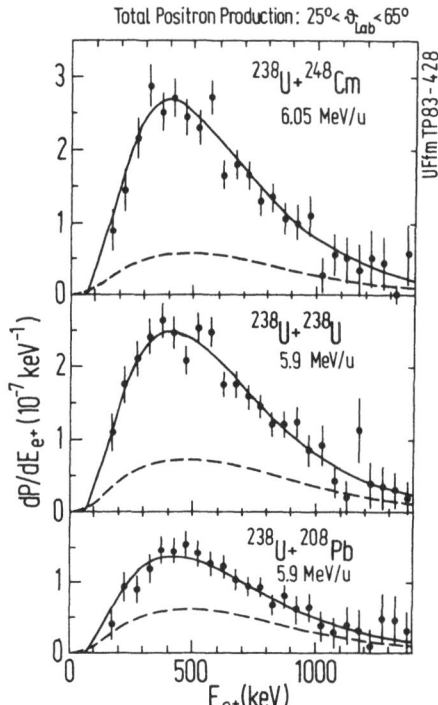

Figure 38. Total positron production spectra with heavy ions deflected into the window $25° < \theta_{lab} < 65°$ for U+Cm, U+U, U+Pb at 5.9 MeV/amu. The dashed curve indicates the measured background; the full curve is the theory plus background.

A possible source which could cause deviations in the shape of the
positron spectra from the results presented so far will be discussed in
below. To obtain theoretical predictions for positron production
it is essential to include the 'spontaneous' coupling for collisions
where the 1sσ-state joins the lower continuum. If it is left out of the
calculation the resulting positron spectra would be strongly altered:

The 'induced' radial coupling is changed at the same time as the spontaneous coupling becomes important. Both contributions add up coherently and cannot be observed separately. A promising strategy to get a clear qualitative signature for the diving process is to try to *modify the time structure* and to select *heavy-ion collisions with prolonged nuclear contact* time. Such nuclear reactions are expected to occur at energies close to or above the Coulomb barrier. The nuclear delay time T should provide a handle to distinguish super-critical systems.

Using a schematic model for the trajectory U. Müller, Th. de Reus, J. Reinhardt et al. [36] have performed coupled channel calculations for the four heavy-ion collision systems Pb+Pb, Pb+U, U+U, and U+Cm, corresponding to Z_{united} =164, 174, 184, and 188, respectively. Independently of assumptions on the incoming and outgoing path, of dissipation of nuclear kinetic energy or angular momentum during the reaction, and of the position of the initial Fermi level, all positron spectra exhibit the following features:

In *subcritical collision systems* ($Z_T + Z_p \leq 173$) a *delay time T causes modulations* in the positron spectrum with a width $\Delta E = h/T$. In fig. 39a positron spectra are displayed for a Pb+Pb collision (E_{lab}=8.73 MeV/u, b=7.11 fm, F=3sσ, $4p_{1/2}$σ) with delay times T=0 (pure Rutherford scattering), 3*, 6*, and 10*10^{-21} sec. The modulations are due to *interference effects* in much the same way as predicted for the δ-electron spectra in deep inelastic heavy-ion collisions [37].

In addition to the interference patterns an *enhancement of positron production* in time-delayed supercritical collisions is observed, where the binding energy of the lowest bound states exceeds the value $2m_o c^2$.

For *long delay times* a *distinct peak* in the positron spectrum is found
at the location of the supercritical bound state resonance (binding
energy minus $2m_o c^2$) due to the spontaneous pair-creation mechanism.
A detailed analysis of the spectra reveals that this peak emerges
gradually as $Z_u = Z_T + Z_p$ exceeds Z_{cr}. Positron spectra for the supercritical system U+U ($E_{lab} = 7.35$ MeV/u, b=3.72 fm, F=3) are shown in
fig. 39b. With increasing delay time the position of the maximum
drifts slowly from the kinematic maximum to the 'resonance energy',
which depends on the combined charge, the separation of the two nuclei
and on the nuclear charge distribution.

However, for any chosen set of experimental parameters, the nuclear
reaction time T may not (and will not) be sharp but distributed over
a certain

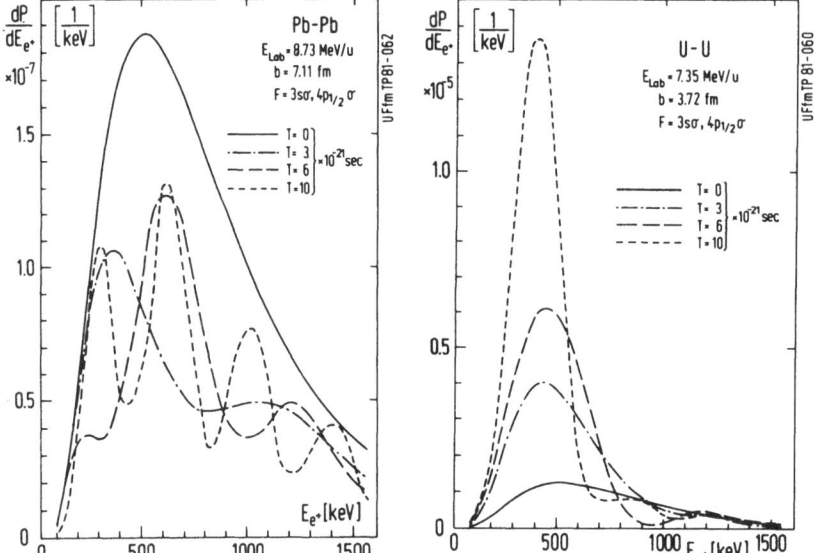

Figure 39: Spectra of positrons created in subcritical (part a)
and supercritical (part b) heavy-ion collisions assuming grazing
Coulomb trajectory (full lines) and nuclear reactions leading to
delay times T=3*, 6*, and 10×10^{-21} sec, respectively, using a schematic model for the trajectory [36]. Whereas for the lighter collision systems modulations in the positron spectra are present, a
distinct peak at the 'resonance' energy $E_{1s\sigma}(R_{min})$ builds up for
systems with $Z_u > 173$.

range with a time distribution function $f(T)$. As an assumption we took a Gaussian centered at $\bar{T}$

$$f(T) = \frac{1}{2\pi\tau} \exp \left[-\frac{(T-\bar{T})^2}{2\tau^2} \right].$$

(9)

The resulting positron spectra for parameters $\bar{T} = 16 \cdot 10^{-21} \text{sec}$ and $\tau = 0$ and $\tau = 2*10^{-21} \text{sec}$ are displayed in fig. 40 for a head-on U+U collision at $E_{lab} = 5.9$ MeV/u. If we consider the 'subcritical' p-states only (part a) we observe that the oscillations disappear already for $\tau = 2*10^{-21} \text{sec}$. Thus we conclude that the *appearance of several oscillations in the spectrum can be expected only for sufficiently sharp nuclear reaction times if the system is subcritical.* Also in the total spectrum (part b) the oscillations are damped out for increasing τ. But most striking is the invariance of the dominant nonvanishing first peak, which originates from the spontaneous part of the positron production mechanisms.

A similar effect is expected in the continuum spectrum of *quasimolecular X-rays.* Calculations for the Pb+Pb system (fig. 41) show that a *line emerges at the united atom K transition energy* if the delay time becomes sufficiently long [38]. The results for the U+U system will be similar, but there is a formidable background from nuclear γ-rays which probably makes an experimental observation very difficult.

The results described so far were obtained within a schematic model for the nuclear motion, which facilitates a systematic study of the *time delay effect* and allows for an investigation of the conceptually interesting limit of large sticking times. To analyse a given experiment, however, the employed nuclear trajectories should be consistent with the elastic and inelastic heavy-ion scattering data.

Many reaction models with different degrees of refinement have been discussed in the literature. We have calculated trajectories with the macroscopic friction model of Schmidt et al. [39], which includes nuclear neck formation. Strong deviations from Coulomb trajectories are found and an energy loss up to ~30% (for b~0) can be obtained. The change of the positron spectrum for collisions with varying degree of nuclear contact is demonstrated in fig. 42. Part (a) shows the modified U+U-trajectories (R(t) for several orbital angular momenta from ℓ=0 head-on to ℓ=400ℏ near grazing collisions. The corresponding positron spectra, part (b), show a gradual enhancement at E_{e+} = 600 keV. As expected a longer delay time ΔT leads to increased positron production in the s-channel.

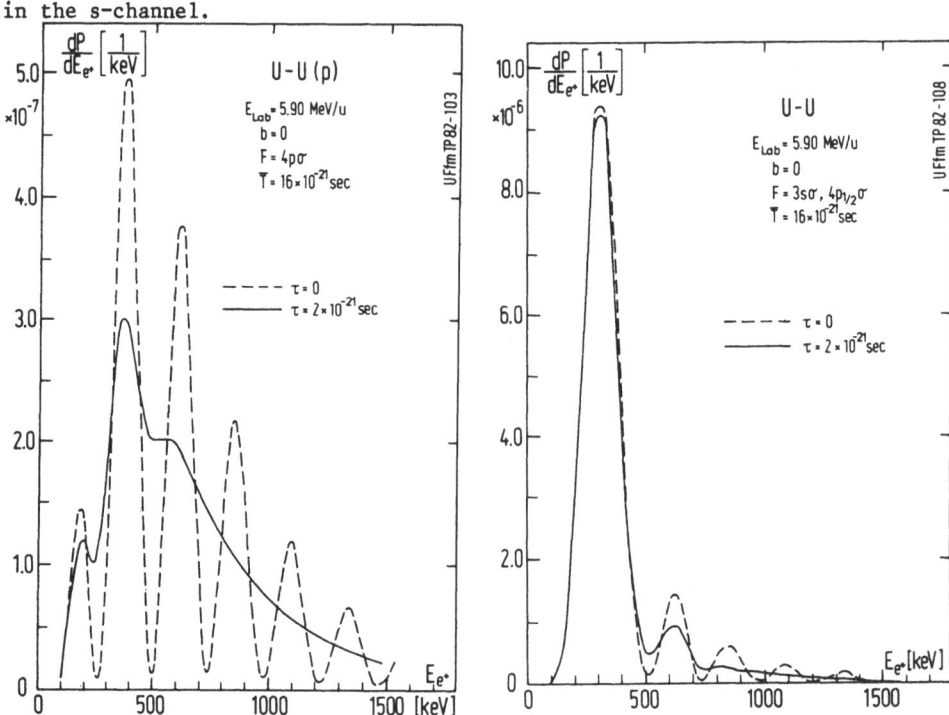

Figure 40: Differential positron production probability versus kinetic positron energy E_{e+} in a central U+U collision at E_{lab}=5.9 MeV/u, assuming a Gaussian nuclear reaction time distribution centered at $\bar{T}$ with a width τ. (a) p-states only, representing a subcritical system, (b) including also the contribution of the s-states. Most striking is the appearance of the first pronounced peak which originates from spontaneous positron production.

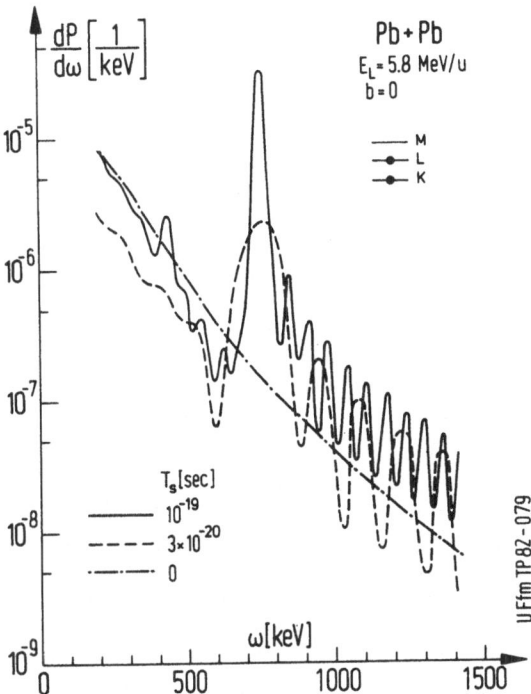

Figure 41: Quasimolecular X-rays of the Pb+Pb system for various sticking times.

On the other hand the change in kinematics causes a drift to lower kinetic energies in the $p_{1/2}$-partial wave spectra due to destructive interference. Both effects taken together lead to an enhancement of the maximum and a drift towards lower energies also in the total spectrum.

Measurements by Backe, Kankeleit et al.[40] seem to indicate such tendencies: In U+U and U+Cm collisions at energies above the Coulomb barrier positron spectra have been measured in coincidence with fission fragments in order to get a signature for close nuclear contact. The analysis shows an enhancement of dP/dE_{e^+} at lower kinetic energies in

qualitative agreement with Fig. 42b. For a quantitative comparison
one has to integrate the impact parameter-dependent positron spectra
over all values of b which lead to a nuclear reaction, weighted by the
corresponding probability w(b) to induce nuclear fission. Performing
the integration with a weight factor w(b) = 1 for b<b_{gr} and w(b)=0
elsewhere, there remains an energy shift of ~50 keV in the experimental data in comparison with the theoretical curves, which might
be due to electron screening effects. Furthermore, as mentioned above,
the theoretical values have to be reduced by an overall factor ~20%.
Fig. 42c shows the experimental data for U+U collisions at E_{lab}=5.9
MeV/u, 7.5 MeV/u, and 8.4 MeV/u, in comparison with theoretical results
excluding electron screening, but reduced by the factor mentioned above.
Dashed lines indicate pure Rutherford scattering trajectories, whereas
the solid lines display spectra calculated with the modified trajectories
of fig. 42a. For an even better agreement longer delay times
($\Delta T \gtrsim 2*10^{-21}$ sec) may be needed. Further investigations along these
lines seem to be very promising, both for establishing the mechanism of
positron production and deducing the nuclear reaction time scale.

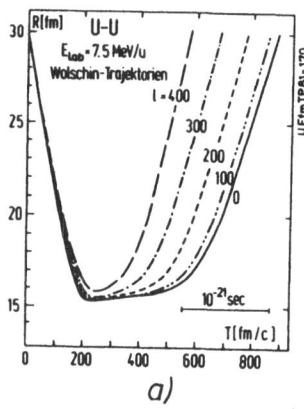

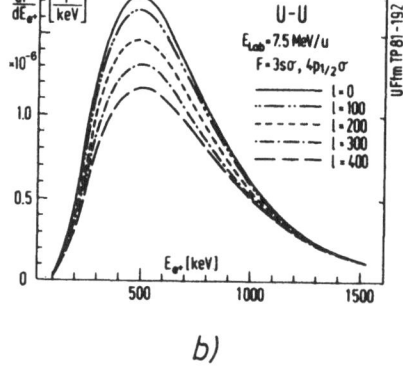

a)   b)

Figure 42a) : Nuclear trajectories calculated in the friction model [39] for
7.5 MeV/u U+U collisions at various
values of the orbital angular momentum
ℓ between 0 and 400 ħ.

Figure 42b) : Energy spectra of
positrons emitted in the collisions
shown in a). The results for the
angular momentum channels s and
$P_{1/2}$ have been added.

183

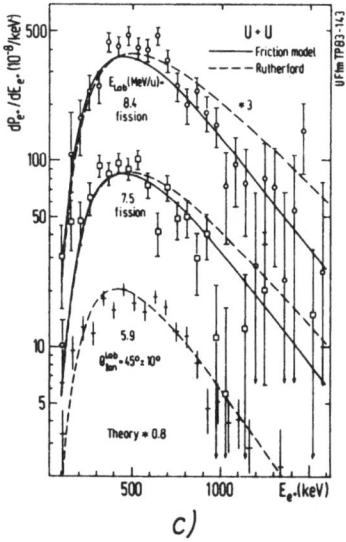

Figure 42c): Comparison of
theoretical predictions for
U+U at E_{lab}=5.9 MeV/u, 7.5
MeV/u, and 8.4 MeV/u, with
experimental data of Backe,
Kankeleit et al. [40], as
described in the text.

As another interesting theoretical problem one might ask about
the presence of *nuclear collisions with very long reaction times*. What
would positron spectra look like if, at a given scattering angle, a
superposition of Rutherford scattering and long-lasting nuclear reaction
is assumed? If, for the sake of simplicity, nuclear scattering with
definite delay time T is assumed, a ratio q<<1 can be used as a measure
of the relative cross section of reactions leading to long contact times,
as compared with the Rutherford cross section. As positron production
is very strongly enhanced for reaction times larger than 10^{-20} sec, a
peak superimposed on the smooth spectrum of positrons emitted in the
much more frequent 'distant' Coulomb collisions could emerge. A rough
estimate shows that for long delay times a peak may be prominent even
if the differential reaction cross section is less than 1% of the
Rutherford cross section.

To obtain the full shape of the positron spectrum, an assumption about the angular distribution $d\sigma^N/d\theta$ of the nuclear reaction component is needed. We shall discuss two extreme simplified cases : (i) isotropic break-up of the compound system, (ii) focussing of the reaction fragments into a narrow angular window in the C.M. system. If the reaction products were emitted isotropically, a line should, in principle, be observable in the positron spectrum at all ion scattering angles. However, it would be most pronounced at $\theta_{c.m.}=90°$, being suppressed at other angles relative to the elastic scattering cross section. fig.43 shows spectra of positrons, reduced by a factor ~2/3, for 5.79 MeV/u U+U collisions for several fixed scattering angles. A long lived nuclear reaction component (dashed lines) with $T = 4*10^{-20}$ sec and admixture $q=2\cdot4*10^{-3}$ (at $\theta_{c.m.}=90°$) has been added under the assumption of an isotropic distribution in the reaction plane.

If the emitted reaction fragments are focussed under a certain scattering angle, a detailed quantitative analysis depends strongly on the strength of the focussing effect. Such an effect can be produced, if the life-times T of the giant nuclear molecule or compound system depend strongly on the angular momentum ℓ carried into the reaction (see section XIV). This is not an unreasonable possibility since it is evident that T must vanish for very large values of ℓ. A detailed investigation of this scenario would require a quantum mechanical treatment of the nuclear motion together with a partial wave analysis of the scattering cross-section (see section XIV).

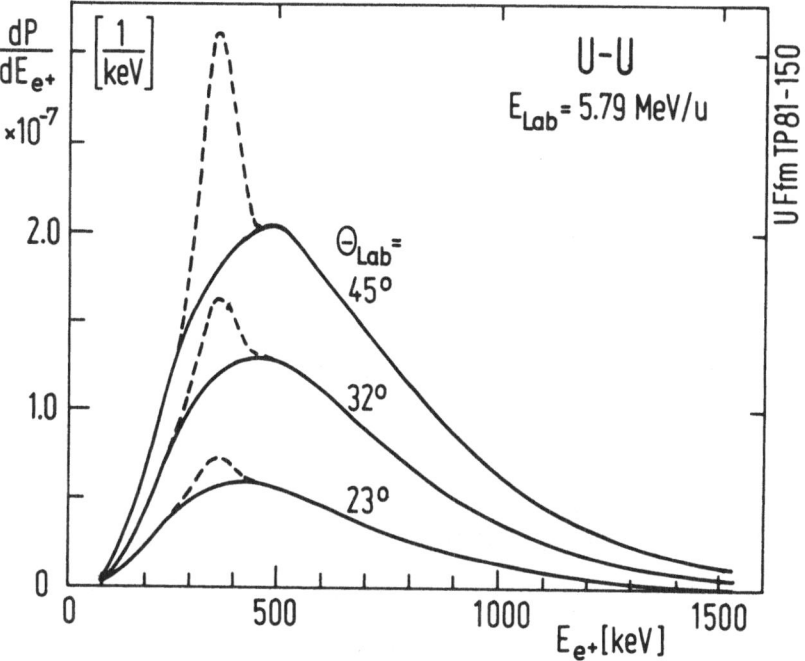

Figure 43: Spectra of positrons emitted in 5.79 MeV/u U+U collisions in coincidence with a scattered nucleus for three selected lab ion angles. The fully drawn curves are calculated assuming Rutherford scattering only. The dashed lines show the effect of an additional nuclear reaction with a lifetime $T = 4*10^{-20}$sec. A relative fraction of $2.4*10^{-3}$ reactions per elastically scattered ion (at 45°) has been assumed.

Two experimental groups, one headed by P. Kienle [41], the other one by J. Greenberg and D. Schwalm [26], during the last years have performed experiments with U+Th, U+U, and U+Cm at energies close to the Coulomb barrier. Contrary to the results of Backe, Kankeleit, et al. [35], their positron spectra show remarkable structures. C. Kozhuharov, P. Kienle et al. [41] measured at 5.8 MeV/u beam energy positron spectra in coincidence with ions scattered into various angular windows. Their analysis shows *sharp maxima in the spectra* that are most pronounced under various laboratory scattering angles (Fig. 43). The U+U-spectra are compared with the Pb+Th-spectra, in fig. 44. While the former

system is overcritical, the latter is undercritical and should,

accordingly, not show line structures (except for Raman-lines - see

later). This, indeed, seems to be the case. In the U+U measurement

the position of the peak was found to be located at ~300 keV with

a width of about ~90 keV, for U+Th the effect is similar. After

substraction of a smooth background the number of positrons per de-

tected ion emitted in the peak is roughly 10^{-5} at θ~45°. If the

observed structure is of quasimolecular origin, it must be produced

in very long-lasting nuclear reactions because of the small width.

If one compares the experimental width with spectra from coupled

channel calculations based on the schematic sticking model a minimum

value T~$4*10^{-20}$sec is required. Due to additional broadening effects

(such as Doppler broadening) the reaction time T would have to be

even longer. According to our calculations the probability for positron

production in a delayed (T=$4*10^{-20}$sec) collision should be ~$4.7*10^{-3}$,

which must be compared with the observed probability ~10^{-5}. Thus

a fraction of q = P_{e+} (exp)/P_{e+}(theor) $\cong 2*10^{-3}$ delayed collisions per

elastically scattered ion is sufficient to produce the observed effect

(at 45°). This number should serve only for general orientation,

since it depends on the details of the model. The most *striking*

structures were detected in the experiment of H. Bokemeyer, J.S.

Greenberg et al. [42] in U+U and U+Cm collisions. For details we refer

to the conference at Florence 1983. A comparison of the theory

with the U+Cm-experiment is shown in fig. 46. The fit is excellent.

Two parameters, the width Γ, and the area (cross section) under the

resonance are fitted. They correspond respectively to the mean lifetime

of the giant system and to the range of impact parameters Δb leading

to fusion (sticking) of the giant system. It turns out that Δb~0.2 fm, which means that only very centrally colliding nuclei in the pole-pole

position are forming the giant nuclear molecule; a very plausible result.

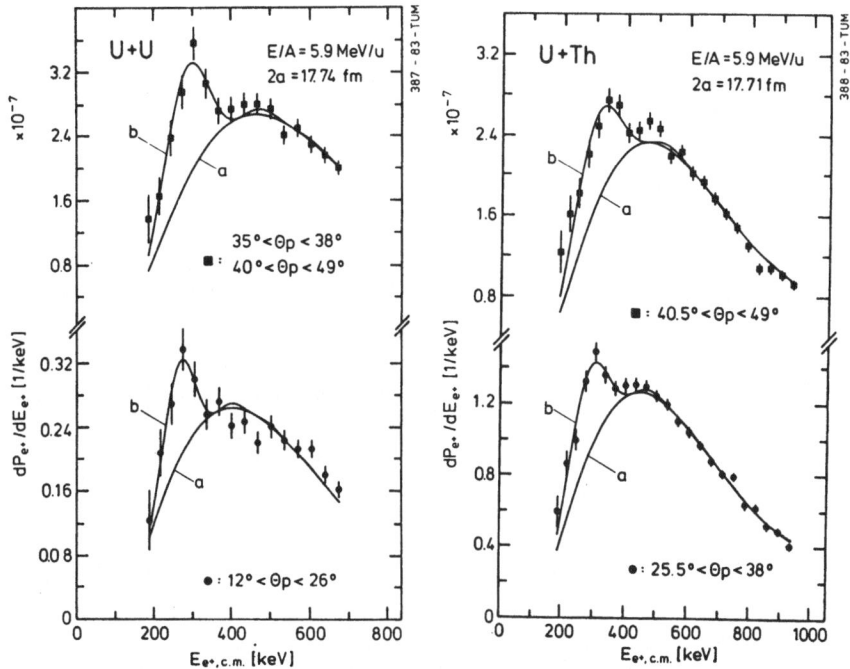

Figure 44: The positron spectra of the Kienle-group [41] for U+U (a)
and U+Th (b) in various angular windows. The line width is somewhat
dependent on the deflection angle of the ion. This is in agreement with
the quantum mechanical model for the heavy ion motion in the giant
molecule-formation. Curve a) indicates the dynamical spectrum according
to theory without shielding and curve b) is a theoretical spectrum with
superposed spontaneous emission (theory of Reinhardt, Müller et al.).

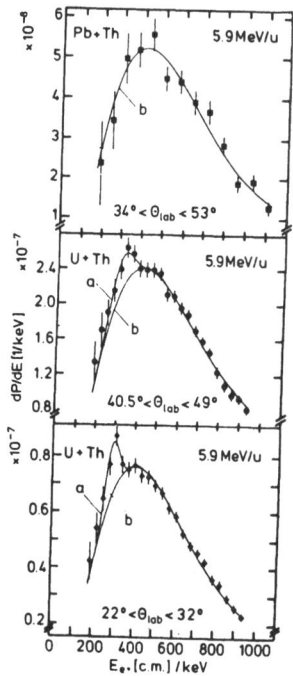

Figure 45: Comparison of the Pb+Th-positron spectra with the U+Th-positron spectra. The former, being under-critical, should not show spontaneous positron lines (except for Raman satellites). This indeed seems to be the case. The data are from the Kienle group. The theory (now with electronic shielding) stems from Reinhardt, Müller et al. [36].

XI. COULD THE LINE STRUCTURE BE OF TRIVIAL ORIGIN?

One might suppose that those peaks are caused by nuclear background processes. However, nuclear transitions of, e.g., multipolarity E1 or E2 should also be observable in the emitted photon spectra provided that proper Doppler shift corrections are performed. This is shown for the U+U-system in fig. 47a and for the U+Cm-system in fig. 47b. In both cases the expected γ-line is indicated if the positron structure would be due to conventional conversion. Obviously, this can be ruled out. Whether the peaks can be caused by E0-processes will be discussed later. On the other hand, if a (sharply) focussed nuclear reaction takes place it is not suprising that an experiment not triggering for the optimal kinematic conditions might smear out any evidence for structure.

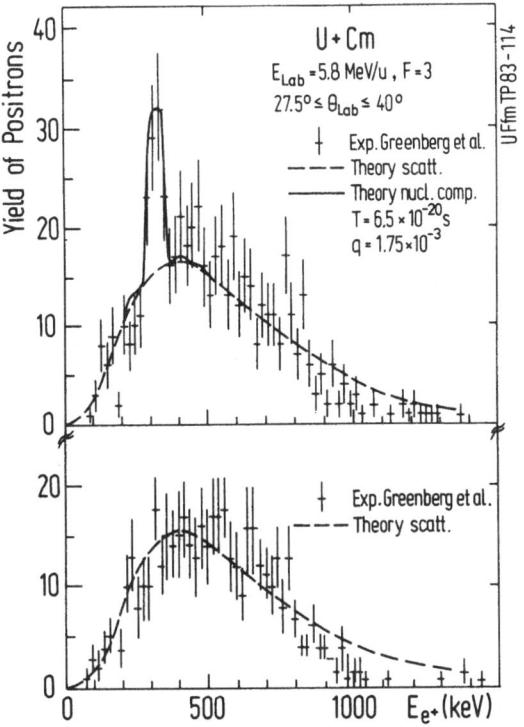

Figure 46: The measurements of H. Bokemeyer, J.S. Greenberg, D. Schwalm et al. [42] for U+Cm positron spectra at 5.8 MeV/u show a remarkable line structure when selecting backward scattered ions (upper part). Theoretical results (in arbitrary units) are compared with the experimental positron yield. The line structure can be explained by a nuclear reaction component with a delay time $T \sim 6.5 * 10^{-20}$ s and an admixture-ratio of $q \sim 1.75 * 10^{-3}$. The lower part displays the positron spectrum for more forward scattered ions, which is in good agreement with theoretical results considering Rutherford scattering only.

Further investigations are needed to settle this question. In fact, a recently *observed threshold effect, supports further the emerging picture that the colliding nuclei have not amalgamate* for a rather long time T to form a giant nuclear system (see section VIII). The observed phenomena seem indeed to be caused by reactions with a very long time scale. This has *far reaching consequences for the physics of nuclear systems in the giant region* (Z~160-190). More about those aspects can be found in section XIII.

The giant nuclear compound system very likely is *not a static object*. Its internal dynamics may influence the spectrum of emitted positrons. Clearly, a rigorous treatment of such effects cannot be based on the semiclassical approximation, but requires a fully quantum mechanical reaction theory for the nuclear scattering (see section XIV). Let us attempt to understand possible consequences in the framework of a very simple (and probably oversimplified) model.

Imagine a classical picture for a nuclear excitation of the super-critical compound system in the spirit of the correspondence principle. The internuclear separation R(t), or better: the *quadrupole deformation of the dinuclear system is supposed to oscillate* around the distance of closest approach R_o of the Coulomb trajectory for a fixed duration T

$$R = R_o \ (1-\alpha_o \ \sin(2\pi\nu t)) \ , \tag{1}$$

where $h\nu$ is the quantum mechanical oscillator energy.

In fig. 48 positron spectra are plotted for a central U+U collision at E_{lab}=6.2 MeV/u. In order to demonstrate the qualitative effect we have restricted

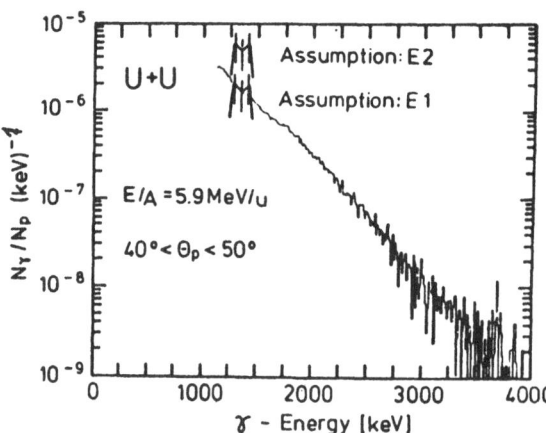

Figure 47a: The γ-ray spectrum for U+U according to Kienle et al. [41] (a) and for U+Cm according to Greenberg et al. [42] (b,c). The expected γ-lines are shown, if the positron line structure would be due to conversion.

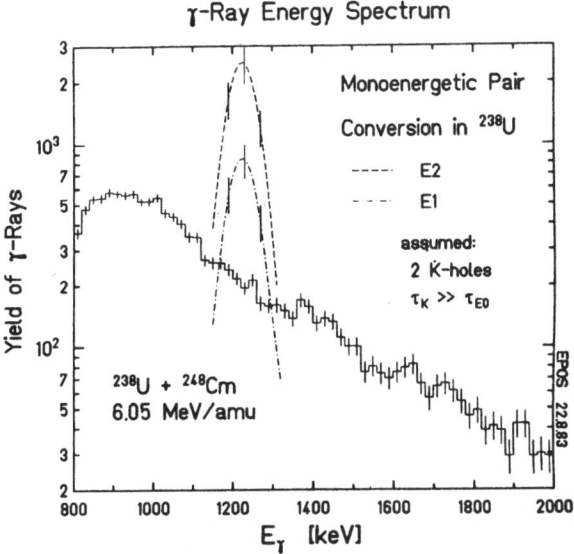

Figure 47b) : Gamma ray spectrum from ≤6.05 MeV/amu uranium and curium collisions measured in 1981. The data shown are coincident with scattered particles with the same kinematic conditions as yield the 320 keV peak in the positron energy spectrum. The expected gamma-ray peak is plotted for the assumption that the positron structure is due to monoenergetic pair conversion of an E1 or E2 transition in the uranium nucleus. We have assumed 2 K-holes at the moment of conversion. The calculated gamma peak includes Doppler shift, peak efficiency and detector resolution. (After Backe, Bokemeyer, Greenberg, Schwalm, Cowan, Schweppe, Vincent, et al.).

the calculations to couplings between s-states only. The parameters in eq. (1) are fixed by $\alpha_o = .25$ and $\nu = .125*10^{21} sec^{-1}$ corresponding to an oscillator energy $h\nu = .25$ keV. Various sticking periods between T=0 and $T=108*10^{-21}$ sec are considered in Fig. 48a. Fig. 48b shows on a linear scale the computed positron spectrum for $T=36*10^{-21}$ sec. As in figs. 39b, 40b we find the dominant 'spontaneous peak'. But in addition a second pronounced peak appears at about $E_{e^+} = 800$ keV. This reflects the fact that part of the vibrational energy of the dinuclear system is transferred to the emitted positron.

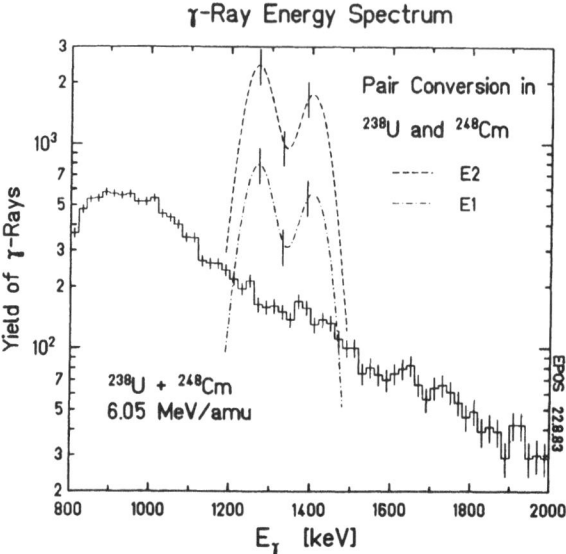

Figure 47c): Gamma ray spectrum from $\leq$6.05 MeV/amu uranium and curium
collisions measured in 1981. The data shown are coincident with scattered
particles with the same kinematic conditions as yield the 320 keV peak
in the positron energy spectrum. The expected gamma-ray peak is plotted
for the assumption that the positron structure is due to nuclear pair
conversion of an E2 or E2 transition in either the uranium or curium
nuclei. The calculated gamma peak includes Doppler shift, peak
efficiency and detector resolution. Note that the positron peak
linewidth strongly argues against the pair conversion process. (After
Backe, Bethge Bokemeyer, Greenberg, Cowan, Schweppe, Vincent et al.).
In order to fully appreciate this discussion see the next subsection.

This is nothing but the classical analogue of a *pair conversion*

process in the supercritical nuclear compound system. If the nuclear

system is vibrationally excited, it may decay to the ground state

transferring its excitation energy to an electron–positron pair. The

electron occupies the vacant 1s–state and the positron carries the energy

balance. Without the nuclear de-excitation this would be the spontaneous

positron creation process, but with it the positron line is shifted by the

nuclear excitation energy. We shall discuss this process further within

the context of pair-conversion processes in the next section. Already

here we can speculate that it might be possible *in the future to observe*

these Raman-satellites and perhaps also *Stokes-satellites* (fig.49). A

spectroscopy of collective states of the giant nuclear system could,

indeed, be within reach.

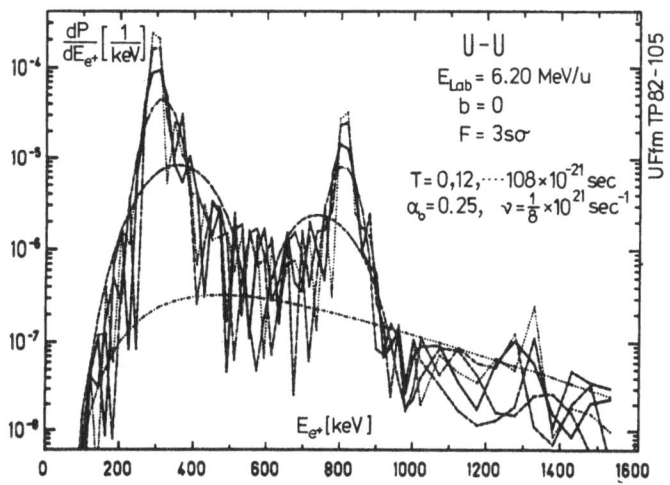

Figure 48: Positron spectra in an U+U head-on collision calculated under the assumption that the distance between the nuclear centres oscillates around the distance of closest approach R_0 of the Coulomb trajectory (cf. eq. (1)). Nuclear reaction times $T=0,12,36,60,84$ and $108*10^{-21}$ sec are considered. The longest duration T corresponds to the most pronounced 'spontaneous peak', etc. Note the logarithmic scale in part a).

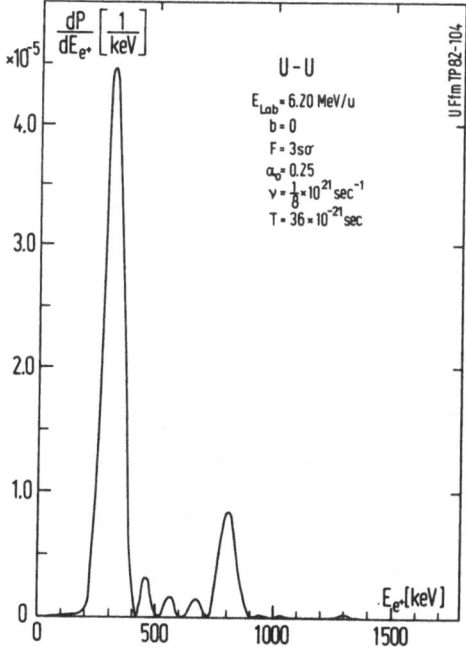

XII. CONVERSION PROCESSES IN SINGLE ATOMS AND IN SUPER-CRITICAL COMPOUND
      SYSTEMS

In collisions of very heavy ions with E_{lab} >3 MeV/u both nuclei are

Coulomb excited. Transfer reactions or even deep inelastic nuclear

reactions can take place which lead to additional excitations of the

nuclei. This internal excitation energy may be carried away by a photon

or may be transferred to a bound electron or to an electron of the

negative energy continuum, which leads to ionization and electron-positron

pair creation, respectively. The latter process requires nuclear

transition energies ω larger than twice the electron rest mass. Nuclear

E0-transitions are characterized by the absence of single photon emissions,

because a photon must carry at least one unit of angular momentum. As

mentioned before such processes form the main source of non-atomic positrons,

and they have to be well understood [43], if one wants to draw firm con-

clusions about the presence or absence of spontaneous pair creation.

The basic processes under investigation are depicted schematically in

fig. 50. The nucleus which makes an E0-transition is labelled by its

initial and final state angular momenta J_i, $J_f = J_i$ and eigenenergies E_i,

$E_f = E_i - \omega$. Process a) describes the *electron-positron pair creation*. An

electron of the negative energy continuum ($\varepsilon = -E < -mc^2$) with Dirac quantum

number κ is lifted to the positive energy continuum. The final state

energy obviously amounts to $E' = \varepsilon + \omega$ whereas the angular momentum quantum

number remains unchanged. Since neither the initial state energy nor the

final state energy is fixed one expects a continuous energy distribution

for the emitted positrons. Process b) indicates the *conversion of a K-*

shell electron (n=1, ℓ=0, j = $\frac{1}{2}$, κ=-1) with energy eigenvalue $E_{1s_{1/2}}$. Thus

bound states with definite energies are involved. Energy conservation then

simply causes monoenergetic lepton emission for a fixed nuclear transition

energy ω.

Process c) symbolizes *monoenergetic positron production*. Here an electron of the negative energy continuum is excited to a bound state, e.g. to the $1s_{1/2}^-$ state. This represents a rather rare process, which can be neglected.

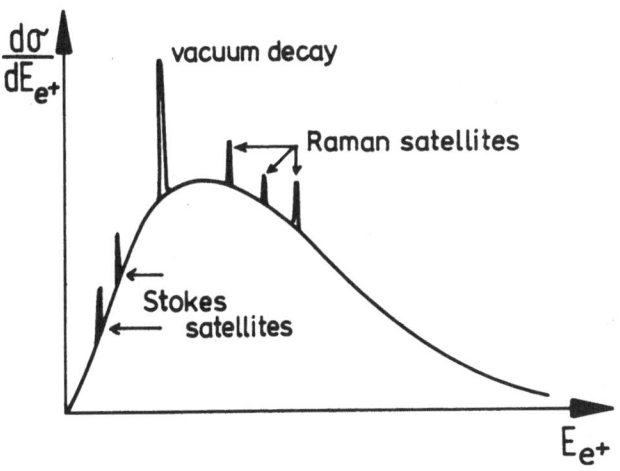

Figure 49: Schematic positron spectrum from a long living giant nuclear system. The expected Raman- and Stokes lines are due to excited (mostly collective) states of the giant nuclear system.

Thus we focus our attention on

i) this *pair conversion coefficient* β, defined as the ratio of the pair production probability (process a)) compared with that of photon emission for a specific nuclear transition with energy ω. Since the energy of the electron and the positron takes continuous values we may express β also as integral of the differential pair conversion coefficient dβ/dE. The lower bound of the integral is determined by the rest mass of the electron, which corresponds to vanishing kinetic energy, while the upper bound is given by the nuclear transition energy ω minus m_e.

ii) the *conversion coefficient* α, defined as the ratio of the probabilities of inner-shell vacancy formation (process b)) and photon emission. In particular this mechanism is important for low energy nuclear transitions.

iii) the *ratio* η *of the two conversion probabilities* for electron-

positron pair creation and for the ionization of bound state electrons.

This ratio is completely determined by the density of the electron

wavefunctions at the nuclear origin, thus being independent of the

nuclear wavefunction.

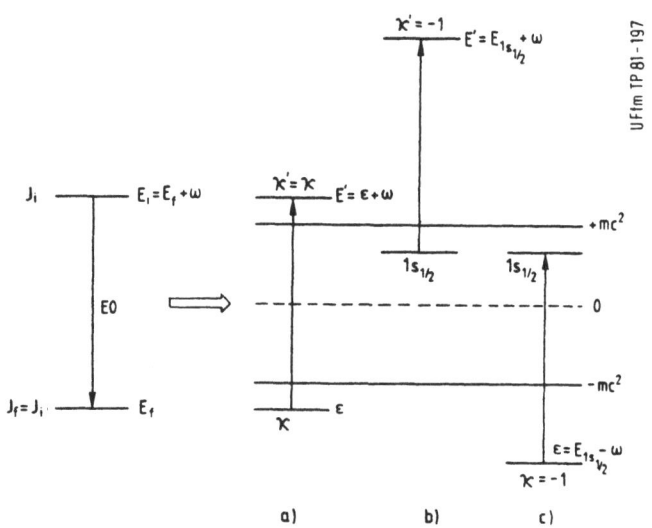

Figure 50: Schematic representation of electron conversion processes accompanying nuclear E0-transition from a state $\{E_i, J_i\}$
to a state $\{E_f=E_i- \omega, J_f=J_i\}$.
a) Electron-positron pair production leading to a continuous
energy distribution of positrons and electrons. b) Conversion
of K-shell electrons – a monoenergetic electron-production mechanism. c) Monoenergetic positron-production – a negligible
process (after Schlüter et. a. [43]).

We computed the differential conversion coefficient dβ/dE for

nuclear E1 and E2 transitions. For the nucleus $_{92}U$ the energy distribu-

tions of emitted positrons is shown in fig. 51a. Nuclear transition

energies of 1323 keV, 1423 keV, 1523 keV, and 1623 keV are considered.

Fig. 51b shows the equivalent differential conversion coefficient dη/dE

for nuclear E0-transitions. As bound state only the atomic K-shell has been taken into account. The conversion probability of higher bound

states is at least one order of magnitude smaller.

Now we can further discuss the possibility whether the observed

structures in positron spectra may originate from nuclear E0-transitions.

One convincing argument against this interpretation is related to the

shape of the e$^+$-energy distribution. According to fig. 51b the halfwidth

of the spectra should be at least 150 keV. However, the observed

structure is much narrower. The second argument is

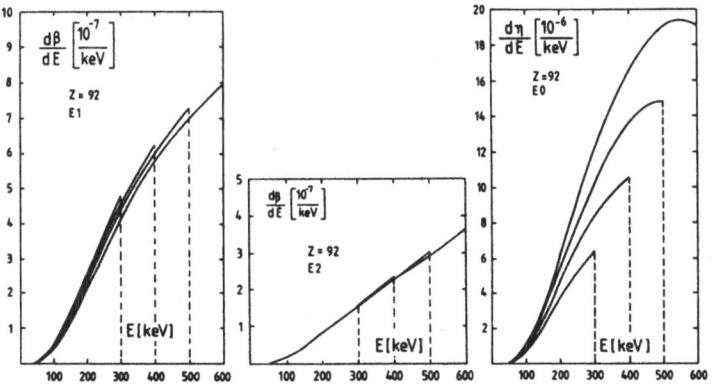

Figure 51 : a) Differential conversion coefficient dβ/dE with respect to the kinetic positron energy E for nuclear E1 - and E2-transitions in $_{92}$U. Nuclear transition energies ω = 1323 keV, 1423 keV, 1523 keV, and 1623 keV are considered, corresponding to maximum kinetic positron energies of E_{max}=300 keV, 400 keV, 500 keV, and 600 keV, respectively. b) Differential conversion probability ratio dη/dE with respect to the kinetic positron energy E for nuclear E0-transitions in U. The same transition energies as in part a).

connected with the energy distribution of the emitted δ-electrons. It

was shown in ref. [43] that, if the observed structures are caused by

nuclear E0-transitions, *one should also observe a distinct peak in the*

δ-electron distribution. Such a peak does not exist, as recent measure-

ments of the Kienle-Kozhuharov-and of the Greenberg-Schwalm- group reveal

(fig. 53) (see conference, Florence 1983). Since *also the*

γ-ray-spectrum (fig 47) does not show any substantial structure which

could eventually be connected with the positron line, *we can conclude, that the sharp positron line structure comes from the decay of the vacuum around a long-living giant nuclear system.* Another analysis of the U-Cm-positron spectra by the Greenberg-group is shown in figure 54. Here the opposite view is taken where various line-fits according to the processes shown in fig. 50 are tried out. Obviously the interpretation of the structure as a spontaneous emission line has by far the highest confidence level.

We now turn to the discussion of electron-positron pair conversion in supercritical compound systems. A supercritical nucleus Z=184 which undergoes a transition with $\hbar\omega > 2m_0c^2$ during the nuclear reaction period T may transfer this excitation energy to one of the electrons in the negative energy continuum. The remaining hole represents a positron. But also the K-shell electron can be lifted

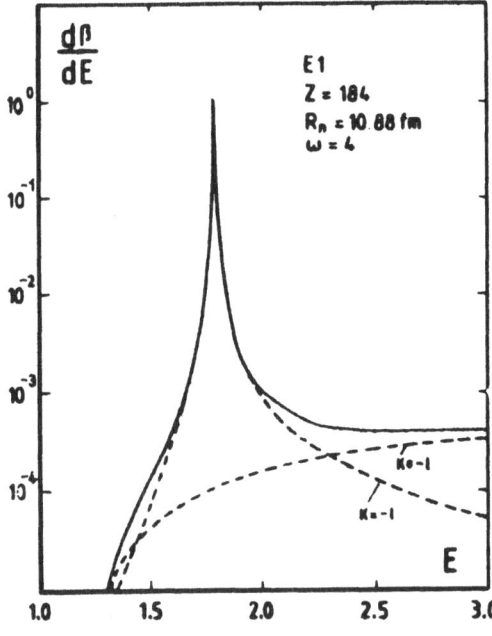

Figure 52: Differential pair conversion coefficient $d\beta/dE$ as function of the positron energy E. Nuclear transitions with $\omega=4$ mc^2 and of multipolarity E1 in a supercritical nucleus Z=184 with a radius R_n=10.88 fm are considered. The dashed curves denote the various contributions of the electron angular momentum states.

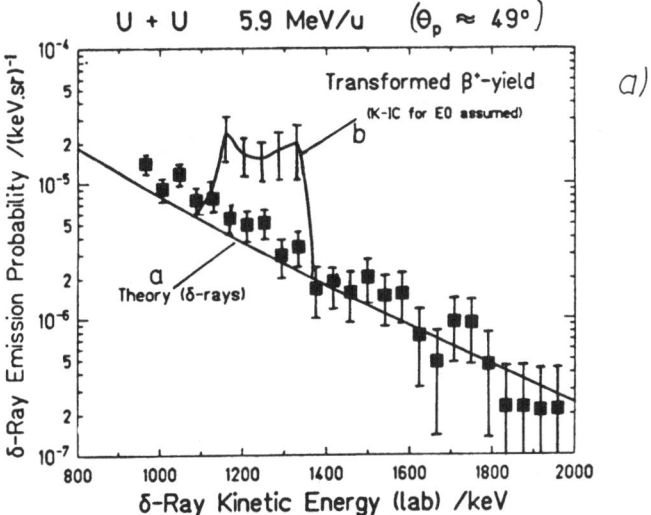

Figure 53: a) The δ-electron spectrum for U+U according to Kienle et al. [41]. The full curve shows the theory of Soff, Reinhardt et al. If the observed line structure in the positron would be due to E0-conversion, the indicated bumps in the δ-electron spectra should be seen. Obviously the δ-spectra are smooth. Consequently the *positron structures must come from the giant nuclear system.* b) Electron spectrum from 6.13 MeV/amu uranium and curium collisions measured June, 1983. The data are coincident with scattered particles in the kinematic region which Yields a 316 keV peak in the positron energy spectrum. The expected internal conversion electron peak is plotted over the data assuming the positron structure is due to pair conversion (PC) or monoenergetic pair conversion (MPC) of an E0 transition in the uranium nucleus. (Note: The positron linewidth argues against the PC process). The plotted peaks are calculated for 1 K-hole at the moment of E0 conversion. In order to explain both the positron peak intensity and the structureless electron spectrum, there must be at least 1.85 K-holes at the moment of E0 conversion (95% efficiency and detector resolution (after Backe, Bokemeyer, Cowan, Greenberg, Schweppe, Schwalm et al.).

to the upper continuum. If T is longer than the spontaneous decay width of the K-shell resonance the K-vacancy will be filled again leading to spontaneous positron emission. This is a sort of *"positron-gun" firing several shots,* the energy of which is supported by the excitation energy of the giant nuclear system.

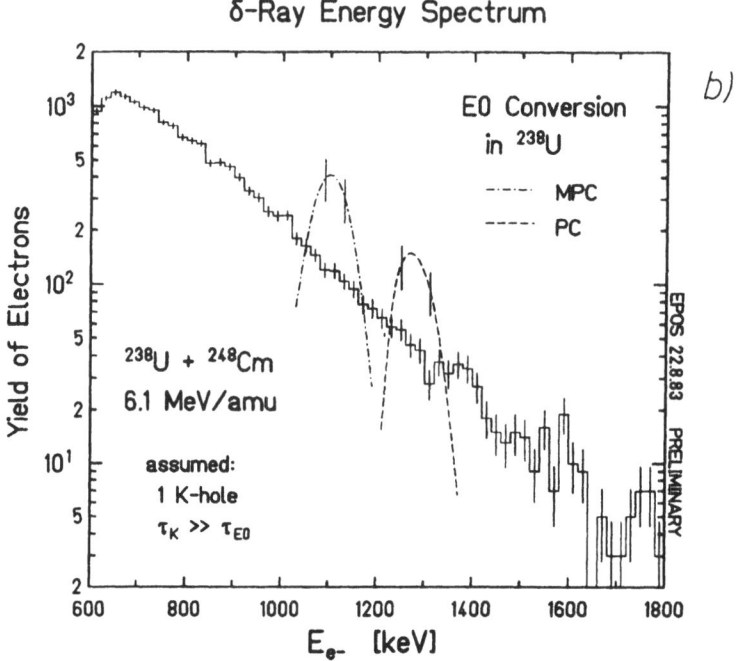

δ-Ray Energy Spectrum

For the nuclear charge distribution a homogeneously charged sphere with a radius R_n=10.88 fm has been assumed. In fig. 52 we show the differential pair conversion coefficient dβ/dE as function of the positron energy E. Nuclear transitions with $\omega=4m_0c^2$ and of multipolarity E1 are considered. The appearance of the pronounced peak at E=E$_{res}$ is striking. The dashed curves represent the various contributions of electron states with κ=-1 and κ≠-1. As expected, the resonance shows up only in the (κ=-1)-channel. Similar ratios dβ/dE are obtained for other multipolarities. Further details can be found in ref. [43].

We now come back to the conversion process in the nuclear compound system in the presence of a K-vacancy, which was discussed at the end of the previous section in the framework of a classical model.

In ordinary stable nuclei this process is rather slow, having transition times in the order of $\gtrsim 10^{-12}$ sec (corresponding to partial decay widths $\Gamma_{conv} < 10^{-3}$ eV). The contraction of bound and continuum

wavefunctions in superheavy atoms strongly enhances this process. This
is most prominent in electric monopole (E0-) conversion, where the width
to a good approximation is given by the simple expression

$$\Gamma_{conv}(i \to f, E0) = 2\pi \left| \frac{1}{6} e^2 \phi_f(0) \phi_i(0) R^2 \rho \right|^2 \tag{2}$$

with the nuclear E0-transition matrix element

$$\rho = \sum_p \int \Psi_f (r_p/R)^2 \Psi_i dV. \tag{3}$$

Γ_{conv} is proportional to the electron (positron) densities in the
initial and final state at the origin. A simple inspection of these
densities assuming ϕ_i to be a state in the positron continuum and ϕ_f
the 1s-state leads to the observation that $\Gamma_{conv}(E0)$ increases by
more than five orders of magnitude when going from a single U nucleus
to the combined Z=184-system, assuming constant ρ.

Depending on the strength of the nuclear matrix element this
means that the width of the conversion process can approach the
spontaneous decay width within one order of magnitude.

In the case of a subcritical charge, $Z < Z_{cr}$, positron lines could be
produced by monoenergetic conversion of excited nuclear states filling
a hole in, e.g., the 1s-, $2p_{1/2}$-, etc. inner shell levels. The width
of the line again would be inversely proportional to the lifetime T of
the nuclear system, provided this is shorter than the decay time of the
excited state. In the same way for supercritical Z the spontaneous decay
of the 1s-resonance could be accompanied by a conversion process
leading to one (or more) additional weaker lines, the relative intensity
being determined by the ratio $\Gamma_{spont}/\Gamma_{conv}$. Here also a new physical
phenomenon is possible, as indicated in the right part of fig. 55.
In an inverse conversion process the nucleus may take up energy released
by the filling of a 1s-hole. The emitted positron has an energy reduced
by the absorbed amount E_N.

Under favourable circumstances, therefore, 'sidebands' might appear in the positron spectrum (Raman- and Stokes- lines - see fig. 49), *allowing for a spectroscopy of the compound system*. To gain a full understanding of these processes it will be necessary to develop a theory accounting for both the classical and quantal aspects of the combined electron and nucleus system in a heavy ion collision.

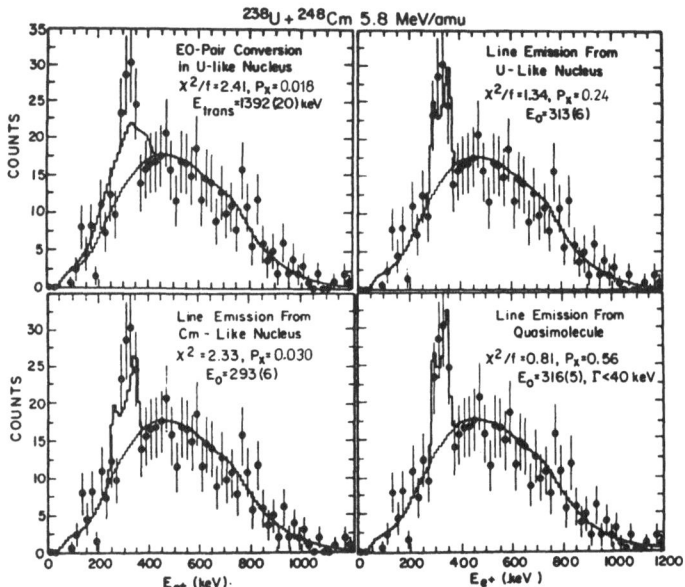

Figure 54: Various fits to the positron line structure according to the possible various interpretations. If the positron line stems from the giant nuclear system (vacuum decay) by far the highest confidence ($P_X = 0.56$) is reached - see lower right corner. (According to the Greenberg–Schwalm–group).

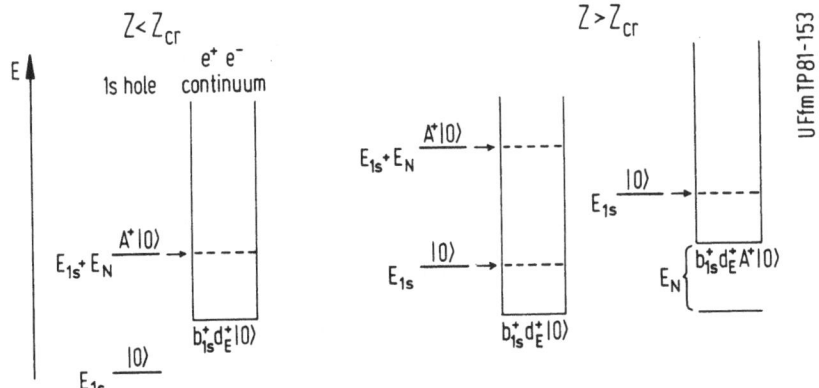

UFfmTP81-153

Figure 55: Monoenergetic pair conversion filling a hole in the 1s-state, induced by a nuclear transition with energy E_N, in a subcritical (left) and supercritical (centre) system. In the latter case also the inverse process is possible, where the nucleus becomes excited while a positron with reduced energy is emitted (right).

XIII. ON THE EXISTENCE OF GIANT NUCLEI AND GIANT NUCLEAR MOLECULES

The question for us nuclear physicists is how it can come about that two such very heavy nuclei can stick together for a time longer than 5×10^{-20} sec., probably even longer than 10^{-19} sec. These are the typical times deduced from the sharp line structure of the positron spectra. At first this seems so unlikely, that one would like to dismiss such a proposal right away. I would like to present some ideas we have worked out recently with Martin Seiwert, Nagwa Abou, Neise, Joachim Maruhn (Frankfurt a.M.) and Volker Oberacker (Vanderbilt University, Nashville, TN) [44]. The problem is to calculate the nucleus-nucleus interaction potential. We first followed the path known from literature and first described many years ago by Scheid and myself, namely to calculate the folding potential of two approaching deformed nuclei. Using the Y3-M interaction V(1,2) whose parameters were fitted to medium heavy elastic ion-ion scattering by Satchler; one calculates

$$V(R) = \int \rho(\vec{r}_1, \vec{R}) \, V(1,2) \rho(\vec{r}_2, \vec{R}) d^3\vec{r}_1 d^3\vec{r}_2 \; . \tag{1}$$

The densities $\rho(r,R,c,t)$ are deformed Fermi densities; c is the radius parameter $c=c_0(1 + \beta\,Y_{20}(\theta,\phi)$ and t is the surface thickness. A typical result is shown in fig. 56 for U+U and various orientations of the two Uranium nuclei. Clearly, one can believe such potentials only up to half density overlap, indicated by arrows in fig. 56; the extremely negative binding at small distances is unrealistic. The question now arises how to take the wrong parts out of the potential. To proceed we recall two facts: First, note that for a homogenous charge distribution $\bar\rho(r_1)$ the integral

$$\int \bar\rho(\vec{r}_1) v(\vec{r}_1\vec{r}_2)\bar\rho(\vec{r}_2) = a_v\,\frac{4\pi}{3}R^3 + a_s\,4\pi R^2 ,$$

where

$$a_v = 2\pi\rho_0^2 \sum_{i=1}^{2} v_i\mu_i^2 - \frac{V_3}{2}\rho_0^2 = -3.884\left[\frac{\text{MeV}}{\text{fm}^3}\right],$$

and

$$a_s = -\pi\rho_0^2 \sum_{i=1}^{2} v_i\mu_i^3 = 1.399\left[\frac{\text{MeV}}{\text{fm}^3}\right] \tag{2}$$

i.e. one gets a volume and a surface term which are functions of the force parameters. Second, remember that we have examples already in physics how to get rid of wrong results: *We renormalize*. This is done so in quantum electrodynamics and, perhaps more familiar to us, in the calculation of the shell corrections. In the latter case we renormalize by subtracting from the single particle sum a smoothed sum, i.e.

$$\Delta_{sc}(R) = \sum_i \varepsilon_i(R) - \sum_i \tilde\varepsilon_i(R,\gamma). \tag{3}$$

Here γ is a more or less phenomenological smoothing parameter. The shell correction $\Delta_{sc}(R)$ is then added to the "average potential" calculated in the empirical liquid drop model. In this way we ensure that the binding properties (Q-values) are properly contained in the potential.

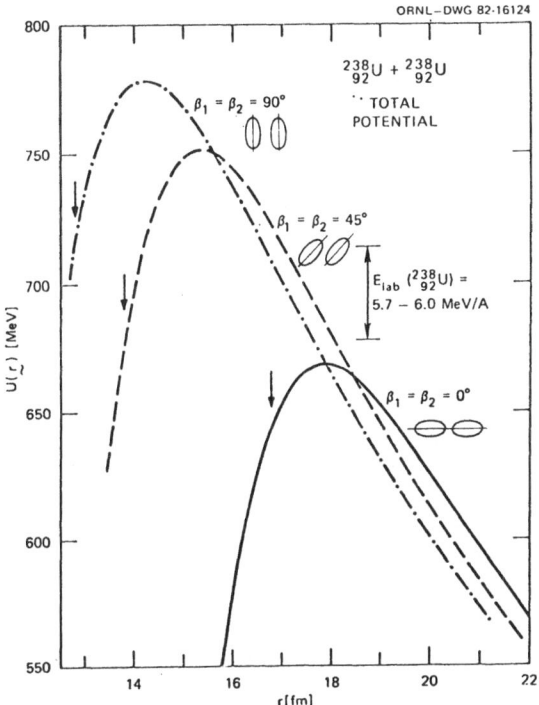

ORNL—DWG 82-16124

Figure 56 : Folding
potentials of two
deformed Uranium nuclei
for various orientations
The arrows indicate
halfdensity overlap.

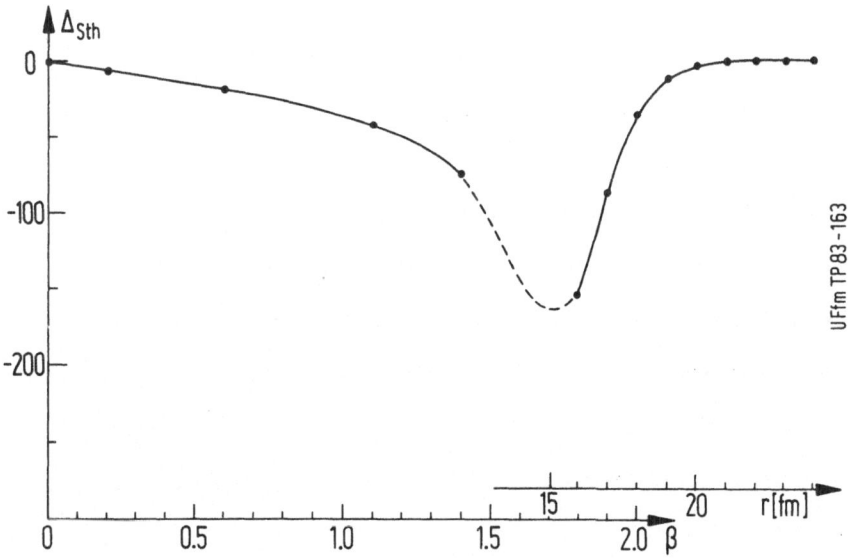

Figure 57: Surface thickness correction for U+U as a function of
distance.

We no now the same with the folding potential by calculating

$$\Delta_{sth}(R) = \int \rho(1,R)V(1,2)\rho(2,R)\ d\tau_1 d\tau_2 - \alpha \int \bar{\rho}(1,R)V(1,2)\bar{\rho}(2,R)d\tau_1 d\tau_2. \quad (4)$$

We call Δ_{sth} *surface thickness correction*. Note that we have taken the volume and surface terms out of the folding potential and simply keep the effect of the surface thickness. If the surface thickness vanishes, $\Delta_{sth}(R) \to 0$. The factor α is determined such that

$$\Delta_{sth}(R) \xrightarrow[\lim R \to 0]{} 0.$$

The typical result for $\Delta_{st}(R)$ is shown in fig. 57.

Adding this to the liquid drop potential, as in the well known shell correction method, the potentials for U+U of fig. 58 result. Again, various orientations are shown. Obviously *binding pockets of molecular type* appear. They have the following interesting properties: 1) For the head-on-configuration the potential pocket is about 20 MeV deep (this depends, of course, on the strength and range of the interaction, which is taken from Satchler's fits, as stated above). It lies precisely at the energy where the positron experiments are carried out, i.e. in the vicinity of 5.8 MeV/A. 2) For other orientations the pocket appears at higher energy, is deeper, and the outer barrier is narrower. The rise in energy is essentially a Coulomb energy effect. In the non-aligned orientations of fig. 58 the nuclei approach closer and hence the Coulomb energy rises (see fig. 59). At those orientations where the nuclear touch is especially intensive (this is because of the quadrupole and hexadecupole deformation of the Uranium nuclei), the potential pockets are considerably deeper and the outer barrier becomes narrower (see fig. 60, where this effect is once more stressed). Clearly this effect can be called *nuclear cohesion*.

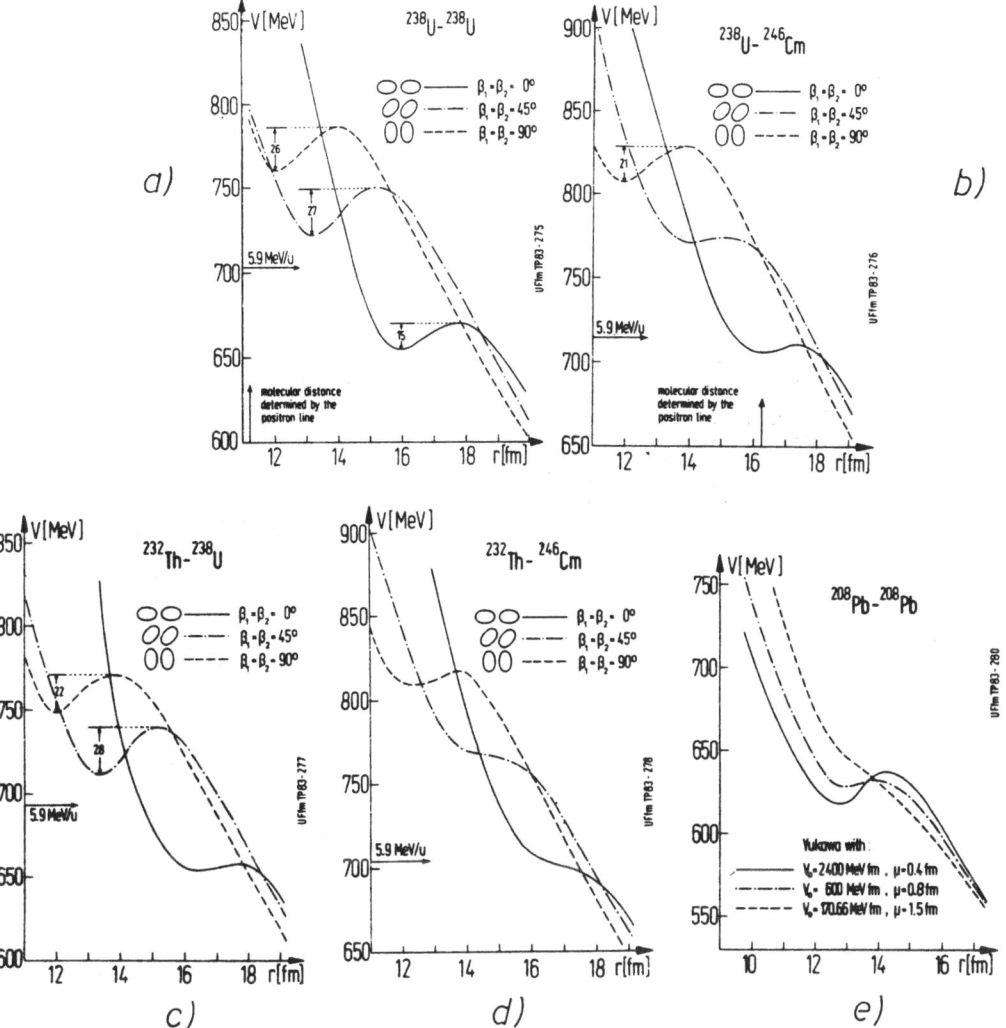

Figure 58: Nucleus-Nucleus potentials with surface thickness corrections for various orientations of the colliding nuclei. Several combinations leading to the giant nucleus domain are presented. The calculations were carried out by Martin Seiwert.

We are now led to the picture that two nuclei form a nuclear molecule of the type illustrated in figs. 59-61. Butterfly and belly dancer modes appear; also relative vibrations of the two nuclei against each other (β-vibrations of the giant system). In the latter case the distance R is oscillating. The former case has similarity with γ-vibrations of the giant system. They represent the remnants of the free rotations which the nuclei would have at large distance. At small distance only those "hindered" rotations survive. There are also the modes of the individual β- and γ-vibrations of the individual two nuclei. Hence we see that the spectrum of collective modes of the giant molecules is extremely rich. At the barrier, where the positron experiments are carried out,

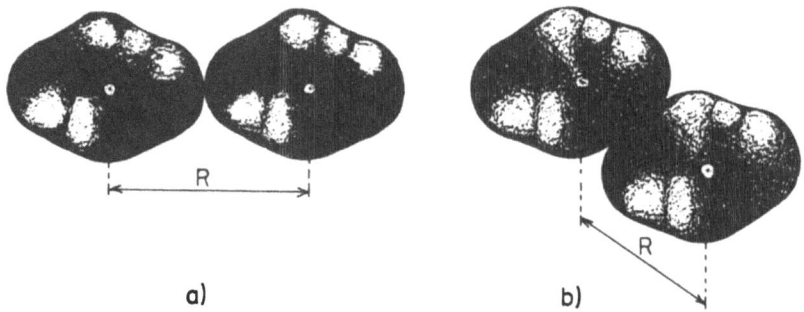

a)   b)

Figure 59: (a) Stretched U+U (b) U+U bended against each other. In the latter case the two nuclei are coming closer to each other; the Coulomb energy in case (b) is certainly higher than in case (a). Because of the quadrupole and hexadecupole deformation of the U-nuclei there are special orientations where the nuclear touching is especially intense.

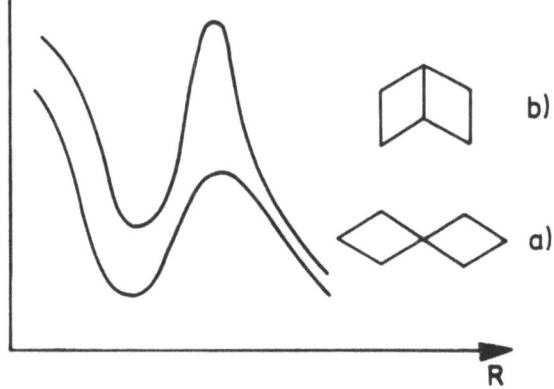

Figure 60: Schematic explanation of *nuclear cohesion*. In orientation (b) the nuclear cohesion force is much more active than in configuration (a).

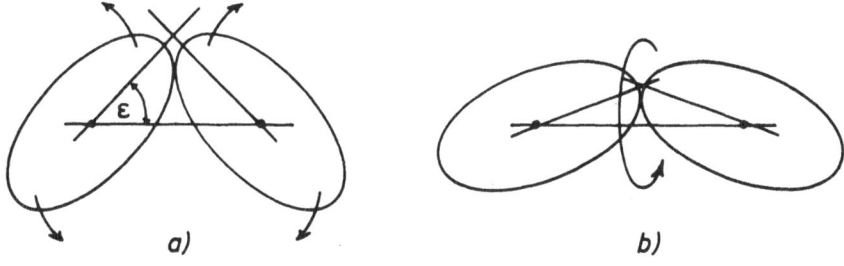

Figure 61: (a) Butterfly- and (b) belly dancer modes of a giant molecule consisting of two deformed nuclei. Their dynamical properties will be theoretically investigated in greater detail in the next section.

we have a highly dense collective spectrum with probably overlapping levels; thus giving rise to the large time delay observed (see fig.62). Because the moment of inertia is so large, the rotational bands are quenched; i.e. the first 2^+-state has an energy level of only a few keV. Similarly the β -vibrations have low energy of the order of 100-800 keV. Peter Hess (Frankfurt) and

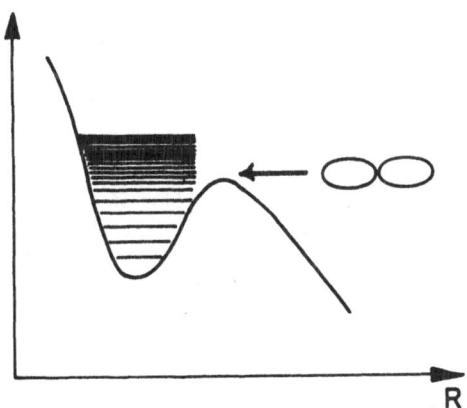

Figure 62: Schematic drawing of the spectrum revealing the high density of collective states of the giant molecule near the barrier.

T. Pinkston (Vanderbilt) investigated these spectra. Their results

will be described below. The richness of the level

structure we know for light nuclear molecules like the Si+Si-system

may have qualitatively a similar origin.

I would like to mention again that a number of these levels can

possibly be investigated experimentally in the future by positron

spectroscopy. The positron line due to spontaneous vacuum decay may

acquire satellites which result from the deexcitation of the giant

molecule (Raman-lines), so that spectra of the type qualitatively shown

in fig. 49 may arise [36].

Due to the time delay of two sticking nuclei the δ-electron spectrum

should show oscillations (fig. 63), because the incoming and outgoing

amplitudes for δ-electron creation now do interfere with a phase propor-

tional to the sticking time. The oscillation energy ΔE can be directly

related to the delay time $\Delta \tau$ by the simple relation

$$\Delta \tau = \frac{\Delta E}{\hbar} \, .$$
(5)

As mentioned already earlier, W. *König* (Heidelberg) reported some time

ago at the Schleching and Regensburg meetings that he found oscillations

of this kind in the deeply inelastic reaction of I+Au. He measured

coincidences of the δ-electrons with projectile and target in a certain

Q-value window. Furthermore the γ-spectrum has been measured and the

electrons stemming from conversion were subtracted. The thus resulting

δ-electron spectrum showed weak oscillations, which indicate a delay time

of the order of $\Delta \tau = 10^{-20}$ sec. These reported results are preliminary.

They could be the beginning of an interesting experimental endeavour;

bringing quantitative (absolute) time scales into nuclear reactions.

Actually, the "atomic clock" should work in the range between 10^{-19} and

10^{-21} sec.

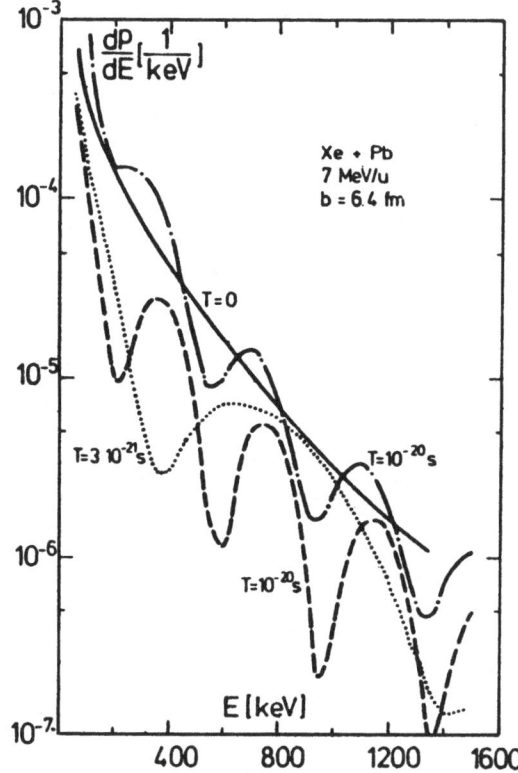

Figure 63: δ -electron spectrum without (straight line) and with time delay [37].

It is remarkable how close the calculated pocket barrier for the aligned position of the U-U or U-Cm system is compared to the energies of 5.7-6.1 MeV/A for the U-beams, at which the positron resonances have been observed. Too far under the barrier the nuclei undergo only Coulomb-deflections. At the Coulomb barrier only the "nose-nose" orientations can overcome the barrier and the positron line should be most pronounced. Too high above the barrier the lifetime of the giant system becomes smaller, and also many more orientations with rather different distances of closest approach lead to a smearing out of the positron line structure. This, is, indeed, what has first been observed by Kienle et al. for the U+U-system and by Greenberg, Schwalm and their associates for the U+Cm-system (see fig. 64). This *threshold-effect gives further evidence for the existence of a nuclear pocket.* The calculated U+Cm-pocket gives exactly the position where the dominant

positron line is observed (320 keV). This is not so for the U+U-system,

where the present calculations indicate that the spontaneous positron

line should be around 180 keV. What is then the observed structure at

300 keV in the U+U spectrum and also the ones appearing at higher

energies? Well, one could imagine that those lines are Raman-lines,

reflecting the excitations of the giant molecular system. Their strong

intensity speaks, at the present

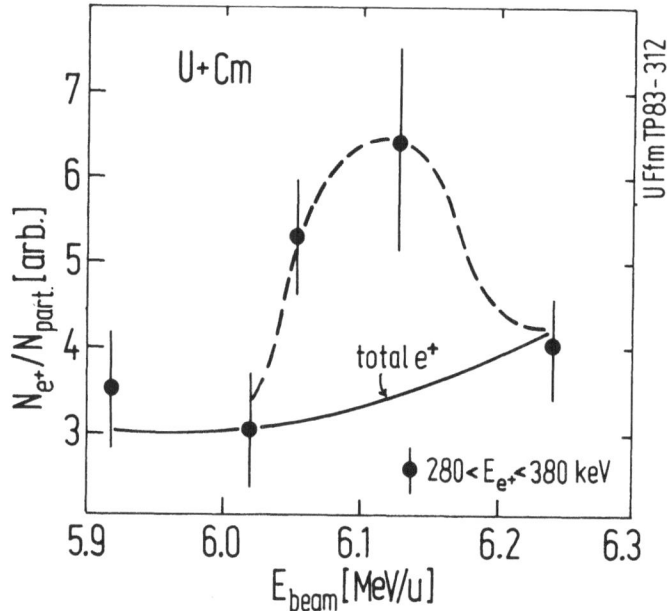

Figure 64: The threshold effect of the appearance of the positron
line structure as a function of the projectile energy [42] indicates
a nuclear pocket.

level of our understanding against such an interpretation. It could also

be possible that there exist other, more spherical quasibound configura-

tions (isomers) of the giant systems, which then would emit spontaneously

positrons at higher energies. The standard, homogenously charged giant

compound nuclei, stabilized due to shell corrections are expected at an

energy of 900 MeV, which is about 180 MeV too high to be considered as a

213

possible candidate for those positron emitters (see fig. 65). However,

due to the strong Coulomb-repulsion it is not unlikely that rather

hollow or torus-like structures could exist as

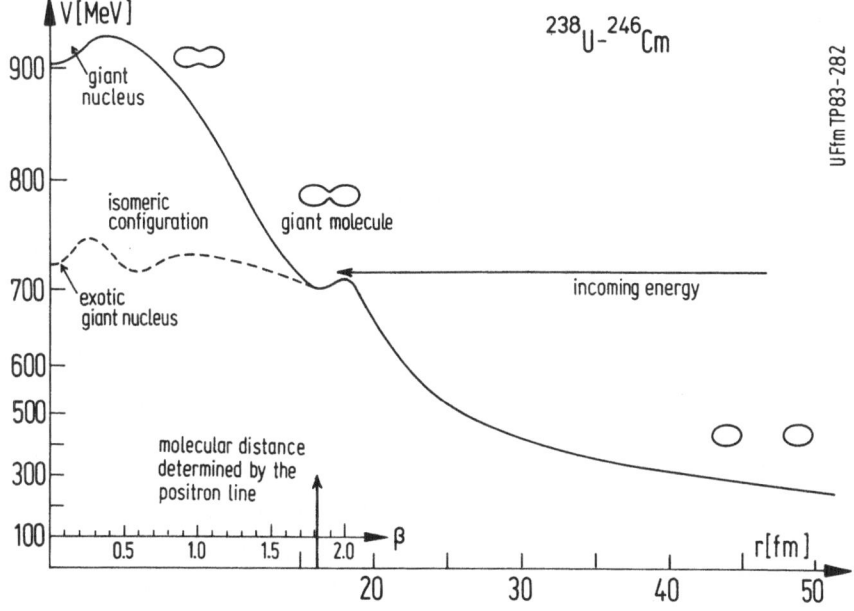

Figure 65 : Potential with giant molecular and giant nucleus
pocket for U+U. The dashed line shows speculations on more
spherical exotic giant structures.

giant nuclear systems. Such speculations are indicated by dashed

lines in fig. 65. Future detailed experiments and also more theore-

tical studies have to clarify these conjectures. It becomes clear,

however, that the *spontaneous positrons*, being carriers of one of the

most fundamental messages, namely the decay of the neutral into a

charged vacuum, *can also be messengers of a most fascinating*

future of nuclear physics; namely of giant nuclei and molecules.

XIV. EXTENSIONS OF THE SEMI-CLASSICAL APPROACH. THE INTERPLAY
      BETWEEN REACTION-DYNAMICS AND POSITRON SPECTROSCOPY

Recently U. Heinz, J. Reinhardt, et al. extended the theory

towards a quantum mechanical treatment of the nuclear relative motion,

with the aim to incorporate in a quantum mechanical way the nuclear

physics responsible for the sticking of the two nuclei, and to gain

insight into the quantum mechanical origin of the classical nuclear

time delay. A very general framework, suitable for the description

of any nuclear and atomic interference phenomena, was derived [45].

Within that framework it was shown that under the assumption that

the atomic and nuclear excitations can be localized in different

spatial regimes (small and large internuclear distances), the *atomic*

excitation amplitude from the two-center eigenstate i to state f in

collisions *with nuclear sticking* can be written as [46]

$$a_{i \to f}^{\ell} (\infty, \infty) = - \sum_n a_{i \to n}^{-\ell}(\infty, R_m) \, e^{2i\delta_\ell^n} \, a_{n \to f}^{+}(R_m, \infty) \, . \quad (1)$$

if the scattering occurs with angular momentum 1. Here $a_{i \to n}^{-\ell} (\infty, R_m)$

is the atomic excitation amplitude on the way in from $R = \infty$ to $R = R_m$

(where R_m is a matching radius of nuclear dimensions), and $a_{n \to f}^{+\ell}(R_m, \infty)$

is the atomic amplitude on the way out from R_m to ∞. *The $a^{\pm}$ can be*

determined semiclassically with sufficient accuracy. δ_ℓ^n is a nuclear

phase shift, describing the nuclear scattering at distances $R < R_m$;

its dependence on the intermediate electronic channel n is via the

available energy in that channel at R_m:

$$\delta_\ell^n \equiv \delta_\ell \, (E - \varepsilon_n(R_m)). \quad (2)$$

These can in principle be determined from the internuclear potential.

We consider the case of a pocket in that potential (as in fig. 65) which

can support many vibrational and rotational states. Then we can

parameterize δ_ℓ as a superposition of Breit-Wigner shaped resonances in a set of, say, N rotational bands on top of vibrational states with energies E_n (n=1,..,), whose energies are $E_{n\ell} = E_n + \gamma(\ell(\ell+1))$ (γ is the rotational constant, about 0.7 keV for the U-U quasimolecule), and whose width we call $\Gamma_{n\ell}$:

$$S_\ell(E) \equiv e^{2i\delta_\ell(E)} = \prod_{n=1}^{N} \frac{E - E_{n\ell} - \frac{i}{2}\Gamma_{n\ell}}{E - E_{n\ell} + \frac{i}{2}\Gamma_{n\ell}} \qquad (3)$$

These resonances will be very dense, and in experiment one will always average over many resonances (e.g. the GSI beam has as an energy spread of the order of 10 MeV). The thus *averaged excitation cross-section* with (26) takes the form

$$\frac{d\sigma_{i \to f}}{d\Omega_N} \simeq \frac{1}{4k^2} \sum_{nn'} a^-_{i \to n'} a^+_{n' \to f} a^{-*}_{i \to n} a^{+*}_{n \to f} \cdot$$

$$\cdot \sum_{\ell\ell'} (2\ell+1)(2\ell'+) P_{\ell'}(\cos\theta) e^{2i(\sigma_{\ell'} - \sigma_\ell)} \qquad (4)$$

$$\cdot \langle S_{\ell'}(E - \epsilon_{n'}) S_\ell^*(E - \epsilon_n) \rangle \cdot$$

σ_ℓ is the Coulomb phase shift, which is slowly varying with beam energy and can be taken out of the energy average. Also the $a^\pm$ are weakly dependent on E and ℓ and have thus been taken out of the ℓ-sum and energy average. $\langle S_\ell S_{\ell'}^* \rangle$ defines a nuclear autocorrelation function; its Fourier transform with respect to the difference in the energy arguments can be interpreted as a distribution of nuclear delay times [47]. It can be analytically calculated from the model [46], using

$$\langle f(E) \rangle \equiv \frac{I}{\pi} \int_{-\infty}^{\infty} \frac{d\epsilon f(\epsilon)}{(\epsilon-E)^2 + I^2} \quad ; \quad (I \sim 10 \text{ , eV}) \qquad (5)$$

as the prescription for the energy average. The result for the excitation cross section can be written as [46]

$$\frac{d\sigma i\rightarrow f}{d\Omega N} = \int_0^\infty dT \ |a_{i\rightarrow f,T}|^2 \ \frac{d\sigma_{delayed}}{d\Omega_N}(\theta,T) \tag{6}$$

It *separates incoherently* into a *direct part*, due to pure Coulomb-scattering without time delay, and a *delayed part due to resonance scattering*. In the latter contribution $|a_{i\rightarrow f},T|^2$ is the semiclassical excitation probability corresponding to a sharp classical delay time T. It is weighted with a different delayed nuclear cross section for every T [46] :

$$\frac{d\sigma_{delayed}}{d\Omega_N}(\theta_1,T) = \frac{1}{4k^2}\left[\sum_\ell (2\ell+1) \ e^{2i\sigma_\ell} \ P_\ell(\cos\theta)\alpha_\ell^E(T)\right]^2 , \tag{7}$$

where $\alpha_\ell^E(T)$ results from the model (3) as

$$\alpha_\ell^E(T) = \sum_{n=1}^N \frac{\sqrt{2I}\,\Gamma_n}{E-E_{n\ell}+iI} \ e^{-iE_{n\ell}T} \ e^{-\frac{1}{2}\Gamma_{n\ell}T} \ .$$

$$\cdot \prod_{m\neq n}^N \frac{E_{n\ell}-E_{m\ell}-\frac{i}{2}(\Gamma_{n\ell}+\Gamma_{m\ell})}{E_{n\ell}-E_{m\ell}-\frac{i}{2}(\Gamma_{n\ell}-\Gamma_{m\ell})} \ . \tag{8}$$

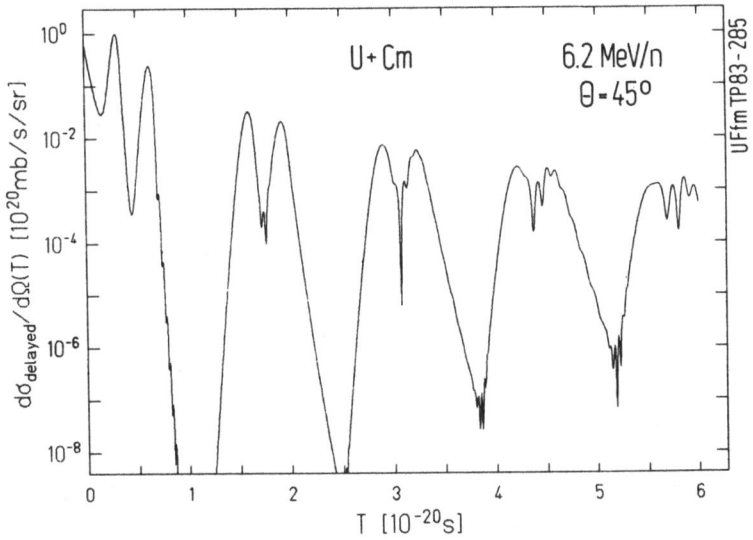

Figure 66: Delayed nuclear cross section as a function of sticking time at Θ=45° for one band of rotational states. The U+Cm system at 6.2 MeV/N is considered. A typical light-house effect (damped) can be recognized.

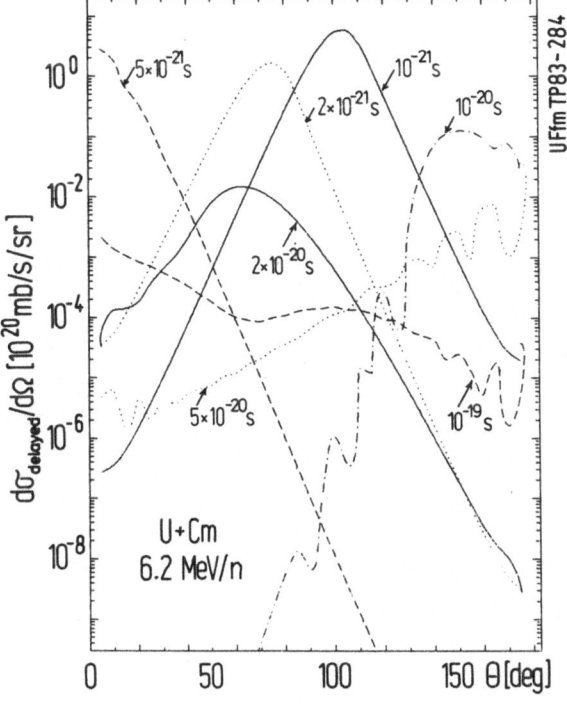

Figure 67: Angular distribution of delayed nuclear cross section for different sticking times of the U+Cm-system at 6.2 MeV/N.

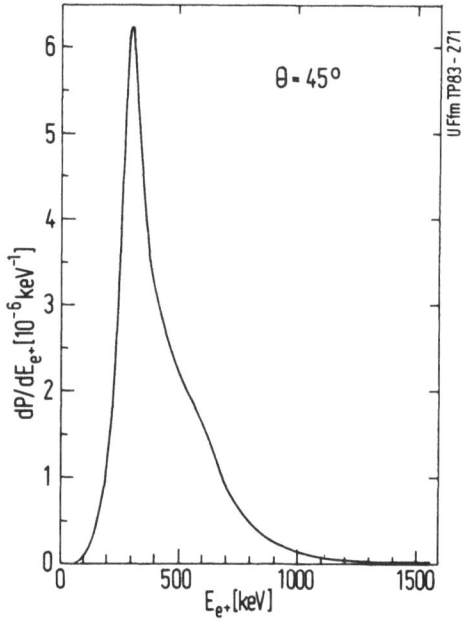

$\theta = 45°$

UFfm TP83-271

Figure 68: Positron spectrum
for the delayed nuclear cross
section of fig. 43. The different lines denote different
upper limits for the time-inte-
gral. The highest line corresponds to the positrons created in all events with sticking
times $\leq 6*10^{-20}$sec. The curves
are normalized to the delayed
nuclear cross section up to
$T=6*10^{-20}$sec.

In figs. (66-69) we show the results from a sample calculation for U-Cm

collisions at 6.2 MeV/N beam energy (E_{CM}=750 MeV) with the assumption

that the maximum of the potential barrier is at V_{max}=725 MeV, and that

the pocket supports one rotational band with band head energy 8 MeV below

the barrier. The widths have been computed by the Hill-Wheeler formula

for a parabolic potential barrier. In fig. 66 we show the time distri-

bution for the delayed nuclear cross section [32] at θ =45°. The

regular peaks can be interpreted as a lighthouse effect generated by a

nuclear molecule rotating with a rather well defined mean angular

momentum $\bar{\ell}$ =226$\hbar$ and decaying after a different number of revolutions

under the chosen scattering angle. The width of the state with

$\bar{\ell}$=226h is 24 keV, explaining the possibility of many revolutions. The

position of the peaks can be explained by observing that pure Coulomb

scattering (T=0) corresponds to θ_c (ℓ=226) $\cong$151°. In fig. 67 we show the angular distribution of the nuclear delayed cross section for fixed

times. One sees very *pronounced dependence on the scattering angle.*

The position of the peaks (which are sharp for small T, but are smeared

out for long T) is consistent with the lighthouse interpretation. Fig.

68 shows a corresponding positron spectrum. The long-time part of the

time-distribution [32] generates a well defined positron peak at the energy

of the spontaneous decay line, whose width has the correct order of magni-

tude (a few tens of keV).

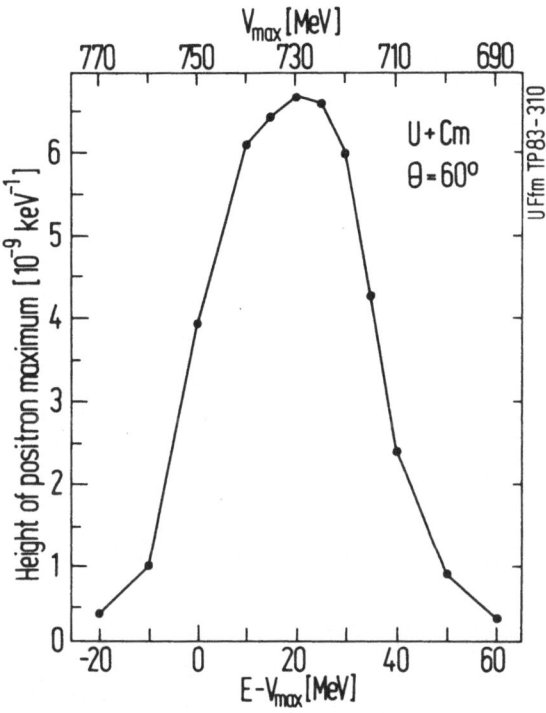

Figure 69: Excitation
function for the sponta-
neous positrons from de-
layed collisions. This
calculation involves 10
bands of rotational states
with band head energies
1,2..., 10 MeV below the
potential barrier. Where-
as the absolute number
for dP/dE may not be
taken too seriously, the
characteristic change of
the height of the positron
peak as a function of
beam energy is a system-
atic feature of the model.
The calculation is by Heinz,
Reinhardt et al. [46].

The simple picture of figs. 68, 67 gets washed out, if one considers

bands of rotational states, due to interferences. What remains, though,

is the *strong dependence on the nuclear scattering angle,* and *a positron*

peak with reasonable widths at some angles. This is in qualitative

agreement with the reported dependence of the experimental positron

spectra on the nuclear scattering angle. One also observes a *quite*

sensitive dependence of the spectra on the beam energy, which is also

supported by experiment [41, 42] (fig. 64). However, problems with the
absolute cross section for the spontaneous positrons, as compared with
the experiments, still remain. They can probably be solved by
including channels similar to those known from the double-resonance-
mechanism in ordinary nuclear molecular reactions.

XV SUMMARY AND OUTLOOK

We shall summarize our presentation with the following statements:

1) The *vacuum structure is most fundamental* for the understanding
   of the physical world.

2) In *overcritical external gauge* fields the vacuum undergoes
   massive changes;   the neutral *vacuum decays* into a charged
   vacuum by emission of antiparticles.   The particle creation
   process will continue until either the potential difference in
   the gaugefield is reduced or the Pauli principle prevents
   further particle creation.

3) If the Dirac field occupies a new ground state, then we speak
   of *dynamically broken symmetry*. The new ground state is called
   supercritical vacuum.   (Charged Vacuum).

4) Symmetry breaking in this sense does not occur in strong gravitational fields (without torsion) because gravity does not
   distinguish between different kinds of particles nor between
   particles and antiparticles.   Therefore, globally static gravitational fields can never lead to spontanous pair creation. This
   very property of the gravitational interaction is also responsible for the inevitable breakdown of global time-independence
   when the strength of gravitational fields exceeds a certain
   limit, so that an event horizon is formed (see [50-54]). The
   Schwarzschild radius separates then a region of static field

from a region where spacetime is intrinsically time dependent. This dependence leads to paircreation, but this process is of basically different (dynamic) nature compared to the case of supercritical electric fields [50].

5) Due to the presence of an event horizon it may be understood as being of topological origin as one has discussed for the rather trivial case of Rindler space [50].

6) *Supercritical gluo-electric fields* in deformed bags lead to spontanous $q\bar{q}$-creation connected with the fission of the bag. This *helps to understand how confinement works.* [48]

7) *Quantum electrodynamics of strong, supercritical* fields plays a distinguished role as the one example that is amenable to *tests in the laboratory.*

8) For $Z > Z_{cr} \approx 173$ the neutral *electron-positron vacuum decays* into a charged vacuum.

9) Important for the test in the laboratory is that in collisions of very heavy ions $(Z_1 + Z_2 > Z_{cr})$ superheavy *(giant) electronic quasimolecules* are formed. The *Two-Center-Dirac-Equation* and the corresponding correlation diagram constitute the theoretical basis of these quasimolecules.

10) The quasimolecules have been tested experimentally through the study of quasimolecular x-rays, inner shell (particularly K-) vacancy production and δ-electron production. The agreement between theory and many experiments is quantitative.

11) The positron production spectrum in a heavy ion collision consists of a *dynamical* and a *spontaneous component.* For Rutherford trajectories the spontaneous component, though important, can hardly be distinguished from the dynamical

spectrum. The latter ("shake off" of vacuum polarization) increases with a very high power of the total charge, namely $(Z_1+Z_2)^{20}$.

12) All the theoretically *predicted features of the dynamical positron spectrum have been experimentally confirmed.*

13) In order to obtain signals for the vacuum decay (spontaneous positrons) a new idea had to be invoked, namely the *formation of giant nuclear systems* (sticking of nuclei, time-delay, formation of giant molecules or other exotic structures like hollow nuclei or toroidal nuclei etc.).

14) Line structures in the positron spectrum have indeed been experimentally veryfied by Kienle et al. and by Greenberg, Schwalm et al.

15) The *positron line-structures are experimentally shown to stem from the giant nuclear system,* because a) the main structures appear only in overcritical systems, b) the δ - electron spectrum is smooth, c) the x-ray-spectrum is smooth, d) a typical threshold effect as a function of the ion energy is observed, e) the width is very small (≤ 30 keV).

16) *Nuclear pockets in the interaction* of very heavy ions seem indeed to be theoretically possible. For deformed nuclei they are orientation dependent. The predicted energy for the spontaneous emission line agrees with experiment for U+Cm, but not yet for U+U, U+Th.

17) Additional *Raman-type structures* due to supercritical conversion are theoretically predicted and likely to be already seen in the experimental positron spectrum. This opens up the possibility of a spectroscopy of the *giant systems*, which

may appear in various isomeric forms.

18) A first *quantum mechanical theory reproduces* (in a yet
schematic approach) all *essential experimental features* and
helps to understand the observations. In particular it is
predicted that those states of the giant nuclear system are
most dominant, which are at (or slightly under) the
outer barrier.

19) The formation of giant nuclear molecules with old (β-,γ-
vibrations, rotations) and new type (butterfly, belly-dancer)
modes of excitations are certainly the one species of giant
system, which immediately explains the observed break-up of
that system close to the entrance configuration.

20) We conclude, that *the vacuum decay in supercritical electric
fields of giant nuclei has been observed*. This is the
conclusion which can be drawn of the presently existing
experimental facts. The spontaneous positrons will in future
serve as a tool for a spectroscopy of the giant systems, which
may, indeed have lifetimes $\tau \gg 10^{-19}$ sec.

Thus a newly discovered fundamental process of field theory
helps to make an *equally basic discovery in nuclear physics*
and opens the possibility for a most exciting future consisting
in the identification of the wealth of structure nature hides
in giant nuclear systems. Information coming from clusters of
nuclear matter practically twice as large as available in the
present periodic system can shed new light on our understanding
of nuclear matter and be quite valuable for astrophysics.

References

1) G. Soff, J. Reinhardt, B. Müller and W. Greiner, Phys.Rev. Lett.
    <u>38</u> (1977) 592.

2) For a quasi-recent review see: J. Reinhardt and W. Greiner, Heavy
    Ion Atomic Physics, appearing in Heavy Ion Sciences, ed. D.A. Bromley,
    Plenum Press, in print. See also J. Reinhardt and W. Greiner,
    Reports on Progress in Physics <u>40</u> (1977) 219.

3) H. Backe, L. Handschug, F. Hessberger, E. Kankeleit, L. Richter,
    F. Weik, R. Willwater, H. Bokemeyer, P. Vincent, Y. Nakayama,
    and J.S. Greenberg, Phys. Rev. Lett. <u>40</u> (1978) 1443.

4) C. Kozhuharov, P. Kienle, E. Berdermann, H. Bokemeyer, J.S.Greenberg,
    Y. Nakayama, P. Vincent, H. Backe, L. Handschug, and E. Kankeleit,
    Phys. Rev. Lett. 42 (1979) 376.

5) W. Heisenberg and H. Euler, Z. Physik, 98:714 (1936).

6) V. Weisskopf, Phys. Z., 34:1 (1933).

7) L. I. Schiff, H. Snyder and J. Weinberg, Phys. Rev.,
    57:315 (1940).
    I. Pomeranchuk and J. Smorodinsky, J. Phys. USSR, 9:97 (1945).
    J. Schwinger, Phys. Rev., 82:664 (1951).
    V.V. Vorokov and N.N. Kolesnikov, Sov. Phys. JETP 12:136 (1961).
    F. Beck, H. Steinwedel and G. Süßmann, Z. Physik 171 (1963) 189
    A. I. Akhiezer and V.B. Berestetskii, Quantum Electrodynamics,
    Interscience Publishers, New York (1965).

8) W. Pieper and W. Greiner, Z. für Physik <u>218</u> (1969) 327.

9) B. Müller, H. Peitz, J. Rafelski, and W. Greiner, Phys.
    Rev. Letters, 28:1235 (1972).

10) B. Müller, J. Rafelski, and W. Greiner, Z. f. Physik, 257:62 (1972).

11) Y.B. Zeldovich and V.S. Popov, Soviet Phys. Uspekhi, 14:673 (1972).

13) W. Greiner (ed.), Quantum Electrodynamics of Strong Fields,
    Plenum Press (1983).

14) U. Fano, Phys. Rev. 124 (1961) 1866.

15) W.L. Wang and C.M. Shakin, Phys. Lett. 32B (1970) 421.

16) J. Reinhardt, B. Müller and W. Greiner, Phys. Rev. A24 (1981) 103.

17) J. Rafelski, B. Müller and W. Greiner, Nucl. Phys. B38 (1974) 585.

18) B. Müller, J. Rafelski and W. Greiner, Nuovo Cimento 18A (1973) 551.

19) W. Greiner, B. Müller and J. Rafelski, Quantum Electrodynamics
    of Strong Fields, Springer-Verlag, Berlin 1984.

20) J. Schneinger, Phys. Rev. 94 (1954) 1362.

21) P. Gärtner, U. Heinz, B. Müller and W. Greiner, Z. Physik A300 (1981) 143.

22) P. Gärtner, J. Reinhardt, B. Müller and W. Greiner, Phys. Lett. 95B (1980) 181.

23) G. Soff, J. Rafelski and W. Greiner, Phys. Rev. A7 (1973) 903.

24) J. Rafelski, L.P. Fülcher and W. Greiner, Phys. Rev. Letters 27 (1971) 958.

25) S.S. Gershtein and Y.B. Zeldovich, JETP 30 (1970) 358.

26) B. Müller and W. Greiner, Zeitschr. f. Naturforschung 31a (1976) 1.

27) W.E. Meyerhof, T.K. Saylor, and R. Anholt, Phys. Rev. A12 (1975) 2641.

28) See P. Vincent in: Quantum Electrodynamics of Strong Fields, ed. W. Greiner, Plenum Press, 1983, p. 359.

29) D. Liesen, P. Armbruster, F. Bosch, S. Hagmann, P.H. Mokler, H.J. Wollersheim, H. Schmidt-Böcking, R. Schuch, and J.B. Wilhelmy, Phys. Rev. Lett. 44 (1980) 983.

30) F. Güttner, W. Koenig, B. Martin, B. Povh, H. Skapa, J. Soltani, Th. Walcher, F. Bosch, and C. Kozhuharov, Z. Phys. A304 (1982) 207;
C. Kozhuharov in: Physics of Electronic and Atomic Collisions, ed. S. Datz, 1982, p. 179.

31) T.H.J. de Reus, J. Reinhardt, B. Müller, W. Greiner, G. Soff, and U. Müller, J. Phys. B: Atom. Mol. Phys. (in press);
G. Soff, J. Reinhardt, B. Müller, and W. Greiner, Z. Physik A294 (1980) 137.

32) J. Reinhardt, B. Müller, and W. Greiner, Phys. Rev. A24 (1981) 103.

33) T. Tomoda and H.A. Weidenmüller, Phys. Rev. A26 (1982) 162.

34) J. Rafelski, B. Müller, and W. Greiner, Z. Physik A285 (1978) 49.

35) H. Backe, W. Bonin, E. Kankeleit, M. Krämer, R. Krieg, V. Metag, P. Senger, N. Trautmann, F. Weik, and J.B. Wilhelmy in: Quantum Electrodynamics of Strong Fields, ed. W. Greiner, Plenum Press, 1983, p. 107.

36) J. Reinhardt, U. Müller, B. Müller, and W. Greiner, Z. Physik A303 (1981) 173;
U. Müller, G. Soff, T. de Reus, J. Reinhardt, B. Müller, and W. Greiner, Z. Physik A (in press).

37) G. Soff, J. Reinhardt, B. Müller, and W. Greiner, Phys. Rev. Lett. 43 (1979) 1981.

38) J. Kirsch, B. Müller, and W. Greiner, Phys. Lett. 94A (1983) 151.

39) R. Schmidt, V.D. Toneev, and G. Wolschin, Nucl. Phys. A311
(1978) 247.

40) H. Backe, P. Senger, W. Bonin, E. Kankeleit, M. Krämer,
R. Krieg, V. Metag, N. Trautmann, and J.B. Wilhelmy, Phys. Rev.
Lett. 50 (1983) 1838.

41) P. Kienle in: Quantum Electrodynamics of Strong Fields, ed.
W. Greiner, Plenum Press, 1983, p. 293; E. Berdermann, M. Clemente,
P. Kienle, H. Tsertos, W. Wagner, F. Bosch, C. Kozhuharov, and
W. Koenig, GSI Scientific Report 83-1, 147; see also 82-1,
138 and 81-2, 128; : Physics Letters, January 1984.

42) H. Bokemeyer, K. Bethge, H. Folger, J.S. Greenberg, H. Grein,
A. Gruppe, S. Ito, R. Schule, D. Schwalm, J. Schweppe, N. Trautmann,
P. Vincent, M. Waldschmidt in: Quantum Electrodynamics of Strong
Fields, ed. W. Greiner Plenum Press, 1983, p. 273; H. Bokemeyer,
H. Folger, H. Grein, Y. Kido, T. Cowan, J.S. Greenberg, J. Schweppe,
K. Bethge, A. Gruppe, R. Merten, Th. Odenweller, K.E. Stiebing,
D. Schwalm, P. Vincent, and N. Trautmann, GSI Scientific Report
83-1, 146 and 82-1, 139; : Phys. Rev. Lett. 51 (1983) 2261.

43) G. Soff, P. Schlüter, and W. Greiner, Z. Physik A303 (1981) 189;
P. Schlüter, Th. de Reus, J. Reinhardt, B. Müller, and G. Soff,
Z. Physik A (in press).

44) M.J. Rhoades-Brown, V.E. Oberacker, M. Seiwert, and W. Greiner
Z. Physik A310 (1983).

45) U. Heinz, B. Müller, and W. Greiner, Ann. Phys. (in press).

46) U. Heinz, J. Reinhardt, B. Müller, W. Greiner, and W.T. Pinkston,
Quantum Mechanics of the Time Structure in Heavy Ion Collisions
with Nuclear Contact, to be published.

47) J. Reinhardt, B. Müller, W. Greiner, and U. Müller, Phys. Rev. A
(in press).

48) D. Vasak, K. -H. Wietschorke, B. Müller, and W. Greiner, Z. Physik
C (in press).

49) A. Chodos, R.L. Jaffe, K. Johnson, C.B. Thorn, and V. Weisskopf,
Phys. Rev. D9 (1974) 3471.

50) M. Soffel, B. Müller, and W. Greiner, Phys. Rep. 85 (1982) 51.

52) S. Hawking, Nature 248 (1974) 30.

53) P.C. Davies, S. Fulling, and W. Unruh, Phys. Rev. D13 (1976) 2720.

54) B. de Witt, Phys. Rep. 19C (1975) 6.

ON THE STRUCTURE OF GIANT NUCLEAR MOLECULES

W. Greiner
Institut für Theoretische Physik
J.W. Goethe Universität Frankfurt, West Germany

In cooperation with P.O. Hess

(1) INTRODUCTION

In recent experiments[1] spontanous positron emission in heavy ion collisions[2] have been measured. The analysis[1,2] of the data indicated that the two nuclei stick together for a time of about 10^{-19}sec or longer. Furthermore few nucleons where transfered only, so that they keep their identity approximately. M.Seiwert et al.[3] described the formation of a nuclear molecule where deformation effects play an important role. In their model the nuclei approach in different orientations relative to each other. When they touch the interplay of Coulomb repulsion and nuclear attraction forms a potential bag. The nuclear interaction is simulated by surface interaction, known as proximity approach[4] To get contact, different energies are necessary for distinct orientations. In extreme cases, i.e. pole-pole and equator-equator orientations, the coulomb energy at contact is smallest and biggest due to different distances. Though the nuclear cohesion is strongest in the equator-equator case due to bigger overlap in surface, the absolute potential minimum will be at the pole-pole orientation. To explain the content of the spontanous positron peak a dense distribution of states has to be assumed.

From the preceeding observations we got the idea to develop a simple model for heavy nuclear molecules. In the pole-pole orientation we can observe different kinds of exitations: Vibration of the individual nuclei, relative vibration , butterfly and belly dancer motion. The last-named represent new collective modes of nuclear molecules not examined up to know! Particular the relative vibration of two spherical nuclei and the interplay with individual exitations of one or both clusters has been a very fruitful idea. This socalled double resonance mechanism[5] can explain a large amount of nuclear structure of light systems in terms of the nuclear molecular picture. Mass transfer takes place in general, but we will not consider it here because the nuclei keep nearly their identity. For simplicity no β and γ vibrations are considered. The simplified model is applied to the system ^{238}U-^{238}U. For the more general approach we refer to a forthcoming publication.

(II) DEFINITION OF VARIABLES

For a nuclear molecule we consider 13 elementary degrees of freedom: ϑ_j^i (i=1,2; j=1,2,3) for each nucleus; r, ϑ, φ giving the relative distance and the orientation of the relative vector $\vec{r}$; $a_o^i = \beta_i \cos\gamma_i$, $a_2^i = (\beta_i/\sqrt{2})\sin\gamma_i$ (i=1,2) the vibration degrees of freedom, where β_1, γ_i are the usual deformation variables. More suitable is the choice of a molecular frame in which the relative vector $\vec{r}$ lies along the z_m- axis and m refers to the molecular frame (see fig.1). The fist and second Euler angle ϑ_1, ϑ_2 give the orientation of the z_m- axis and by definition of $\vec{r}$ for that system.i.e. $\vartheta=\vartheta_2$ and $\varphi=\vartheta_1$. If at least one of the nuclei, let us call it no.1, is strongly deformed, e.g. prolate, the third Euler angle can be fixed. We require that the z_1- axis, along which the moment of inertia is lowest, lies in the (x_m,z_m)- plane. The system of principal axes of the nuclei in general does not coincide with the molecular one. We have yet to rotate by ψ_1 to the z_1- axis and by ϕ_1 around this axis to reach the principle system of nucleus no.1. The orientation of nucleus no.2 to the molecular frame is prescribed by the Euler angles χ_2, ψ_2, ϕ_2 (see fig.1). The quadrupole variables α_{2m}^i (i=1,2) are related to a_K^i (K=-2,0,2) via

$$\alpha_{2m}^{1} = \sum_{k} D_{mk}^{2*}(\vartheta_j^1)\,a_k^1 = \sum_{m',k} D_{mm'}^{2*}(\vartheta_j^1)\,D_{m'k}^{2*}(0,\varphi_1,\phi_1) \quad (1a)$$

$$\alpha_{2m}^{2} = \sum_{k} D_{mk}^{2*}(\vartheta_j^2)\,a_k^2 = \sum_{m',k} D_{mm'}^{2*}(\vartheta_j^2)\,D_{m'k}^{2*}(x_2,\varphi_2,\phi_2) \quad (1.b)$$

(1.a-b) provides us with a relation between the old variables ϑ_j^i, a_o^i, a_2^i and the new ones, namely ϑ_1, ϑ_2, ϑ_3, φ_i, ϕ_i, x_2, a_o^i, a_2^i. In this letter we restrict us to the pole-pole orientation which corresponds to freeze the variable x_2. Exitations with $x_2 \neq 0$ are assumed to lie at high energy. We do not include β and γ vibrations which simplifies (1.a-b) enormously. α_{2m}^i does not depend on ϕ_1 and ϕ_2 because $a_o^i = \beta_{oi}$ is different from zero only. Therefore the relevant variables are ϑ_1, ϑ_2, ϑ_3, φ_1, φ_2 and r.

(III) COORDINATE SYMMETRIES

In general one is confronted with ambiguities after having defined an internal system. For example in a quadrupole deformed nucleus exist 24 possible choices for a system of principal axes. This is extensively discussed in the book of Eisenberg and Greiner[6] where three fundamental operations $\hat{R}_k$ (k=1,2,3) are given. With them all those can be generated relating all internal systems. $\hat{R}_k$ act on the components $(\bar{x},\bar{y},\bar{z})$ of a vector on the intrinsic system as

$$\hat{R}_1(\bar{x},\bar{y},\bar{z}) = (\bar{x},-\bar{y},-\bar{z})$$

$$\hat{R}_2(\bar{x},\bar{y},\bar{z}) = (\bar{y},-\bar{x},\bar{z})$$

$$\hat{R}_3(\bar{x},\bar{y},\bar{z}) = (\bar{y},\bar{z},\bar{x})$$

In a nuclear molecule we have to consider three different internal systems, namely the molecular frame and the two principal axes systems. In the following we illustrade the procedure. $\vec{x}$ is a vector in space while $\vec{x}_m$ and $\vec{x}_p$ are the same ones in the molecular frame and system of principal axes respectively. They are related via

$$\vec{x} = [\hat{D}_1 \, \hat{m}_1] \cdot [\hat{m}_1^{-1} \, \hat{D}_2 \, \hat{m}_2] \cdot [\hat{m}_2^{-1} \, \vec{x}_p] = \hat{D}_1' \, \hat{D}_2' \, \vec{x}_p' = \hat{D}_1' \, \vec{x}_m' \qquad (2)$$

where $\hat{D}_1$ prescribes the rotation from $\vec{x}$ to $\vec{x}_m$ and $\hat{D}_2$ from $\vec{x}_m$ to $\vec{x}_p$. $\hat{m}_1$ and $\hat{m}_2$ are rotations leading to equivalent systems $\vec{x}_m = \hat{m}_1 \, \vec{x}_m'$ and $\vec{x}_p = \hat{m}_2 \, \vec{x}_p'$. $\hat{D}_1'$ and $\hat{D}_2'$ rotate to the new frames. (2) provides us with a relation of the old orientation angles to the new ones, knowing the transformation $\hat{m}_1$ and $\hat{m}_2$.

The intrinsic system has yet to be defined more uniquely in order to reduce the ambiguity. Let us restrict to a symmetric system of nuclear molecules. The axes in the molecular frame are already defined uniquely, except for their orientations. The ambiguity in the choice of principal axes can be reduced, requireing that the unit vector along the z_i - axis has to have a positive component on the z_m- axis. Therefore only the combinations $\hat{R}_{1p} \, \hat{R}_{1m}$, $\hat{R}_{2m}^{A2}$, $\hat{R}_{2p} \, \hat{R}_{2m}^{A2}$ and powers of them are allowed. $\hat{R}_{kp}$ and $\hat{R}_{km}$ (k=1,2) act respectively on the principal axes and molecular frame! In table I the action on the variables and components of a vector in different systems is given. Though the simplified model does not depend on ϕ_i, it is given for completeness. Because a symmetric system is considered, i.e. $\Psi_1 = -\Psi_2 = \varepsilon$, we have yet to symmetrize with respect to the indices 1 and 2. Let us call this operator P(1⟷2), which changes Ψ_1 to Ψ_2, i.e. ε to $-\varepsilon$.

(IV) THE MODEL HAMILTON FUNCTION AND ITS QUANTIZATION

As the classical kinetic energy we choose

$$T = \frac{B}{2} \sum_{i=1}^{2} \sum_{m=-2}^{+2} (-1)^m \dot{\alpha}_{2m}^i \dot{\alpha}_{2-m}^i + \frac{M}{2} \dot{\vec{r}}^2 \qquad (3)$$

The first term gives the contribution in exitation for each nucleus, while the last term describes the relative motion. μ is the reduced mass. Substituting $\dot{\alpha}_{2m}^i$, $\vec{r}$ and its time derivatives into (3) we get after a set of manipulations, descibed extensively in ref.6,

$$T = \sum_{\ell} \Theta_{\ell\ell} \, \omega_{\ell}'^2 + \frac{1}{2} \Theta_\varepsilon \dot{\varepsilon}^2 + \frac{M}{2} \dot{r}^2 \qquad (4)$$

with

$$\Theta_{11} = 6 B \beta_0^2 \cos^2\varepsilon + \mu r^2 \qquad (4.a)$$

$$\Theta_{22} = 6 B \beta_0^2 \qquad\quad + \mu r^2 \qquad (4.b)$$

$$\Theta_{33} = 6 B \beta_0^2 \sin^2\varepsilon \qquad (4.c)$$

$$\Theta_\varepsilon = 6 B \beta_0^2 \qquad (4.d)$$

ω_ℓ' (ℓ =1,2,3) are the angular velocities with respect to the ℓ 'th axis in the intrinsic frame. β_0 gives the ground state deformation. The first term in (4) is the rotational contribution to the kinetic energy. Note that all off diagonal elements of the moment of inertia tensor $\Theta_{\hbar\ell}$ ($\hbar \neq \ell$) vanish! The second term describes the butterfly motion and the third one the relative kinetic energy. The belly dancer mode is hidden in the rotational part. It corresponds to a rotation around the z_m- axis.

As a potential we choose

$$V = \frac{C_\varepsilon}{2} \varepsilon^2 + \frac{C_r}{2} (r - r_0)^2 \qquad (5)$$

C_ξ and C_r give the strength of the potential for the butterfly and relative degree of freedom respectively. In principle the term εr can also appear. For simplicity we do not consider it.

The quantization is carried out in the same manner as prescribed in
ref.6 including a change of volume element. The proceedure is called
Pauli quantization and is nothing else as quantizing the Hamilton function in cartesian components and transforming it afterwards to curvilinear coordinates. In the volume element appears the factor $\sqrt{g}$. g
is the determinant of the metric tensor appearing in the classical ex-
pression $2Tdt^2 = \sum_{\ell m} g_{\ell m} dq_\ell dq_m$. q_ℓ denote the curvilinear coordinates.
When $\hat{L}_k^{'}$ (k=1,2,3) are the components of angular momentum operator
in the molecular frame, we obtain for the Hamiltonian

$$\hat{H}' = \frac{\hbar^2 \hat{L}_1^{'2}}{2\,\theta_{11}} + \frac{\hbar^2 \hat{L}_2^{'2}}{2\,\theta_{22}} + \frac{\hbar^2 \hat{L}_3^{'2}}{2\,\theta_{33}} - \frac{\hbar^2}{2\,\theta_\xi}\frac{\partial^2}{\partial\varepsilon^2} - \frac{\hbar^2}{2\mu}\frac{\partial^2}{\partial r^2} +$$
$$+ \frac{C_\xi}{2}\varepsilon^2 + \frac{C_r}{2}(r-r_0)^2 - \frac{\hbar^2}{48\,\beta\beta_0^2 \sin^2\xi}(1+\sin^2\varepsilon) \tag{6}$$

(V) SOLUTION OF THE SCHRÖDINGER EQUATION

The Hamiltonian is expanded in ε and $\bar{r} = r - r_0$, assuming small values
of ε and r. Taking the lowest order part $\hat{H}_0$ we have to solve the eigenvalue problem of

$$H = \frac{\hbar^2(\hat{L}^2 - \hat{L}_3^{'2})}{2\,\theta_0} + \frac{\hbar^2\hat{L}_3^{'2}}{12\,\beta\beta_0^2\varepsilon^2} - \frac{\hbar^2}{12\,\beta\beta_0^2}\frac{\partial^2}{\partial\varepsilon^2} - \frac{\hbar^2}{48\,\beta\beta_0^2\varepsilon^2} +$$

$$+ \frac{C_\varepsilon}{2} \varepsilon^2 - \frac{\hbar^2}{2\mu} \frac{\partial^2}{\partial F^2} + \frac{C_r}{2} \bar{F}^2 \qquad (7)$$

with $\quad \Theta_0 = 6 B \beta_0^2 + \mu r_0^2$

(7) has the same structure as the Hamiltonian of the Rotation-Vibration- Model (RVM)[6,7]. Indeed the solution is similar to the RVM except to one change: The projection K of the angular momentum I on the z - axis has to be changed to 2K! The reason lies in the different numerical factor of the terms $\sim \hbar^2 \hat{L}_3^{'2}$ and $\sim \hbar^2$. The ratio is 4 while in the RVM it is 1. Therefore we have to multiply the nominator and denominator of the term $\sim \hbar^2 \hat{L}_3^{'2}$ by 4, which gives $4K^2$. The solution after symmetrisation with respect to the operators in table I is given by

$$|I M K n_\varepsilon n_r\rangle = \left(\frac{(2I+1)}{16\pi^2(1+\delta_{K0})} \right)^{\frac{1}{2}} \left(D_{MK}^{I*}(\vartheta_i) + (-1)^I D_{M-K}^{I*}(\vartheta_i) \right) \chi_{2K, n_\varepsilon}(\varepsilon) g_{n_r}(F)$$

$$(8)$$

With χ_{2K, n_ε} given in ref.6 and g_{n_r} , which is a one dimensional harmonic oscillator. The energy is given by

$$E = \frac{\hbar^2}{2\Theta_0} [I(I+1) - K^2] + \hbar \sqrt{\frac{C_r}{\mu}} + \hbar \sqrt{\frac{C_\varepsilon}{6 B \beta_0^2}} (|K| + 2n_\varepsilon + 1) \qquad (9)$$

with

$$K = 0, 2, 4, \dots$$

$$I = 0, 2, 4, \dots \quad \text{if } K = 0$$

$$I = K, K+1, K+2, \dots \text{ if } K \neq 0$$

Applying P(1⟷2) to the solution (8) leads to the condition of even K. The structure of the spectrum (9) suggests to interpret the r-

and ε- motion as β- and γ- vibrations respectively for the giant nuclear molecule.

(VI) APPLICATION TO THE MOLECULAR SYSTEM

$^{238}U-^{238}U$

The system $^{238}U-^{238}U$ was examined by M. Seiwert et al.[3]. We take the parameters C and C from their work, while the others are given by the RVM. In this sense the model is parameterfree! For the factors in the energy we got: $\frac{\hbar^2}{2\theta_0} = 0.57*10^{-3}MeV$, $C_\gamma = 30MeV$, $C_\varepsilon = 279Mev/rad^2$. In fig.2a a typical spectrum is shown. Only bandheads are drawn. Upon each band head there is a dense rotational structure. Each band is classified by the quantum numbers $(K, n_\varepsilon, n_\gamma)$. For example (200)I=2 and (010)I=0 correspond to a pure belly dancer and butterfly mode respectively. Furthermore bandheads $(2n_\varepsilon, 0, 0)$, $(0, n_\varepsilon, 0)$ are nearly degenerate due to the small contribution of the rotational part.

In fig 2b the angular momentum distribution , within 10MeV window, is shown. The window starts at the barrier. 10MeV is the uncertainty in energy of the heavy ion beam used in experiment. Only states are taken into account whose width is neither too small nor too broad $(10^{-6} \lesssim \Gamma_z \lesssim 10^6)$. We proceed in accordance to the work of U. Heinz[8], using the Hill-Wheeler formula[9]. It is interesting to note that the angular momentum distribution is peaked around an average value I=100-120$\hbar$ and that the total number of states contributing is of the order 1000!

(VII) Conclusion

A very simplified model for nuclear molecules was presented. The general version will be given in a forthcoming paper. Here only the main ideas are outlined. New collective modes are introduced, such as butterfly and belly dancer motion. Finally the model is applied to the symmetric system ^{238}U-^{238}U. The crude energy spectrum and the angular momentum distribution have been determined. The latter showed a peaked structure generated by approximately 1000 states. The model has yet some uncertainties as the correct values for C_ξ and C_r. The model from which these parameters are deduced has yet some uncertainties. We would like to stress that the model not only can be applied to giant systems but also to light ones, as C-C and Mg-Mg. Calculations for C_ξ have yet to be carried out for such systems.

TABLE I

	$\bar{x}_i$	$\bar{y}_i$	$\bar{z}_i$	x_m	y_m	z_m	ϑ_1	ϑ_2	ϑ_3	φ_i	ϕ_i
$R^i_{1p}\,\hat{R}_{1m}$	$\bar{x}_i$	$-\bar{y}_i$	$-\bar{z}_i$	x_m	$-y_m$	$-z_m$	$\vartheta_1+\pi$	$\pi-\vartheta_2$	$-\vartheta_3$	$-\varphi_i$	$-\phi_i$
$\mathbb{1}\,(\hat{R}_{2m})^2$	$\bar{x}_i$	$\bar{y}_i$	$\bar{z}_i$	$-x_m$	$-y_m$	z_m	ϑ_1	ϑ_2	$\vartheta_3+\pi$	$-\varphi_i$	$\phi_i+\pi$
$\hat{R}^i_{2p}(\hat{R}_{2m})^2$	$\bar{y}_i$	$-\bar{x}_i$	$\bar{z}_i$	$-x_m$	$-y_m$	z_m	ϑ_1	ϑ_2	$\vartheta_3+\pi$	$-\varphi_i$	$\phi_i-\frac{\pi}{2}$

References:

1 M.Clemente, E.Berdermann, P.Kienle, H.Tsertos, W.Wagner, F.Bosch, C.Kozhuharov and W.Koenig; Proc of the Intern. Conf on Nuclear Physics, Florence, Aug.29- Sept.3, 1983, p.693; H.Bokemeyer, H.Folger, H.Grein, T.Cowan, J.S.Greenberg, J.Schweppe, K.Bethge, A.Gruppe, K.E.Stiebing, D.Schwalm, P.Vincent and N.Trautmann; ibid., p.694

2 J.Reinhardt, U.Müller, B.Müller and W.Greiner; Z.Phys. A303, (1981) 173

3 M.Seiwert, W.T.Pinkston and W.Greiner; to be published

4 J.Blocki, J.Randup, W.J.Swiatecki, C.F.Tsang; Annals of Phys. 105 (1977), 427

5 W.Scheid, W.Greiner, R.Lemmer; Phys Rev. Lett. 25, (1970) 176

6 J.M.Eisenberg, W.Greiner; Nuclear Theory I, North- Holland- Publishing- Company, Amsterdam (1975)

7 A.Faessler, W.Greiner; Z.Phys. 168 (1962) 425; A.Faessler, W.Greiner; Z.Phys. 170 (1962) 105; A.Faessler, W.Greiner; Z.Phys. 177 (1964) 190; A.Faessler, W.Greiner, R.K.Sheline; Nucl. Phys. 80 (1965) 417

8 U.Heinz; Habilitation thesis, University Frankfurt/Main (1983)

9 M.S.Child, "Molecular Collision Theory", Academic Press, New york (1974)

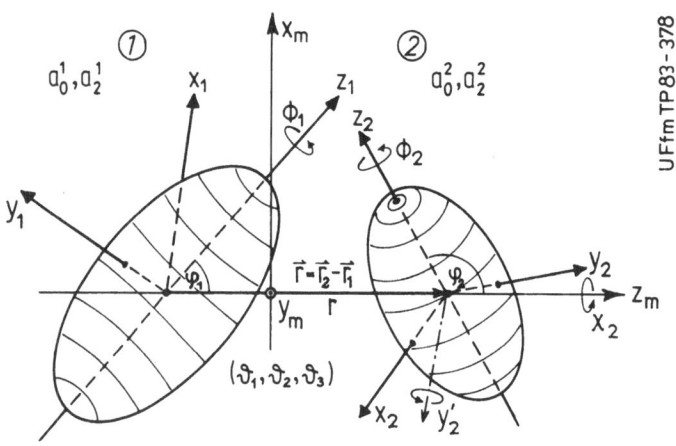

Fig.1: Definition of coordinates in the molecular frame with arbitrary orientations of the nuclei with respect to each other.

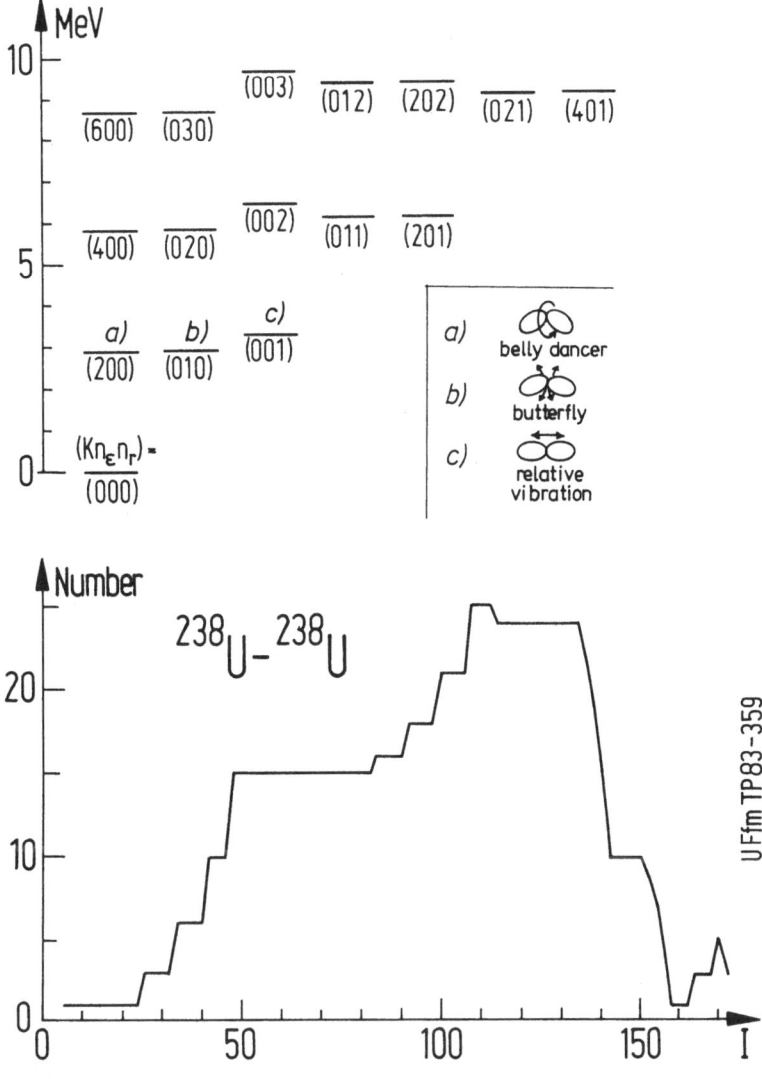

Fig.2: In 2.a a typical energy spectrum is plotted. The system investigated is ^{238}U- ^{238}U. In 2.b the distribution of angular momentum
states in a 10 MeV window is given for that system. The 10MeV window starts at the barrier.

PION BREMSSTRAHLUNG

IN SUBTHRESHOLD HEAVY ION COLLISIONS

W. Greiner
Institute für Theoretische Physik
J.W. Goethe Universität Frankfurt, West Germany

In cooperation with

D. Vasak and B. Müller

Subthreshold pion production has become of considerable interest during recent years, particularly in connection with with the search for cumulative (or cooperative) effects. The idea to consider a bremsstrahlung-type mechanism for the production of secondary particles in hadronic collisions is not new[1]. If this model can be shown to describe pion production in nuclear collisions, it could be useful source of information about the time development of the reaction and the spin-isospin structure of the colliding nuclei[2].

In the following we apply the bremsstrahlung model to collisions of equal nuclei far below the single nucleon-nucleon threshold for pion production. In this approach the pion field is treated in semiclassical approximation and the recoil of the collective source, the nucleus, propagating along a given trajectory, is neglected. Deceleration gives rise to a shake-off of the pionic cloud surrounding the nucleus. The inclusive differential cross-section for radiation of pions with a selected charge in this schematic model[2] reads

$$\frac{d^2\sigma_\pi}{d\Omega\,dE} = \frac{\sigma_0 A^2 g_0^2}{3\,(4\pi)^3}\,\frac{p}{M^2} \tag{1}$$

$$\times \sum_{s_P,s_T}\left|\sum_{i=P,T}\int_{-\infty}^{+\infty}\frac{d}{dt}\left[\frac{\tilde{s}(\hat{p}_i)(\sigma_i\cdot p)}{(u_i\cdot p)}\right]\exp\,i(Et-\vec{p}\cdot\vec{R}_i(t))\,dt\right|^2$$

where E is the pion energy, $p = (E^2 - m^2)^{1/2}$, and Ω the pion solid angle. For equal nuclei, the effective cross-section $\sigma_0 = 4\pi R^2$ with R being the half-density radius of the colliding nuclei. M = 930 MeV is the mass of a bound nucleon, m = 135 MeV that of a neutral pion, $\rho_0 = .17\ fm^{-3}$ is the normal nuclear density and g_0 is the pion-nucleon coupling constant, $g_0^2/4\pi = 14$. The cross-section is averaged over the direction of the spin vectors s_i of both participating nuclei. $\tilde{p}(\hat{p})$ is the Fourier-transform of the Gaussian nuclear density $\rho(r) = \eta\rho_0\exp(-(r/2a)^2)$ with the Lorentz-boosted momentum $\hat{p} = [|\vec{p}|^2 - (\vec{p}\cdot\vec{R}(t))^2]^{1/2}$, the compression $\eta = \rho/\rho_0$ and $1/a = 2(\rho/A)\sqrt{\pi}$. Finally, $\vec{R}_i(t)$ are the linear trajectories of the projectile (i=P) and the target (i=T) in the c.m. system, parametrized by the deceleration time τ. We express τ as a unique multiple of the "passing time" $\tau_s = R/(2\gamma_{in}v_{in})$, where v_{in} is the initial c.m. velocity and $\gamma_{in} =$

$(1-v_{in}^2)^{-1/2}$; i.e., $\tau = \nu\tau_s$. u_i^μ are the corresponding four-velocities and σ_i^μ the spin vectors boosted from the particle's rest frame into the c.m. system. For more details cf. ref. 2.

A short remark to the rôle of the spin density is appropriate: In the schematic approach of ref. 2 it was assumed that the spin vectors of the projectile and target are not correlated at all. This assumption lead to incoherent addition of the pion yields from both nuclei. On the other hand, in the hydrodynamical picture of heavy ion collisions the nucleons in the reaction zone, consisting of hot compressed nuclear matter, do not "remember" their origin. Thus there is only one spin direction, and the pion yields should be added coherently. While incoherent addition leads to enhanced pion production at $90°$ in the c.m. system, in case of coherence a negative interference sideways leaves a forward-backward enhancement in the angular distribution and gives a smaller total pion yield. In fact, a nonzero source of pion radiation is a consequence of spin fluctuations[3], which appear preferably in the reaction zone, where the individual nucleon-nucleon collisions take place and lead to occupation of vacant states. Seen in this light, the coherent addition is the natural one.

In order to avoid complications caused by the Coulomb force when looking at charged pion data[4] we compare our calculations (cf. fig. 1) with recent experiments on neutral pion production[5,6,7]. The pion angular distributions from a C + C reaction[5] measured by the GSI-group at the CERN-SC at 84, 74 and 60 MeV/n exhibit a forward-backward behaviour and thus favour the coherent addition of the pion yields from the projec-tile and target nuclei: in fig. 1(a) the experimental pion angular distribution[5] is compared with our results. The dashed curve in the upper energy cut is obtained by incoherent addition (here the deceler-ation parameter has been adjusted to $\nu = 0.55$). The solid lines show the results from coherent addition. In this case the parameter $\nu = 0.38$ is somewhat smaller to compensate for the negative interference mentioned above. In the framework of our model we conclude that the spin distrib-utions of the participating nuclei must be closely correlated or, in other words, that deceleration, creation of compressed nuclear matter and pion production are simultaneous processes. Therefore, to be consistent, the compression has also been estimated by taking the values from hydrodynamical calculations[8]. Note that hydrodynamics is questionable for nuclei as small as carbon and for energies as low as 20 MeV/n.

Because the deceleration by the Coulomb force is very long range, it essentially does not contribute to pion radiation, but takes away a considerable part of the available scattering energy for heavy nuclei. We thus subtract the energy $E_{coul} = Z_P Z_T \alpha / 2R$ from the bombarding energy E_{lab}.

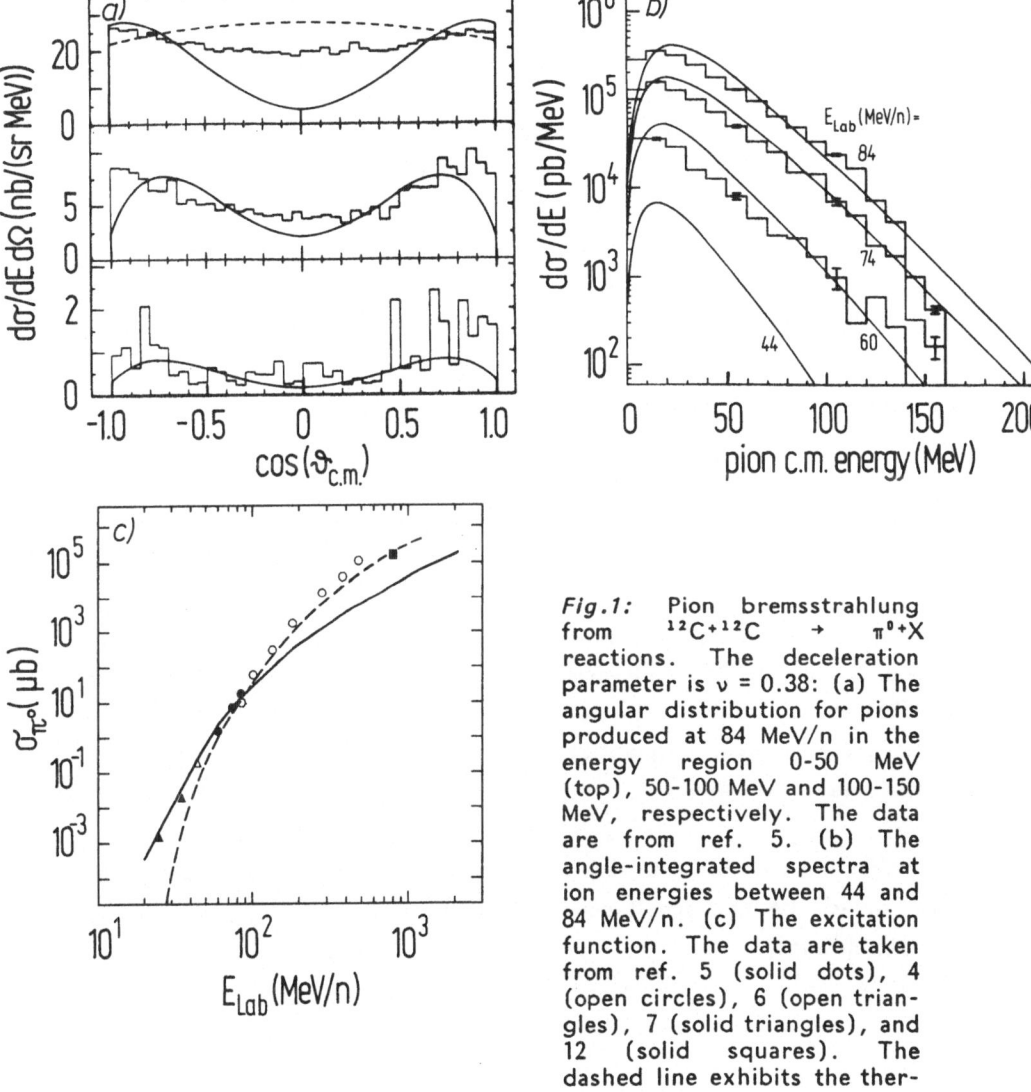

Fig.1: Pion bremsstrahlung from $^{12}C + ^{12}C \rightarrow \pi^0 + X$ reactions. The deceleration parameter is $\nu = 0.38$: (a) The angular distribution for pions produced at 84 MeV/n in the energy region 0-50 MeV (top), 50-100 MeV and 100-150 MeV, respectively. The data are from ref. 5. (b) The angle-integrated spectra at ion energies between 44 and 84 MeV/n. (c) The excitation function. The data are taken from ref. 5 (solid dots), 4 (open circles), 6 (open triangles), 7 (solid triangles), and 12 (solid squares). The dashed line exhibits the thermal pions.

In fig. 1(b) the experimental[5] and theoretical angle-integrated spectra are shown to coincide in shape as well as in magnitude. The exponential decay of the cross-sections with growing pion energy reflects the influence of the gaussian form-factor, i.e., the shape of the reaction zone[9]. In fig. 1(c) the pion excitation function is shown. The parameter ν is kept fixed at the value 0.38. At higher bombarding energies the pion bremsstrahlung underestimates considerably the experimental yield[4,10]. This is not surprising, since we expect there other production mechanisms like the thermal production from highly excited isobars[11] to contribute to the total pion yield. Moreover, the deceleration parameter may differ from the value $\nu = 0.38$ used at low energies. Still, we find excellent agreement with the pion production from Ar + Ca at 44 MeV/n taken at GANIL[6] and the very recent preliminary data[7] for N + Ni collisions (measured after our present calculations were already available) with the 35 MeV/n beam of the MSU superconducting cyclotron by the Stony Brook-Oak Ridge-GSI collaboration[7] (the experimentally determined[5,7] A-dependence is used to renormalize the yields to be comparable with the carbon data). For completeness, an upper bound for the thermal pion yield (dashed line) has been estimated in the shock-wave model[10] with a hard equation of state (Fermi gas) in which the temperatures and the densities are higher than in the hydrodynamical model:

$$\sigma_\pi^{therm} = \sigma_0 \, \frac{4\pi}{3} R_c^3 \int_0^\infty \left[\exp(E/T) - 1 \right]^{-1} \frac{d^3p}{(2\pi)^3}$$

R_c is the half-density radius and T the temperature of the compressed compound nucleus. Thermally created pions are obviously negligible in the very low energy regime. The sum of the thermal and bremsstrahlung pion yields gives an overall quantitative description of the available experimental data.

An interesting possibility to test the time development of the collision process would be a simultaneous measurement of pion and γ-ray bremsstrahlung[6].

A classical formula, similar to (1), holds also for photons[12]:

$$\frac{d^2\sigma_\gamma}{d\Omega\, d\omega} = \frac{\sigma_0 Z^2 \alpha}{4\pi^2 \omega} \left| \sum_{i=P,T} \int_{-\infty}^{+\infty} \frac{d}{dt} \frac{\vec{n} \times \vec{v}_i}{1 - \vec{n} \cdot \vec{v}_i} \exp\left[i\omega(t - \vec{n} \cdot \vec{R}_i) \right] dt \right|^2$$

where $\alpha = 1/137$, ω is the photon frequency and n the direction of emission. Until now this formula has been applied to low ion energy collisions[13] where also experimental data[14] exist. Our results, using the same nuclear trajectory as for pion production, are presented in fig. 2. If the sources of electromagnetic radiation move on a straight line the yield in the forward direction is zero as it is at 90° (c.m.) for symmetric systems. These minima are clearly seen in the differential cross-section in the laboratory system (fig. 2(a)). A significant intensity of high energy photons (25 MeV$\leq\omega\leq$150 MeV) is predicted (cf. fig. 2(b)).

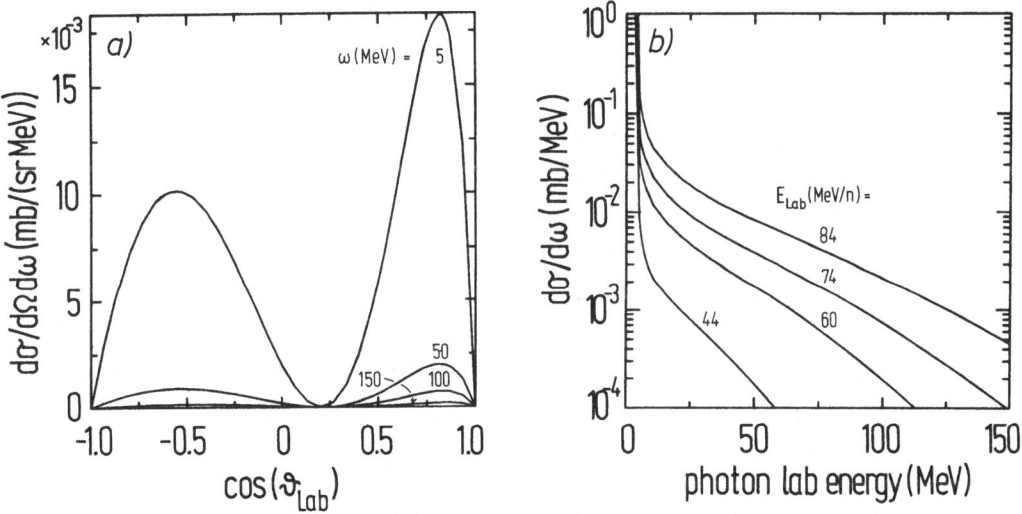

Fig.2: The γ-ray bremsstrahlung from $^{12}C + ^{12}C \to \gamma + X$ reactions: (a) The double-differential cross-section at 84 MeV/n in the lab system. (b) The angle-integrated γ-ray spectra in the lab-system.

We conclude that

1. In particular at low bombarding energies the bremsstrahlung mechanism can help to understand the (symmetric) pion production cross-section. Within our bremsstrahlung model the π^0-data of refs. 5,6 and 7 can be explained with the unique deceleration parameter $\nu = 0.38$. The deceleration is similar to that obtained from nuclear hydrodynamics[9]. Other pion producing mechanisms[11,15] are supressed in this energy regime. The influence of pion reabsorbtion can be neglected for nuclei as small as carbon; it will be important only for the high momentum tails of the spectra[16].

2. The forward-backward peaking of the experimental angular distributions is quantitatively explained, if the pion radiation from the projectile and target are coherently added, supporting the cooperative nature of subthreshold pion production.

3. Information about the time-development of the collision process is also contained in electromagnetic processes. We therefore suggest that besides γ-radiation also processes such as electron and positron emission[17] can be used to check the underlying collision dynamics.

We thank G. Buchwald, G. Graebner, J. Maruhn and M. Uhlig for providing us with their results. We also acknowledge discussions on the bremsstrahlung problem with P. Braun-Munzinger, E. Grosse, J. Julien, Ch. Michel, H. Noll, F. Obenshain and P. Paul and are grateful that they made their data available to us prior to publication.

References:

[1]H. Gemmel and H. A. Kastrup, Nucl. Phys. B14(1969), 566

[2]D. Vasak, B. Müller and W. Greiner, Phys. Scr. 22(1980), 25;
D. Vasak, H. Stöcker, B. Müller and W. Greiner, Phys. Lett. 93B(1980), 243

[3]In this point our model is not entirely a classical model: spin has no classical counterpart. The fluctuations are in principle calculable only in a full microscopic treatment of the collision. A reasonable approximation could also be obtained by inclusion of the spin-isospin degrees of freedom in a hydrodynamical calculation. Here we define an "effective" spin representing the spin density averaged over the space-time history of the fluctuations.

[4]W. Benenson et al., Phys. Rev. Lett. 43(1979),683 and 44(1980),54E;
J. Julien: Subthreshold Pion Production by Heavy Ions, paper presented at the 3rd. International Conference on Nuclear Reaction Mechanisms, Varenna, 1982;
B. Jakobsson: Proc. of the Nordic Meeting on Nuclear Physics, Fuglso, Denmark, 1982;
T. Johansson et al., Phys. Rev. Lett. 48(1982), 732;
J. P. Sullivan et al., Phys. Rev. 25C(1982), 1499

[5]H. Noll et al.: GSI Scientific Report, 1982, p. 32;
E. Grosse, Proc. of the International Workshop on Gross Properties of Nuclei and Nuclear Excitations XI, Hirschegg, Austria, 1983, p.65 and private communication;
Ch. Michel, Proc. of the XXI International Winter Meeting on Nuclear Physics, Bormio, Italy, 1983, p.539

[6]H. Noll et al., Proc. of the International Conference on Nuclear Physics, Florence, Italy, 1983, p. 682

[7]P. Braun-Munzinger et al.: Pion Production in Heavy Ion Collisions at $E_{lab}/A = 35$ MeV, to be published

[8]G. Buchwald, G. Graebner, J. Theis, J. A. Maruhn, W. Greiner and H. Stöcker, Phys. Rev. C28(1983), 1119 , and private communication

[9]The parameter p_o in the form factor $A\exp(-p^2(1-v_i^2\cos^2\vartheta)/p_o^2)$ has the form $p_o \sim 1/R_I$ where R_I is the size of the reaction zone. In this model $R_I \sim R_c$, i.e., equal to the half-density width of the compressed nucleus.

[10]G. F. Chapline, M. H. Johnson, E. Teller and M. S. Weiss, Phys. Rev. D8(1973), 4302;
W. Scheid, H. Müller and W. Greiner,Phys. Rev. Lett. 32(1974), 741;
H. Stöcker, G. Graebner, J. A. Maruhn, and W. Greiner, Phys. Lett. 95B(1980), 192

[11]S. Nagamiya et al., Phys. Rev. C24(1981), 971

[12]J. D. Jackson: Classical Electrodynamics, J. Wiley & Sons, New York, 1975

[13] J. Reinhardt, G. Soff and W. Greiner, Z. Phys. A276(1976), 285

[14] H. P. Trautvetter, J. S. Greenberg and P. Vincent, Phys. Rev. Lett. 37(1976), 202

[15] G. F. Bertsch, Phys. Rev. C15(1977), 713

[16] J. M. Eisenberg, D. S. Koltun: Theory of Meson Interactions in Nuclei, J. Wiley & Sons, New York, 1980, p. 117

[17] G. Soff, J. Reinhardt, B. Müller and W. Greiner, Z. Phys. A294(1980), 137, and private communication

PIONS AND OTHER HADRONIC DEGREES
OF FREEDOM IN NUCLEI[*]

W. Weise
Institute of Theoretical Physics
University of Regensburg
D-8400 Regensburg, FR Germany

---

[*] Work supported in part by Bundesministerium für Forschung und Technologie
(grant MEP-33-REA) and by Deutsche Forschungsgemeinschaft (grant We 655/7-6).

LECTURE 1

1. Introduction

The traditional picture of the nucleus in low energy nuclear physics is that of an interacting many-body system of structureless, pointlike protons and neutrons. Here low energy
nuclear physics is understood to be the region of excitation energies ΔE smaller than
the Fermi energy ($\varepsilon_F \simeq 30 - 40$ MeV) and momentum transfers $\Delta q \lesssim 1/R$, where R is the
nuclear radius.

The situation changes as ΔE and/or Δq is increased up to several hundreds of MeV, the
domain of intermediate energy physics. At this point explicit mesonic degrees of freedom
become directly visible. The pion, in particular, is of fundamental importance. With
its small mass of $m_\pi = 140$ MeV it is by far the lightest of all mesons. It is the
generator of the long range nucleon-nucleon interaction. The pion Compton wavelength,
$\lambdabar_\pi = \hbar/m_\pi c = 1.4$ fm, defines the length scale of nuclear physics.

As mesons become important, nucleons begin to reveal their intrinsic structure. Inse-
parably connected with pionic degrees of freedom is the role of the $\Delta(1232)$, the spin
3/2-isospin 3/2 isobar reached from the nucleon by a strong spin-isospin transition at
an excitation energy $\Delta E = M_\Delta - M \simeq 300$ MeV, the Δ-nucleon mass difference.

In these lectures, the position will be taken that the nucleus consists of nucleons and
their excited states (primarily the $\Delta(1232)$) which communicate by exchange of mesons
(in particular: the pion). Such a description has turned out to be quite successful in
correlating various phenomena and data at intermediate energies, remarkably though
without the need, so far, for explicit reference to underlying quark degrees of freedom.
This progress has gone parallel with the similarly successful meson exchange phenomenology of nucleon-nucleon forces at long and intermediate distances ($r \gtrsim 0.8$ fm). A sur-
vey of the rapid experimental and theoretical progress in meson-nuclear physics can be
obtained by consulting the conference proceedings [1] and [2], and recent reviews in
ref. [3-6].

While there may not be a need for explicitly invoking quark degrees of freedom in nuclei
up to a few hundred MeV of excitation energy, there is an obvious necessity to understand the phenomenological input into nuclear forces from a more fundamental (quark-gluon
dynamical) point of view. Attempts to establish relationships of this kind are still
at their very beginning, but there is little doubt that activities in this direction
will constitute a substantial part of intermediate energy physics research in coming
years. Some of the developments will be touched in these lectures, though not at a
very detailed level.

2. The Nucleon-Nucleon Interaction

2.1 Survey: Mesons and the Nuclear Force

The nucleon-nucleon interaction has been a problem of fundamental interest and challenge ever since Yukawa's pioneering work in 1935. The problem is still unsolved: it is yet impossible to derive nuclear forces directly from Quantum Chromodynamics, the theory of strong interactions. However, over the years, meson exchange models have established a highly successful phenomenology.

A schematic picture of the nucleon-nucleon potential in the $^1S_0$ state is shown in Fig. 1. At distances of the order of the pion Compton wave length and beyond, the one-pion exchange interaction dominates. At intermediate distances two-pion exchange mechanisms become important. The lowest angular momentum carried by the exchanged pion pair is $J^\pi = 0^+$, together with isospin $I = 0$ in accordance with the symmetry of the $(\pi\pi)$ state. The corresponding $(\pi\pi)$ mass spectrum has a broad distribution. In one-boson exchange models, this is usually prametrized in terms of an effective "σ" boson with a mass between 400 and 600 MeV.

Furthermore, two interacting pions in a $J^\pi = 1^-$ and isospin $I = 1$ state resonate strongly to form the ρ meson with a mass $m_\rho = 770$ MeV.

Down to about $r \gtrsim 0.8$ fm, two-pion exchange processes can be treated rather accurately using dispersion relation methods, such as in the Paris [7] or Stony Brook [8] NN-interaction, or in refined versions of the Bonn potential [9]. At shorter distances $(r \lesssim 0.8$ fm), the understanding of the NN force is more or less on phenomenological grounds only. In a one-boson exchange description (e.g. of the Bonn [10] or Nijmegen [11] groups), the short-range repulsion is simulated by exchange of a strongly coupled ω meson ($J^\pi = 1^-$, $I = 0$) with a mass $m_\omega = 783$ MeV.

Both ρ and ω exchange take place primarily at distances comparable to their Compton wavelengths $m_\rho^{-1} \simeq m_\omega^{-1} \sim 1/4$ fm, which is the same order of magnitude as the nucleon size itself. It is therefore difficult to imagine how a ρ or ω meson can travel freely between two nucleons. One has to expect that there is a massive influence of finite-size cutoffs. In any case, one probably has to interpret these short-range vector meson exchanges as phenomenological representations of complex mechanisms taking place at the level of quarks and gluons, once two nucleons approach each other at distances so small that their quark cores most likely overlap.

Nevertheless, the one- and two-boson exchange phenomenology provides a quantitatively successful description of NN scattering data and deuteron properties. We summarize properties of the exchanged mesons and meson-nucleon coupling constants in table 1. The coupling constants refer to meson-nucleon effective Lagrangians of the following types:

$$\text{Scalar:} \qquad \mathcal{L}_s = g_s \, \bar{\psi}(x) \, \psi(x) \, \phi_s(x) , \qquad\qquad (2.1a)$$

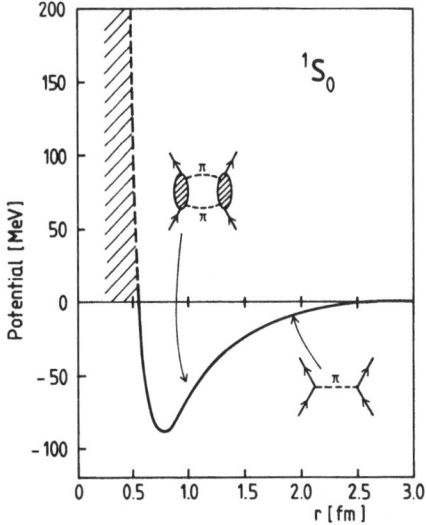

Figure 1: Schematic picture of the NN interaction in the $^1S_0$ channel

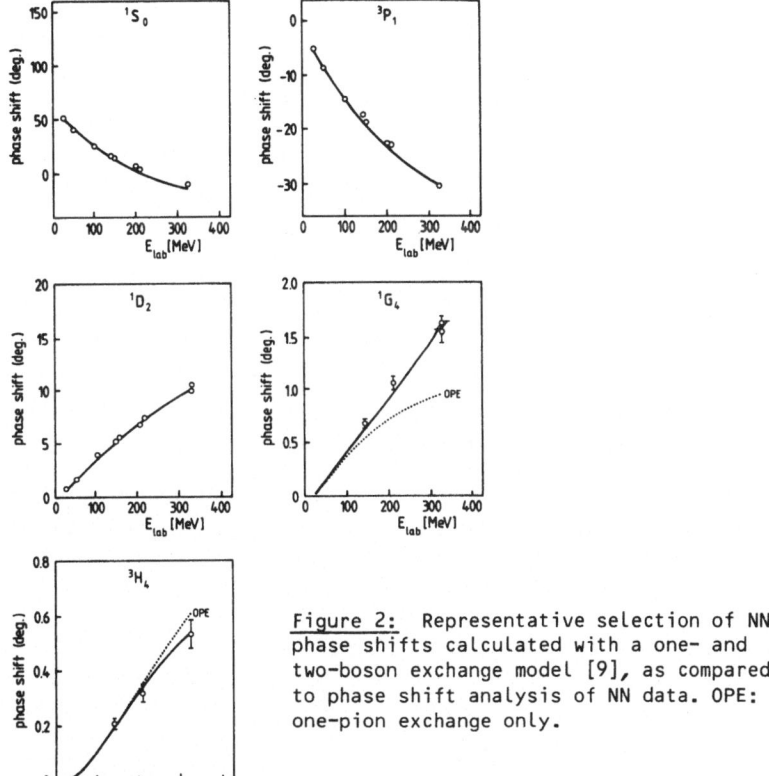

Figure 2: Representative selection of NN phase shifts calculated with a one- and two-boson exchange model [9], as compared to phase shift analysis of NN data. OPE: one-pion exchange only.

Pseudoscalar: $\mathcal{L}_P = g_P \bar{\psi}(x) i \gamma_5 \psi(x) \phi_P(x)$, (2.1b)

Vector: $\mathcal{L}_V = [g_V \bar{\psi}(x) \gamma_\mu \psi(x) + (g_T/2M) \bar{\psi}(x)\sigma_{\mu\nu} \psi(x) \partial^\nu] V^\mu(x)$. (2.1c)

Here $\psi(x)$ are the nucleon Dirac spinor fields, and we follow the Bjorken and Drell conventions for metric and Dirac-γ matrices [+]. The ϕ_s, ϕ_p and V^μ refer to scalar, pseudoscalar and vector meson fields. For isovector mesons the isospin dependence enters in the form $\vec{\tau} \cdot \vec{\phi}$ or $\vec{\tau} \cdot \vec{V}^\mu$, respectively, where $\vec{\tau} = (\tau^1, \tau^2, \tau^3)$ are the three Pauli isospin matrices for nucleons.

meson	J^π	I	mass m[MeV]	$g^2/4\pi$ Bonn	$g^2/4\pi$ GK
$\pi^\pm$	0^-	1	139.6	14.4	14.3
π^0			135.0		
η	0^-	0	548.8	4.95	0
ρ	1^-	1	770	0.48 (6.0)	0.55 (6.1)
ω	1^-	0	783	10.6	8.1 ± 1.5

table 1: Properties and coupling constants of mesons commonly used in Boson exchange models of the NN interaction. The Bonn [9] results refer to vertex functions modified by monopole form factors

$$F(q^2) = (\Lambda^2 - m^2)/(\Lambda^2 - q^2) \;,\; q^2 = q_0^2 - \vec{q}^2,$$

with λ = 1.5 GeV. Also shown are the coupling constants obtained by a dispersion theoretic analysis of Grein and Kroll (GK) [12]. For the vector mesons, the coupling constant $g_V^2/4\pi$ is given and the ratio g_T/g_V shown in parantheses (this ratio is small for the ω meson).

Fig. 2 shows a representative selection of nucleon-nucleon phase shifts in low and higher partial waves calculated with the recent one- and two-boson exchange interaction of the Bonn group [9]. This calculation includes a selected set not only of ($\pi\pi$) exchange, but also ($\pi\rho$), ($\pi\sigma$) and ($\pi\omega$) exchange processes. The results obtained with the Paris potential [7] are of similar quality.

Note that the higher partial waves up to laboratory energies $E_{lab} \simeq 100$ MeV are dominated by one-pion exchange, because of their peripheral nature.

[+] We use conventions such that

$$\gamma_0 = \begin{pmatrix} 1 & 0 \\ 0 & -1 \end{pmatrix}, \; \vec{\gamma} = \begin{pmatrix} 0 & \vec{\sigma} \\ -\vec{\sigma} & 0 \end{pmatrix}, \; \gamma_5 = \begin{pmatrix} 0 & 1 \\ 1 & 0 \end{pmatrix}, \; \{\gamma_\mu, \gamma_\nu\} = 2g_{\mu\nu} \text{ with } g_{00} = 1,$$

$$g_{ij} = -\delta_{ij}; \; \bar{\psi} = \psi^+ \gamma_0$$

2.2 Reminder of the One-Pion Exchange interaction

The best known part of the nuclear force is one-pion exchange (OPE). It is the prototype of spin-isospin dependent interactions and plays a most important role in all
subsequent discussions.

For a static, pointlike nucleon, the pion-nucleon interaction Hamiltonian derived from
eq. (2.1a) by a non-relativistic reduction is

$$H_{\pi NN} = \frac{f}{m_\pi} \, \vec{\sigma} \cdot \vec{\nabla} \, \vec{\tau} \cdot \vec{\phi}(\vec{r}), \tag{2.2}$$

where $\vec{\sigma}$ and $\vec{\tau}$ are nucleon spin and isospin and $\vec{\phi}(\vec{r})$ is the isovector pionfield. Second
order perturbation theory with $H_{\pi NN}$ gives the static one-pion exchange (OPE) potential
(see Fig. 3). In momentum space

$$V_\pi(\vec{q}) = -\frac{f^2}{m_\pi^2} \frac{\vec{\sigma_1} \cdot \vec{q} \; \vec{\sigma_2} \cdot \vec{q}}{\vec{q}^2 + m_\pi^2} \, \vec{\tau_1} \cdot \vec{\tau_2}, \tag{2.3}$$

where $\vec{q}$ is the momentum transfer carried by the exchanged pion. The coupling constant
is

$$\frac{f}{m_\pi} = \frac{g_{\pi NN}}{2M}, \quad \frac{g_{\pi NN}^2}{4\pi} \simeq 14 \qquad \text{(see table 1)} \tag{2.4}$$

where m_π and M are the pion and the nucleon mass, respectively, i.e. $f \simeq 1$. The V_π of
eq. (2.3) can be split into a spin-spin and tensor piece,

$$V_\pi(\vec{q}) = -\frac{1}{3} \frac{f^2}{m_\pi^2} \left[\left(1 - \frac{m_\pi^2}{q^2 + m_\pi^2}\right) \vec{\sigma_1} \cdot \vec{\sigma_2} + \frac{\vec{q}^2}{\vec{q}^2 + m_\pi^2} S_{12}(\hat{q}) \right] \vec{\tau_1} \cdot \vec{\tau_2}, \tag{2.5}$$

with

$$S_{12}(\hat{q}) = 3 \, \vec{\sigma_1} \cdot \hat{q} \, \vec{\sigma_2} \cdot \hat{q} - \vec{\sigma_1} \cdot \vec{\sigma_2}, \tag{2.6}$$

where $\hat{q} = \vec{q}/|\vec{q}|$. In r-space, one obtains the familiar form:

$$V_\pi(\vec{r}) = \frac{f^2}{m_\pi^2} \, \vec{\sigma_1} \cdot \vec{\nabla} \, \vec{\sigma_2} \cdot \vec{\nabla} \, \frac{e^{-m_\pi r}}{4\pi r} \, \vec{\tau_1} \cdot \vec{\tau_2}$$

$$= -\frac{1}{3} \frac{f^2}{4\pi} \left\{ \left[\frac{e^{-m_\pi r}}{r} - \frac{4\pi}{m_\pi^2} \delta^3(\vec{r}) \right] \vec{\sigma_1} \cdot \vec{\sigma_2} + \right. \tag{2.7}$$

$$\left. \left(1 + \frac{3}{m_\pi r} + \frac{3}{m_\pi^2 r^2} \right) \frac{e^{-m_\pi r}}{r} S_{12}(\hat{r}) \right\} \vec{\tau_1} \cdot \vec{\tau_2}.$$

The characteristic feature of OPE is its strong tensor force. The δ-function piece is
obviously an artifact of the assumed pointlike nature of the nucleon source. Nucleons
are, of course, far from being pointlike objects, and we shall examine how their size
and intrinsic structure modifies the properties of OPE at short distances.

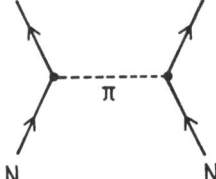

<u>Figure 3:</u>   Static one-pion exchange interaction between nucleons

2.3 Isovector Two-Pion Exchange

At shorter distances, the spin-isospin dependent nucleon-nucleon interaction receives contributions from the exchange of two interacting pions in the channel with ($J^\pi = 1^-$, I = 1), the one carrying the quantum numbers of a ρ meson. (See Fig. 4)

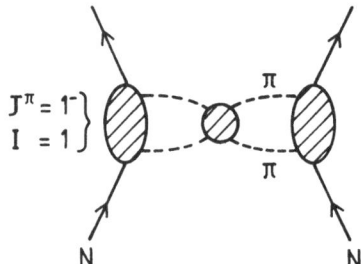

<u>Figure 4:</u>   Exchange of a ($\pi\pi$) pair coupled to ($J^\pi = 1^-$, I = 1) including ρ exchange.

If the ($\pi\pi$) mass distribution is approximated by a single δ-function located at m_ρ = 770 MeV, and for infinitely heavy, pointlike nucleons, the ρ exchange interaction obtained by non-relativistic reduction from eq. (2.1b) becomes:

$$V_\rho(\vec{q}) = - \frac{f_\rho^2}{m_\rho^2} \frac{(\vec{\sigma_1} \times \vec{q}) \cdot (\vec{\sigma_2} \times \vec{q})}{\vec{q}^2 + m_\rho^2} \; \vec{\tau_1} \cdot \vec{\tau_2} \; . \tag{2.8}$$

We note that the $\vec{\sigma} \times \vec{q}$ type interaction comes from the dominant tensor coupling $\bar{\psi} \, \sigma_{\mu\nu} \, \psi \, \partial^\nu \rho^\mu$ of the ρ meson to the nucleon. Here $f_\rho / m_\rho = (g_T)_{\rho NN}/2M$. Empirically, one finds $f_\rho^2/m_\rho^2 \simeq 2 \; f_\pi^2/m_\pi^2 \simeq 2m_\pi^{-2}$. In r-space,

$$V_\rho(\vec{r}) = \frac{1}{3} \frac{f_\rho^2}{4\pi} \left\{ 2 \left[\frac{e^{-m_\rho r}}{r} - \frac{4\pi}{m_\rho^2} \, \delta^3(\vec{r}) \right] \vec{\sigma_1} \cdot \vec{\sigma_2} - \right.$$

$$\left. - \left(1 + \frac{3}{m_\rho r} + \frac{3}{m_\rho^2 r^2} \right) \frac{e^{-m_\rho r}}{r} \, S_{12}(\hat{r}) \right\} \vec{\tau_1} \cdot \vec{\tau_2} \; . \tag{2.9}$$

The ρ exchange tensor force has opposite sign as compared to π exchange and therefore tends to reduce the pathologically strong OPE tensor force at short distance. However, a picture like this is probably only of limited relevance, since the ρ meson Compton wavelength m_ρ^{-1} is comparable to the nucleon size, as mentioned before. That is, one has to expect that V_ρ is cut down massively by form factors.

LECTURE 2

3. Pion-Nucleon Coupling in Relativistic Quark Models

Given the fact that nucleons have their own intrinsic quark structure, it is necessary to address the question why a description of nuclei in terms of nucleon quasi particles and mesons instead of quarks is successful even at momentum transfers where one would expect the size of nucleons to play a substantial role. There is, of course, no satisfactory answer to this question. It is nevertheless useful to obtain some insight into the relevant length scales involved in pion-nucleon interactions, and in particular to see how the magnitude of the phenomenological pion-nucleon coupling constant $g_{\pi NN}$ can be related to the underlying quark dynamics.

3.1 Facts from QCD

Non-strange hadrons are composed uf u- and d-quarks which form a flavour-SU(2) (isospin) doublet. In this flavour subsector, the QCD Lagrangian is

$$\mathcal{L}_{QCD} = \bar{q}(x) \left[i \gamma_\mu D^\mu - m \right] q(x) - \frac{1}{4} F_{\mu\nu}^a(x) F_a^{\mu\nu}(x), \tag{3.1}$$

where $q(x)$ are the quark fields and m is the mass matrix:

$$q(x) = \begin{pmatrix} q_u(x) \\ q_d(x) \end{pmatrix} , \quad m = \begin{pmatrix} m_u & 0 \\ 0 & m_d \end{pmatrix} . \tag{3.2}$$

Here

$$D_\mu \equiv \partial_\mu - ig \frac{\lambda_a}{2} G_\mu^a(x), \tag{3.3}$$

where $G_\mu^a(x)$ is the gluon field with color indices a = 1, ..., 8; $F_{\mu\nu}^a$ is the corresponding field tensor, and λ_a are the SU(3) color matrices.

Now, there are many hints that the (current) quark masses m_u and m_d are very small compared to typical hadron masses. The important point is that for $m_u = m_d = 0$, $\mathcal{L}_{QCD}$ of eq. (3.1) is invariant under the chiral transformation

$$q(x) \rightarrow e^{i\gamma_5 \vec{\tau} \cdot \vec{\theta}} q(x) , \quad \bar{q}(x) \rightarrow \bar{q}(x) e^{i\gamma_5 \vec{\tau} \cdot \vec{\theta}} . \tag{3.4}$$

That is, chiral symmetry is a fundamental symmetry of QCD with massless quarks. This symmetry combines the conservation of helicity for massless, free Fermions, with the (u,d) iso-doublet structure of the quark fields.

Invariance under the chiral transformation, eq. (3.4), implies that the quark axial
current,

$$\vec{A}_\mu(x) = \bar{q}(x)\,\gamma_\mu\,\gamma_5\,\frac{\vec{\tau}}{2}\,q(x) \tag{3.5}$$

is conserved for free quarks, i.e.

$$\partial^\mu \vec{A}_\mu(x) = 0 . \tag{3.6}$$

On the other hand, the solutions of the equations of motion derived from $\mathcal{L}_{QCD}$ are expected to generate confinement for individual quarks. Once confinement sets in, chiral
symmetry is necessarily broken. To illustrate this, consider for example a single,
massless quark whose motion is partly confined by a reflecting wall. Reflection at the
wall implies that the quark momentum changes from $\vec{p}$ to $-\vec{p}$, whereas the quark spin
remains unaffected. Thus the helicity $h = \vec{\sigma}\cdot\vec{p}/|\vec{p}|$ changes sign, i.e. the quark wave-
function is not an eigenfunction of helicity any more. In more general terms, chiral
symmetry is spontaneously (or rather: dynamically) broken. This can be cast into simple
phenomenological terms as shown in the following section.

3.2 Confining Potentials and Chiral Symmetry breaking

The phenomenology of confined quarks has been developed quite successfully in terms of
Bag Models [13] and their extension to incorporate Chiral Symmetry [14-16]. We shall
follow here a slightly different path, though with a similar physical picture in mind,
by assuming that non-strange baryons are composed of massless u- and d-quarks confined
by a scalar potential M(r) [17,18]. This potential is to be interpreted as the mean
field experienced by individual quarks and generated by the confining forces which are
probably due to non-perturbative gluon interactions. Soliton models [19,20] simulate
these degrees of freedom in terms of a scalar soliton field $\sigma(r)$, so that the local
quark mass becomes $M(r) = g\sigma(r)$, where g is a coupling constant. The quark Hamiltonian
in such a picture is ($\vec{\alpha} = \gamma_0\vec{\gamma}$, $\beta = \gamma_0$):

$$H = \vec{\alpha}\cdot\vec{p} + \beta M(r) , \tag{3.7}$$

and the quark fields $q(x) = q(\vec{r},t)$ satisfy the Dirac equation

$$\left[i\gamma_\mu\partial^\mu - M(r)\right] q(\vec{r},t) = 0 . \tag{3.8}$$

The confining potential M(r) should have some of the qualitative features suggested
by QCD, assuming that M(r) represents a mean field primarily of gluonic origin: in
the hadron center, M(r) should be small, so as to allow quarks to move freely, in
accordance with asymptotic freedom. Towards the surface, M(r) should grow rapidly to
yield confinement. Absolute confinement requires $M(r) \to \infty$ beyond some distance from
the hadron center.

An ansatz for M(r) can be made as a power series in r, or simply by a single power law
$M(r) = cr^n$. For such potentials and the Dirac equation eq. (3.8) a virial theorem can
be derived [21]: The potential energy,

$$E_{pot} = \int d^3r \; <q^+ \, \gamma_o \, M(r) \, q >$$

is related to the total energy E in a given quark orbit by

$$E_{pot} = \frac{E}{n+1} \; . \qquad (3.9)$$

For n = 3 the confining potential $M(r) = cr^3$ essentially replaces the volume part of the energy in the standard MIT bag model, where the energy per quark is

$$\frac{E(R)}{N} = \frac{x}{R} + \frac{4\pi}{3N} \, BR^3 \; . \qquad (3.10)$$

The first term in eq. (3.10) is to be interpreted as the quark kinetic energy, with x = 2.04 for the lowest $s_{1/2}$ orbit. The condition dE/dR = 0 implies that the volume term, $(4\pi/3N)BR^3$, is 1/4 of the total energy, just as for the r^3-potential. The parameter c for n = 3 plays the role of an energy density, which we expect to be of the order of 1 GeV/fm^3.

Consider now the axial current of a single quark satisfying the Dirac equation, eq. (3.8). We take the divergence and find, using the Dirac equation:

$$\partial^\mu A_\mu^\lambda (x) = M(r) \; \bar{q}(x) \, i\gamma_5 \, \tau^\lambda q(x) \; . \qquad (3.11)$$

This result tell us that the breaking of chiral symmetry, measured by the nonzero divergence of the axial current, is directly related to the confining potential. The limit of free, massless quarks would be obtained with $M(r) \equiv 0$. The right hand side of eq. (3.11) acts as a pseudoscalar-isovector source function. This source function obviously peaks at the baryon surface, since M(r) rises like a power, whereas the quark wave functions q(r) decrease exponentially beyond a distance comparable to the baryon size.

3.3 Introducing the Goldstone Pion

If QCD has an underlying SU(2) x SU(2) chiral symmetry, then the dynamical breaking of this symmetry by confinement at the quark level must be restored by a compensating field carrying the quantum numbers of a pion. The Goldstone theorem requires the existence of such a Boson field with zero mass. To demonstrate this, one generalizes the axial current,

$$A_\mu^\lambda (x) = \bar{q}(x) \gamma_\mu \gamma_5 \frac{\tau^\lambda}{2} q(x) - f_\pi \, \partial_\mu \, \phi^\lambda(x) \quad \text{+ terms non-linear in } \phi^\lambda \qquad (3.12)$$

by introducing the pseudoscalar-isovector field $\phi^\lambda(x)$ just mentioned. Here f_π is a constant. Restoring chiral symmetry means to require that the divergence of eq. (3.12) vanishes.

Suppose now that we can omit the terms non-linear in ϕ^λ as a first approximation [16].

Then together with eq. (3.11), the condition $\partial^\mu A^\lambda_\mu = 0$ implies the following field equation for ϕ^λ:

$$f_\pi \,\square\, \phi^\lambda(x) = M(r)\,\bar{q}(x)\,i\gamma_5\,\tau^\lambda q(x).$$

(3.13)

The suggestion is, of course, to identify ϕ^λ with the pion. This pion has zero mass according to eq. (3.13). We refer to it as the Goldstone Boson associated with the breaking of chiral symmetry at the quark level.

The step from a conserved axial current to PCAC can be made by introducing a finite pion mass, m_π = 140 MeV. Furthermore, f_π should be identified with the pion decay constant, f_π = 93 MeV, since the pionic part of the axial current determines the decay rate for $\pi \rightarrow \mu\nu$. Eq. (3.13) is then replaced by

$$(\square + m_\pi^2)\,\phi^\lambda(x) = \frac{M(r)}{f_\pi}\,\bar{q}(x)\,i\gamma_5\,\tau^\lambda q(x),$$

(3.14)

and the divergence of the axial current becomes

$$\partial^\mu A^\lambda_\mu(x) = m_\pi^2\,f_\pi\,\phi^\lambda(x).$$

(3.15)

In chiral bag models, the source function on the right hand side of eq. (3.14) is proportional to a δ-function at the bag boundary.

The pion is introduced here on purely phenomenological grounds, as in chiral bag models. There is no obvious relation to the pion as a bound $q\bar{q}$ pair at this level. A more profound approach can be based on the Nambu and Jona-Lasinio model [22]. This model starts from a chiral invariant effective Lagrangian for massless quarks and demonstrates that if the quarks acquire a non-zero effective mass by sufficiently strong self-interactions, at which point chiral symmetry is spontaneously broken, a bound quark-antiquark mode carrying pion quantum numbers develops with zero mass. The physical pion mass is then obtained by starting from finite, but small quark masses of order 10 MeV. The pion in such a picture is a coherent superposition of $q\bar{q}$ states [23] and has properties analogous to low-lying collective particle-hole states in many body systems. The pion core must be small ($r_\pi \lesssim 0.4$ fm) in order to obtain the correct decay constant f_π [24,25].

The very special nature of the pion as compared to other mesons is clearly one of the most fundamental aspects of nuclear forces, although we cannot go into further details here. Some interesting features of pion-nucleon dynamics can however be discussed already at the present level.

3.4 Pion-Nucleon Coupling and the Axial Form Factor

Suppose that nucleons are described by three massless [+] quarks occupying the lowest orbit of the confining potential M(r). Eq. (3.13) tells that the coupling of a pion to quarks in the nucleon is given by the source function

$$J_5^\lambda(x) = \frac{M(r)}{f_\pi} \sum_{j=1}^{3} \bar{q}_j(x) \, i\gamma_5 \tau^\lambda q_j(x). \tag{3.16}$$

In the static limit, we define a pion-nucleon form factor $G_{\pi NN}(q^2)$ by

$$-i \, G_{\pi NN}(q^2) \langle \vec{\sigma} \cdot \vec{q} \, \tau^\lambda \rangle = 2M \langle N | \int d^3 r \, e^{i\vec{q}\cdot\vec{r}} J_5^\lambda | N \rangle, \tag{3.17}$$

where $\langle \vec{\sigma}\tau^\lambda \rangle$ refers to matrix elements taken with nucleon spin and isospin operators, and $|N\rangle$ is the SU(4) three-quark wave function of the nucleon. The πN coupling constant is

$$g_{\pi NN} = G_{\pi NN}(q^2=0). \tag{3.18}$$

(It is actually defined as $G_{\pi NN}(q^2 = m_\pi^2)$, but we ignore this minor detail.)

Another form factor of interest is the one related to the quark axial current. The axial form factor measures the spin distribution of quarks inside the nucleon. At momentum transfers $q^2 \ll 4M^2$, it is given by

$$G_A(q^2) \langle \sigma^i \tau^\lambda/2 \rangle = \langle N | \int d^3 r \, e^{i\vec{q}\cdot\vec{r}} A_i^\lambda(\vec{r}) | N \rangle, \tag{3.19}$$

where A_i^λ is one of the three-vector components ($i = 1,2,3$) of the $A_\mu^\lambda = \bar{q} \, \gamma_\mu \gamma_5 (\tau^\lambda/2) q$, summed over the three valence quarks. The $G_A(q^2)$ is normalized according to

$$g_A = G_A(q^2=0), \tag{3.20}$$

where g_A is the axial charge. (Empirically, $g_A = 1.26$). Now, it can be shown [25,26] that $G_A(q^2)$ does not receive contributions from a pionic term proportional to $f_\pi \partial_\mu \phi$ of the axial current as long as ϕ is a continuous function. This makes $G_A(q^2)$ a particularly suitable quantity to discuss the quark core size. For a confining potential $M(r) = cr^3$ with $c \simeq 1$ GeV/fm^3, we find the result, Fig. 5. The axial charge comes out to be $g_A = 1.21$. Center-of-mass corrections, obtained by projection of the quark momenta onto good total momentum, turn out to be small, if the projection procedure is constrained by the gauge invariance requirement for the corresponding electromagnetic current [26]. The rms radius associated with $G_A(q^2)$ is $\langle r^2 \rangle^{1/2} \simeq 0.6$ fm.

It is straightforward to show by using the Dirac equation that $g_{\pi NN}$ and g_A are connected by the Goldberger-Treiman relation,

---

[+] We could add at this point small current quark masses of about 10 MeV, consistent with a finite, but small pion mass.

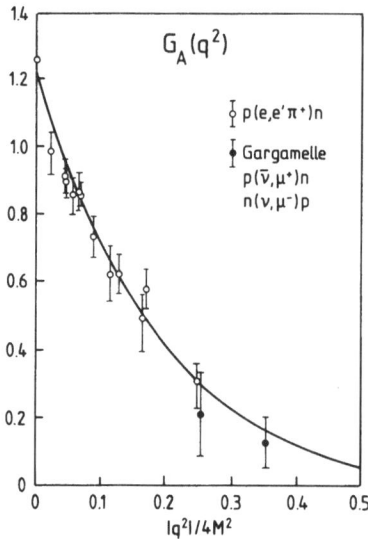

$G_A(q^2)$

Figure 5: The nucleon axial form factor calculated for three quarks confined in the potential $M(r) = cr^3$ with $c = 0.95$ GeV/fm^3 [26].

$$g_{\pi NN} = \frac{M}{f_\pi} g_A . \tag{3.21}$$

For $g_A = 1.21$ obtained with the cr^3 potential, $g_{\pi NN} \simeq 12$ results, to be compared with the empirical value $g_{\pi NN} \simeq 13$.

The pion-nucleon source function is shown in Fig. 6. It exhibits the characteristic surface peaking. The resulting form factor $G_{\pi NN}(q^2)$ [26] is slightly softer than the axial form factor $G_A(q^2)$. Similar conclusions have been drawn in ref. [25].

Unlike the nucleon electromagnetic form factors, $G_A(q^2)$ receives practically no contribution from the pion cloud in this model. Effects from 3π states could be present in principle, but they would probably change the picture very little, the dominant contribution in this channel being the A_1 with a mass of no less than 1.3 GeV.

In chiral quark models, the difference in radius between the axial form factor, which measures essentially the spin distribution within the nucleon, and the charge radius $\langle r_c^2 \rangle^{1/2} = 0.83$ fm is assigned to the charged pion cloud surrounding the quark core. We present in Fig. 7 the results of such a calculation [17] where the quark core is the same as used to obtain $G_A(q^2)$ of Fig. 5. The calculation includes approximate center-of-mass corrections. It shows that the proton charge radius is in fact determined largely by the pion cloud which represents about 1/3 of the total charge.

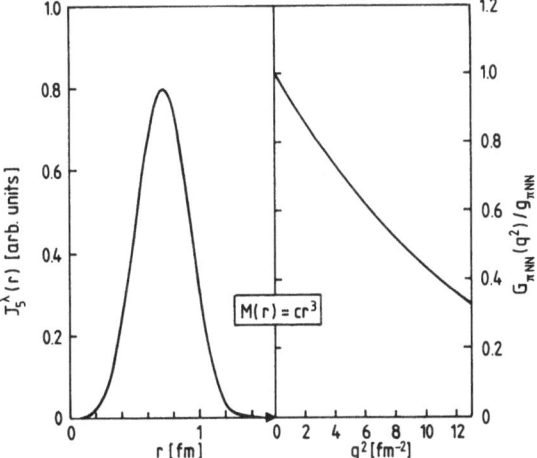

<u>Figure 6:</u> Pion-nucleon source function $\langle N|J_5^\lambda(r)|N\rangle$ (left) and the corresponding pion-nucleon form factor $G_{\pi NN}(q^2)$ evaluated according to eqs.(3.16,17) with a confining potential $M(r) = cr^3$ with $c = 0.95$ GeV/fm^3.

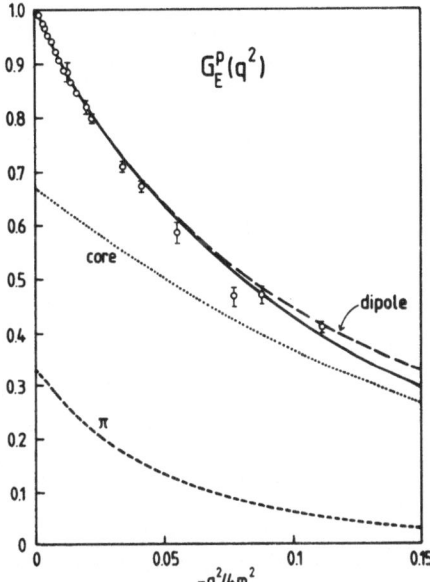

<u>Figure 7:</u> Proton charge form factor calculated according to ref. [27]. The contribution of the quark core and of the pion cloud are shown separately; the sum of both is compared to data. The quark core is the same as the one giving the $G_A(q^2)$ of Fig. 5.

3.5 Constraints on the OPE Tensor Force: the deuteron asymptotic D/S-ratio

A quark core of about 1/2 fm radius will introduce substantial modifications as compared to OPE with pointlike nucleon sources. The static OPE potential with form factors becomes

$$V_\pi(\vec{q}) = - \frac{G_{\pi NN}^2(q^2)}{4 M^2} \frac{\vec{\sigma}_1 \cdot \vec{q} \; \vec{\sigma}_2 \cdot \vec{q}}{\vec{q}^2 + m_\pi^2} \; \vec{\tau}_1 \cdot \vec{\tau}_2 \; . \tag{3.22}$$

For $G_{\pi NN}(q^2)$ as obtained from a quark core following the preceeding discussion, we show the resulting tensor potential in Fig. 8, for a core radius $\langle r^2 \rangle^{1/2} = 0.5$ fm. The finite size of the core effectively weakens the tensor force, by an amount determined by the rms radius.

One of the best possibilities to examine the tensor force is by investigating the asymptotic D/S-ratio in the deuteron. We follow here the discussion of ref. [28]. The D/S-ratio η is defined in terms of the asymptotic S- and D-state components of the deuteron wave function (u(r) and w(r), respectively) as follows:

$$u(r) \xrightarrow[r \to \infty]{} N e^{-\alpha r}, \quad w(r) \xrightarrow[r \to \infty]{} \eta N \left(1 + \frac{3}{\alpha r} + \frac{3}{\alpha^2 r^2}\right) e^{-\alpha r}, \tag{3.23}$$

where $\alpha^2 = \varepsilon M$ and ε is the deuteron binding energy. The value of η is determined to such high accuracy that it allows for a detailed test of the tensor potential at

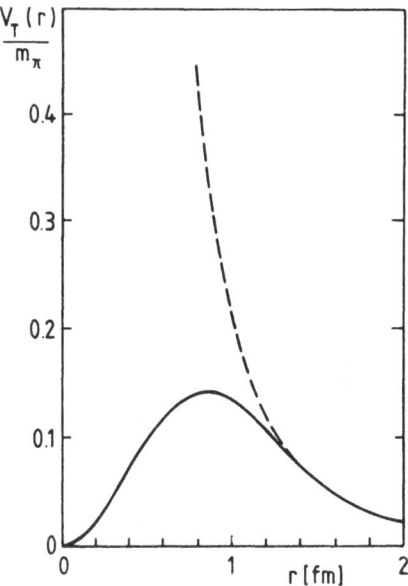

Figure 8: The one-pion exchange tensor potential with point-like nucleons (dashed curve) and modified by form factors $G_{\pi NN}(q^2)$ calculated with a quark confining potential M(r) = cr^3, c = 1 GeV/fm^3 (solid curve).

distances $r \gtrsim 0.6$ fm, the unknown short distance behaviour being suppressed [28].

We show in Fig. 9 a calculation of η following the method of Ericson and Rosa-Clot, using the quark core πN form factor $G_{\pi NN}(q^2)$ as input, and varying the quark core density radius $<r^2>^{1/2}$. The result indicates that the measured value of η sets an upper limit to $<r^2>^{1/2}$ of about 0.6 fm. This result does not depend on the precise form of $G_{\pi NN}(q^2)$, the essential parameter being just $<r^2>$ [29]. It may well be, of course, that $G_{\pi NN}(q^2)$ in such an analysis represents a variety of complicated short-distance processes, so that the immediate relation to the quark core size is obscured. In any case, the data tell that deviations from pointlike OPE should effectively not extend beyond a distance $r \gtrsim 0.6$ fm.

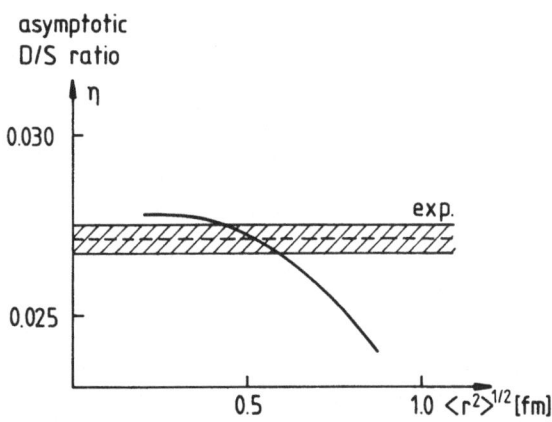

Figure 9: Deuteron asymptotic D/S-ratio calculated according as in [28], with $G_{\pi NN}(q^2)$ from eq. (3.17). The radius $<r^2>^{1/2}$ refers to the density distribution of the quark core as calculated with confining potential $M(r) = cr^3$ (see text).

3.6 Nucleonic Spin-Isospin Transitions, the $\Delta(1232)$, and $\pi N\Delta$ -Coupling

Three u or d quarks in a $(1s_{1/2})^3$ configuration of a bag or confining potential can be coupled either to a spin 1/2, isospin 1/2 state (the nucleon) or a spin 3/2, isospin 3/2 state (the Δ). The mass splitting between N(938) and $\Delta(1232)$ is due to spin-dependent residual forces. The most important mechanisms contributing to this splitting are supposed to be chromomagnetic interactions from gluon exchange and the spin-isospin dependent self-interactions of the nucleon or Δ via the surrounding pion cloud (see e.g. refs. [15,16]).

Such a model, together with the pion-quark interaction developed before, necessarily implies a strong $\pi N\Delta$ coupling. The reason is that the transition from nucleon to $\Delta(1232)$ is made by a single spin-isospin flip on one of the quarks, without changing their spatial $(1s)^3$ configuration.

Given the pion-quark source function J_5^λ of eq. (3.16), we can define a $\pi N\Delta$ transition form factor $G_{\pi N\Delta}(q^2)$ by

$$-i\, G_{\pi N\Delta}(q^2)\, \langle \vec{S}^+ \cdot \vec{q}\; T_\lambda^+ \rangle = 2\sqrt{MM_\Delta}\, \langle \Delta | \int d^3 r\, e^{i\vec{q}\cdot\vec{r}} J_5^\lambda | \Delta \rangle, \quad (3.24)$$

where $|N\rangle$ and $|\Delta\rangle$ are the quark model wave functions of nucleon and $\Delta(1232)$, respectively, and the matrix element $\langle \vec{S}^+\vec{T}^+ \rangle$ on the left hand side refers to transition matrix elements taken between spin-isospin - 1/2 and spin-isospin - 3/2 states. The transition operators are defined as

$$\langle \tfrac{3}{2}\, m_\Delta\, |\, S_\mu^+\, |\, \tfrac{1}{2}\, m_N \rangle = (\tfrac{3}{2}\, m_\Delta\, |\, 1\mu\, \tfrac{1}{2}\, m_N),$$
$$\langle \tfrac{3}{2}\, \tau_\Delta\, |\, T_\mu^+\, |\, \tfrac{1}{2}\, \tau_N \rangle = (\tfrac{3}{2}\, \tau_\Delta\, |\, 1\mu\, \tfrac{1}{2}\, \tau_N) \qquad (3.25)$$

A property of the spin transition operators which will be used frequently in practical applications is

$$S_i\, S_j^+ = \delta_{ij} - \tfrac{1}{3}\, \sigma_i\, \sigma_j, \qquad (3.25a)$$

where i and j denote Cartesian components (i = 1,2,3). Since the orbital parts of the three-quark-wave functions $|N\rangle$ and $|\Delta\rangle$ are the same as long as the N-Δ mass splitting is treated in first order perturbation theory, the ratio of $G_{\pi N\Delta}$ and $G_{\pi NN}$ is determined entirely by spin-isospin coupling coefficients, i.e. by the underlying SU(2) x SU(2) symmetry of the problem. More precisely [30]:

$$\frac{\langle \Delta | J_5^\lambda | N \rangle}{\langle N | J_5^\lambda | N \rangle} = \sqrt{\frac{72}{25}}. \qquad (3.26)$$

The resulting model of πNN and $\pi N\Delta$ couplings can be summarized in terms of the interaction Hamiltonians

$$H_{\pi NN} = i\, \frac{f(q^2)}{m_\pi}\, \vec{\sigma}\cdot\vec{q}\; \vec{\tau}\cdot\vec{\phi}, \qquad (3.27)$$

$$H_{\pi N\Delta} = i\, \frac{f_\Delta(q^2)}{m_\pi}\, \vec{S}^+\cdot\vec{q}\; \vec{T}^+\cdot\vec{\phi} + h.c., \qquad (3.28)$$

where

$$f(q^2) = \frac{m_\pi}{2M}\, G_{\pi NN}(q^2),$$
$$f_\Delta(q^2) = \frac{m_\pi}{2\sqrt{MM_\Delta}}\, G_{\pi N\Delta}(q^2), \qquad (3.29)$$

with $f(q^2 = 0) \equiv f = 1$, $f_\Delta^2/f^2 = 72/25$. This gives already a reasonable description of p-wave pion-nucleon scattering and one-pion exchange forces. More detailed quantitative agreement can be obtained by adding relativistic corrections and readjusting f_Δ to the Chew-Low value, $f_\Delta = 2$ [5], the model we shall adopt in many-body applications.

3.7 Summary

Chiral symmetry is a fundamental symmetry of QCD with massless u and d quarks. Confinement necessarily breaks chiral symmetry at the quark level and implies the existence of a massless Goldstone pion. The finite pion mass is probably related to finite, but small current quark masses m_u, m_d of order 10 MeV. Chiral bag or confinement potential models are a convenient way to introduce the coupling of pions to quarks at the baryon surface. In such models, the πNN or $\pi N\Delta$ form factor can be calculated; for a cr^3 confining potential it turns out to be well approximated by

$$f(q^2) = \sqrt{\frac{25}{72}}\, f_\Delta(q^2) = f\, exp\left[-\frac{\vec{q}^2}{\Lambda^2}\right],$$ (3.30)

where Λ^{-1} is related to the quark core rms radius. For example, one obtains $\Lambda = 750$ MeV for $<r^2>^{1/2} = 0.5$ fm. Such relatively soft pion-nucleon form factors are consistent with the nucleon axial form factor (although the experimental uncertainties are unfortunately too large at present to draw more quantitative conclusions). It would be difficult, however, to accomodate such a Λ with the existing one-boson exchange phenomenology of the nucleon-nucleon interaction [10] which suggests cutoffs Λ of about 1 GeV or larger. To establish connections between boson exchange models and the underlying quark dynamics at short distance remains as a key problem.

LECTURE 3

4. Virtual Pions in Nuclei

4.1 Pion Exchange Currents

Some of the strongest evidence for pionic degrees of freedom in nuclei comes from investigations with electromagnetic probes. The exchange of virtual charged pions between nucleons contributes genuine two-body pieces to the total nuclear current. Effects of these socalled exchange currents have been studied in great detail [31,32], especially for simple systems like the deuteron.

We recall that the static pion-nucleon interaction Hamiltonian (with pointlike nucleons) is

$$H_{\pi NN} = \frac{f}{m_\pi}\, (\vec{\sigma}\cdot\vec{\nabla})\, (\vec{\tau}\cdot\vec{\phi}).$$ (4.1)

Now, the rule for introducing the photon field in a gauge invariant way is to replace

$$\vec{\nabla} \rightarrow \vec{\nabla} \mp ie\vec{A},$$ (4.2)

where $\vec{A}$ is the vector potential, and $\pm$ refers to a π^+ or π^- meson, respectively. This generates an interaction of the form

$$H_{\gamma\pi N} = \frac{f}{m_\pi}\, ie\,\sqrt{2}\; \tau_\pm\, \phi_\mp\, \vec{\sigma}\cdot\vec{A}. \tag{4.3}$$

Furthermore, the pion itself carries a current

$$\vec{J}_\pi = ie\left[\phi_+ \vec{\nabla}\phi_- - \phi_- \vec{\nabla}\phi_+\right], \tag{4.4}$$

which interacts with the photon, the corresponding term of the interaction being

$$H_{\gamma\pi} = \vec{J}_\pi\cdot\vec{A}. \tag{4.5}$$

To lowest order in the two-body one-pion exchange interaction, perturbation theory with $H_{int} = H_{\pi NN} + H_{\gamma\pi N} + H_{\gamma\pi}$ generates the Feynman diagrams shown in Fig. 10.

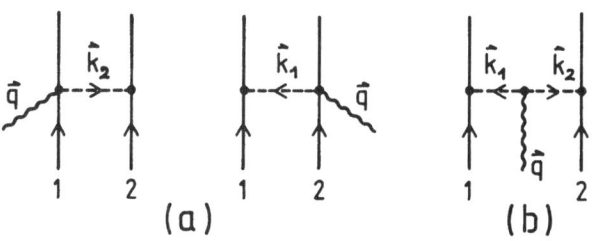

Figure 10:   Two-body exchange currents from OPE, (a) Pair current; (b) pionic current.

The two diagrams (a) contain the point interaction $H_{\gamma\pi N}$ together with $H_{\pi N}$, connected by the propagator of the exchanged pion. This part is called the pair current because it derives a non-relativistic reduction of a process which relativistically (with γ_5 coupling) involves the virtual excitation of a particle-antiparticle pair. The corresponding two-nucleon current is

$$\vec{J}_{pair}(\vec{k}_1,\vec{k}_2) = -ie\left(\frac{f^2}{m_\pi^2}\right)\left[\vec{\tau}(1)\times\vec{\tau}(2)\right]_3 \left\{\frac{\vec{\sigma}_1(\vec{\sigma}_2\cdot\vec{k}_2)}{k_2^2+m_\pi^2} - \frac{\vec{\sigma}_2(\vec{\sigma}_1\cdot\vec{k}_1)}{k_1^2+m_\pi^2}\right\}. \tag{4.6}$$

The isospin dependence reflects the fact that the nucleon isospins always come in the combination $\tau_-(1)\tau_+(2) - \tau_+(1)\tau_-(2) = 2i\left[\vec{\tau}(1)\times\vec{\tau}(2)\right]_3$.

The term Fig. 10b represents the coupling of the photon to the exchanged pion. The pion current, eq. (4.4), gives a factor $(\vec{k}_1 - \vec{k}_2)$. The πNN vertex $H_{\pi NN}$ appears twice. There are two pion propagators, one for each pion of momentum $\vec{k}_1$ and $\vec{k}_2$, respectively:

$$\vec{J}_\pi(\vec{k_1},\vec{k_2}) = ie\left(\frac{f^2}{m_\pi^2}\right)\left[\vec{\tau}^{(1)}\times\vec{\tau}^{(2)}\right]_3 \frac{\vec{\sigma_1}\cdot\vec{k_1}\,(\vec{k_1}-\vec{k_2})\,\vec{\sigma_2}\cdot\vec{k_2}}{(\vec{k_1}^2+m_\pi^2)(\vec{k_2}^2+m_\pi^2)}\,. \qquad (4.7)$$

Note that the sum of pionic and pair current, if multiplied by $\vec{q} = \vec{k}_1 + \vec{k}_2$, becomes

$$\vec{q}\cdot(\vec{J}_{pair}+\vec{J}_\pi) =$$

$$= -ie\left(\frac{f^2}{m_\pi^2}\right)\left[\vec{\tau}^{(1)}\times\vec{\tau}^{(2)}\right]_3\left\{\frac{\vec{\sigma_1}\cdot\vec{k_1}\,\vec{\sigma_2}\cdot\vec{k_1}}{\vec{k_1}^2+m_\pi^2} - \frac{\vec{\sigma_1}\cdot\vec{k_2}\,\vec{\sigma_2}\cdot\vec{k_2}}{\vec{k_2}^2+m_\pi^2}\right\} \qquad (4.8)$$

This demonstrates that the pionic and pair currents have always to be taken together in order to satisfy current conservation (or equivalently, gauge invariance).

It is useful also to present the currents in r-space

$$\vec{J}(\vec{x},\vec{r_1},\vec{r_2}) = \int\frac{d^3k_1}{(2\pi)^3}\frac{d^3k_2}{(2\pi)^3}\,e^{i\vec{k_1}\cdot(\vec{r_1}-\vec{x})}\,e^{i\vec{k_2}\cdot(\vec{r_2}-\vec{x})}\,\vec{J}(\vec{k_1},\vec{k_2})\,. \qquad (4.9)$$

This yields

$$\vec{J}_{pair}(\vec{x},\vec{r_1},\vec{r_2}) = -e\left(\frac{f^2}{4\pi}\right)\left[\vec{\tau}^{(1)}\times\vec{\tau}^{(2)}\right]_3 Y_1(m_\pi r)\,.$$

$$\left[\vec{\sigma_1}(\vec{\sigma_2}\cdot\hat{r})\,\delta^3(\vec{r_1}-\vec{x}) + \vec{\sigma_2}(\vec{\sigma_1}\cdot\hat{r})\,\delta^3(\vec{r_2}-\vec{x})\right]\,; \qquad (4.10a)$$

$$\vec{J}_\pi(\vec{x},\vec{r_1},\vec{r_2}) = -e\left(\frac{f^2}{4\pi}\right)\left[\vec{\tau}^{(1)}\times\vec{\tau}^{(2)}\right]_3\,.$$

$$(\vec{\nabla_1}-\vec{\nabla_2})(\vec{\sigma_1}\cdot\vec{\nabla_1})(\vec{\sigma_2}\cdot\vec{\nabla_2})\,Y_0(m_\pi|\vec{r_1}-\vec{x}|)\,Y_0(m_\pi|\vec{r_2}-\vec{x}|)\,. \qquad (4.10b)$$

This illustrates most clearly the way in which the photon couples to the charged pion at the point $\vec{x}$ between the two nucleons located at points $\vec{r}_1$ and $\vec{r}_2$. Here $Y_0(z) = \frac{e^{-z}}{z}$ and $Y_1(z) = (1 + \frac{1}{z})\,Y_0(z)$.

The above expressions are obtained for pointlike nucleons and pions. At high momentum transfers, modifications due to the finite size of nucleons and pions should be introduced. This is not a trivial procedure since the introduction of form factors has to be done such that gauge invariance is strictly satisfied.

4.2 The Exchange Magnetic Moment

Pion exchange currents have substantial influence on magnetic properties of nuclei in general, and on magnetic moments in particular. We recall that the magnetic moment density is related to the total current $\vec{J}$ by

$$\vec{\mu}(\vec{x}) = \frac{1}{2}\,\vec{x}\times\vec{J}(\vec{x})\,, \qquad (4.11)$$

where $\vec{J}$ contains both single-nucleon currents and two-body exchange currents,

$$\vec{J}(\vec{x}) = \sum_{i=1}^{A} \vec{J_i}(\vec{x}) + \vec{J}_{ex}(\vec{x}).$$ (4.12)

The exchange magnetic moment is given by

$$\vec{\mu}_{ex} = \frac{1}{2} \int d^3x \left[\vec{x} \times \vec{J}_{ex}(\vec{x}) \right].$$ (4.13)

By far the dominant contribution comes from the pair current. Using eq. (4.10a) one obtains

$$\vec{\mu}_{ex}(1,2) = \frac{ef^2}{8\pi} \left[\vec{\tau}(1) \times \vec{\tau}(2) \right]_3 \cdot$$

$$\cdot \left\{ [\vec{\sigma_1} \times \vec{r_1}](\vec{\sigma_2} \cdot \hat{r}) + (\vec{\sigma_1} \cdot \hat{r})[\vec{\sigma_2} \times \vec{r_2}] \right\} Y_1(m_\pi r),$$ (4.14)

where $\vec{r} = \vec{r_1} - \vec{r_2}$. The exchange magnetic moment gives a characteristic correction to g_ℓ, the orbital g-factor of a nucleon in a nucleus. For example, an odd proton can exchange a π^+ with neutrons in the core. The corresponding correction $\delta g_\ell(p)$ is positive. Similarly, if the odd particle is a neutron, the corresponding $\delta g_\ell(n)$ arises from a π^- exchanged with protons in the core. Hence $\delta g_\ell(n)$ is negative, and these considerations lead to

$$\frac{\delta g_\ell(p)}{\delta g_\ell(n)} = -\frac{N}{Z}$$ (4.15)

for pion exchange corrections to g_ℓ. The above arguments are summarized in the form [33]:

$$\delta g_\ell = \zeta \tau_3 ,$$ (4.16)

where actual calculations yield $\zeta = 0.1 - 0.2$. The isovector character of δg_ℓ is of course related to the isovector nature of the pionic current. The δg_ℓ due to pion exchange effects is a substantial fraction of the measured δg_ℓ in various nuclei (see T. Yamazaki in ref. [33]), but a detailed discussion of higher order configuration mixing effects is necessary to make the discussion quantitative.

4.3 The $\Delta(1232)$ Exchange Current

The $\Delta(1232)$ is reached from the nucleon by a spin-flip isovector transition which can be induced either by a pion or by the isovector M1 part of a photon field. The $\gamma N\Delta$ coupling is of the type

$$H_{\gamma N\Delta} = i \frac{f_{\gamma N\Delta}}{m_\pi} (\vec{S}^+ \times \vec{q}) \cdot \vec{A} \; T_3^+ ,$$ (4.17)

where $\vec{A}$ is the electromagnetic vector potential and $\vec{S}^+$, $\vec{T}^+$ refer to the $(1/2 \to 3/2)$ spin and isospin transition operators, eq. (3.25). The coupling strength can be determined from the photoproduction of neutral pions ($\gamma N \to \pi^0 N$) which is strongly dominated by the M1 excitation of the $\Delta(1232)$ via eq. (4.17). One obtains $f_{\gamma N\Delta} = 0.116$ and notices that the following ratio holds to good accuracy:

$$\frac{f_{\gamma N \Delta}}{f_{\pi N \Delta}} = \frac{e}{g_{\pi N N}} \mu_v \tag{4.18}$$

where $g_{\pi NN}^2/4\pi = 14.4$ and μ_v is the isovector magnetic moment of the nucleon,

$$\mu_v = \tfrac{1}{2}(\mu_p - \mu_n),$$

with $\mu_p = 2.79$ and $\mu_n = -1.91$. Eq. (4.18) represents the ratio of electromagnetic to strong interaction scales, the additional μ_v arising because of the isovector magnetic dipole nature of the transition.

The virtual excitation of a $\Delta(1232)$ followed by pion exchange contributes to the two-body exchange current, as shown in Fig. 11. Following standard rules, one obtains for this part of the current:

$$\vec{J}_\Delta(\vec{q}, \vec{k}_1, \vec{k}_2) = ie\,\frac{4}{9}\,\frac{f\,f_{\pi N \Delta}\,f_{\gamma N \Delta}}{m_\pi^3\,(M_\Delta - M)} \cdot$$

$$\cdot \left\{ \left[4\,\tau_3(1)\,\frac{\vec{\sigma}_1 \cdot \vec{k}_1\,(\vec{k}_1 \times \vec{q})}{\vec{k}_1^2 + m_\pi^2} + (1 \leftrightarrow 2) - \right.\right. \tag{4.19}$$

$$\left.\left. - [\vec{\tau}(1) \times \vec{\tau}(2)]_3 \left[\frac{(\vec{\sigma}_1 \times \vec{k}_1)\,\vec{\sigma}_2 \cdot \vec{k}_2}{\vec{k}_2^2 + m_\pi^2} - (1 \leftrightarrow 2) \right] \times \vec{q} \right\} .$$

The ΔN mass difference $M_\Delta - M$ appearing in the denominator corresponds to virtual Δ-excitation after photon absorption. The reverse ordering is also possible as shown in Fig. 11, but involves the energy denominator in the form $(M_\Delta + M)^{-1}$ and is therefore suppressed. The interpretation of the isospin structure of the different terms in eq. (4.19) is straightforward. The first two terms which do not change the charge of either nucleon 1 or nucleon 2 correspond to the exchange of a π^0. The other terms proportional to $[\vec{\tau}(1) \times \vec{\tau}(2)]_3$ represent the exchange of charged $(\pi^\pm)$ pions.

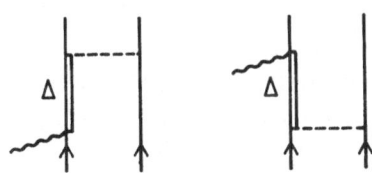

Figure 11:   Exchange current involving the virtual excitation of a $\Delta(1232)$.

4.4. Exchange Currents in Few-Nucleon Systems

4.4.1 Deuteron Electrodisintegration

A classical example of evidence for pion exchange currents is the backward inelastic electron-deuteron scattering process

$$e + d \rightarrow e' + p + n$$

near threshold. The basic process is the same as in the np $\rightarrow$ dγ reaction at threshold, but now taken at high momentum transfers. There the only possibility for the final pn pair is to be in a $^1S_0$ state which is then reached from the $^3S_1 - ^3D_1$ deuteron ground state by an M1 transition.

In the one-photon approximation close to threshold, the double differential cross section at large momentum transfers is given by

$$\frac{d^2\sigma}{dE'd\Omega} = C(\theta,E)\left\{\hat{k}'\cdot\vec{J}_{fi}^* \; \hat{k}\cdot\vec{J}_{fi} + \hat{k}\cdot\vec{J}_{fi}^* \; \hat{k}'\cdot\vec{J}_{fi} + 2\,|\vec{J}_{fi}|^2 \sin^2\frac{\theta}{2}\right\}, \qquad (4.20)$$

where $C(\theta,E)$ contains the Mott cross section and kinematic factors, and $(E,\vec{k})$, $(E',\vec{k}')$ are the in- and outgoing electron four-momenta.

The matrix elements $\vec{J}_{fi}$ receive contributions from one-body (impulse approximation) and two-body (exchange) currents. The one-body transition form factor has a character-istic minimum at $q^2 \simeq 12$ fm^{-2} due to interference between $^3D - ^1S$ and $^3S - ^1S$ transi-tions. The matrix elements due to different parts of the current operator are shown in Fig. 12, taken from ref. [34]. We note that the pair current $\vec{J}_{pair}$ dominates at large momentum transfers, with negligible additional effects from the pionic current $\vec{J}_\pi$; the $\Delta(1232)$ current $\vec{J}_\Delta$ also contributes, but much less than the pair current.

The comparison with data taken at Saclay [35] is shown in Fig. 12, where pionic exchange currents are incorporated following ref. [36]. Similar results have been obtained in ref. [37].

The agreement of the theory with data at relatively low q^2 is essentially a consequence of chiral symmetry and soft pion theorems which are implicit in the $q^2 \rightarrow 0$ limit of the exchange current. The interesting feature is the validity of this simplest possible description of exchange currents exclusively in terms of pions even at large momentum transfers. It seems then that the short range nuclear forces suppress short distance corrections to this picture rather efficiently.

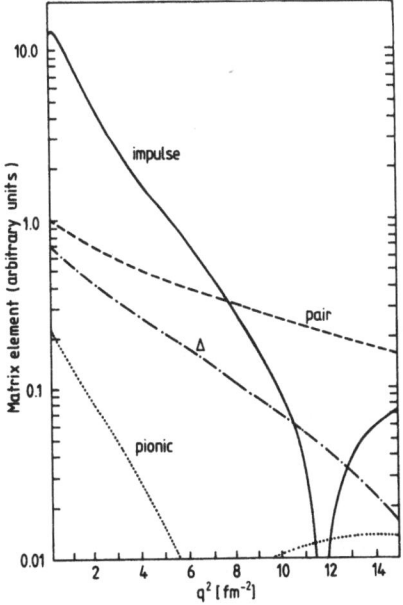

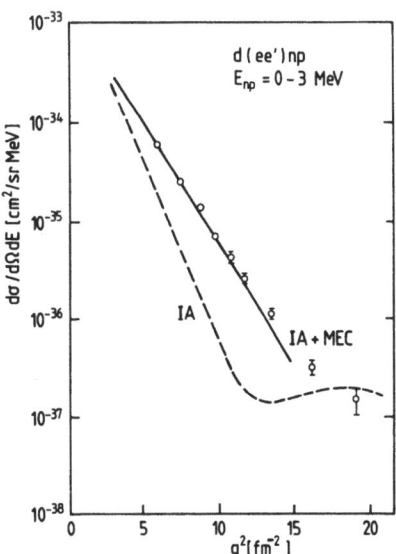

Figure 12: Contributions to the deuteron electrodisintegration matrix element [34] from pair, pionic and Δ(1232) exchange currents as compared to the impulse approximation.

Figure 13: Differential cross section for d(e,e´)np close to threshold. Dashed curve: impulse approximation; solid curve: inclusion of pair current; calculations: [36,37]; data from [35].

4.4.2 The ^{3}He Magnetic Form Factor

Another example of the influence of mesonic exchange currents is the magnetic form factor of ^{3}He. From our previous discussion, one naturally expects that the dominant M1 structure of the pair current should show up most pronouncedly in magnetic observables. This is demonstrated here for the form factor related to the spin distribution of nucleons in ^{3}He, measured by high resolution elastic electron scattering at backward angles.

This form factor is related to the magnetic moment density $\vec{\mu}(\vec{x})$ of eqs. (4.11, 4.12) by

$$F_M(q^2) = \int d^3x \, e^{i\vec{q}\cdot\vec{x}} \langle \mu_z(\vec{x}) \rangle, \qquad (4.21)$$

where $\langle\vec{\mu}\rangle$ is the expectation value taken with the ^{3}He ground state. Theoretical uncertainties in the treatment of the three-nucleon wave function are considerable greater than in the deuteron case. Nevertheless, there is general agreement that one-body currents alone evaluated with different types of three-body wave functions (Faddeev, Variational or "realistic" phenomenological wave functions) badly fail [38] when confronted with data [39]. With inclusion [40] of pion exchange currents of the form discussed previously except for modifications due to nucleon form factors, the result is shown in Fig. 14, in remarkable agreement with data.

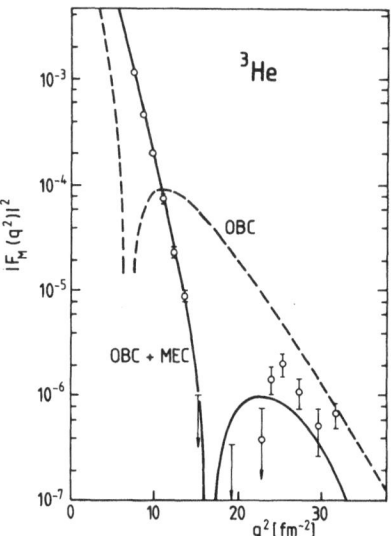

Figure 14: Magnetic form factor of ^{3}He. The dashed curve is calculated with one-body currents only. The solid line includes the effect of meson exchange currents [40]; data from [39].

<u>LECTURE 4</u>

5. Pion-Nucleon Scattering

In order to gain more insight into the role of the $\Delta(1232)$, and also for later purposes in treating pion-nucleus scattering, it is necessary to discuss pion-nucleon scattering, in some detail.

5.1 πN Scattering Amplitude

We write the πN elastic scattering amplitude as $f(\vec{q}',\vec{q})$, where $\vec{q}$ and $\vec{q}'$ denote in- and outgoing center-of-mass (c.m.) momenta. The differential cross section is given by $\frac{d\sigma}{d\Omega} = 1/2 \sum |f|^2$, where the summation is over nucleon spins. The partial wave decomposition of $f(\vec{q}',\vec{q})$ is

$$f(\vec{q}\,';\vec{q}) = \sum_I Q_I \left\{ \sum_\ell \left[(\ell+1) f_{I\ell_+}(E) + \ell f_{I\ell_-}(E) \right] P_\ell(\cos\theta) \right.$$
$$\left. - i\,\vec{\sigma}\cdot(\hat{q}\,'\times\hat{q}) \sum_\ell \left[f_{I\ell_+}(E) - f_{I\ell_-}(E) \right] P_\ell'(\cos\theta), \right. \tag{5.1}$$

where E is the c.m. energy, $\cos\theta = \hat{q}\cdot\hat{q}\,'$, Q_I projects onto the possible isospins $I = \frac{1}{2}, \frac{3}{2}$ and $\ell_\pm$ refers to channels with total angular momentum $j = \ell \pm 1/2$, respectively. The partial wave f_α ($\alpha = I\ell j$) are related to the phase shifts δ_α by

$$2iq\, f_\alpha(E) = \left[S_\alpha(E) - 1 \right], \quad S_\alpha = e^{2i\delta_\alpha}. \tag{5.2}$$

A useful quantity to work with is the K matrix,

$$K_\alpha = \frac{1}{q}\tan\delta_\alpha, \quad S_\alpha = \frac{1+iqK_\alpha}{1-iqK_\alpha}, \quad f_\alpha = \frac{K_\alpha}{1-iqK_\alpha}. \tag{5.3}$$

The threshold behaviour is described by scattering lengths (a_{2I}) for s-waves and scattering volumes ($a_{2I,2j}$) for p-waves, defined by

$$a_\alpha = \lim_{q\to 0} K_\alpha / q^{2\ell}. \tag{5.4}$$

We summarize their experimental values in table 2 :

s-wave scattering lengths $[m_\pi^{-1}]$		p-wave scattering volumes $[m_\pi^{-3}]$	
a_1	0.173 ± 0.003	a_{11}	-0.081 ± 0.002
a_3	-0.101 ± 0.004	a_{13}	-0.030 ± 0.002
		a_{31}	-0.045 ± 0.002
		a_{33}	-0.214 ± 0.002

table 2: Empirical πN s-wave scattering lengths and p-wave scattering volumes, taken from ref. [41].

Note that there is a strong cancellation in the s-wave isospin-even combination $a_1 + 2a_3$. We mention that this combination vanishes in soft pion theories. Note also that the only strong channel in the p-wave is the one with spin and isospin 3/2, where a_{33} indicates strong attraction which, as we already know, supports the formation of the $\Delta(1232)$ resonance.

5.2 The Isobar Model of p-wave πN Scattering

To illustrate how the $\Delta(1232)$ enters in the p-wave πN amplitude, let us derive f_{33} in a model with nucleon, $\Delta(1232)$ and pion-baryon couplings given by eqs. (3.27 - 28). Furthermore, the simplifying assumption of static nucleons will be made. The K matrix, still keeping its operator structure in terms of nucleon spins and isospins, is then derived as follows [5,30]:

$$\langle \vec{q}'j|K|\vec{q}i\rangle = \frac{f^2}{4\pi m_\pi^2} \left[\frac{\vec{\sigma}\cdot\vec{q}'\,\vec{\sigma}\cdot\vec{q}}{-\omega}\,\tau_j\tau_i + \frac{\vec{\sigma}\cdot\vec{q}\,\vec{\sigma}\cdot\vec{q}'}{\omega}\,\tau_i\tau_j \right]$$
$$+ \frac{f_\Delta^2}{4\pi m_\pi^2} \left[\frac{\vec{S}\cdot\vec{q}'\,\vec{S}^\dagger\cdot\vec{q}}{\omega_\Delta-\omega}\,T_jT_i^+ + \frac{\vec{S}\cdot\vec{q}\,\vec{S}^\dagger\cdot\vec{q}'}{\omega_\Delta+\omega}\,T_iT_j^+ \right], \qquad (5.5)$$

where ω is the pion c.m. energy ($E = M +\omega$ in the static limit). Eq. (5.5) summarizes all direct and crossed terms with nucleon and $\Delta(1232)$ intermediate states. Note that in the K matrix, the poles appear on the real axis at the physical masses of the inter-mediate particles in the absence of inelasticities. The crossing symmetry is evident in eq.(5.5) by examining the invariance under the replacements $\omega \leftrightarrow -\omega$, $\vec{q} \leftrightarrow -\vec{q}'$ and $i \leftrightarrow j$, where i and j refer to the pion (cartesian) isospin indices. The energy denominator related to the $\Delta(1232)$ contains the NΔ mass difference $\omega_\Delta = M_\Delta - M = 2.1\ m_\pi$.

Projecting into the P_{33} channel and neglecting the small crossed Δ-isobar term propor-tional to $[\omega_\Delta + \omega]^{-1}$, we obtain

$$K_{33}(\omega) = \frac{1}{q}\tan\delta_{33} = \frac{1}{3}\frac{q^2}{4\pi m_\pi^2}\left[\frac{4f^2}{\omega} + \frac{f_\Delta^2}{\omega_\Delta-\omega}\right]. \qquad (5.6)$$

If the Chew-Low value $f_\Delta = 2f$ is used, we find

$$f_{33}(\omega) = \frac{1}{q}e^{i\delta_{33}}\sin\delta_{33} = \frac{\frac{1}{3}(f_\Delta^2/4\pi m_\pi^2)\,q^2\,(\omega_\Delta/\omega)}{\omega_\Delta-\omega-i\,\Gamma_\Delta(\omega)/2},$$

$$\Gamma_\Delta = \frac{2}{3}\frac{f_\Delta^2}{4\pi}\frac{q^3}{m_\pi^2}\left(\frac{\omega_\Delta}{\omega}\right). \qquad (5.7)$$

Here Γ_Δ is the $\Delta \to \pi N$ decay width. At resonance ($\omega = \omega_\Delta$) where $q = 1.64\ m_\pi$, one obtains $\Gamma_\Delta \simeq 130$ MeV which is not far from the experimental width ($\Gamma_\Delta = 115 \pm 5$ MeV).

For a quantitative description of the p-wave πN phase shifts, the static limit is not accurate enough. Relativistic kinematics has to be treated appropriately, and inclusion of the N*(1470) in the spin-isospin 1/2 channel is necessary to reproduce the correct energy dependence of the P_{11} phase shift. Details are given in ref. [5]. Relativistic corrections can effectively be absorbed in the πNN and πNΔ coupling Hamiltonians, eqs. (3.27 - 28) by multiplying a factor $(2M/M + E)^{1/2}$ and $(2M_\Delta/M_\Delta + E)^{1/2}$, respectively. The experimental $\Delta \to \pi N$ width is then obtained with $f_\Delta^2/4\pi = 0.37$.

The p-wave phase shifts obtained in such a refined isobar model are shown in Fig. 15. We conclude that a p-wave π-nucleon K matrix with masses of the free nucleon, Δ and N*, is an appropriate starting point for subsequent discussions of pion-nucleus interactions.

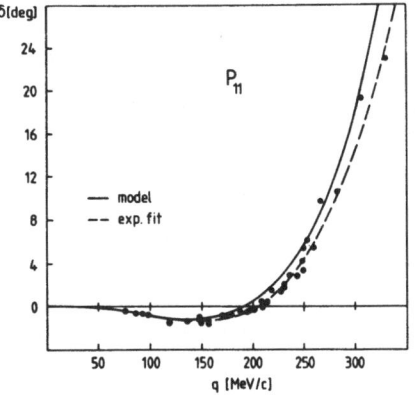

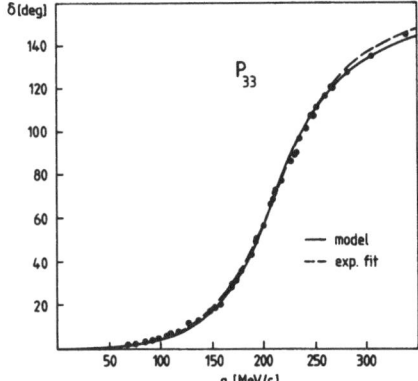

Figure 15: The p-wave πN phase shifts obtained from the isobar model K matrix including relativistic corrections, as described in ref. [5] (solid curve). The dashed curve is a fit to the experimental data.

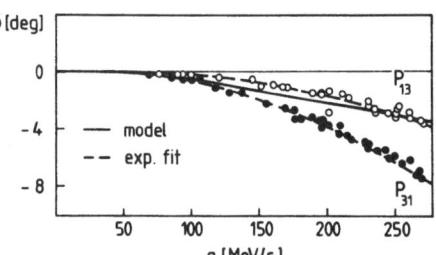

LECTURE 5

6. Pions in Nuclear Matter

6.1 Introduction: the Pion Propagator

This section is intended to prepare some of the basic concepts for the treatment of the pion-nucleus many-body problem, following ref. [5]. The framework to start with is defined by a Hamiltonian

$$H = H_N + H_\Delta + H_\pi + H_{\pi NN} + H_{\pi N\Delta} , \tag{6.1}$$

where H_N, H_Δ and H_π correspond to free nucleons, $\Delta(1232)$ and pions, $H_{\pi NN}$ and $H_{\pi N\Delta}$ are the π-nucleon and πNΔ-coupling Hamiltonians, eq. (3.27-28).

The pion field in the medium obeys the equation

$$\left(\Box + m_\pi^2\right) \vec{\phi}(\vec{r},t) = \vec{J}_5(\vec{r},t), \tag{6.2}$$

where $\vec{J}_5$ is the isovector, pseudoscalar pion source function. In the absence of sources, the free pion field is written in second quantized form as

$$\phi^\lambda(\vec{r},t) = \int \frac{d^3q}{(2\pi)^{3/2}} \; \frac{1}{\sqrt{2\omega}} \left[a_{q\lambda} e^{-i(\omega t - \vec{q}\cdot\vec{r})} + a_{q\lambda}^+ e^{i(\omega t - \vec{q}\cdot\vec{r})} \right], \quad (6.3)$$

where $\omega = \omega(\vec{q}) = \sqrt{\vec{q}^2 + m_\pi^2}$ and λ denotes isospin. For a pion in a nuclear medium, the spectrum $\omega(\vec{q})$ will change due to the interactions with the medium.

The pion propagator D is defined by

$$i\,\delta_{\lambda\lambda'}\, D(\vec{r},\vec{r}';t) = \langle 0 | T\, \phi^\lambda(\vec{r},t)\, \phi^\lambda(\vec{r}',0) | 0 \rangle, \quad (6.4)$$

where T denotes the time-ordered product, and the vacuum refers to the many-body ground state. It is convenient to work in momentum space and express D in terms of the free pion propagator.

$$D_0(\omega,\vec{q}) = \int \frac{d^4x}{(2\pi)^4} \; e^{iq\cdot x} D_0(\vec{r},0;t) = \left[\omega^2 - \vec{q}^2 - m_\pi^2 + i\delta \right]^{-1}, \quad (6.5)$$

by

$$D(\omega,\vec{q}) = D_0(\omega,\vec{q}) + D_0(\omega,\vec{q})\, \Pi(\omega,\vec{q}) D(\omega,\vec{q}) \quad (6.6)$$

$$= \left[\omega^2 - \vec{q}^2 - m_\pi^2 - \Pi(\omega,\vec{q}) + i\delta \right]^{-1}.$$

This defines the pion self-energy $\Pi(\omega,\vec{q})$ which can be interpreted as the "potential" experienced by the pion due to the interactions with the medium.

The singularities of $D(\omega,\vec{q})$ determine the spectrum $\omega(\vec{q})$ of excitations carrying the quantum numbers of the pion. The pionic response function, or pseudoscalar-isovector current correlation function, is defined by

$$\delta_{\lambda\lambda'}\, R(\vec{r},\vec{r}';t) = \langle 0 | T\, J_5^\lambda(\vec{r},t)\, J_5^{\lambda'}(\vec{r}',0) | 0 \rangle. \quad (6.7)$$

In momentum space, the response function is obtained from the pion self-energy by

$$R(\omega,\vec{q}) = \Pi(\omega,\vec{q}) + \Pi(\omega,\vec{q})\, D_0(\omega,\vec{q})\, R(\omega,\vec{q})$$

$$= \Pi(\omega,\vec{q}) + \Pi(\omega,\vec{q})\, D(\omega,\vec{q})\, \Pi(\omega,\vec{q}). \quad (6.8)$$

The quantity of primary interest is clearly the pion self-energy $\Pi(\omega,\vec{q})$ which summarizes all (irreducible) interactions of the pion with the medium.

6.2. Pion-Selfenergy and related quantities

We consider now the pionic response of an infinite medium with equal number of protons and neutrons. The pion field has energy ω and momentum $\vec{q}$. Following the discussion of previous chapters, we expect that the pion field will polarize the medium primarily via the p-wave πN interaction by exciting either nucleon-hole or $\Delta(1232)$-hole states.

A first order picture of the response is then given by the pion self-energy $\Pi^{(0)}$ illustrated in Fig. 16. We write it, somewhat schematically, as

$$\Pi^{(0)}(\omega,\vec{q}) = -\sum_{Nh} \frac{|\langle\pi(q)|H_{\pi NN}|Nh\rangle|^2}{\varepsilon_N - \varepsilon_h - \omega} + \begin{array}{l}\text{crossed}\\\text{term}\end{array}$$

$$-\sum_{\Delta h} \frac{|\langle\pi(q)|H_{\pi N\Delta}|\Delta h\rangle|^2}{\varepsilon_\Delta - \varepsilon_h - \omega} + \begin{array}{l}\text{crossed}\\\text{term}\end{array} \quad . \tag{6.9}$$

Here $|Nh\rangle$ and $|\Delta h\rangle$ denote a nucleon-hole or Δ-hole state carrying pion quantum numbers, and ε_Δ, ε_N, ε_h are the corresponding single particle energies.

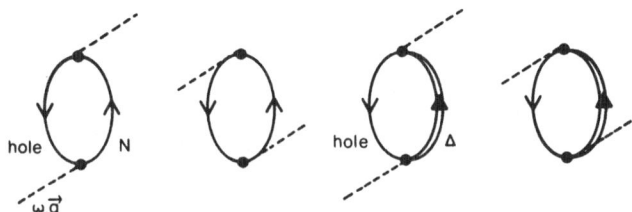

<u>Figure 16</u>: Lowest order p-wave pion selfenergy through nucleon-hole and $\Delta(1232)$-hole excitations.

We rewrite eq. (6.9) as

$$\Pi^{(0)}(\omega,\vec{q}) = -\vec{q}^{\,2}\,\overset{\circ}{\chi}(\omega,\vec{q}) , \tag{6.10}$$

where the factor $\vec{q}^{\,2}$ is due to the p-wave nature of the interaction. This defines the lowest order pionic susceptibility $\overset{\circ}{\chi}$ of the medium. Because of the spin-dependence of the underlying interaction it is useful to point out [42] the analogy between the pionic response problem and the one encountered in magnetic materials. In that sense, $\overset{\circ}{\chi}$ has a "diamagnetic" part $\overset{\circ}{\chi}_\Delta$ which involves high-lying Δ-hole excitations and a "paramagnetic" component $\overset{\circ}{\chi}_N$ related to low lying nucleon-hole excitations.

The explicit form of $\overset{\circ}{\chi} = \overset{\circ}{\chi}_N + \overset{\circ}{\chi}_\Delta$ for symmetric nuclear matter is as follows:

$$\overset{\circ}{\chi}_N(\omega,\vec{q}) = \frac{4f^2(q^2)}{m_\pi^2} \int \frac{d^3p}{(2\pi)^3} \frac{n(\vec{p})[1 - n(\vec{p}+\vec{q})]}{\varepsilon(\vec{p}+\vec{q}) - \varepsilon(\vec{p}) - \omega - i\delta} + \begin{array}{l}\text{crossed}\\\text{term}\end{array} \tag{6.11}$$

where the factor of 4 comes from the spin-isospin sum, $n(\vec{p})$ is unity for $|\vec{p}| \leq k_F$, the Fermi momentum, and zero otherwise, and $\varepsilon(\vec{p}) = \vec{p}^{\,2}/2M^*$, where M^* is the nucleon effective mass. The crossed term is obtained from the direct one by the replacements $\omega \to -\omega$ and $\vec{q} \to -\vec{q}$. The integral eq. (6.11) can be worked out analytically. The result is given in ref. [43]. For large pion frequencies ω and low nuclear densities ρ, one finds

$$\overset{\circ}{\chi}_N(\omega,q) \simeq \frac{f^2(q^2)}{m_\pi^2}\left[\frac{\rho}{q^2/2M^*-\omega} + \frac{\rho}{q^2/2M^*+\omega}\right]. \tag{6.12}$$

Note that $\overset{\circ}{\chi}_N(\omega,q) \to 0$ as $M^* \to \infty$, so that in this limit the pionic response for symmetric nuclear matter is completely determined by $\Delta(1232)$ excitations. This appliés, in particular, for $\omega \gtrsim m_\pi$, the kinematic region of pionic atoms and low energy pion-nucleus scattering. At $\omega = 0$ and $q < k_F$, one finds

$$\overset{\circ}{\chi}_N(\omega=0,q) = \frac{f^2(q^2)}{m_\pi^2}\frac{2M^*k_F}{\pi^2}\left[1 - \frac{1}{12}\frac{q^2}{k_F^2} + \dots\right]. \tag{6.13}$$

To leading order, $\overset{\circ}{\chi}_N(\omega = 0, \vec{q} \to 0)$ is then determined by the density of states at the Fermi surface, $2M^*k_F/\pi^2$.

The $\Delta(1232)$-hole susceptibility $\overset{\circ}{\chi}_\Delta$ is

$$\overset{\circ}{\chi}_\Delta(\omega,q) = \frac{16}{9}\frac{f_\Delta^2(q^2)}{m_\pi^2}\int\frac{d^3p}{(2\pi)^3}\frac{n(\vec{p})}{\mathcal{E}_\Delta(\vec{p}+\vec{q})-\mathcal{E}(\vec{p})-\omega} + \begin{array}{c}\text{crossed}\\\text{terms}\end{array}. \tag{6.14}$$

In the absence of Δ-interactions other than those responsible for free $\Delta \to \pi N$ decay,

$$\mathcal{E}_\Delta(\omega,\vec{p}) = M_\Delta - M + \frac{\vec{p}^2}{2M_\Delta} - \frac{i}{2}\Gamma_\Delta(\omega). \tag{6.15}$$

At large pion energies ($\omega \gg q^2/2M$), one obtains

$$\overset{\circ}{\chi}_\Delta(\omega,q) \simeq \frac{4}{9}\frac{f_\Delta^2(q^2)}{m_\pi^2}\left[\frac{\rho}{\omega_\Delta-\omega-i\Gamma_\Delta(\omega)/2} + \frac{\rho}{\omega_\Delta+\omega}\right], \tag{6.16}$$

where $\omega_\Delta = M_\Delta - M = 2.1\ m_\pi$. At $\omega = 0$, $\overset{\circ}{\chi}_N$ dominates the pionic susceptibility, but

$$\overset{\circ}{\chi}_\Delta(\omega=0,q) = \frac{8}{9}\frac{f_\Delta^2(q^2)}{m_\pi^2}\frac{\rho}{\omega_\Delta} \tag{6.17}$$

still gives a contribution of about 35 % to the total $\overset{\circ}{\chi}$.

6.3 The Optical Potential and the Diamesic Function

For $\omega \gtrsim m_\pi$, the "pion optics" domain explored by pion elastic scattering, it is convenient to work with an optical potential. The lowest order p-wave optical potential is related to the pion self-energy simply by

$$U_{opt}^{(o)}(\omega,q) = \frac{1}{2\omega}\Pi^{(o)}(\omega,q) = -\frac{4\pi}{2\omega}q^2c_o\rho. \tag{6.18}$$

The quantity c_o is the spin-isospin averaged scattering volume; at threshold ($\omega = m_\pi$),

$$c_o = \frac{1}{3}\left[4a_{33} + 2a_{13} + 2a_{31} + a_{11}\right] = 0.21\ m_\pi^{-3} \tag{6.19}$$

in terms of the p-wave scattering volumes, table 2. Recalling from eqs. (6.12, 6.16) that only $\overset{\circ}{\chi}_\Delta$ contributes at $\omega = m_\pi$, $q \to 0$, one finds

$$c_0(\omega = m_\pi, q = 0) = \frac{8}{9} \frac{f_\Delta^2}{4\pi m_\pi^2} \frac{\omega_\Delta}{\omega_\Delta^2 - m_\pi^2}, \tag{6.20}$$

which is quite close to eq. (6.19).

The response function in this simplest possible model is obtained by iteration of $\Pi^{(0)}$ with pure one-pion exchange:

$$R(\omega, q) = \Pi^{(0)}(\omega, q) + \Pi^{(0)}(\omega, q) \frac{1}{\omega^2 - q^2 - m_\pi^2} R(\omega, q). \tag{6.21}$$

It is convenient to summarize the medium effects in a diamesic function $\varepsilon(\omega, q)$ [5,44]:

$$R(\omega, q) = \frac{\Pi^{(0)}(\omega, q)}{\varepsilon(\omega, q)}. \tag{6.22}$$

From eqs. (6.21), (6.10) one obtains the standard RPA result

$$\varepsilon(\omega, q) = 1 + \frac{q^2}{\omega^2 - q^2 - m_\pi^2} \overset{\circ}{\chi}(\omega, q). \tag{6.23}$$

The zeros of ε determine the spectrum $\omega(q)$ of pion-like excitations.

In the optical part of the spectrum, a complex index of refraction n for pions can be defined by

$$n^2 = \frac{q^2}{\omega^2 - m_\pi^2} = \frac{1}{1 - \chi(\omega, q)}. \tag{6.24}$$

The pion scattering T-matrix is simply given by $T = R/2\omega$. This T matrix, with eq. (6.21, 22) represents the multiple scattering expansion with the first order p-wave optical potential.

The properties of the diamesic function $\varepsilon(\omega, q)$ have been discussed in great detail in the literature. One particularly interesting question is whether the nuclear medium can act as an amplifier for the pion field at no cost of energy. To find this out, let us discuss the diamesic function at zero frequency. By examining eq. (6.23) one sees that $\varepsilon(\omega = 0, q)$ can approach values small compared to unity at sufficiently large q and sufficiently high density. In fact, as $\varepsilon \to 0$ for $\omega = 0$ a pionic soft mode develops which indicates a phase transition into a pion condensate [45, 46].

For the simple model described above the particle-hole interaction which drives pionic modes is entirely given by one-pion exchange. The OPE tensor force is sufficiently attractive at high momentum transfers ($q_c \sim 2\text{-}3\ m_\pi$) so that in the absence of repulsive correlations, the condition $\varepsilon \to 0$ is actually met at critical densities ρ_c which are only a fraction of nuclear matter density, $\rho_0 = 0.17\ \text{fm}^{-3}$. Obviously, the picture is much oversimplified up to this point, since there seem to be no traces of critical pionic phenomena in nuclei.

LECTURE 6

6.4 Spin-Isospin dependent Particle-Hole Interaction

We proceed now to discuss spin-isospin correlations other than one-pion exchange. For later purposes, we shall not only consider here pionic modes of excitation, those driven by spin-longitudinal interactions of the type $\vec{\sigma}_1 \cdot \vec{q} \; \vec{\sigma}_2 \cdot \vec{q}$, but also spin-transverse ones. The particle-hole interaction in spin-isospin excitation channels must be of the general form (in momentum space, and in direct particle-hole channels):

$$V_{\sigma\tau}(\omega,\vec{q}) = \left[W_\ell(\omega,q)\, \vec{\sigma}_1 \cdot \hat{q}\, \vec{\sigma}_2 \cdot \hat{q} + W_t(\omega,q)(\vec{\sigma}_1 \times \hat{q}) \cdot (\vec{\sigma}_2 \times \hat{q}) \right] \vec{\tau}_1 \cdot \vec{\tau}_2 . \tag{6.25}$$

We ignore spin-orbit interactions which are empirically small in isovector channels. Note that an equivalent formulation of eq. (6.25) is:

$$V_{\sigma\tau} = \left\{ \left[\tfrac{1}{3} W_\ell + \tfrac{2}{3} W_t \right] \vec{\sigma}_1 \cdot \vec{\sigma}_2 + \left[\tfrac{1}{3} W_\ell - \tfrac{1}{3} W_t \right] S_{12}(\hat{q}) \right\} \vec{\tau}_1 \cdot \vec{\tau}_2 . \tag{6.26}$$

Eqs. (6.25, 26) have been written down for nucleons; for Δ's the appropriate replacements $\vec{\sigma} \to \vec{S}^+$ and $\vec{\tau} \to \vec{T}^+$ have to be made.

The prototype of the longitudinal interaction W_ℓ is one-pion exchange. As in the previous section, for large energy transfers ω, one has to go beyond the static approximation and include retardation:

$$W_\pi(\omega,q) = \frac{f^2(q^2)}{m_\pi^2} \; \frac{q^2}{\omega^2 - q^2 - m_\pi^2 + i\delta} \; , \tag{6.27}$$

with f replaced by f_Δ if N-Δ transitions are involved. This is how far one can go in a model based on the Hamiltonian, eq. (6.1). In the hierarchy of interactions, W_π of eq. (6.27) represents the well established long-range part; the less well established shorter range contributions will now have to be discussed in some detail. In fact, much of the presently ongoing debate is about uncertainties at the level of short-range spin-isospin correlations.

The prototype of the transverse coupling interaction W_t is isovector two-pion exchange, usually in its simplified ρ-exchange version, following eq. (2.8):

$$W_\rho(\omega,q) = \frac{f_\rho^2}{m_\rho^2} \; \frac{q^2}{\omega^2 - q^2 - m_\rho^2} \; . \tag{6.28}$$

We have mentioned previously that the short range behaviour of eq. (6.28) is too simplistic because of strong cutoff corrections at high momentum transfers [47] which act differently in spin-spin and tensor channels. In actual calculations, such cutoffs have been introduced phenomenologically [48,49]. In any case, one expects the prototype interactions to be accompanied by screening effects at short distances from several possible sources:

(a) repulsive short-range correlations, or alternatively: effects from quark-gluon dynamics at short distances;

(b) many-body vertex corrections, e.g. from exchange terms.

Within the framework of Landau-Migdal theory [50,51], these screening effects are altogether summarized in terms of a phenomenological repulsive Fermi-liquid interaction,

$$g' \vec{\sigma}_1 \cdot \vec{\sigma}_2 \ \vec{\tau}_1 \cdot \vec{\tau}_2 \ , \tag{6.29}$$

to be added to the prototype interactions. Actually, $g' = g'(\omega, q)$ is a function of ω and q as well; if the underlying interactions are short ranged, one expects this dependence to be smooth.

Additional tensor correlation pieces of the type $h'(\omega, q) \ S_{12}(\hat{q})\vec{\tau}_1 \cdot \vec{\tau}_2$ have also been discussed in the literature and found to be small compared to the leading pieces from π and ρ exchange [53] but the question might have to be reopened in the light of more complete many-body calculations [54].

The magnitude of g' will turn out to be of crucial importance in all subsequent discussions. Note that g' acts in the same way in longitudinal and transverse channels, since $\vec{\sigma}_1 \cdot \vec{\sigma}_2 = \vec{\sigma}_1 \cdot \hat{q} \ \vec{\sigma}_2 \cdot \hat{q} + (\vec{\sigma}_1 \times \hat{q}) \cdot (\vec{\sigma}_2 \times \hat{q})$. Hence for the static interaction $V_{\sigma\tau}(\omega = 0)$, the following ansatz results:

$$W_\ell (q^2) = g' - \frac{f^2}{m_\pi^2} \frac{q^2}{q^2 + m_\pi^2} \ ; \quad W_t(q^2) = g' - \frac{f_\rho^2}{m_\rho^2} \frac{q^2}{q^2 + m_\rho^2} \ . \tag{6.30}$$

The static long-wavelength spin-isospin response is entirely determined by g', since

$$V_{\sigma\tau}(\omega = 0, q = 0) = g' \vec{\sigma}_1 \cdot \vec{\sigma}_2 \ \vec{\tau}_1 \cdot \vec{\tau}_2 \ , \tag{6.31}$$

and g' is related, up to a constant, to the Landau-Migdal Fermi liquid parameter G_0'. If this identification is made, g' already includes exchange terms of the particle-hole interaction by definition, so that only direct particle-hole matrix elements should be calculated with $V_{\sigma\tau}$.

Information about g' for nucleons is obtained from investigations of various magnetic nuclear properties within the Landau-Migdal framework. Commonly accepted values are [+] $g' = 0.6 - 0.7$ [52,53]. Reaction matrix calculations starting from realistic nucleon-nucleon interactions come close to such numbers, and they also show that g' is a smooth function of momentum transfer q [54]. At a more phenomenological level [5], similar values of g' can be obtained in a model where $V_{\sigma\tau}$ is made of π and ρ exchange times a two-body correlation function which approximates Brueckner reaction matrix calculations.

6.5 The Lorentz-Lorenz Correction, and g′ from a Chiral Bag Model point of view

The Lorentz-Lorenz (LL) correction [55] has always been subject to a great deal of interest as a prototype many-body effect in pion-nucleus interactions. The classical

---

[+] We give values of g' in pionic units, i.e. in units of $(f/m_\pi)^2$.

LL effect is equivalent to $g' = 1/3$; it simply corresponds to the removal of the δ-function piece from V_π of eq. (27) in the presence of a repulsive core [56]. This effect can be given a simple interpretation [57] from the point of view of chiral bag models. In the simplest version of such models, the space inside the bag is occupied by quarks, but no pions, while the pion field exists outside the bag and joins to the quark axial current at the boundary such as to make it continuous across the boundary. The axial current is

$$\vec{A}^\lambda = \bar{q}\, \vec{\gamma}\, \gamma_5 \frac{\tau^\lambda}{2}\, q\; \theta(R-r) + f_\pi\, \vec{\nabla}\, \phi^\lambda\; \theta(r-R), \tag{6.32}$$

where the q are quark fields and R is the bag radius. For zero pion mass, the axial current is conserved, $\vec{\nabla} \cdot \vec{A} = 0$, which imples that on the bag boundary,

$$\bar{q}\, \vec{\gamma}\, \gamma_5 \frac{\tau^\lambda}{2}\, q \cdot d\vec{S} = f_\pi\, \vec{\nabla}\, \phi^\lambda \cdot d\vec{S} \tag{6.33}$$

for each surface element normal to the bag sphere.

Let us consider now a medium of well separated bags. We wish to rederive the classical Lorentz-Lorenz correction assuming that the average distance between the bags is large at the boundary of a given bag differs from the field in free space by the spin-isospin polarizability of the medium. Therefore, in the space between bags, the pion field equation $\nabla^2 \phi = 0$ is modified in first order according to $\vec{\nabla}(1-\chi)\vec{\nabla}\phi = 0$, where χ is the pionic susceptibility, eq. (6.10). Consequently the boundary condition, eq. (6.33), becomes

$$\frac{1}{f_\pi}\, \bar{q}\, \vec{\gamma}\, \gamma_5 \frac{\tau^\lambda}{2}\, q \cdot d\vec{S} = (1-\chi)\, \vec{\nabla}\, \phi^\lambda \cdot d\vec{S}. \tag{6.34}$$

The local field correction, $\delta\vec{\nabla}\phi \cdot d\vec{S} = -\chi \vec{\nabla}\phi \cdot d\vec{S}$, has to be averaged over angles which gives a factor 1/3. Performing the correction to all orders yields the local field at the bag boundary

$$\phi_{loc} = \frac{\phi_0}{1 + \frac{1}{3}\chi} \tag{6.35}$$

where ϕ_0 is the pion field in free space. The factor $(1 + \chi/3)^{-1}$ is equivalent to a Landau-Migdal parameter $g' = 1/3$, the classical Lorentz-Lorenz correction, obtained here simply because of the total screening of medium polarization effects from the baryon interior due to the confining forces keeping the quarks inside. Eq. (6.35) still holds if $q\bar{q}$ components carrying pion quantum numbers are allowed inside the bag, such as in the Cloudy Bag model (A.W. Thomas [16]), as long as the confining boundary provides a perfect screening of the baryon interior against the spin-isospin polarization effects in the many-body medium surrounding this baryon.

The significance of this simple model is that it obviously provides a universal g' for both nucleons and Δ-isobars (after separating trivial spin-isospin factors), thereby maintaining the underlying SU(4) symmetry of the problem. We regard this g' as the "minimal" one, expected to be present in any nuclear medium with non-overlapping bags.

Additional contributions to g' are expected to come from short-range dynamics involving overlapping bags.

In discussions of the Landau-Migdal parameter g'_Δ for Δ's, two different positions are presently taken:

(a) the universality hypothesis, based in one or another form on the above quark model considerations and their underlying spin-isospin SU(4) symmetry. In this picture, $g'_\Delta = g'_N$ is assumed;

(b) arguments based on one-boson exchange NΔ-interactions with phenomenological cutoffs, but otherwise treating nucleon and Δ(1232) as elementary particles. In such models, exchange terms of the short-range NΔ-interactions tend to cancel direct terms [58], such that the resulting g'_Δ is substantially smaller than g'_N.

Both approaches are probably too simple to be realistic. The universality argument may well hold for the classical LL correction, $g' = 1/3$, which represents one half of the empirical g' for nucleons. The question is whether the other half, e.g. from overlapping bags, follows the same universality rule. On the other hand, the treatment of the short-distance exchange terms of the NΔ force as if N and Δ were elementary particles is similarly questionable. In a situation like this, we shall strictly follow the Landau-Migdal framework and treat g'_Δ as phenomenological parameter.

6.6 Summary: Spectrum of Pion-like Excitations in Nuclear Matter

With incorporation of correlations other than one-pion exchange, pion self-energy is turned into

$$\Pi = \frac{-q^2 \overset{\circ}{\chi}}{1 + g' \overset{\circ}{\chi}} , \qquad (6.36)$$

and the diamesic function becomes

$$\mathcal{E}(\omega, q) = 1 + \left[\frac{q^2}{\omega^2 - q^2 - m_\pi^2} + g' \right] \overset{\circ}{\chi} . \qquad (6.37)$$

The presence of a positive (repulsive) g' raises the critical density ρ_{crit} for a pionic instability. We show this in Fig. 17. For $g' = 0.6 - 0.7$, ρ_{crit} is raised beyond three to five times nuclear matter density, which moves both pion condensates and its precursors far away from experimentally explorable domains.

To summarize this section, we present in Fig. 18 a schematic picture of the spectrum $\omega(q)$ of pionic excitations in nuclear matter determined by the condition $\varepsilon(\omega, q) = 0$, i.e. by the singularities of the pion-nuclear response function $R(\omega, q)$. The following domains are of particular interest:

(I) The area of low-energy π-nucleus scattering and pionic atoms. The issue there is to learn about many-body corrections to the first order pion-nucleus optical potential (such as the Lorentz-Lorenz correction and other effects).

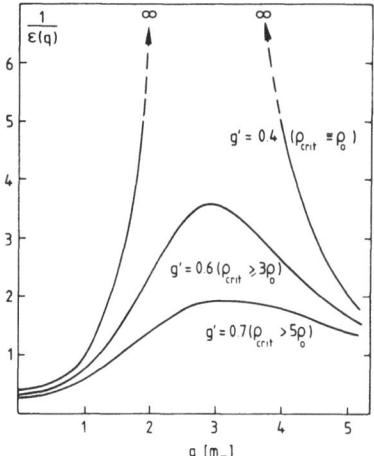

Figure 17: The inverse static diamesic function $\varepsilon^{-1}(\omega = 0, q)$. Shown is the dependence on the Landau parameter g'.

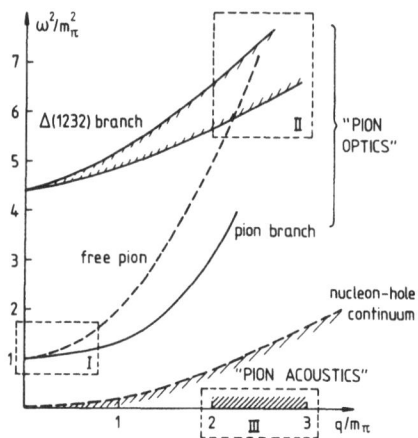

Figure 18: Schematic picture of the various pionic modes of excitation in nuclear matter, as explained in the text.

(II) The region of $\Delta(1232)$ propagation. Here the interest is in the interactions of the Δ with surrounding nucleons, resulting in a change of its mass and decay width (indicated by the broad band associated with the $\Delta(1232)$ branch in Fig. 18).

(III) The domain of possible pionic soft-modes embedded into the continuum of low frequency pion-like nucleon-hole excitations. For sufficiently large g', such soft mode behaviour is essentially ruled out, as mentioned before.

(IV) The low frequency, long wavelength limit where according to Fig. 17 one expects of spin-isospin dependent response. The amount of quenching is determined by $g'(\omega = 0, q = 0)$.

LECTURE 7

7. Nuclear Spin-Isospin Response

7.1 Introduction

We wish now to follow the conceptual framework developed for nuclear matter and adapt it to finite nuclei. At the same time we generalize the scheme to describe spin-isospin response problems not only of pionic type, but also with spin-transverse operators.

Following previous discussions, we recall that there exist two basic types of spin-iso-spin excitation mechanisms in nuclei:

(1) Nucleonic transitions involving the excitation of the Δ(1232) by spin-isospin flip
at the quark level (see Fig. 19a). In many-body language, these are Δ-hole excitations.
(2) Nuclear spin-isospin transitions involving spin-flip, ΔT = 1 nucleon-hole excitations
(see Fig. 19b).

At first sight, these two types of excitations would appear to be rather unrelated,
because of the very different energy scales involved: the scale for Δ hole excitations
is determined by the Δ-N mass difference of about 300 MeV, whereas the characteristic
energy of the relevant nucleon-hole states is given by the energy spacing between $j_>$
and $j_<$ spin-orbit partners or their analogues, typically of order 10 MeV.

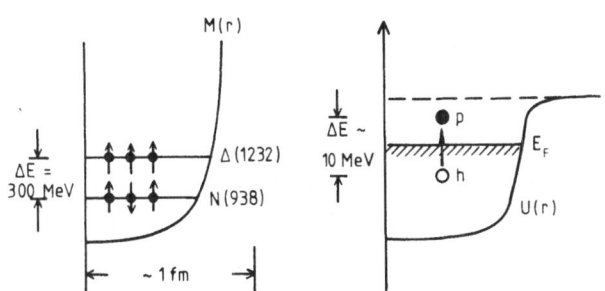

Figure 19: Illustration of spin-isospin excitation mechanisms
(a) N→Δ transition at the quark level;
(b) Isovector-spinflip nucleon-hole excitation

On the other hand, one must keep in mind that N-Δ transitions are strong, following
developments in previous chapters, and that the suppression by a large energy denominator can be partly compensated by a large coupling strength. One expects therefore
that there will be a coupling between nucleon-hole and Δ-hole modes, even at low
energy. The degree to which this coupling occurs will depend on the strength of N-Δ
interactions.

The main problem in discussing the role of the Δ(1232) in low energy nuclear spin-isospin transitions is to discriminate its effects from standard core polarization induced
by strong tensor correlations [59]. These core polarization mechanisms involve virtual
nuclear excitations at similar energy scale (i.e. several hundred MeV) as Δ-hole
excitations, hence one expects that both effects have to be discussed at the same level.

7.2 Spin-Isospin Response Function

We shall consider two basic types of spin-isospin operators :

$$F_\ell = \sum_{j=1}^{A} \vec{\sigma}_j \cdot \hat{q} \; \tau_j^\lambda \; e^{i\vec{q}\cdot\vec{r}_j}, \qquad \text{(longitudinal)} \quad (7.1)$$

$$F_t = \sum_{j=1}^{A} (\vec{\sigma}_j \times \hat{q})_z \; \tau_j^\lambda \; e^{i\vec{q}\cdot\vec{r}_j}. \qquad \text{(transverse)} \quad (7.2)$$

The terms "longitudinal" and "transverse" indicate the preferred alignment between spin and momentum transfer q. For N-Δ-transitions the replacements $\vec{\sigma} \to \vec{S}^+$ and $\vec{\tau} \to \vec{T}^+$ are to be made. Excitations driven by F_ℓ are called "pion-like" because of their pseudoscalar-isovector nature. Transverse excitations via F_t are encountered in magnetic isovector-spin transitions. The (p,n) and (p,p') reactions involve combinations of both F_ℓ and F_t.

It is convenient to introduce the response function (or susceptibility) of the nuclear many-body system with respect to a perturbing spin-isospin operator F, in close analogy with developments in section 6.1:

$$R_F(\vec{q}',\vec{q};\omega) = \sum_n \left\{ \frac{<o|F^+(\vec{q}')|n><n|F(\vec{q})|0>}{\omega - E_n + i\delta} - \frac{<o|F^+(-\vec{q})|n><n|F(-\vec{q}')|0>}{\omega + E_n} \right\},$$

(7.3)

where $|0>$ and $|n>$ are the nuclear ground state and excited states, respectively, and E_n are the excitation energies. For a finite nucleus, the in- and outgoing momenta $\vec{q}$ and $\vec{q}'$ can be different, unlike the situation in nuclear matter.

The strength function is defined by

$$S_F(\vec{q},\omega) = -\frac{1}{\pi} Im\, R_F(\vec{q},\vec{q};\omega) = \sum_n \delta(\omega - E_n)|<n|F(\vec{q})|0>|^2.$$

(7.4)

In scattering experiments like (e,e'), (p,p') or (p,n), the differential cross section is proportional to S_F:

$$\frac{d^2\sigma}{d\Omega\, dE} \propto S_F(\vec{q},\omega).$$

(7.5)

In pion or photon scattering, the scattering amplitude is directly proportional to R_F. For example, the pion-nucleus amplitude f is obtained by identifying $F \equiv \delta H = H_{\pi NN} + H_{\pi N\Delta}$:

$$f(\theta,\omega) = -\frac{1}{4\pi} \sum_n \frac{<o|\delta H^+(\vec{q}')|n><n|\delta H(\vec{q})|0>}{\omega - E_n + i\delta} + \text{crossed term},$$

(7.6)

and the total cross section is

$$\sigma(\omega) = \frac{4\pi}{q} Im\, f(\theta=0,\omega).$$

(7.7)

7.3 First Order Response

The elementary mode of nuclear polarization is given by particle-hole (either nucleon-hole or Δ-hole) excitations $|ph>$ coupled to the appropriate quantum numbers:

$|Nh> = |(\text{nucleon-hole})\, J^\pi, \Delta T = 1>,$

$|\Delta h> = |(\Delta(1232)\text{-hole})\, J^\pi, \Delta T = 1>.$

The lowest order response is described by:

$$\overset{\circ}{R}_F = \langle o | F^+ G_o(\omega) F | o \rangle, \tag{7.8}$$

where

$$G_o = G_o^N + G_o^\Delta .$$

$$\tag{7.9}$$

Here

$$G_o^N(\omega) = \sum_{Nh} |Nh\rangle \left[\frac{1}{\omega - \varepsilon_N + \varepsilon_h + i\delta} - \frac{1}{\omega + \varepsilon_N - \varepsilon_h - i\delta} \right] \langle Nh|$$

is the nucleon-hole Green's function. We now choose ε_N and ε_h, the single particle energies as eigenvalues of an appropriate Hartree-Fock Hamiltonian H_o. The Δ-hole Green's function is

$$G_o^\Delta(\omega) = \sum_{\Delta h} |\Delta h\rangle \left[\frac{1}{\omega - E_{\Delta h}(\omega)} - \frac{1}{\omega + E_{\Delta h}(-\omega)} \right] \langle \Delta h|, \tag{7.10}$$

where the Δ-hole energy is a complex function of the excitation energy ω, as in eq. (6.15), but modified by possible Hartree-Fock type binding effects.

7.4 Response Function in the RPA approximation

The next step beyond single particle response is by introduction of the spin-isospin particle-hole interaction, see section 6.4. The frequently used RPA approximation corresponds to iterating the process, Fig. 20, to all orders:

$$R_F^{RPA} = \langle o | F^+ [G_o(\omega) + G_o(\omega) V G_o(\omega) + \dots] F | o \rangle$$

$$= \langle o | F^+ \frac{G_o(\omega)}{1 - V G_o(\omega)} F | o \rangle . \tag{7.11}$$

For pion-nucleus scattering, we recall that this is equivalent to summing the multiple scattering series with the first order optical potential.

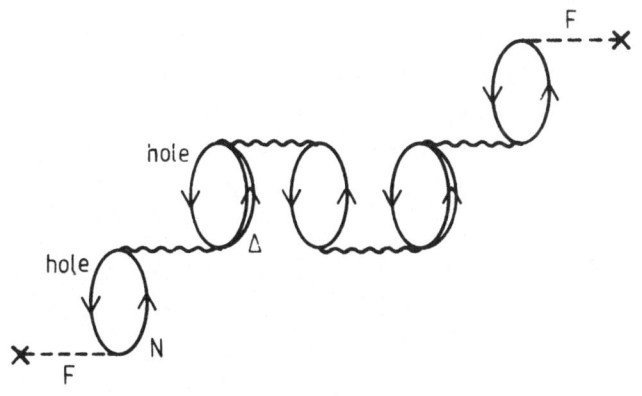

<u>Figure 20:</u> Graphical illustration of the response function within RPA

7.5 Two-Particle-Two-Hole Contributions

The one-particle-one-hole RPA framework described above is not sufficient to provide a realistic description of the response function at large energy transfers. The 1p1h states couple strongly to the 2p2h continuum. As we shall see, this is the leading mechanism for pion absorption in nuclei which proceeds mainly on nucleon pairs.

Such corrections provide an additional width and shift to the nucleon-hole or Δ-hole states. To illustrate this, let $|ph>$ be a particle-hole state and $|\lambda> = |2N2h>$ a two-nucleon-two-hole continuum state. We denote by V the interaction which couples these states. Second order perturbation theory leads to a complex shift of the particle-hole energy

$$\delta E_{ph} = \sum_\lambda \frac{|<ph|V|\lambda>|^2}{E_o - E_\lambda - i\delta} , \qquad (7.12)$$

where the sum over λ is understood as an integration over E_λ, and E_o is the starting energy, given essentially by the energy transfer ω to the nucleus. If E_o is larger than the two-nucleon emission threshold, we obtain an energy dependent width

$$<ph|\Gamma_{2p2h}|ph> = -2 \, Im \, \delta E_{ph} . \qquad (7.13)$$

This width contributes to the imaginary part of the response function.

Later on we shall discuss the influence of such absorptive mechanisms on the pion-nucleus cross section. Here we would like to give an impression of the relative size of 2p2h contributions to the response function by examining the transverse structure function $S_T(\omega,q)$ derived from inclusive inelastic electron scattering.

The differential cross section for single arm (e,e′) measurement with one-photon exchange is

$$\frac{d^2\sigma}{d\Omega \, dE'} = K \sigma_{MOTT} \left\{ \left(\frac{q_\mu^2}{q^2}\right)^2 S_L(\omega,q) + \left[\left(\frac{q_\mu^2}{q^2}\right) + \tan^2\frac{\theta}{2}\right] S_T(\omega,q) \right\}, \qquad (7.14)$$

where K is a kinematical factor and $q_\mu^2 = \omega^2 - \vec{q}^2$. At backward angles, the transverse structure function S_T can be separated from the longitudinal one, S_L. This S_T measures the response due to the interaction with the nuclear current and spin-magnetization. The latter one involves operators of the type $\vec{\sigma} \times \vec{q}$, as discussed before.

Fig. 21 shows $S_T(\omega,q)$ deduced for ^{56}Fe [60] and compared with a Fermi gas RPA plus 2p2h calculation in the local density approximation [61]. The calculation covers the nucleon-hole excitation region and shows the substantial influence of 2p2h contributions in filling the minimum around $\omega \simeq 200$ MeV. At larger energy transfers, S_T rises again towards the $\Delta(1232)$ region.

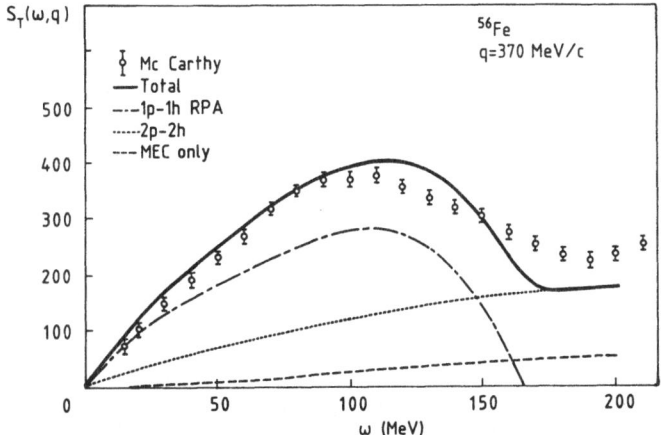

$S_T(\omega, q)$

Mc Carthy
— Total
—·— 1p-1h RPA
······ 2p-2h
——— MEC only

^{56}Fe
q=370 MeV/c

ω (MeV)

<u>Figure 21:</u>   Transverse structure function deduced from inelastic electron
scattering on ^{56}Fe [60]. The calculations [61] show the response
in 1p1h RPA, 2p2h contributions, and the effect of meson exchange
currents (MEC).

LECTURE 8

8. Pion-Nucleus Scattering and Related Processes

Pion-nucleus scattering (or: pion optics) divides naturally into the low energy region
$(0 < T_\pi \lesssim 80$ MeV) and the $\Delta(1232)$ resonance region (80 MeV $\lesssim T_\pi \lesssim 400$ MeV), where T_π
is the pion kinetic energy. The characteristic features can be illustrated already by
considering the pion mean free path in nuclear matter, $\ell = (\rho\sigma)^{-1}$, where ρ is the density and σ the isospin averaged πN cross section. In the Δ resonance region, one obtains
$\ell \simeq 0.5$ fm; the interaction of a pion with the nucleus is therefore surface dominated.
This is to be seen in contrast with low energy scattering. There the πN interaction is
weak. The mean free path is several fm, of the order of the nuclear size or larger; the
π-nucleus interaction takes place all over the nuclear volume, and specific many-body
corrections are important.

8.1 Low-Energy π-Nucleus Elastic Scattering

The starting point of the theory of low energy π-nucleus interactions is the optical
potential, eq. (6.18). In the local density approximation, $\rho \to \rho(\vec{r})$, the first order
potential becomes

$$2\omega\, U_{opt}^{(o)}(\omega, \vec{r}) = -4\pi \left[b_o(\omega)\, \rho(\vec{r}) - c_o(\omega)\, \vec{\nabla}\rho(\vec{r})\vec{\nabla} \right], \tag{8.1}$$

where we have added an s-wave part to the leading p-wave interaction. At threshold,
$b_o = 1/3(a_1 + 2a_3)$.

However, first order potentials of this type fail badly when confronted with pionic
atom data and low energy cross section. Corrections of higher order in density are
required. The essential ones have already been introduced: those related to pion absorption, primarily through coupling to the two-particle-two-hole continuum, and the
Lorentz-Lorenz correction.

A consistent description of both pionic atom and low energy scattering can be obtained
with the following potential [55,62] (for spherical N = Z nuclei):

$$U_{opt} = U_{opt}^{s} + U_{opt}^{p} , \qquad (8.2)$$

where

$$2\omega \, U_{opt}^{s} \, (\omega, r) = - 4\pi \left[b_{o} (\omega) \rho(r) + B_{o} (\omega) \rho^{2}(r) \right], \qquad (8.3a)$$

$$2\omega \, U_{opt}^{p} \, (\omega, \vec{r}) = - 4\pi \, \vec{\nabla} \frac{c(r) + C(r)}{1 + 4\pi \, g'[c(r) + C(r)]} \, \vec{\nabla} , \qquad (8.3b)$$

where $\qquad c(r) = c_{o}(\omega) \rho(r) \quad , \quad C(r) = C_{o}(\omega) \rho^{2}(r) , \qquad (8.3c)$

and we have omitted for simplicity kinematic corrections of order ω/M which are non-
negligible in practical calculations.

A typical calculation is presented in Fig. 22 using the parameters in table 3 [62]
which also reproduce the shifts and widths of pionic atom levels. The main point to
summarize the results is that many-body corrections of order ρ^2 are important. There
is a strong correlation between Re C_o, which can be interpreted as the dispersive
shift due to absorption effects together with binding effects, and the parameter g'
representing the Lorentz-Lorenz correction. It is therefore not possible to determine
either one of these parameters directly from pion elastic scattering.

The imaginary part of the optical potential at threshold and at low energy is entirely
determined by pion absorption. The absorption cross section is

$$\sigma_{abs} = - (2\omega/q) <\phi_{q}| \, Im \, U| \, \phi_{q}>,$$

where q is the pion momentum and ϕ_q are the pion distorted waves evaluated with the
full U_{opt}. Given Im B_o and Im C_o as in table 3, σ_{abs} comes out to be sizeable, typically
more than 1/2 of the total cross sections at low energy. Both B_o and C_o have been
evaluated in microscopic models, assuming that pion absorption takes place primarily
on nucleon pairs. The calculations can then be constrained by the measurements of the
$\pi d \leftrightarrow NN$ process.

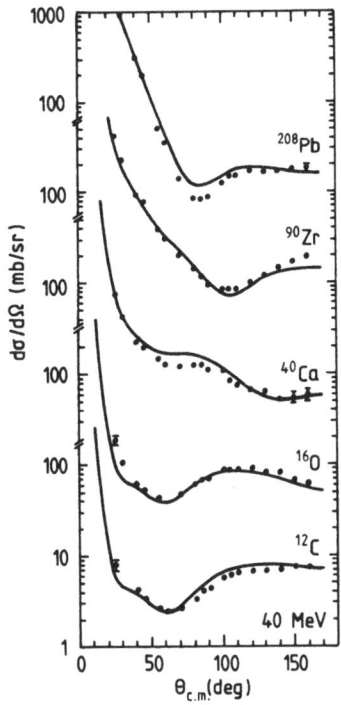

$b_0 [m_\pi^{-1}]$		-0.03	
Re $B_0 [m_\pi^{-4}]$		0.0	
Im $B_0 [m_\pi^{-4}]$		0.05	
$c_0 [m_\pi^{-3}]$		0.23	0.23
g'		0.47	0.60
Re $C_0 [m_\pi^{-6}]$		0.04	0.14
Im $C_0 [m_\pi^{-6}]$		0.12	0.17

table 3: Best fit values of optical potential parameters (eq. (8.3) adjusted to pionic atom data. The two alternative sets of p-wave parameters give equivalent fits.

Figure 22: Differential cross sections for π-nucleus elastic scattering at $T_\pi = 40$ MeV. The parameters are those of table 3 and include a smooth variation from pionic atoms to 40 MeV scattering.

8.2 Scattering in the Δ-Resonance region

A characteristic feature of pion-nucleus scattering around $T_\pi = 160$ MeV is the strong diffractive structure of the differential cross section dσ/dΩ, indicating that a large part of the scattering at resonance is simply determined by geometry. In fact, because of the Δ(1232) resonance in the elementary πN amplitude, the imaginary part of the first order (Born) scattering amplitude is very large at these energies. The scattering process is qualitatively similar to scattering from a black sphere. By analogy with optics, diffraction at the edge of the sphere of radius R leads to a characteristic pattern in dσ/dΩ, the first minimum appearing roughly at the angle where $qR \sim \pi$.

This picture, though qualitatively correct, is too primitive however when it comes to a more quantitative discussion. The precision and abundance of data for some selected nuclei is sufficient to allow for a partial wave analysis. A prototype nucleus is ^{16}O which we shall now examine in more detail.

We denote by F(θ,ω) the $\pi^{16}O$ scattering amplitude where ω is the pion energy in the π-nucleus c.m. system and dσ/dΩ = $|F|^2$. For targets with zero spin such as ^{16}O, we have the partial wave decomposition

$$F(\theta,\omega) = \sum_{J=0}^{\infty} (2J+1) F_J(\omega) P_J(\cos\theta). \qquad (8.4)$$

that the Δ(1232), even in a nuclear environment, can be treated as quasiparticle, like the nucleon itself, specified by an effective mass and width.

The question whether such a picture works, and where are its limits, can be regarded as one of the primary motivations for investigating pion-nucleus scattering in the Δ region.

The Δ-hole model combines the basic requirement of a good first order input with a proper many-body framework for systematic improvements in the treatments of higher order corrections. The framework is that of the response function developed in Chapter 7.

The lowest order response, or optical potential, through Δ-hole intermediate states is

$$2\omega \langle \vec{q}' | U_{opt}^{(o)}(\omega) | \vec{q} \rangle = \langle \vec{q}' | H_{\pi N\Delta}^{+} \, G_o^{\Delta}(\omega) \, H_{\pi N\Delta} | \vec{q} \rangle \tag{8.7}$$

where $\vec{q}$ and $\vec{q}'$ denote in- and outgoing pion momenta, and $G_o^{\Delta}(\omega)$ is the first order Δ-hole Green's function, eq.(7.10). Iteration of G_o^{Δ} to all orders with the Δ-hole interaction $V_{\Delta h}$, i.e. the Δ-hole analogue of the spin-isospin interaction $V_{\sigma\tau}$ of eq. (6.25), gives the RPA Δ-hole Green's function G^{Δ}, which is further modified by coupling of the Δ-hole states to the two-particle-two-hole continuum etc. to account for absorptive damping. Once this is done, the pion-nucleus T-matrix is given in the form (see eq. (7.6)):

$$2\omega \langle \vec{q}' | T(\omega) | \vec{q} \rangle = \langle \vec{q}' | H_{\pi N\Delta}^{+} \, G^{\Delta}(\omega) \, H_{\pi N\Delta} | \vec{q} \rangle \tag{8.8}$$
$$= -4\pi \, F(\theta,\omega).$$

The full Δ-hole Green's function is a sum of terms, each characterized by a specific Δ-hole angular momentum J^{π}.

The actual calculations [5,64 - 66] all follow essentially the same basic RPA approach to the response function. They differ in the detailed treatment of Δ-hole interactions and in their (either microscopic [5,66] or phenomenological [64]) incorporation of important couplings to two-nucleon-two-hole continuum states, the ones relevant for the description of pion absorption channels.

8.4 Δ-Hole Doorway States

We proceed now to present an example of such a calculation. We point out that in the partial wave expansion of the scattering amplitude, eq. (8.4), the $J^{\pi} = 0^{+}, 2^{-}, 3^{+}, \ldots$ etc. coincides with the angular momentum and parity of the Δ-hole excitation modes. These modes are shown in Fig. 24 for $\pi^{16}O$ in terms of the partial cross sections,

$$\sigma_J(\omega) = \frac{4\pi}{q} (2J+1) \, Im \, F_J(\omega). \tag{8.9}$$

The complex F_J have been obtained for all partial waves $0 \leq J \leq 8$ in ref. [63]. Except for the $J^\pi = 0^-$ partial wave which supposedly carries a strong portion of non-resonant s-wave interactions, all partial waves with $J \leq 5$ show a resonant behaviour. However, there is a massive damping in all partial waves: the inelasticities η_J, defined by

$$\eta_J = |S_J| \qquad \text{where} \quad S_J = 1 + 2iq\, F_J , \qquad (8.5)$$

q being the π-nucleus c.m. momentum, come down as far as $\eta_J \simeq 0.2$ for $J \leq 3$ in the kinetic energy region 100 MeV $\leq T_\pi \leq$ 180 MeV. The total cross section,

$$\sigma_{tot}(\omega) = \frac{4\pi}{q}\, Im\, F_J\,(\theta = 0, \omega), \qquad (8.6)$$

and the inelastic cross section for π^{16}O derived from this analysis are presented in Fig. 23.

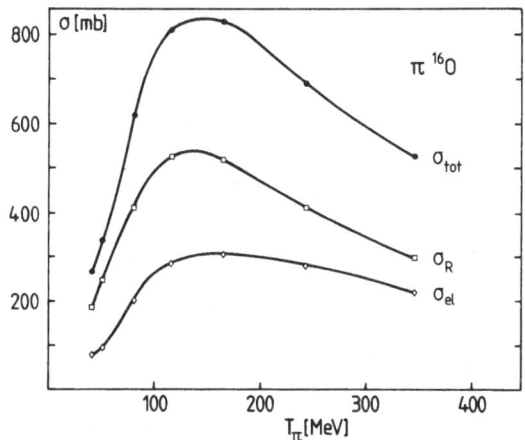

Figure 23: Total, elastic and reaction cross sections for π^{16}O scattering (with Coulomb corrections removed) taken from the analysis of ref. [63].

About one half of the reaction cross section turns out to be related to pion absorption. The major part of the other half comes from inelastic scattering processes with knockout of one or more nucleons (such as quasifree scattering, $(\pi, \pi'N)$).

The large reactive content of the total pion-nucleus cross section is an important feature. In models where the scattering process is described by the excitation and subsequent propagation of a $\Delta(1232)$ inside the nucleus, this implies that there must be a substantial damping width experienced by the $\Delta(1232)$ in a many-body environment.

8.3 The Δ-Hole Model

The dominance of the $\Delta(1232)$ in the pion-nucleon spin-isospin-3/2 channel suggests that the basic mode of excitation at pion kinetic energies between 100 - 300 MeV is the creation of Δ-hole pairs. This is the asumption behind the Δ-hole model. It asserts

The result can be cast into a simple form by identifying each strength distribution for a given J^π with essentially one or two Δ-hole doorway states $|d_J\rangle$:

$$2\omega \langle \vec{q}'|T(\omega)|\vec{q}\rangle = \sum_{d_J} \frac{\langle \vec{q}'|H_{\pi N\Delta}^+|d_J\rangle\langle \tilde{d}_J|H_{\pi N\Delta}|\vec{q}\rangle}{\omega - E_J(\omega) + i\Gamma_J(\omega)/2} \ . \qquad (8.10)$$

Each Δ-hole mode of given J^π has a resonant structure, the position of the maximum moving upward with increasing J. The width of each mode is determined by several

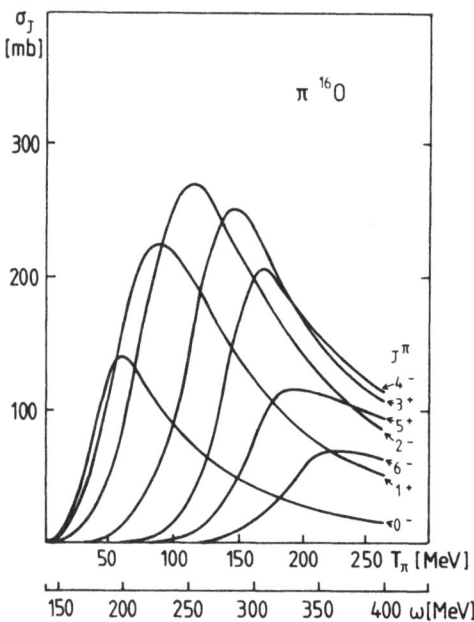

Figure 24: Distribution of Δ-hole strength as seen in pion-nucleus scattering from ^{16}O. Shown is the result of a microscopic Δ-hole calculation [5,66].

factors: (i) the $\Delta \to \pi N$ decay width, partly quenched by Pauli blocking effects, and very importantly, (ii) the absorptive channels arising from the coupling of Δ-hole to the 2N2h continuum. A special feature of the longitudinal spin-isospin response such as it is studied in pion-nucleus scattering, is the downward shift in energy of Δ-hole peaks in lower partial waves. This shift comes mainly because of the strong attraction from non-static one-pion exchange in the (direct) Δ-hole interaction, or equivalently, from the coherent multiple scattering of the pion through the nucleus. The downward shift is partly reduced by the repulsive Δ-hole interaction proportional to g'_Δ, and further influenced by dispersive shifts from absorptive channels. The latter effect is one of the reasons why it is not possible to determine g' directly from Δ-nucleus scattering. Nevertheless the Δ-hole model has been remarkably successful in its capacity to treat genuine many-body corrections to the propagation of the $\Delta(1232)$

inside a nucleus. In fact, the Δ-hole strength distributions, Fig. 24, are very close to those obtained from the partial wave analysis [63] except for the $J^\pi = 0^-$ partial wave which requires a careful treatment of s-wave πN interactions not incorporated in $H_{\pi N\Delta}$

The downward shift of Δ-hole strength in low partial waves observed in π-nucleus scattering is absent in photonuclear cross sections in the Δ-region. This is easily understood because of the dominant transverse $\gamma N\Delta$ coupling proportional to $\vec{S}^+ \times \vec{q}$ which suppresses the direct OPE Δ-hole interaction, the latter involving longitudinal operators of the type $\vec{S}^+ \cdot \vec{q}$ [67].

8.5 The Δ-Nuclear Effective Potential

One can think of all the above mentioned effects on the $\Delta(1232)$ propagation inside the nucleus as being described by a complex optical potential for the Δ. This potential will in general be non-local, because of the large distances over which the excited nuclear many-body system can propagate in space. An equivalent local potential of the form

$$V_\Delta(E,r) = W_o(E) \frac{\rho(r)}{\rho_o} + 2\vec{\ell}_\Delta \cdot \vec{s}_\Delta \, V_{LS}(r) \tag{8.11}$$

has been used successfully [64] in systematically reproducing elastic pion-nucleus data together with total absorption cross sections. (In such an approach, the Pauli blocking terms (Fock terms) are usually calculated explicitly so that their real and imaginary parts do not appear in V_Δ). The result obtained for ^{12}C and ^{16}O is

$$W_o \cong (-30 - i\,40)\ \text{MeV}, \tag{8.12}$$

almost independent on pion energy in the range $T_\pi = 100 - 250$ MeV.

Non-localities enter, at least partly, through the spin-orbit term, where $\vec{\ell}_\Delta$ and $\vec{s}_\Delta$ are the orbital angular momentum and the spin of the propagating $\Delta(1232)$. With the parametrization

$$V_{LS} = V_o^{LS} \mu r^2 e^{-\mu r^2} \tag{8.13}$$

and $\mu = 0.3$ fm^{-2} the following value has been obtained for ^{16}O:

$$V_o^{LS} = (-10 - i\,4)\ \text{MeV}. \tag{8.14}$$

The size of the spin-orbit interaction for Δ's is therefore roughly comparable with that for nucleons. With inclusion of a spin-orbit term, the fit to differential cross sections improves systematically, as Fig. 25 shows. In the absence of V_{LS} the required W_o is strongly energy dependent.

The phenomenological Δ-nuclear potential is sometimes called spreading potential in order to emphasize the spreading of strength from the Δ-hole doorway states into other inelastic channels. Its significance is that it sets the scale for the size of many-body effects related to absorption and other reactive channels. A reactive width of $\Gamma_R \sim 80$ MeV for the central partial waves ($J \leq 3$ at $T_\pi = 160$ MeV) emphasizes the importance of the partial Δ decay into channels other than $\Delta \rightarrow \pi N$ inside the nucleus. The imaginary part of W_0 is actually quite well reproduced by coupling the Δ-hole states to the two-nucleon-two-hole continuum in a model constrained by the $\pi d \rightarrow pn$ absorption amplitude [5].

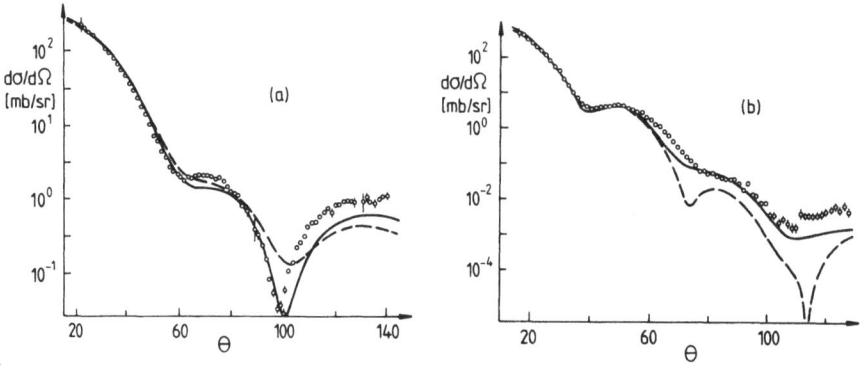

Figure 25: Δ-hole model calculation of π^{16}O differential cross sections (Horikawa et al. [64]) showing the influence of the Δ-nucleus spin-orbit interaction (solid curve) as compared to $V_{LS} = 0$ and W_0 readjusted (dashed curve). (a) $T_\pi = 114$ MeV; (b) $T_\pi = 240$ MeV.

LECTURE 9

9. Gamow-Teller and Magnetic Isovector Transitions

Substantial experimental progress has been made in the past few years, in the systema-tic exploration of Gamow-Teller (GT) strength by (p,n) [68] and (^{3}He, t) [69] processes, and in the investigation of magnetic transitions in a variety of nuclei using high resolution inelastic electron scattering. Both the GT strength and the magnetic tran-sitions of low multipolarity are observed to be systematically quenched as compared to shell model expectations. Part, but not all of the quenching can be attributed to standard nuclear core polarization. The interesting question is then to what extent the unexplained parts of the quenching can be interpreted as signatures of sub-nucleonic effects, such as polarization involving virtual Δ excitations.

The (p,n) process takes advantage of the fact that at energies 100 - 200 MeV of the incoming proton, the spin-isospin dependent part of the nucleon-nucleon interaction dominates strongly over the purely isospin dependent one, thus favouring GT($\sigma\tau$) tran-sitions over Fermi (τ) transitions. The same is true for the (^{3}He,t) reaction.

9.1 Quenching of Gamow-Teller Transitions

GT transitions are related to the nuclear axial current operator

$$\vec{A}^{\pm}(\vec{r}) = g_A \sum_{i=1}^{A} \vec{\sigma_i}\, \tau_i^{\pm}\, \delta^3(\vec{r}-\vec{r_i}).$$ (9.1)

The total GT strength as seen in (p,n) reactions at forward angles can be compared with the sum rule [68]

$$\frac{1}{g_A^2} \sum_n \left\{ |\langle n|\vec{A}^+(\vec{q}=0)|0\rangle|^2 - |\langle n|\vec{A}^-(\vec{q}=0)|0\rangle|^2 \right\} = 3(N-Z)$$ (9.2)

where the intermediate states |n> are purely nucleonic ones. Only $\vec{\sigma}\tau^+$ contributes to (p,n) reactions, but the τ^- part is strongly suppressed by the Pauli principle for nuclei with large neutron excess. The actual strength observed is greatly reduced as compared to eq. 9.2 . If reexpressed in terms of an effective g_A^{eff}, the ratio (g_A^{eff}/g_A) is obtained as shown in Fig. 26 [68]. The fraction of GT strength observed as compared to the sum rule is (65 $\pm$ 5)%. This includes a careful consideration of background subtractions from energies above the actual GT resonance state [71].

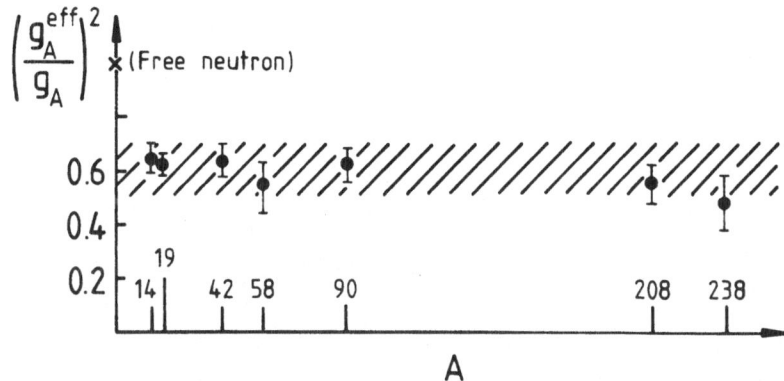

Figure 26: Fraction of Gamow-Teller sum rule strength observed in (p,n) reactions [68].

The systematics of the quenching of the axial charge g_A in nuclei for a wide range of mass numbers is a remarkable feature. We note that with g_A = 1.26 for a free neutron, the effective g_A in nuclei appears to be

$$g_A^{eff} = 1.02 \pm 0.03.$$ (9.3)

Similar conclusions are drawn by a systematic analysis of magnetic moments and beta decays of mirror nuclei [72].

It is instructive to look back at the GT sum rule from a quark model point of view [73]. We recall that eq. (9.2) is derived for a nucleus consisting of nucleons only. The

$\Delta(1232)$ can be introduced by writing the sumrule starting at the quark level from an axial current

$$\vec{A}^{\pm} = \sum_{Q=1}^{3A} \vec{\sigma}_Q \tau_Q^{\pm} , \qquad (9.3)$$

where the $\vec{\sigma}\tau^{\pm}$ now operate on the individual u and d quarks. The corresponding quark sum rule is

$$\sum_n \left\{ |<n| \sum_{Q=1}^{3A} \vec{\sigma}_Q \tau_Q^+ |0>|^2 - |<n| \sum_{Q=1}^{3A} \vec{\sigma}_Q \tau_Q^- |0>|^2 \right\} = 3(N-Z). \quad (9.4)$$

Note that the summation over intermediate states $|n>$ necessarily involves both nucleon and Δ intermediate states, since these states just truncate all possible quark spin-isospin transitions at zero momentum transfer. If the summation over $|n>$ is restricted to nucleons only, the result is $(25/3)(N-Z)$ on the right hand side of eq. (68). By comparison with eq. (9.2), this reveals $g_A = 5/3$, the familiar constituent quark model value for g_A. We note the reduction of the quark sum rule, with Δ's included, over the nucleon sum rule (excluding Δ's) by a factor 9/25. The discussion cannot be applied to realistic situations since the model is obviously oversimplied, giving $g_A = 5/3$ instead of 1.26. Nevertheless, such considerations are useful since they illustrate the connections between nucleons and Δ's in spin-isospin excitation channels. If looked at from a quark sum rule point of view, the presence of Δ's reduces the GT sum rule, eq. (9.2), by effectively replacing g_A by $g_A^{eff} = 1$, in surprising coincidence with eq. (9.3).

9.2 Quenching of Isovector Magnetic Spin Transitions

Isovector magnetic spin transitions are related to the nuclear spin current

$$\vec{J}_M(\vec{r}) = g_M \sum_i (\vec{\sigma}_i \times \vec{\nabla}) \tau_i^3 \delta^3(\vec{r}-\vec{r}_i) . \qquad (9.5)$$

A similar systematics as in GT transitions is found in M1 and M2 transitions, if the observed strength is reexpressed in terms of an effective spin-g-factor, $g_M^{eff} = \gamma g_M$ in eq. (9.3) [70]. The quenching factor γ is shown in Fig. 27. This is not a model independent evaluation, since an RPA calculation has been used for reference (no sum rule of the simple type, eq. (9.2), exists for magnetic transitions).

The quenching of M1 and M2 strength by a factor $\gamma^2 \simeq 1/2$ (except for light nuclei) raises the question about common quenching mechanisms for both g_A and g_M, the corresponding operators being obtained from each other just by an isospin rotation.

Before going into a more detailed discussion of spin-isospin quenching mechanisms, it is interesting to recall the magnetic moments situation. Arima [73] has repeatedly emphasized that the systematics of renormalization effects observed in magnetic moments over a wide range of nuclei can be well accounted for by standard core polarization and tensor correlations, with only little room left for polarization effects involving the $\Delta(1232)$. In that respect, it is important to note [74 ,75] that there is a

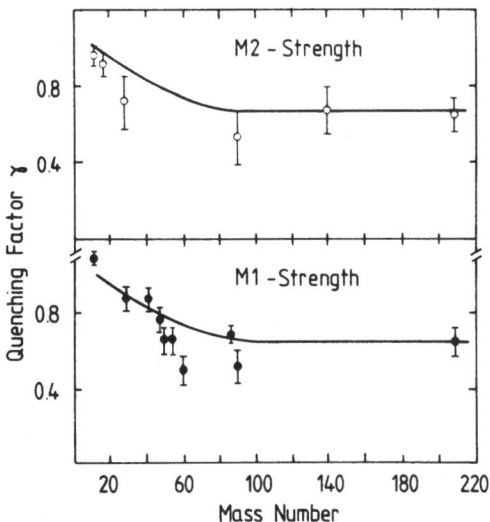

Figure 27: Quenching factor $\gamma = g_M^{eff}/g_M$ associated with the reduced M1 and M2 strength observed in (e,e') experiments (from ref. [70]).

substantial difference between magnetic moments and M1 transitions, as far as the role of the $\Delta(1232)$ is concerned. The effective magnetic dipole operator can be written as

$$\vec{\mu}_{eff} = \frac{1}{2} g_s^{eff} \vec{\sigma} + g_\ell^{eff} \vec{\ell} + \sqrt{\frac{\pi}{2}} \, g_p \, [\vec{\sigma} \times Y_2]^{[1]} , \qquad (9.6)$$

where the g-factors stand for

$$g = g^{(o)} + g^{(1)} \tau_3 . \qquad (9.7)$$

The contribution of eq. (9.6) to magnetic moments and reduced M1 transition matrix elements is (e.g. [75]):

$$\mu \, (j = \ell + \tfrac{1}{2}) = \frac{1}{2} \left\{ g_s^{eff} + 2\ell g_\ell^{eff} + \frac{\ell}{2\ell+3} \, g_p \right\} , \qquad (9.8a)$$

$$\mu \, (j = \ell - \tfrac{1}{2}) = -\frac{1}{2} \left(\frac{2\ell-1}{2\ell+1} \right) \left\{ g_s^{eff} - 2(\ell+1) g_\ell^{eff} + \frac{\ell+1}{2\ell-1} \, g_p \right\} , \qquad (9.8b)$$

$$\langle j' \| \vec{\mu}_{eff} \| j \rangle = (-)^{\ell + \frac{1}{2} - j'} \left[\frac{2j+1}{2(2\ell+1)} \right]^{\frac{1}{2}} \left(-g_s^{eff} + g_\ell^{eff} + \frac{1}{4} g_p \right) . \qquad (9.8c)$$

The point is that contributions from g_s and g_p appear with the same sign in diagonal matrix elements, but with opposite sign in M1 transition matrix elements. Now, Δ-hole polarization effects are shown [74,75] to contribute with opposite signs to g_s and g_p, which means that $\Delta(1232)$ effects interfere destructively in magnetic moments, whereas they add coherently in B(M1) values. Thus the place to look for possible Δ degrees of freedom is in spin transitions rather than moments.

6.3 Renormalization of Spin-Isospin Operators in Nuclei

We proceed now with a discussion of the possible sources for the quenching of spin-isospin strength in the low energy, long wavelength limit. Given a repulsive spin-isospin dependent particle-hole interaction at low momentum transfers (recall eq. (6.31)), a substantial reduction of spin-isospin strength, as compared with independent particle model exceptions, is already obtained within standard RPA with nucleons only. This is seen by recalling from eq. (7.11) that RPA introduces a "quenching factor" ε^{-1}, with

$$\varepsilon = 1 - V_{\sigma\tau} \, G_o^N \tag{9.9}$$

in the response function. For finite nuclei, this ε is a non-local operator [5]. Its analogue in nuclear matter, where longitudinal and transverse spin-isospin channels decouple, is a simple function of ω and q which is often referred to as the longitudinal or transverse diamesic function.

For transitions to specific final states, the screening of the perturbing operator F by the RPA polarization cloud introduces an effective operator,

$$F_{eff} = \varepsilon^{-1} F = \frac{F}{1 - V_{\sigma\tau} \, G_o^N} \tag{9.10}$$

Since G_o^N at $\omega = 0$ goes like

$$G_o^N(\omega = 0) = -2 \sum_{Nh} \frac{|Nh\rangle\langle Nh|}{\varepsilon_N - \varepsilon_h} \tag{9.11}$$

and $V_{\sigma\tau}(\omega = q = 0) = g' \, \vec{\sigma}_1 \cdot \vec{\sigma}_2 \, \vec{\tau}_1 \cdot \vec{\tau}_2$ with $g' > 0$, the reduction of F_{eff} with respect to F is obvious. Part of this RPA screening can simply be interpreted as an effect of ground state correlations, as shown in Fig. 28. The situation is quite different at high momentum transfers, especially in longitudinal ($\vec{\sigma} \cdot \vec{q} \, \tau^\lambda$ type) channels. As q increases, the attraction from the OPE part of $V_{\sigma\tau}$ sets in, and screening may be turned into antiscreening, depending on the effects of cutoffs in πNN vertex form factors.

In practical calculations, one usually truncates the particle-hole basis (the model space, or P space). Any polarization effect outside that model space (involving the residual Q space, P + Q = 1) tends to introduce additional quenching. For example, Bertsch and Hamamoto [77] find in a perturbative calculation that there is a strong mixing of the Gamow-Teller resonance with high-lying 2p2h configurations, so that a

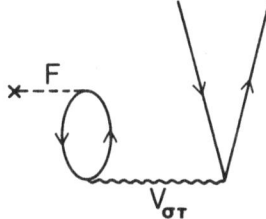

Figure 28: Quenching of spin-isospin transitions by RPA type ground state correlations. Note that by Pauli principle considerations, this affects M1 transitions but not GT transitions.

large fraction of the GT strength in ^{90}Zr is moved up to the continuum between 10 - 45 MeV, by mechanisms illustrated in Fig. 29. The tensor part of the effective interaction becomes very important in these mixings, a fact also pointed out in [73-75] tensor force effects. The Bertsch and Hamamoto result, namely that about half of the GT strength is moved to the 2p2h continuum, seems to be an overestimate, though. A careful reanalysis [76] of the GT background places upper limits ($\lesssim$ 20 %) for the strength moved into the continuum between 20 - 40 MeV.

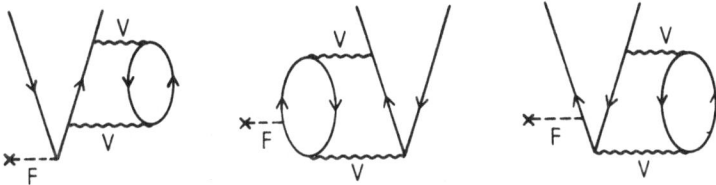

Figure 29: Three out of many diagrams involving mixings of 1p1h states with high lying 2p2h excitations.

6.4 Δ-hole induced Screening of Spin-Isospin Operators

In addition to the screening due to conventional nuclear polarization mechanisms, we expect that virtual Δ-hole excitations contribute to the quenching factor ε. We have demonstrated the existence of broad Δ-hole states at excitation energies around 300 MeV. The question is now to what extent virtual Δ-hole excitations participate in the nuclear spin-isospin response even at low energy. Suppose that all relevant conventional nucleon degrees of freedom are treated explicitly in a sufficiently large model space (P-space), such that the remaining Q-space contains all polarization effects where intrinsic N → Δ transitions are involved. Within RPA reduced to P space, the effective spin-isospin operators incorporating Δ-hole screening effects (see Fig. 30) will now be

$$F_{eff} = \mathcal{E}_{\Delta}^{-1} F, \qquad \mathcal{E}_{\Delta} = 1 - V G_{o}^{\Delta}, \qquad (9.12)$$

where $V = V(\Delta h)$ is the Δ-hole interaction and G_{o}^{Δ} is the Δ-hole Green's function, eq. (7.10)

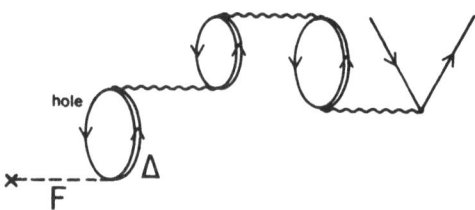

Figure 30: Screening of spin-isospin operators by virtual Δ-hole excitations within RPA.

It is instructive to discuss ε_{Δ} in the static long-wavelength limit for nuclear matter. The Δ-hole interaction in this limit becomes

$$V(\Delta h) = g_{\Delta}' \; \vec{S}_1 \cdot \vec{S}_2^+ \; \vec{T}_1 \cdot \vec{T}_2^+, \qquad (9.13)$$

where g_{Δ}' is the relevant Landau-Migdal parameter derived from the ΔN-interaction. The G_{o}^{Δ} at $\omega = 0$ is proportional to $\rho/(M_{\Delta}-M_N)$, where ρ is the nuclear density and $M_{\Delta} - M_N$ is the ΔN mass difference. Carrying out spin-isospin sums, one obtains

$$\mathcal{E}_{\Delta} = 1 + g_{\Delta}' \, \lambda \, \frac{\rho}{\rho_o}, \qquad (9.14)$$

where the density is given in units of nuclear matter density, $\rho_o = 0.17 \; \text{fm}^{-3}$, and the constant λ is

$$\lambda = \frac{8}{9} \left(\frac{72}{25} \right) \frac{\rho_o}{M_{\Delta} - M} \; m_{\pi}^{-2}. \qquad (9.15)$$

The factor 72/25 is obtained if one assumes the SU(4) scaling between NN and NΔ spin-isospin transitions, in which case $\lambda \simeq 0.6$. (In the Chew-Low model, the 72/25 would be replaced by 4 and λ would be increased correspondingly.) The Δ-induced quenching factor is seen to be determined by g_{Δ}' in the long wavelength limit. For example, with $g_{\Delta}' = 0.5$ one obtains $\varepsilon_{\Delta} = 1.3$ at nuclear matter density, $\rho = \rho_o$. This quenching is obviously common to both GT and magnetic spin transitions. That is, the effective axial vector coupling constant and isovector spin g-factor become:

$$g_A^{eff}/g_A = g_M^{eff}/g_M = \left[1 + \lambda g_{\Delta}' \, \rho/\rho_o \right]^{-1}. \qquad (9.16)$$

This Δ-hole induced screening of spin-isospin transitions has been discussed widely in the literature [78].

In finite nuclei, ε_{Δ} becomes a non-local operator, as discussed before, and calculations are usually performed keeping the full finite range structure of the Δ-hole interaction, including one-pion exchange and ρ exchange. The non-local and finite range effects have two consequences, namely that the Δ-induced quenching effect depends on the nuclear mass number (quenching is less for light nuclei) and on the angular momentum J of the state considered (less quenching for large J; see ref. [5]).

The essential parameter governing the Δ-hole screening is g'_Δ. We have already mentioned that, unlike g' for nucleons, g'_Δ is subject to considerable uncertainty. In many-body schemes which start from a boson exchange model of the NN $\rightarrow$ NΔ or NΔ $\rightarrow$ NΔ interaction, exchange terms (Fig. 31b) tend to cancel direct terms (Fig. 31a) of the Δ-hole inter-action [79], the cancellation being most effective in short range pieces, like exchange. The resulting g'_Δ would be small, about 0.3, hence Δ-hole quenching would not be substantial. In fact, the cancellation is complete for a zero range interaction. However, recent estimates [80] indicate that one has to carry on with the question of exchange terms along the lines of ref. [54] to include the induced interaction (Fig. 31c). In fact, diagrams (b) and (c) (taken to all orders) of Fig. 31 tend to cancel largely among themselves, leaving Fig. 31(a) as the dominant piece. In any case, this is just a limited set out of many more diagrams, and one has to raise the question how far the standard many-body framework with "elementary" nucleons and Δ exchange terms can be pushed at short distances. The Landau-Migdal framework avoids these problems by operating with the direct particle-hole interaction, Fig. 31(a) only, and assigning a phenomenological g'_Δ, including exchange, to this channel. As mentioned before, we shall strictly maintain this philosophy in the following.

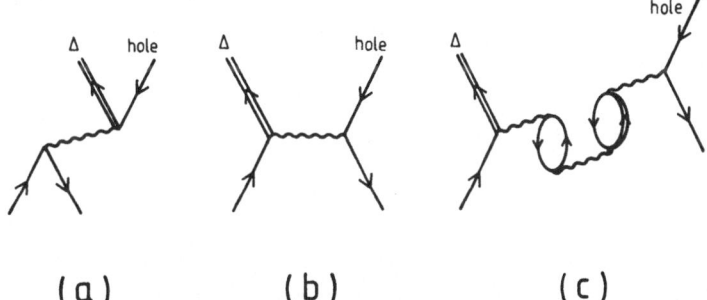

(a)   (b)   (c)

Figure 31: Direct (a) and exchange (b) pieces of the Δ-hole transition interaction. The exchange terms are screened by higher order diagrams of the type (c), the induced interaction in this channel.

LECTURE 10

9.5 A specific example: M1 Transition to the 10.2 MeV State in ^{48}Ca

This state, seen first in (e,e') scattering [81] is by now one of the best studied examples of quenched M1 strength. According to McGrory and Wildenthal [82], this state has a relatively simple shell model structure dominated by a $f_{5/2}f_{7/2}^{-1}$ neutron-hole configuration. The wave function obtained from a full fp-shell model calculation (which defines the model space, or P space) is

$$|^{48}Ca;1^+> = 0.89\,|\nu f_{5/2}\,f_{7/2}^{-1};1^+> + 0.11\,|\nu f_{7/2}\,f_{5/2}^{-1};1^+> , \qquad (9.17)$$

plus additional small admixtures of more complicated configurations. The dominant neutron-hole component makes this a favourable case for studying renormalization effects of the spin g factor. A pure $f_{7/2} \rightarrow f_{5/2}$ single particle transition using the unrenormalized value for g_s gives B(M1)↑ = 12 μ_N^2, whereas the experimental value is $(3.9 \pm 0.3)\mu_N^2$ [81]. Using the wave function, eq. (9.17), the B(M1)↑ comes down to $7.3\mu_N^2$. A major fraction of this quenching comes from 2p-2h ground state correlations of the type shown in Fig. 28. Such 2p-2h correlations are also incorporated in standard RPA calculations (Suzuki, Krewald and Speth, 1981), where B(M1)↑ = $8\mu_N^2$ is found. The additional quenching of about $1\mu_N^2$ is due to more complicated many-particle-many-hole effects not present in RPA.

The effect of pionic exchange currents is small, but acts to increase the effective g_s by $\lesssim 10$ % (Kohno and Sprung [78]). This discussion indicates that subtle cancellations are involved (Towner and Khanna [74]). It shows also, however, that it is difficult to obtain a B(M1) much less than 7-$8\mu_N^2$ from ground state correlations and mesonic exchange currents. Another factor 1.5 - 2 reduction is still required.

Now, if g'_Δ is sufficiently large, Δ-hole screening is a candidate for supplying a good fraction of the remaining quenching. This is shown in Fig. 32 (Härting et al. [78]; see also ref. [83]), where the Δ-hole screening (on top of the McGrory-Wildenthal pf-shell model space) has been calculated with a Δ-hole force consisting of π and ρ exchange plus a Landau-Migdal zero range interaction proportional to g'_Δ, the parameter which has been varied. The full non-local structure of the diamesic function ε_Δ as well as the proper angular momentum projection is kept in this calculation. The Chew-Low ratio $f_\Delta/f = 2$ has been used here. (For comparison with calculations using the constituent quark model value $f_\Delta/f = \sqrt{72/25}$, multiply g'_Δ in Fig. 14 by a factor 1.4). Note that for finite nuclei such as ^{48}Ca, there is a mixing of transverse and longitudinal parts of the Δ-hole interaction even though the probing M1 field is purely transverse. As a consequence the attraction from OPE reduces somewhat the quenching from g'_Δ alone, an effect observable in the limit $g'_\Delta = 0$.

Next, we wish to consider the M1 form factor of the same 10.2 MeV state in ^{48}Ca, which has been measured by Steffen et al. [84]. We do this in several steps, starting from

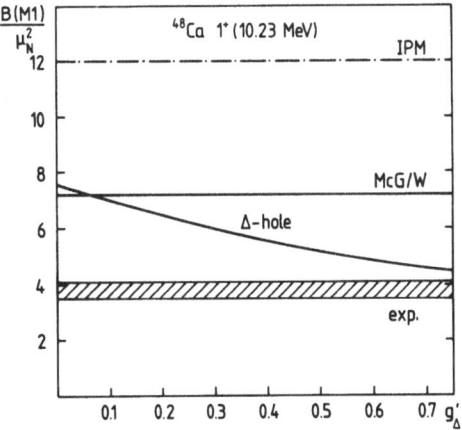

Figure 32: B(M1)↑ value for ^{48}Ca(1^+). IPM: result for pure $(f_{5/2} f_{7/2}^{-1})$ neutron-hole configuration. McG/W: result of full fp-shell model calculation (McGrory and Wildenthal, [82]). Δ-hole: result including Δ-hole screening in addition to McG/W as a function of the Δ-hole Landau-Migdal parameter g'_Δ.

the McGrory-Wildenthal wave function, eq. (9.17), and introducing Δ-hole screening as in Fig. 30. The calculation here is comparable to the large space RPA calculation of Suzuki et al. [78]). They use a similar Δ-hole interaction, but with inclusion of exchange terms for π and ρ exchange, which is equivalent to choosing a much reduced g'_Δ (in the Landau Fermi Liquid picture). However, they also observe that they have to add in by hand a $\delta g'_\Delta > 0$ in order to fit the energy of the 1^+ state. This $\delta g'_\Delta$ compensates for the reduction of g'_Δ obtained by explicit calculation of exchange terms. In our calculation, exchange terms are systematically omitted, for reasons given earlier.

The full fp-shell model space has the advantage that it includes many-particle-many-hole configurations not present in RPA. But it omits nucleon core polarization effects outside that model space. We have included such effects at least partly by incorporating all RPA type nucleon-hole polarization diagrams outside P-space to all orders. The different steps of the calculation are shown in Fig. 33a. Note that the quenching effect due to Δ-hole and nucleon-hole polarization is q-dependent, reflecting the q-dependence of the Δ-hole interaction from π and ρ exchange.

Meson exchange current effects increase the M1 form factor up to the first maximum by about 10 %. Consequently, for $g'_\Delta = 0.6$, there is still room for an additional renormalization of the isovector spin-g factor by about 10 %. Fig. 33b shows the result [85] when all effects are included, together with a $g_s^{eff} = 0.9\ g_s$. This latter factor may represent, for example, second order core polarization processes of the type, Fig. 29, not included within RPA.

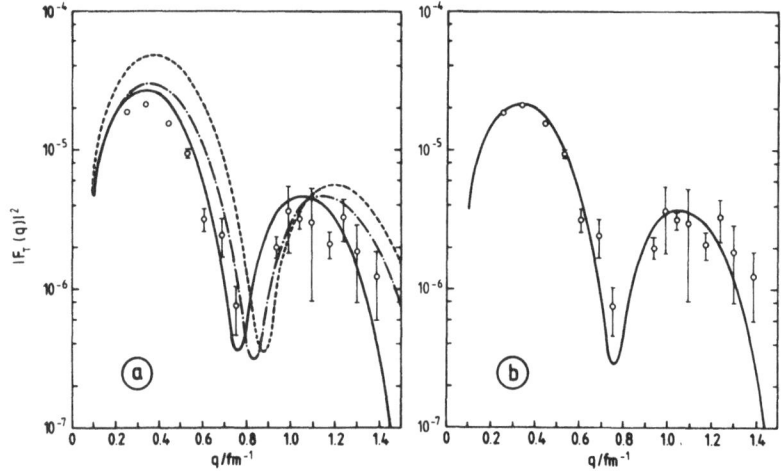

<u>Figure 33:</u> DWBA-calculations of the transverse form factor for ^{48}Ca(e,e')
(1$^+$; 10.2 MeV). (a) Dashed curve: McGrory-Wildenthal fp-shell
model result; dash-dotted curve: effect of Δ-hole screening,
with g_Δ' = 0.6; solid curve: additional effect of RPA-type nucleon-
hole polarization outside the fp-shell; (b) curve obtained from
solid curve of (a) by adding meson exchange currents and using
g_s^{eff} = 0.9 g_s (from ref. [85]; exp. data: ref. [84]).

In summary, the above example gives indications for Δ(1232) induced quenching, but has
also evinced the difficulties in discriminating such effects against "standard"
nuclear properties, such as ground state correlations and core polarization. A Δ-hole
induced reduction of the isovector g_s and g_A by 30 - 35 % can be obtained for values
g_Δ' = 0.5 - 0.6 of the Δ-hole Landau-Migdal parameter. We note that while this Δ-hole
quenching is common to both GT and M1 transitions, ground state correlations act
differently in both cases, namely, they are reduced for GT transitions in neutron rich
nuclei. Meson exchange currents contribute relatively little to the renormalization of
g_s (an increase by less than 10 % for the ^{48}Ca example). The situation here is
different from that in very light nuclei, where pion exchange dominate and Δ(1232)
effects are relatively small (see Chapter 4).

10. Hyperons in Nuclei

10.1 Strangeness Exchange Reactions

We have discussed mechanisms to create a Δ(1232) in a nucleus by pion-induced processes.
The main motivation for doing so was to study interactions of the Δ with surrounding
nucleons. In a similar way, kaon beams have been used to implant Λ and Σ hyperons in
nuclei in order to investigate their interactions with a nuclear environment.

The Λ and Σ have strangeness S = - 1. They are produced in the following strangeness
exchange reactions on nucleons:

$$K^- + N \rightarrow \Lambda + \pi, \tag{10.1}$$

$$K^- + N \rightarrow \Sigma + \pi. \tag{10.2}$$

The (K^-, π^-) reaction on nuclei has been used extensively to form Λ and Σ hypernuclei [87 - 89].

Consider first the Λ production process, eq. (10.1). A particularly interesting feature of the kinematics is that for a K^- momentum of 500 MeV/c, the recoiling Λ is produced with zero momentum (recoilless production, see table 4), if the pion is detected under forward angles. This has important consequences for Λ hypernucleus produced under these

K^- momentum [MeV/c]	0	100	300	500	700	900
Λ momentum [MeV/c]	250	190	70	0	40	80

table 4: Recoil momentum in $K^- n \rightarrow \pi^- \Lambda$ with pions detected at angle 0°.

kinematical conditions. It means that the neutron in a given shell model orbit will be replaced preferentially by a Λ carrying the same orbital quantum numbers.

10.2 Spectroscopy of Λ-Hypernuclei

We consider here the Λ-hypernuclear excitation spectra in ^{12}C and ^{16}O obtained with the (K^-, π^-) reaction and shown in Fig. 34. The spectra are plotted as a function of the mass difference $M_{HY} - M_A$ of the hypernucleus and the target nucleus. Also plotted is the binding energy B_Λ, and the Λ-nuetron mass difference is indicated for orientation.

The interesting point to note is first that the spectrum looks very much like one which would follow from a simple shell model picture: the neutron is removed from the p-shell of carbon and oxigen and replaced by a Λ which occupies any one of the p- or s-shell orbits available to it in an assumed Λ-nucleus average potential. Now, in ^{12}C, only the $p_{3/2}$ neutron shell is occupied, whereas in ^{16}O, a neutron in either $p_{3/2}$ or $p_{1/2}$ orbit can be replaced by a Λ. A comparison of ^{16}O and ^{12}C as in Fig. 34 therefore permits to extract not only the depth of the average Λ-nucleus potential, but also the strength of the Λ-nucleus spin-orbit interaction. A detailed phenomenological analysis [90] yields the following results: if the Λ-nucleus single particle potential is written as

$$V_\Lambda(r) = W_o \frac{\rho(r)}{\rho_o} + V_o^{LS} \frac{d}{dr}\left(\frac{\rho(r)}{\rho_o}\right) \frac{\vec{\ell} \cdot \vec{\sigma}}{r}, \tag{10.3}$$

then

$$W_o = (-32 \pm 2) \, MeV, \tag{10.4}$$

$$V_o^{LS} = (4 \pm 2) \, MeV \, fm^2. \tag{10.5}$$

Thus the central potential depth for a Λ is about half as deep as that of a nucleon, while the Λ spin-orbit coupling is only about 1/4 or less compared to the spin-orbit force of nucleons in nuclei.

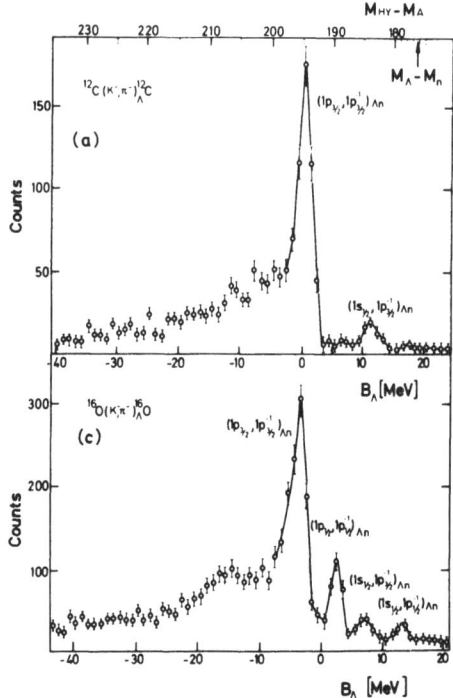

Figure 34: Spectra obtained from the (K^-, π^-) reaction [88] on ^{12}C and ^{16}O at a kaon momentum of 715 MeV.

10.3 The Hyperon-Nucleon Spin-Orbit Interaction

The size of the spin-orbit force is evidently an interesting piece of information, since it reflects properties of the hyperon-nucleon effective interaction at relatively short distances. Several attempts have been made to relate the results, eqs. (10.4-5), to properties of the underlying two-body interaction.

One such approach starts from a relativistic boson exchange model and relates the central potential depth to the spin-orbit force in a Dirac-Hartree-Fock calculation [91,92]. With constraints set by the potential well depth and spin-orbit splitting in nuclei, and with SU(3) applied to hyperon-nucleon interactions, one finds values of W_0 and V_0^{LS} in good agreement with the empirical values.

The smallness of the ΛN spin-orbit force comes as a natural result also in simple quark rearrangement plus gluon exchange models [93]. The Λ is a combination of u, d and s quark in such a way that (ud) couple to a spin and isospin singlet. Therefore the spin-orbit interaction due to exchange of u or d quarks vanishes for the diagonal ΛN → ΛN interaction where no s quark is exchanged. The contribution from ΛN → NΛ exchange processes is small.

However, the same naive quark model predicts a value 4/3 for ratio of ΣN to NN spin-orbit forces in nuclei [93]. In contrast, boson exchange models generally suggest a small Σ-nucleus spin-orbit coupling [94,95]. Experimental data [89] on Σ-hypernuclei can be interpreted assuming an average Σ-nucleus potential depth of about − 30 MeV. Unfortunately, the data seem so far not to be sufficiently accurate to deduce an unambgibuous Σ-nucleus spin-orbit potential; nevertheless there are claims [96] in favour of an interpretation with a large V_{LS} for Σ's.

Whether this apparent discrepancy between meson exchange and simple quark models leads us to the limits of the boson exchange phenomenology is a question of vital importance. In any case, one would wish that Σ-hypernuclear data become available in the future at a level of accuracy such that this problem can be sorted out.

REFERENCES

[1] Proc. 8th Int. Conf. on High Energy Physics and Nuclear Structure, Vancouver 1979, Nucl. Phys. A 335 (1980)

[2] Proc. 9th Int. Conf. on High Energy Physics and Nuclear Structure, Versailles 1981, Nucl. Phys. A 372 (1982)

[3] Mesons in Nuclei, Vol. I-III, M. Rho and D.H. Wilkinson, eds., North-Holland Publ. Co. (1979)

[4] A.W. Thomas and R.H. Landau, Phys. Reports 68 (1980)121

[5] E. Oset, H. Toki and W. Weise, Phys. Reports 83 (1982) 281

[6] F. Lenz and E. Moniz, Adv. in Nucl. Phys. 13 (1983)

[7] M. Lacombe et al., Phys. Rev. C 23 (1981) 2405; R. Vinh Mau, in: Mesons in Nuclei, Vol. I, M. Rho and D. Wilkinson, eds., North-Holland, Amsterdam (1979)

[8] A.D. Jackson, D.O. Riska and B.J. Verwest, Nucl. Phys. A 249 (1975) 397; G.E. Brown and A.D. Jackson, The Nucleon-Nucleon Interaction, North-Holland, Amsterdam (1976)

[9] K. Holinde, Phys. Reports 68 (1981) 121; R. Machleidt, in: Quarks, Mesons and Isobars in Nuclei, R. Gurardiola and A. Polls, eds., World Scientific, Singapore (1983)

[10] K. Erkelenz, Phys. Rep. 13 (1974) 191

[11] M.M. Nagels, T.A. Rijken and J.J. de Swart, in: Few Body Systems and Nuclear Forces, Vol. I (1978), Lecture Notes in Physics, Springer (1978)

[12] W. Grein and P. Kroll, Nucl. Phys. A 338 (1980) 332

[13] A. Chodos, R.L. Jaffe, C.B. Thorn and V. Weisskopf, Phys. Rev. D 9 (1974) 3471; A. Chodos, R.L. Jaffe, K. Johnson and C.B. Thorn, Phys. Rev. D 10 (1974) 2599;

[14] A. Chodos and C.B. Thorn, Phys. Rev. D 12 (1975) 2733

[15] G.E. Brown and M. Rho, Phys. Lett. 82 B (1979) 177; G.E. Brown, M. Rho and V. Vento, Phys. Lett. 84 B (1979) 383; V. Vento et al., Nucl. Phys. A 345 (1980) 413

[16] G.A. Miller, A.W. Thomas and S. Théberge, Phys. Lett. 91 B (1980) 192; Phys. Rev. D 22 (1980) 2823; A.W. Thomas, Adv. in Nucl. Phys. 13 (1983) 1

[17] W. Weise, in: Quarks, Mesons and Isobars in Nuclei, R. Guardiola and A. Polls, eds. World Scientific (1983), p. 146

[18] R. Tegen, R. Brockmann and W. Weise, Z. Physik A 307 (1982) 339; R. Tegen, M. Schedl and W. Weise, Phys. Lett. 125 B (1983) 9

[19] T.D. Lee, Particle Physics and Introduction to Field Theory, Harwood Publ., London (1981) R. Friedberg and T.D. Lee, Phys. Rev. D 18 (1978) 2623

[20] R. Goldflam and L. Wilets, Phys. Rev. C 25 (1982) 1951

[21] M. Brack, Phys. Rev. D 27 (1983) 1950

[22] Y. Nambu and G. Jona-Lasinio, Phys. Rev. 122 (1961) 345, 124 (1961) 246

[23] R. Brockmann, W. Weise and E. Werner, Phys. Lett. 122 B (1983) 201; V. Bernard, R. Brockmann, M. Schaden, W. Weise and E. Werner, Nucl. Phys. A 412 (1984) 349

[24] S.J. Brodsky and G.P. Lepage, Phys. Scripta, 23 (1981) 945; S.J. Brodsky, in: Quarks and Nuclear Forces, Springer Tracts in Mod. Phys. 100 (1982) 81

[25] P.A.M. Guichon, G.A. Miller and A.W. Thomas, Phys. Lett. 124 B (1983) 109

[26] R. Tegen and W. Weise, Z. Physik A 314 (1983) 357

[27] E. Oset, R. Tegen and W. Weise, Univ. Regensburg preprint TPR-84-1 (1984)

[28] T.E.O. Ericson and M. Rosa-Clot, Nucl. Phys. A 405 (1983) 497

[29] J. de Kam, Z. Physik A 310 (1983) 113; M. Schedl, private communication

[30] G.E. Brown and W. Weise, Phys. Reports 22 (1975) 279

[31] M. Chemtob and M. Rho, Nucl. Phys. A 163 (1971) 1

[32] D.O. Riska, in: Mesons in Nuclei, Vol. II, M. Rho and D. Wilkinson, eds., North-Holland (1979)

[33] H. Miyazawa, Progr. Theor. Phys. 6 (1951) 801; J.I. Fujita and M. Ichimura, in: Mesons in Nuclei, M. Rho and D. Wilkinson, eds., North-Holland (1979), p. 625; T. Yamazaki, ibido, p. 651

[34] J.A. Lock and L.L. Foldy, Ann. of Phys. 93 (1973) 276

[35] M. Bernheim et al., Phys. Rev. Lett. 46 (1981) 402

[36] J. Hockert, D.O. Riska, M. Gari and G. Hoffmann, Nucl. Phys. A 217 (1973) 14
[37] H. Arenhövel, Nucl. Phys. A 374 (1982) 521 c
[38] R.A. Brandenburg et al., Phys. Rev. Lett. 32 (1974) 325
[39] J.M. Cavedon et al., contribution 9th Int. Conf. High Energy Physics and Nuclear
     Structure, Versailles (1981)
[40] E. Hadjimichael, R. Bornais and B. Goulard, Phys. Rev. Lett. 48 (1982) 583
[41] G. Höhler, Landolt-Börnstein, Numerical Data ...,
     Vol. I/9b2, p. 279, Springer (1983)
[42] M. Ericson, Ann. of Phys. 63 (1971) 562;
     T.E.O. Ericson, in: Proc. Conf. on Common Problems in Low- and Medium-Energy
     Physics, B. Goulard and F.C. Khanna, eds., Plenum (1979)
[43] A.L. Fetter and J.D. Walecka, Quantum Theory of Many-Particle Systems,
     McGraw-Hill (1971)
[44] N.C. Mukhopadhyay, H. Toki and W. Weise, Phys. Lett. 84 B (1979) 35
[45] A.B. Migdal, Rev. Mod. Phys. 50 (1978) 107
[46] G.E. Brown and W. Weise, Phys. Reports 27 (1976)1
[47] W. Dickhoff, H. Müther and A. Faessler, Phys. Rev. Lett. 49 (1982) 1902
[48] M. Anastasio and G.E. Brown, Nucl. Phys. A 285 (1977) 516
[49] W. Weise, Nucl. Phys. A 278 (1977) 4021
[50] A.B. Migdal, Theory of Finite Fermi Systems, Interscience (1967)
[51] J. Speth, E. Werner and W. Wild, Phys. Reports 33 (1977) 127
[52] J. Speth, V. Klemt, J. Wambach and G.E. Brown, Nucl. Phys. 343 (1980) 382;
     P. Ring and J. Speth, Nucl. Phys. A 235 (1974) 315
[53] S.O. Bäckman, A.D. Jackson and O. Sjöberg, Nucl. Phys. A 321 (1979) 10
[54] W.H. Dickhoff et al., Nucl. Phys. A 369 (1981) 445; Phys. Rev. C 23 (1981) 1154
[55] M. Ericson and T.E.O. Ericson, Ann. of Phys. 36 (1966) 383
[56] S. Borshay, G.E. Brown and M. Rho, Phys. Rev. Lett. 32 (1974) 787
[57] W. Weise, Phys. Lett. 117 B (1982) 150
[58] A. Arima, T. Cheon, K. Shimizu, H. Hyuga and T. Suzuki, Phys. Lett. 122 B (1983)126
[59] A. Arima and H. Hyuga, in: Mesons in Nuclei, Vol. II, M. Rho and D. Wilkinson,
     eds., North Holland (1979)
[60] J.S. McCarthy, Nucl. Phys. A 335 (1980) 27c
[61] W. Alberico, M. Ericson and A. Molinari,
     Nucl. Phys. A 386 (1982) 412, and CERN preprint
[62] K. Stricker, J. Carr and H. McManus, Phys. Rev. C 22 (1980) 2043
[63] S. Ciulli, H. Pilkuhn and H.G. Schlaile, Z. Phys. A 302 (1981) 45
[64] M. Hirata, F. Lenz and K. Yazaki, Ann. of Phys. 109 (1977) 16;
     M. Hirata, J. Koch, F. Lenz and E.J. Moniz, Ann. of Phys. 120 (1979) 205;
     Y. Horika, M. Thies and F. Lenz, Nucl. Phys. A 345 (1980) 386
     F. Lenz, M. Thies and Y. Horikawa, Ann. of Phys. 140 (1982) 266
[65] K. Klingenbeck, M. Dillig and M.G. Huber,
     Phys. Rev. Lett. 41 (1978) 387; Phys. Rev. C 22 (1980) 681
[66] W. Weise, Nucl. Phys. A 278 (1977) 402
     E. Oset and W. Weise, Phys. Lett. 77 B (1978) 159;
     Nucl. Phys. A 319 (1979) 477; A 358 (1981) 163c;
[67] W. Weise, Nucl. Phys. A 358 (1981) 163c;
     J.H. Koch, NIKHEF preprint (1983)
[68] C. Gaarde et al., Nucl. Phys. A 369 (1981) 258;
     C.D. Goodman, Nucl. Phys. A 374 (1982) 241c
[69] C. Ellegard et al., Phys. Rev. Lett. 50 (1983) 1745
[70] A. Richter, Nucl. Phys. A 374 (1982) 177c
[71] F. Osterfeld, Phys. Rev. C 26 (1982) 762
[72] B. Buck and S.M. Perez, Phys. Rev. Lett. 50 (1983) 1975
[73] J. Delorme, M. Ericson and P. Guichon, Phys. Lett. 115 B (1982) 86
[74] A. Arima and H. Hyuga, in: Mesons in Nuclei, Vol. II, M. Rho and D.H. Wilkinson,
     eds., North-Holland (1979)
[75] I.S. Towner and F.C. Khanna, Nucl. Phys. A 399 (1983) 334
[76] R.D. Lawson, Phys. Lett. 125 B (1983) 255
[77] J. Rapaport, Proc. IUCF Workshop on "Interactions between Medium Energy Nucleons
     in Nuclei", preprint, Ohio University (1982)
[78] G.F. Bertsch and I. Hamamoto, Phys. Rev. C 26 (1982) 1323

[79] M. Ericson, A. Figureau and C. Thévenet, Phys. Lett 45 B (1973) 19;
     R. Rho, Nucl. Phys. A 231 (1974) 493;
     K. Ohta and M. Wakamatsu, Nucl. Phys. A 234 (1974) 445;
     I.S. Towner and F.C. Khanna, Phys. Rev. Lett. 42 (1979) 51;
     E. Oset and M. Rho, Phys. Rev. Lett. 42 (1979) 47;
     H. Toki and W. Weise, Phys. Lett. 97 B (1980) 12;
     W. Knüpfer, M. Dillig and A. Richter, Phys. Lett. 95 B (1980) 349;
     G.E. Brown and M. Rho, Nucl. Phys. A 372 (1981) 397;
     A. Bohr and B. Mottelson, Phys. Lett. 100 B (1981) 10;
     A. Härting, W. Weise, H. Toki and A. Richter, Phys. Lett. 104 B (1981) 261;
     T. Suzuki, S. Krewald and J. Speth, Phys. Lett. 107 B (1981) 9;
     M. Kohno and D. Sprung, Phys. Rev. C 25 (1982) 297;
     R.D. Lawson, Phys. Lett. 125 B (1983) 255
[80] A. Arima, T. Cheon, K. Shimizu, H. Hyuga and T. Suzuki,
     Phys. Lett. 122 B (1983) 126
[81] G.E. Brown, K. Nakayama and J. Speth, private communication and preprint
[82] W. Steffen, R. Benz, H. Gräf, A. Richter, E. Spamer, O. Titze and W. Knüpfer,
     Phys. Lett. 95 B (1980) 23
[83] J.B. McGrory and B.H. Wildenthal, Phys. Lett. 103 B (1981) 173
[84] W. Weise, Nucl. Phys. A 396 (1983) 373c
[85] W. Steffen, H. Gräf, A. Richter, A. Härting, W. Weise, U. Deutschmann, G. Lahm
     and R. Neuhausen, Nucl. Phys. A 404 (1983) 413
[86] A. Härting, M. Kohno and W. Weise, Nucl. Phys. A (1984), in print
[87] B. Povh, Ann. Rev. Nucl. Sci 28 (1978) 1
[88] W. Brückner et al., Phys. Lett. 79 B (1978) 157
[89] R. Bertini et al., Phys. Lett. 90 B (1980) 375
[90] A. Bouyssy, Nucl. Phys. A 290 (1977) 324; Phys. Lett. 91 B (1980) 15
[91] R. Brockmann, Phys. Rev. C 18 (1978) 1510;
     R. Brockmann and W. Weise, Phys. Rev. C (1977)
[92] R. Brockmann and W. Weise, Phys. Lett. 69 B (1977) 167;
     Nucl. Phys. A 355 (1981) 365
[93] H.J. Pirner, Phys. Lett. 85 B (1979) 190
[94] R. Brockmann, Phys. Lett. 104 B (1981) 256
[95] C.B. Dover and A. Gal, BNL preprint (1982)
[96] B. Povh, Progr. in Part. and Nucl. Phys. 8 (1982) 325

QUARK MODELS OF HADRONIC INTERACTIONS

L. Wilets
Institute for Nuclear Theory
Department of Physics, FM-15
University of Washington
Seattle, Washington 98195, U.S.A.

PREPARED FOR THE U.S. DEPARTMENT OF ENERGY

1.0 MODELLING QUANTUM CHROMODYNAMICS

1.1 Models Of QCD

Although quantum chromodynamics is generally accepted as the fundamental theory of strongly interacting elementary particles, there exist no exact solutions to the theory. Important general properties of the theory have been derived, such as asymptotic freedom, the running of the coupling constant, and (plausibly) color confinement. A small (but very impressive) amount of information has been extracted from Monte Carlo lattice gauge theory (LGT) calculations. Such results are limited (in their accuracy and the number of properties calculated) by computer size and time considerations. To date, such calculations include only gluonic interactions, and do not allow for quark-antiquark virtual excitations.

At another level of sophistication is modelling. The exact theory is replaced by a mathematically simpler model (or effective theory) which incorporates as many features of the exact theory as possible. A small set of parameters may be fixed by (for example) LGT calculations and experiment. The utility of any model is then determined by its accuracy in describing other experimental data and predicting new phenomena. Thus modelling provides a bridge between fundamental theory and experiment. Comparison with experiment not only tests the model but, more importantly, provides a method to test the fundamental theory.

The MIT bag[1] was one of the earliest and most successful models of QCD, imposing confinement a _priori_ and including perturbatively interactions between quarks and gluons. An evolution of the MIT bag came with the introduction of the chiral[2] and cloudy[3] bags, which treat pions as elementary particles. The pion is an "anomalously" light hadron, and in a chirally invariant QCD should emerge as the massless Goldstone boson. Non-relativistic potential models have also had remarkable success, and have the advantage of being amenable to dynamic calculations.

The soliton model proposed by Friedberg and Lee[4,5] is particularly attractive. It is a covariant field theory and sufficiently general so that, for certain limiting cases of the adjustable parameters, it can describe either the MIT or SLAC (shell) bags. The confinement mechanism appears as a dynamic field. This allows non-static processes, such as bag oscillations and collisions (including bag creation and fission) to be calculated utilizing the well-developed techniques of nuclear many-body theory.

1.2 The Soliton Bag Model Of Friedberg And Lee

In the soliton model, the (effective) Lagrangian density is

$$\mathcal{L} = \mathcal{L}_q + \mathcal{L}_\sigma + \mathcal{L}_{q\sigma} + \mathcal{L}_G + \text{(counter terms, Higgs fields, etc)}, \quad (1.1)$$

where the individual terms have the following interpretation:

$\mathcal{L}_q = \Sigma_f \bar{\psi}_f (\gamma \cdot p - m_f) \psi_f$ describes the quarks as Dirac particles of mass m_f, where f is the flavor. We take $m_u = m_d = 0$.

$\mathcal{L}_\sigma = \frac{1}{2} (\partial \sigma)^2 - U(\sigma)$ describes the scalar soliton field σ, which represents the complex structure of the vacuum, arising from virtual gluons and quark-pairs interacting among themselves. The momentum operator conjugate to σ is $\pi = \dot{\sigma}$, and the two satisfy the canonical equal-time commutation relations

$$[\sigma(\vec{r},t), \pi(\vec{r}',t)] = i\delta^3(\vec{r} - \vec{r}') \quad (1.2)$$

The non-linearity of the soliton field enters through the self-interaction function (see Fig. 1)

$$U(\sigma) = \frac{1}{2} a \sigma^2 + \frac{1}{3!} b \sigma^3 + \frac{1}{4!} c \sigma^4 + B. \quad (1.3)$$

The polynomial terminates in fourth order to ensure renormalizability. $U(0) = B$ is to be identified with the "bag constant" or volume energy density of a cavity. With a suitable adjustment of the constants, the function has two minima, one at $\sigma = 0$, and another, lower minimum, at $\sigma = \sigma_v$. The physical vacuum corresponds to the second minimum, and the constant B is chosen so that $U(\sigma_v) = 0$.

The quarks interact with the soliton field through the term $\mathscr{L}_{q\sigma} = g\bar{\psi}\sigma\psi$. In the presence of (real) quarks, the sum $U(\sigma) + g\bar{\psi}\sigma\psi$ may have a minimum (depending on the parameters) near $\sigma = 0$ (the perturbative vacuum). This leads to a cavity in the σ-field, which is called the "bag."

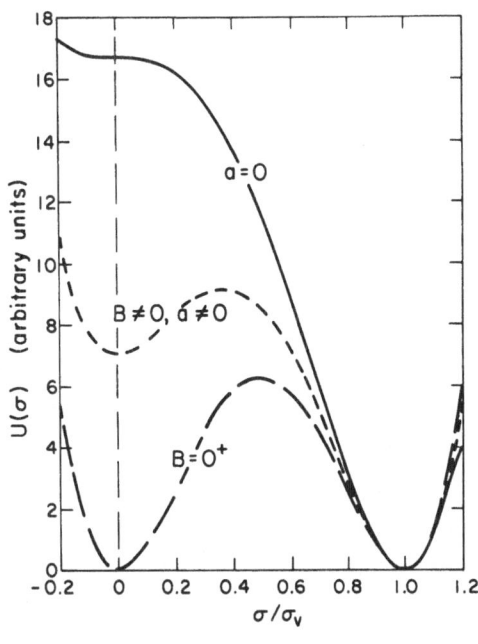

Fig. 1. Three forms for $U(\sigma)$.

Color gluon fields are introduced as in QCD, except that they interact with the soliton field through a dielectric function $\kappa(\sigma)$, chosen such that $\kappa(0) = 1$ and $\kappa(\sigma_v) = 0$. A convenient (but not unique) form is

$$\kappa(\sigma) = (\sigma/\sigma_v - 1)^2 \tag{1.4}$$

The magnetic susceptibility is $\mu = \kappa^{-1}$. The gluon part of the Lagrangian is written

$$\mathscr{L}_G = \frac{-\kappa}{4} F^c_{\mu\nu}F^{c\mu\nu} - g_s\bar{\psi}\gamma_\mu\frac{\lambda^c}{2}A^{c\mu}\psi \tag{1.5}$$

where g_s is the magnitude of the color charge. The strong coupling constant is $\alpha_s = g_s^2/4\pi$. The λ^c are the SU(3) color matrices; $c = 1....8$. The ψ functions have 4 (Dirac) times 3 (color) times n (flavor) components.

The requirements on κ yield color confinement. This can be seen easily if one keeps only terms linear in the gluon field. Then Gauss's law gives

$$\vec{\nabla}\cdot\vec{D}^c = \rho^c , \tag{1.6}$$

where c is the color index. If the total color charge does not vanish within some finite cavity, the D-field will fall off as r^{-2} as $r \rightarrow \infty$, and the color electric energy in medium

$$\frac{1}{2} \int d^3r \ D^2/\kappa(r) \tag{1.7}$$

will be infinite because $\kappa \rightarrow 0$ (exponentially in the model) as $r \rightarrow \infty$.

As long as one calculates diagrams only through order of one gluon exchange, there is no problem of double counting: the soliton field represents at least two-gluon structures. If higher order diagrams are calculated, the coefficients in the effective Lagrangian must be readjusted at each stage to compensate.

There are five parameters in the model: a, b, c, g and α_s. The first four involve only the soliton field and quarks; α_s is the quark-gluon coupling constant. The following key data may be used to help fix these parameters:

1.3 Soliton-quark Parameters

From LGT calculations we have available the following pieces of data.

{1} The bag constant[6] B = (220+20) MeV/fm^3. This is considerably larger than the MIT value[8] of 57 MeV/fm^3.

In the MIT model (if we neglect for the moment gluon effects), the bag energy is given by

$$E = \frac{2.04 \ N_q \ \hbar c}{R} + \frac{4}{3}\pi \ R^3 B \ , \tag{1.8}$$

where N_q is the number of quarks (in the lowest $s_{\frac{1}{2}}$-state). Minimization with respect to R yields

$$R = \left(\frac{2.04 \ N_q \ \hbar c}{4\pi B}\right)^{\frac{1}{4}} \tag{1.9}$$

Because of the appearance of $B^{\frac{1}{4}}$ in the denominator, this is the quantity often quoted. The discrepancy between the LGT and MIT values, in terms of $B^{\frac{1}{4}}$, is then only 40% rather than a factor of 4.

{2} $m_{GB} = (720+40)$ MeV, the mass of the glueball state (0^{++}) [7]. We identify this state with an excitation of the pure soliton field, such that $U''(\sigma_v) = m_{GB}^2$.

The errors quoted in {1} and {2} above are presumably statistical only, and do not account for systematic effects of the lattice size or the omission of dynamical interactions with quarks.

From experiment we select the subset

{3} The mean of the nucleon and delta masses, $\bar{m} = (m_N + m_\Delta)/2 = 1.087$ GeV.

{4} The proton size $\langle r_p^2 \rangle = 0.83$ fm.

{5} The proton magnetic moment, $\mu = 2.7925$ nuclear magnetons.

{6} The ratio of the axial to vector coupling constants, $g_A/g_V = 1.25$.

1.4 The Strong Coupling Constant

α_s is a "running" constant, and is evaluated for the regime of hadronic sizes. To fix it, we may utilize

{7} The delta-nucleon mass difference, $m_\Delta - m_N = 297$ MeV.

{8} The "string constant, which is the coefficient of the linear term in the potential between massive quarks, such as charmed quarks. This has been fit phenomenologically to the charmonium spectrum using a non-relativistic potential model [9] and has also been obtained in LGT calculations [7]; the value is $t \cong 1$ GeV/fm. The string constant can be calculated in the MIT model as follows. The energy per unit length in a uniform tube is

$$E/L = (B + \frac{1}{2} \vec{E}^c \cdot \vec{E}^c) A \qquad (1.10)$$

where A is the cross sectional area. The flux of color electric field emanating from a source and passing through the cross section of the tube is

$$A E^c = \frac{1}{2} Q \lambda^c , \qquad (1.11)$$

hence

$$t \equiv E/L = B A + \frac{1}{8} Q^2 \lambda^c \cdot \lambda^c /A . \qquad (1.12)$$

But $< \lambda^c \cdot \lambda^c > = 16/3$ and $\alpha_s = Q^2/4\pi$. This gives

$$t = B A + \frac{8}{3} \pi \alpha_s / A. \tag{1.13}$$

Minimization with respect to A yields the result

$$t = (32 \pi \alpha_s B/3)^{\frac{1}{2}} \tag{1.14}$$

There is much more data to fit once the parameters are determined. These include hadronic spectra and resonances, decay widths, various reactions (especially nucleon-nucleon scattering), etc. There are many more experimental data of high and low quality, but the data from LGT calculations are still few and of uncertain accuracy.

In adjusting the model parameters, the fitting is dependent on the level of sophistication of the model calculations. Whenever more diagrams are included, one must readjust the model parameters. For example, if we were able to calculate all gluon interactions exactly, then presumably the effects of the soliton field would disappear completely.

2.0 THE MEAN FIELD APPROXIMATION

If we neglect gluon interactions, the Hamiltonian can be written

$$H = \int d^3r \, \{ \tfrac{1}{2}[\pi^2 + |\vec{\nabla}\sigma|^2] + U(\sigma) + \psi^\dagger (\vec{\alpha}\cdot\vec{p}+\beta g\sigma) \, \psi \} . \tag{2.1}$$

As a preliminary step to dynamical calculations, we consider first static solutions to the field equations. The simplest of these is the mean field approximation (MFA). We set

$$\sigma = \sigma_0(\vec{r}) + \sigma_1; \quad \pi = \pi_0 + \pi_1 , \tag{2.2}$$

where $\sigma_0(\vec{r})$ is a time-independent c-number and σ_1 is the quantum fluctuation operator. Because σ_0 is static, $\pi_0 = 0$. σ_1 and π_1 satisfy

$$[\sigma_1(\vec{r},t), \pi_1(\vec{r}',t)] = [\sigma(\vec{r},t), \pi(\vec{r}',t)] = i \, \delta^3(\vec{r}-\vec{r}') \; . \tag{2.3}$$

Similarly, we represent the quark field operators by

$$\psi = \Sigma_k \, C_k \, \psi_k \quad , \tag{2.4}$$

where the ψ_k are a complete set of spinor-color-flavor vectors and

$$\{C_k^\dagger, C_{k'}\} = \delta_{\vec{k},\vec{k}'} \quad . \tag{2.5}$$

In the MFA, we consider a fixed occupation of quarks (3 quarks for nucleons; a quark-antiquark pair for mesons) and neglect the σ_1 field. The energy, relative to the vacuum, is then

$$E = \int d^3r\{\tfrac{1}{2}|\vec{\nabla}\sigma_0|^2 + U(\sigma_0) + \sum_{k\text{-occ}} \psi_k^\dagger(\vec{\alpha}\cdot\vec{p}+\beta g\sigma_0) \, \psi_k\} \; . \tag{2.6}$$

Extremization of E with respect to ψ_k (subject to the normalization constraint) and with respect to σ_0 leads to the coupled set of equations

$$(\vec{\alpha}\cdot\vec{p} + \beta g\sigma_0) \, \psi_k = \varepsilon_k \, \psi_k \; , \tag{2.7a}$$

$$-\nabla^2\sigma_0 + U'(\sigma_0) + g \sum_{k\text{-occ}} \bar{\psi}_k \, \psi_k = 0 \; , \tag{2.7b}$$

where $U' = dU/d\sigma_0$. The first is a linear eigenvalue problem (if σ_0 is given); the second is a non-linear inhomogeneous differential equation (if ψ_k is given). The equations are solved alternately until self-consistency obtains.

The cycling proceeds as follows. Let m be the cycle number. We solve (2.7a) for $\psi_k^{(m)} \{\sigma_0^{(m-1)}\}$. Then we solve (2.7b) for $\sigma_0^{(m)} \{\psi_k^{(m)}\}$. Simple iteration was found not to converge, so a cycle convergence factor f was introduced, so that the replacement $f\sigma_0^{(m)} + (1-f)\sigma_0^{(m-1)} \to \sigma_0^{(m)}$ was made before advancing m. Rapid convergence was achieved for f = 0.5.

There are many techniques for solving (2.7a). I will outline here the method described in ref. [5]. Let $\sigma_0(r)$ be spherically symmetric. Suppressing the cycle index m, the Dirac function is written

$$\psi_k = \begin{pmatrix} u_k(r) \\ i\vec{\sigma}\cdot\hat{r}\ v_k(r) \end{pmatrix} \mathcal{Y}_{km} \ ,$$

(2.8)

where $\mathcal{Y}_{km} = \mathcal{Y}_{jm}^\ell$ is a Pauli spinor. The Dirac quantum number $\kappa = (j+\frac{1}{2})\ (-1)^{j-\ell+\frac{1}{2}}$. Here k stands for the set of all quantum numbers. The u and v satisfy

$$du_k/dr = -(g\sigma_0 + \varepsilon_k)\ v_k \ ,$$

(2.9a)

$$dv_k/dr + 2v_k/r = (-g\sigma_0 + \varepsilon_k)\ u_k \ .$$

(2.9b)

We begin with a previous guess of the eigenvalue, $\varepsilon_k^{(n)}$ and integrate the radial equations outward (satisfying the boundary condition at r = 0) to some match point R. Similarly the equations are integrated inward from some large starting radius to the same point R. Unless $\varepsilon_k^{(n)}$ is a true eigenvalue, it is not possible to match both u and v at R. The functions are scaled so that $u_k(R^-) = u_k(R^+)$. Then an improved estimate for the eigenvalue is

$$\varepsilon_k^{(n+1)} = \frac{\int \psi_k^{(n)\dagger}(\vec{\alpha}\cdot\vec{p} + \beta g\sigma_0)\psi_k^{(n)}d^3r}{\int \psi_k^{(n)\dagger}\psi_k^{(n)}\ d^3r}$$

$$= \varepsilon_k^{(n)} + \frac{u_k^{(n)}(R)\ (v_k^{(n)}(R^+) - v_k^{(n)}(R^-))\ R^2}{\int r^2 dr\ (u_k^{(n)2} + v_k^{(n)2})}$$

(2.10)

The process is iterated until the desired accuracy is achieved before going on to the solution of the σ_0-equation.

The non-linear σ_0 differential equation is solved by the very effective iterative method of Henyey and Wilets[10]. The radial differential equation (2.7b) can be written ($f = r\sigma_0$)

$$d^2f/dr^2 + G(f) = 0 \ .$$

(2.11)

Let $f^{(n)}$ be the approximation resulting from the n-th iteration. Then set

$$f^{(n+1)}(r) = f^{(n)}(r) + y^{(n)}(r) . \qquad (2.12)$$

Then to first order in $y^{(n)}$, (2.11) becomes

$$d^2 y^{(n)}/dr^2 + G'(f^{(n)}) y^{(n)} = - [d^2 f^{(n)}/dr^2 + G(f^{(n)})], \qquad (2.13)$$

which is a linear, inhomogeneous differential equation. $f^{(n)}$ is required to satisfy the proper boundary conditions for each n. In the present case,

$$f^{(n)}(0) = 0, \quad f^{(n)}(r) \rightarrow r\sigma_v \text{ for } r \rightarrow \infty . \qquad (2.14a)$$

Therefore

$$y^{(n)}(0) = 0, \quad y^{(n)}(r) \rightarrow 0 \text{ for } r \rightarrow \infty . \qquad (2.14b)$$

Eq. (2.13) can be solved for $y^{(n)}$ by various, standard techniques. The accuracy of the result, however, is determined by the accuracy of the approximation used to represent the inhomogeneous term $d^2 f^{(n)}/dr^2$. What we have done was to approximate the non-linear differential equation (2.11) by the Noumerov method. With $f_i = f(i\Delta r)$, we set

$$(f_{i+1} - 2 f_i + f_{i-1})/(\Delta r)^2 +$$

$$\frac{1}{12} (G(f_{i+1}) + 10 G(f_i) + G(f_{i-1})) = 0 , \qquad (2.15)$$

then linearize the equations to obtain a band matrix operating on $y_i^{(n)}$. The matrix equation can be solved as follows (drop superscript "n"): Let

$$\alpha_i y_{i+1} + \beta_i y_i + \gamma_i y_{i-1} + H_i = 0 . \qquad (2.16a)$$

Ansatz:

$$y_i = A_i y_{i+1} + B_i . \qquad (2.16b)$$

Then

$$A_i = - \frac{\alpha_i}{\beta_i + \gamma_i A_{i-1}} \quad , \quad B_i = - \frac{\gamma_i B_{i-1} + H_i}{\beta_i + \gamma_i A_{i-1}} \, . \tag{2.16c}$$

(2.16c) gives upward recursion relations for the A_i and B_i. Check boundary conditions: For $y_0 = 0$, we may begin with $A_0 = B_0 = 0$. At some maximum r (I_{max}), there is another boundary condition, for example $y_{Imax} = 0$. Then one can recur downward to obtain all of the y_i.

The numerical procedures work extremely well, and usually converge when a solution exists.

A non-exhaustive search of parameters to fit data {1} through {8} above has been made. It should be stressed again that the parameter set obtained depends on the order of the calculation. The result quoted here was obtained in the mean field approximation, including recoil corrections (which will be discussed below). Up to this point, gluon exchange has not been included. We will return later to interaction of the gluon field.

We find a "reasonable" parameter set which gives a moderately good fit to experimental data:

a = 236.13, b = -11 614, c = 180 000, g = 25.

The structure of the soliton field and the quark wave function for the nucleon are shown in Fig. 2. Other hadronic properties are summarized in the following table, which includes recoil effects discussed below[11,12]:

Table 1.

Bag characteristics for the parameters listed in the text.

	Model	Experiment	LGT
$\langle r_p^2 \rangle^{\frac{1}{2}}$	0.83 fm	0.83	
$\bar{m}$	1,011 MeV	1,087	
μ_p	2.53 nm	2.7925	
g_A/g_V	1.04	1.25	
B	23.6 MeV/fm^3		220
m_{GB}	3,500 MeV		720
$\varepsilon(1s)$	287 MeV		

Our value of B is an order of magnitude smaller than the LGT calculations and somewhat smaller than the MIT value, indicating that surface tension plays an important role in confinement. We could "tolerate" an even smaller value. However, as mentioned above, the number usually quoted is B , which makes discrepancies among the various values appear to be less pronounced. We also find a larger glueball mass than that given by LGT calculations. It must be emphasized that the LGT calculations are incomplete in that $q\bar{q}$ excitations are not included.

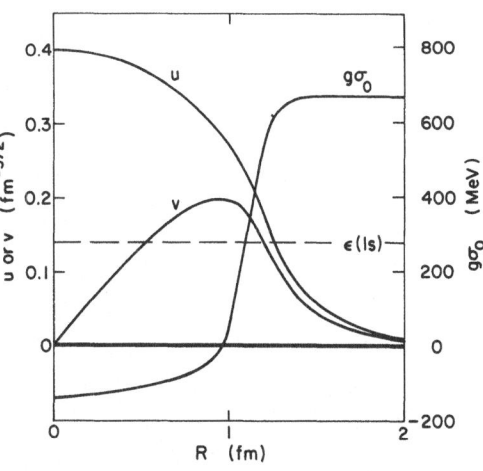

Fig. 2. Numerical results for the parameter set given in text above.

3.0 QUANTUM ALTERNATIVES TO THE MEAN FIELD

For handling the quantum fluctuations σ_1 and also for treating σ_0 as a quantum field, we now turn to the expansion of the field operators σ and π in terms of modes. We consider two approaches: (1) the coherent state and (2) Q-space.

3.1 The Coherent State

Instead of the division $\sigma = \sigma_0 + \sigma_1$, where $\sigma_0(r)$ is a c-number, we rather expand σ and π in a full set of modes,

$$\sigma = \sigma_v + \sum_n \left(\frac{1}{2\omega_n}\right)^{\frac{1}{2}} (a_n^\dagger s_n^* + a_n s_n) ,$$

$$\pi = i \sum_n \left(\frac{\omega_n}{2}\right)^{\frac{1}{2}} (a_n^\dagger s_n^* - a_n s_n) , \qquad (3.1)$$

where the ω_n are as yet undetermined and the $\{s_n\}$ are a complete orthonormal set. $\sigma(\vec{r},t)$ and $\pi(\vec{r},t)$ satisfy the usual equal time commutation relations if

$$[a_n(t), a_{n'}^\dagger(t)] = \delta_{nn'} \ ,$$

$$[a_n(t), a_{n'}(t)] = [a_n^\dagger(t), a_{n'}^\dagger(t)] = 0 \ . \tag{3.2}$$

A "coherent" state in one mode, say $n = 0$, is obtained by the construction

$$|\lambda\rangle = e^{\lambda a_0^\dagger} |0\rangle \ , \tag{3.3}$$

where $a_n|0\rangle = 0$ for all n. The coherent state satisfies

$$a_0|\lambda\rangle = \lambda|\lambda\rangle \tag{3.4}$$

and

$$\langle\lambda|\lambda\rangle = e^{|\lambda|^2} \tag{3.5}$$

There is no loss of generality in choosing λ real. We consider now the case where s_0 is real.

To study the energy of the system relative to the "vacuum" for these modes, $|0\rangle$, we normal order the Hamiltonian, $:H:$. In calculating the expectation value of $:H:$, we will need the following

$$\langle\lambda|: (a_0^\dagger + a_0)^n :|\lambda\rangle = (2\lambda)^n \langle\lambda|\lambda\rangle \ ,$$

$$\langle\lambda|: (a_0^\dagger - a_0)^n :|\lambda\rangle = 0 \text{ for } n > 0 \ . \tag{3.6}$$

Expectation values involving a_i or $a_i^\dagger$ $(i > 0)$ vanish. We write the state vector in product form

$$|\Psi\rangle = |\lambda\rangle|q\rangle / \langle\lambda|\lambda\rangle^{\frac{1}{2}}$$

where $|\lambda\rangle$ is the soliton coherent state and $|q\rangle$ is the normalized quark state vector; let $\lambda = \xi \sqrt{\omega_0}/2$.

$$\langle\Psi|:H:|\Psi\rangle =$$

$$\int d^3r \{ \Sigma_k \ \psi_k^\dagger (\vec{\alpha}\cdot\vec{p} + \beta g\sigma_0)\psi_k + \frac{1}{2}|\nabla\sigma_0|^2 +$$

$$\frac{a'}{2}\sigma_0^2 + \frac{b}{3!}\sigma_0^3 + \frac{c}{4!}\sigma_0^4 + B' \} \ . \tag{3.7}$$

The coefficients a' and B' differ from a and B by (infinite) constants arising from normal ordering. Except for the renormalization of the coefficients a and B, (3.7) is identical with the mean field approximation expression for the energy (2.6) if we identify

$$\sigma_0(\vec{r}) \equiv \xi s_0(\vec{r}) + \sigma_v = <\Psi|\sigma|\Psi> \tag{3.8}$$

with the mean field quantity. Note that the quantities ω_k do not enter explicitly in the final result.

Multiple-mode coherent state wave functions can be constructed by taking a product of the exponential operators:

$$|\lambda_1 \lambda_2 \ldots> = \Pi_n e^{\lambda_n a_n^\dagger} |0> \tag{3.9}$$

but this may also be regarded as a single mode function!

3.2 Q-space

We proceed again with an orthonormal mode analysis, but rather than utilizing annihilation and creation operators, we introduce the operators P and Q :

$$\sigma = \sigma_v + \Sigma\, s_n Q_n \ ,$$
$$\pi = \qquad \Sigma\, s_n P_n \ , \tag{3.10}$$

where

$$[Q_n, P_{n'}] = i\, \delta_{nn'} \ . \tag{3.11}$$

Once again we restrict consideration to one mode, and ignore cross coupling to other modes. We drop the subscript "0". The state vector is now written

$$|\Psi> = \Phi(Q)|q> \tag{3.12}$$

and we use the representation

$$P = -i\, \partial/\partial Q \ . \tag{3.13}$$

Before deriving the equation for (Q), we note that in the one mode approximation this approach differs from the coherent state in that the latter has an explicit form for occupation numbers. A more general form would be some arbitrary function, $F(a^\dagger)|0\rangle$. The Q representation allows for the most general form of F; it gives the exact one mode (Tomanaga) formulation of the problem. In principle, we could normal order the $a \equiv (\omega^{-\frac{1}{2}}Q + i\omega^{-\frac{1}{2}}P)2^{-\frac{1}{2}}$ and $a^\dagger$ operators, but that would lead to a fourth order differential equation for $\Phi(Q)$. Rather, we keep the original ordering, and later subtract the vacuum energy associated with the mode(s) considered.

With $|\Psi\rangle$ given by (3.12), we find

$$\langle\Psi|H|\Psi\rangle =$$

$$\int d^3r\{\sum_k \psi_k^\dagger(\vec{\alpha}\cdot\vec{p} + g\beta(\sigma_v + s\langle Q\rangle))\psi_k + s^2\langle P^2\rangle + |\vec{\nabla}s|^2\langle Q^2\rangle$$

$$+ \langle U(\sigma_v + sQ)\rangle\} , \tag{3.14}$$

where

$$\langle P^2\rangle \equiv -\int \Phi^* \frac{d^2}{dQ^2} \Phi \, dQ ,$$

$$\langle Q^n\rangle \equiv \int \Phi^* Q^n \Phi \, dQ . \tag{3.15}$$

We extremize $\langle H\rangle$ with respect to $\psi_k(\vec{r})$, $s(\vec{r})$ and $\Phi(Q)$, subject to normalization constraints on all three functions to obtain the three coupled equations

$$[\vec{\alpha}\cdot\vec{p} + \beta g(\sigma_v + s\langle Q\rangle - \varepsilon_k]\psi_k(\vec{r}) = 0 , \tag{3.16a}$$

$$[- \langle Q^2\rangle\vec{\nabla}^2 + \langle P^2\rangle + \langle QU'(\sigma_v + sQ)\rangle + \langle Q\rangle g\sum \bar{\psi}_k\psi_k - \lambda]s(\vec{r}) = 0 , \tag{3.16b}$$

$$[\int d^3r(- s^2 \frac{d^2}{dQ^2} + |\vec{\nabla}s|^2Q^2 + U(\sigma_v + sQ) + g\sum_k \bar{\psi}_k s\psi_k Q) -E_Q]\Phi(Q) = 0. \tag{3.16c}$$

These equations have been solved self-consistently. Eqs. (a) and (c) are linear eigenvalue equations. Eq. (c) is a second order differential equation with a "potential" which is a fourth order polynomial in Q. Eq. (b) is a non-linear, inhomogeneous differential equation of the same form as the σ_0 equation (2.7b). In solving (3.16b), however, the Lagrange multiplier λ must be varied until a normalized function s is obtained. The three equations are cycled through (with a convergence factor on s) until self-consistency obtains.

Physical quantities are obtained by subtracting out the no-quark, same one-mode state quantities. Thus the s obtained by solving Eqs. (3.16a-b) with N quarks is now used to solve only Eq. (3.16c) with the term containing $\bar{\psi}_k s \psi_k$. absent. Comparisons have been made with the mean field approximation for the energy and $\langle \sigma \rangle$. The results agree (for the standard parameter set) to about 20%. Another measure of the validity of the mean field approximation is the quantum fluctuation in Q. We find

$$\frac{\langle Q \rangle}{\langle Q^2 \rangle^{\frac{1}{2}}} \simeq 0.78$$

$$(3.17)$$

The results of this section were obtained by Dethier.

4.0 SMALL AMPLITUDE OSCILLATIONS

We return to the mean field representation $\sigma = \sigma_0 + \sigma_1$, $\pi = \pi_1$. The Hamiltonian is given by

$$H = \mathcal{E}_{\sigma_0} + \sum_k{}' \varepsilon_k c_k^\dagger c_k$$
$$+ \int d^3r \{ \tfrac{1}{2}(\pi_1^2 + |\vec{\nabla}\sigma_1|^2 + U''(\sigma_0)\sigma_1^2) + \tfrac{1}{6} U'''(\sigma_0)\sigma_1^3 + \tfrac{1}{24} c\sigma_1^4$$
$$+ g \sum_{k\ell}{}' \bar{\psi}_k \sigma_1 \psi_\ell c_k^\dagger c_\ell \} . \qquad (4.1)$$

where $\mathcal{E}_{\sigma_0} = \int d^3r \, \mathcal{H}_\sigma (\sigma_0(r))$.

The primes on the sums denote subtraction of the same terms for k (and ℓ) occupied. Note the vanishing of terms linear in σ_1, if all quarks are in occupied states. The quantum part of the soliton field can be expanded in an orthonormal set:

$$\sigma_1 = \sum_n \left(\frac{1}{2\omega_n}\right)^{\frac{1}{2}} (a_n^\dagger s_n^* + a_n s_n) , \qquad (4.2a)$$

$$\pi_1 = i \sum_n \left(\frac{\omega_n}{2}\right)^{\frac{1}{2}} (a_n^\dagger s_n^* - a_n s_n) . \qquad (4.2b)$$

The Hamiltonian (4.1) simplifies if we <u>choose</u> $s_n(r)$ and ω_n to satisfy

$$(- \nabla^2 + U''(\sigma_0) - \omega_n^2) \, s_n(\vec{r}) = 0 \; ; \tag{4.3}$$

then

$$H = \mathcal{E}_{\sigma_0} + \Sigma_k' \; \epsilon_k \, c_k^+ c_k + \Sigma_n \, \omega_n (a_n^\dagger a_n + \tfrac{1}{2}) + H_1 + H_3 + H_4 , \tag{4.4}$$

with

$$H_1 = g \sum_{k\ell n}' \int d^3r \; \bar{\psi}_k \, s_n \, \psi_\ell \, (\frac{1}{2\omega_n})^{\frac{1}{2}} a_n + \text{hc.} , \tag{4.5a}$$

$$H_3 = \frac{1}{6} \int d^3r (b + c\sigma_0) \sigma_1^3 , \tag{4.5b}$$

$$H_4 = \frac{c}{24} \int d^3r \; \sigma_1^4 , \tag{4.5c}$$

where σ_1 in H_3 and H_4 is represented by (4.2a). The diagrammatic meaning of these terms is shown in Fig. 3.

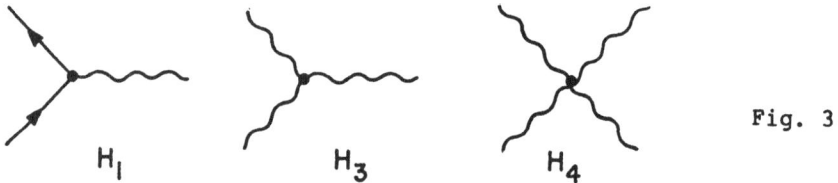

Fig. 3

H_1 H_3 H_4

Eq. (4.3) defines normal modes for oscillations about the mean field solution. Since we have $\sigma_0(r)$ to be spherically symmetric, we can set

$$s_n(\vec{r}) \equiv s_{\ell m n}(\vec{r}) = r^{-1} u_{\ell n}(r) Y_{\ell m}(\theta,\phi) , \tag{4.6}$$

whence

$$\left(- \frac{d^2}{dr^2} + \frac{\ell(\ell+1)}{r^2} + U'' - \omega_{\ell n}^2\right) u_{\ell n} = 0 . \tag{4.7}$$

$U''(\sigma_0(r))$ has a sharp dip in the vicinity of the bag surface, as can be seen in Fig. 4. A rough estimate of the eigenfrequencies can be obtained by setting

$$U'' = U''(\sigma_0(r_0)) + (r-r_0)^2 \Omega^2 , \tag{4.8}$$

where r_0 is the location of the minimum and

$$\Omega^2 = \frac{1}{2} \frac{d^2}{dr^2} U''(\sigma_0(r)) \Big|_{r=0} \tag{4.9}$$

Then

$$\omega_{\ell n}^2 \simeq U''(\sigma_0(r_0)) + \frac{\ell(\ell+1)}{r_0^2} + \Omega(2n+1) . \tag{4.10}$$

For the lowest state, using our standard parameters, we find $\omega_{00} \simeq$ 1400 MeV. The quark and soliton (σ_1) spectra are displayed schematically in Fig. 5.

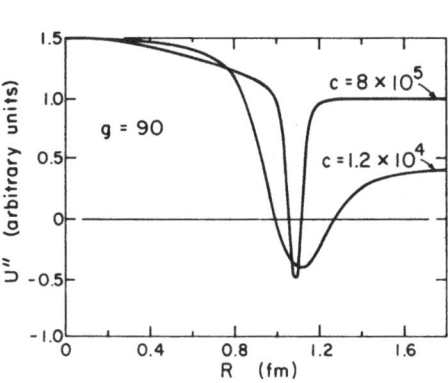

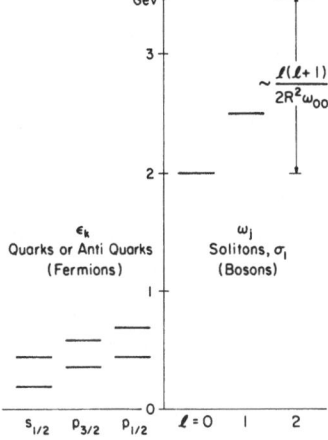

Fig. 4. $U''(\sigma(r))$ showing surface dip (B=0).

Fig. 5. Schematic representation of quark and soliton spectra.

Because of the sharp dip in U'' near the surface, the lowest normal modes are surface modes.

The soliton-quark interaction terms, H_1, couples the σ_1 excitations to quark particle-hole pairs. $q\bar{q}$ virtual excitations are interpreted as giving rise to a "meson" cloud surrounding the bag or, more specifically, the nucleon. I will return to the meson cloud later.

Although we began with a covariant Lagrangian, the MFA destroyed covariance by the selection of a preferred frame. Inclusion of the σ_1 part of the soliton field can restore covariance. For example, the localized MFA bag has $\langle \vec{P}^2 \rangle \neq 0$. A state with P=0 is spread out over all space. Thus $\langle r^2 \rangle$ will increase (ultimately to ∞) as the approximations are improved, and we are alerted to the fact that $\langle r^2 \rangle$ is not a measure of nucleon size. We return to this in Section 6.

The terms H_3 and H_4 involve only σ_1 operators and lead to a restructuring of the soliton (σ_1) spectra. They can be handled in the one mode or uncoupled mode approximations straightforwardly in the Q-space formalism. For present purposes, we will neglect these terms and assume that their effects can be absorbed into the effective parameters. There may, however, be important physical effects in these terms.

5.0 LARGE AMPLITUDE MOTION AND COLLISIONS

5.1 The Generator Coordinate Method.

The method of generator coordinates (GCM) is being applied to large amplitude bag with particular application to N-N scattering. Consider a parameter or set of parameters which describe the static configuration of a system of quarks and the soliton field and let $|\alpha,n$ denote a set of basis states which is complete for any α. A method of obtaining these basis states is described below. The GCM state vector is written

$$|\Psi\rangle = \sum_n \int \phi_n(\alpha) |\alpha,n\rangle d\alpha . \tag{5.1}$$

Since the set is complete for each α, the expansion is overcomplete. In practice, this causes no problem since the sum is truncated to a small number of terms. In what follows, we consider only a single term and suppress n; actually, several configurations may be required. The generalization is straightforward.

The weight function $\phi(\alpha)$ is obtained by extremizing the expectation value of the Hamiltonian

$$\int d\alpha \phi^*(\alpha) \langle \alpha|H|\alpha'\rangle \phi(\alpha') d\alpha'$$

subject to the normalization constraint. Then

$$\int d\alpha' \langle \alpha|H-E|\alpha'\rangle \phi(\alpha') = 0 \tag{5.2}$$

This is the basic GCM integral equation for $\phi(\alpha)$. Depending upon whether the spectrum is discrete or continuous, it is either an eigenvalue or a scattering equation. Although we can work with the integral equation, it is instructive to consider the approximate differential equation which it satisfies. For a system which has well developed collective motion, we expect $\langle\alpha|H-E|\alpha'\rangle$ to fall off rapidly as a function of $\alpha - \alpha'$. To utilize that property, it is convenient to introduce the mean and relative parameters

$$\bar{\alpha} = \frac{1}{2}(\alpha+\alpha') , \quad \delta = (\alpha-\alpha')$$

(5.3)

Then

$$\langle\Psi|H-E|\Psi\rangle = \int d\bar{\alpha}\int d\delta\phi^*(\bar{\alpha}+\frac{1}{2}\delta)\langle\bar{\alpha} + \frac{1}{2}\delta|H-E|\bar{\alpha} - \frac{1}{2}\delta\rangle\phi(\bar{\alpha} - \frac{1}{2}\delta)$$

$$= \int d\bar{\alpha}\int d\delta[\phi^*(\bar{\alpha}) + \frac{1}{2}\delta\phi'^*(\bar{\alpha}) + \frac{1}{8}\delta^2\phi''^*(\alpha) + \ldots] \times$$

$$\langle\bar{\alpha} + \frac{1}{2}\delta|H-E|\bar{\alpha} - \frac{1}{2}\delta\rangle \times$$

$$[\phi(\bar{\alpha}) - \frac{1}{2}\delta\,\phi'(\alpha) + \frac{1}{8}\delta^2\phi''(\alpha) + \ldots] .$$

(5.4)

Now $\langle\bar{\alpha} + \frac{1}{2}\sigma|H - E|\bar{\alpha} - \frac{1}{2}\delta\rangle$ is even in δ. The only integrals which survive are of the form

$$\mathcal{O}_n(\bar{\alpha}) = \int\langle\bar{\alpha} + \frac{1}{2}\delta|\mathcal{O}|\bar{\alpha} - \frac{1}{2}\delta\rangle\delta^n d\delta$$

(5.5)

for n even. Here $\mathcal{O}$ is either H or N = 1. Through order ϕ'' we have

$$\langle\psi|H-E|\psi\rangle =$$

$$\int d\bar{\alpha}\{\phi^*(H_0-E\,N_0)\phi + \frac{1}{4}(H_2-E\,N_2)(-\phi'^*\phi' + \frac{1}{2}\phi^{*''}\phi + \frac{1}{2}\phi^*\phi'')\} .$$

(5.6)

This may be cast into the more familiar form

$$\int\tilde{\phi}^*(-\frac{d}{d\bar{\alpha}}\,\frac{1}{2B(\bar{\alpha})}\,\frac{d}{d\bar{\alpha}} + V(\bar{\alpha}) - E)\tilde{\phi}\,d\bar{\alpha}$$

(5.7)

with

$$V(\bar{\alpha}) = \frac{H_0}{N_0} + \frac{1}{2N_0^{\frac{1}{2}}} \frac{d}{d\bar{\alpha}} (H_2 - EN_2) \frac{d}{d\bar{\alpha}} \frac{1}{N_0^{\frac{1}{2}}} , \tag{5.8a}$$

$$B(\bar{\alpha}) = - \frac{N_0}{H_2 - EN_2} , \tag{5.8b}$$

and

$$\tilde{\phi} = N_0^{\frac{1}{2}} \phi . \tag{5.8c}$$

Below we define α such that $\alpha \longrightarrow r$ as $r \longrightarrow \infty$, where r is (say) the separation of two bags. However, $V(\bar{\alpha})$ has no simple interpretation as a potential until $B(\bar{\alpha})$ is determined!

Following Mosel[13], we now introduce a change of variable from $\bar{\alpha}$ to x such that

$$x = X(\bar{\alpha}) = \int^{\bar{\alpha}} [B(\alpha')/\mu]^{\frac{1}{2}} d\alpha' , \tag{5.9a}$$

$$\tilde{\phi}(\bar{\alpha}) = f(\bar{\alpha})\psi(x) , \tag{5.9b}$$

$$f(x) = \text{const} \ [\mu B(\bar{\alpha})]^{\frac{1}{4}} \tag{5.9c}$$

where $\mu = M/2$ is the reduced nucleon mass. Then variation of Eq.(5.7) yields

$$(- \frac{1}{2\mu} \frac{d^2}{dx^2} + V + V_1 - E)\psi(x) = 0 \tag{5.10}$$

with

$$V_1 = \frac{1}{8} \frac{B''}{B^2} - \frac{1}{32} \frac{(B')^2}{B^3} \tag{5.11}$$

Since we choose α (or $\bar{\alpha}$) so that for separated bags $\alpha \rightarrow r$ (the bag separation) this is consistent with $B \rightarrow \mu$, $\alpha \rightarrow x$.

5.2 Generating The State Vector

We determine $|\alpha\rangle$ by minimizing the expectation value of the total Hamiltonian, $\langle\alpha|H|\alpha\rangle$ with respect to a variational mean field wave function for the quarks and a coherent state wave function for the soliton field subject to a constraint

$$\langle\alpha|Q|\alpha\rangle = Q_0(\alpha) \tag{5.12}$$

where Q is some moment of the quark distribution

$$Q = \int \bar{\psi}\, q(\vec{r})\,\psi\, d^3r \; . \tag{5.13}$$

The constrained mean field equations now assume the form

$$[\vec{\alpha}\cdot\vec{p} + \beta[g\sigma_0(\vec{r}) - \lambda q(\vec{r})] - \varepsilon_k]\psi_k = 0 \; , \tag{5.14a}$$

$$- \nabla^2\sigma_0 + U'(\sigma_0) + g \sum_{k-occ} \bar{\psi}_k\psi_k = 0 \; , \tag{5.14b}$$

where λ is a Lagrange multiplier. Instead of specifying the constraint function $q(\vec{r})$ explicitly and solving the pair of equations (5.14 a&b), it is more physical to specify the function in square brackets

$$[g\sigma_0(\vec{r}) - \lambda q(\vec{r})] \equiv \mathcal{V}(\vec{r}) \; , \tag{5.15}$$

then solve for the $\psi_k(\vec{r})$, and then for $\sigma_0(\vec{r})$; there is no iteration involved. The self consistency is implicit. The constraint can now be "discovered" by solving

$$\lambda q = g\sigma_0 - \mathcal{V}. \tag{5.16}$$

We parameterize $\mathcal{V}(\vec{r}) = \mathcal{V}(\alpha,\vec{r})$ by a folding procedure. We consider two spheres of radius R with centers separated by a distance α , as shown in Fig. 5. For $\alpha >0$, we define a function $\Theta(\vec{r})$ equal to unity if $\vec{r}$ lies in the interior of either sphere and zero otherwise. If $\alpha <0$, we choose $\Theta(r)$ to equal unity only in the intersecting, lens-shaped volume; this gives a natural continuation of α from positive values (prolate shapes) to negative values (oblate shapes). The radii of the spheres are chosen so that the enclosed volume is independent of α, and equals $2 \times (4/3)\ \pi R^3$, where R is the radius of a three quark bag. (The constant volume approximation, which is valid for a relativistic Fermi gas, can be relaxed to, say, minimize the energy as a function of R for each α, or to let R be another shape parameter.) α can assume all values, $-\infty < \alpha < +\infty$, and is equal to the separation of isolated bags for $\alpha > 2R$.

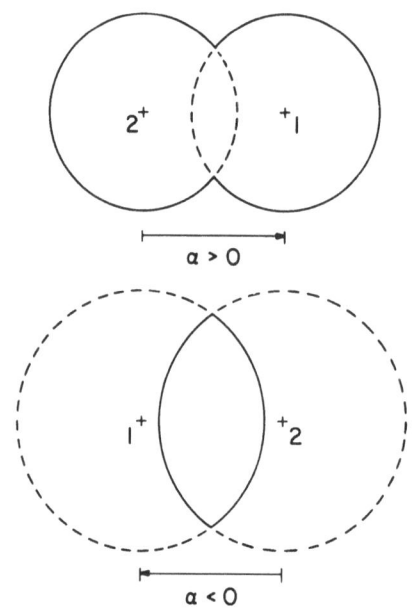

Fig. 5. Geometric shapes used to define $\Theta(\vec{r})$ and α.

Into this geometric shape is folded a Yukawa smoothing function, yielding

$$\mathcal{V}(\vec{r}) = g\sigma_v [1 - \frac{\gamma^2}{4\pi} \int \frac{e^{-\gamma|\vec{r}-\vec{r}'|}}{|\vec{r}-\vec{r}'|} \Theta(\vec{r}')\ d^3\vec{r}'] \qquad (5.17)$$

Note that $\mathcal{V} \to g\sigma_v$ as $r \to \infty$ and $\mathcal{V} \cong 0$ for $\vec{r}$ well inside the geometric volume. It is adjusted to approximate the self consistent, unconstrained spherical solution for isolated bags.

The method of solution of equations (18 a&b) is a generalization of that described in ref. [5]. Here, however, both $\mathcal{V}(\vec{r})$ and $\sigma_0(\vec{r})$ are expanded in terms of even Legendre polynomials, and the quark functions are expanded in terms of Dirac spinors of good quantum number κ. An example of the shape of $\mathcal{V}(\vec{r})$ is shown in Fig. 6. Low lying eigenvalues of the Dirac equation as a function of α are shown in Fig. 7. Parity is a good quantum number for the quark functions, and we note that for well-separated bags the eigenvalues become doubly degenerate with respect to the two parities, corresponding to degenerate left and right states. All of the GCM calculations described here were performed by A. Schuh, and the work is continuing.

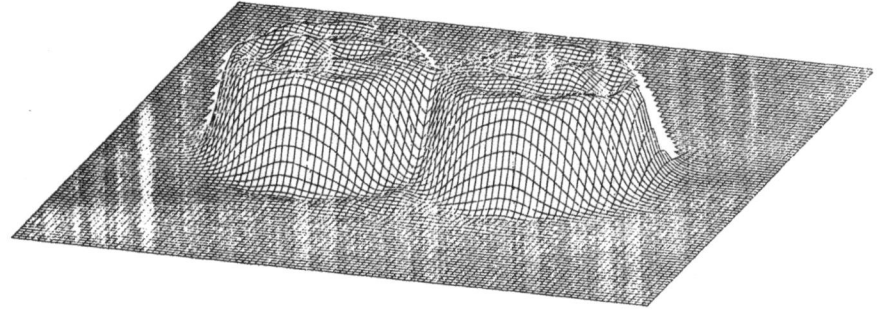

Fig. 6. The function $-\mathcal{V}(r)$ for α = 2 fm and R = 1 fm.

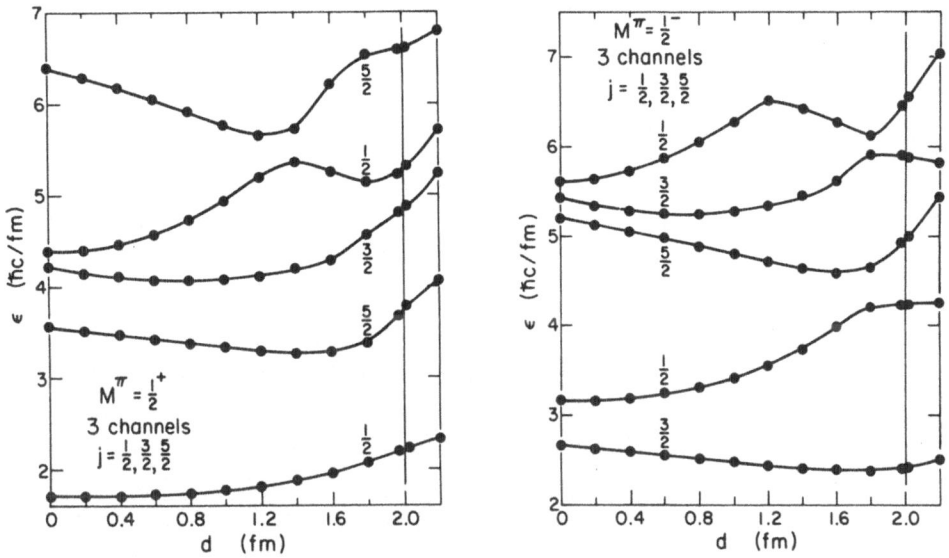

Fig. 7. Low lying quark eigenenergies calculated coupling three κ -states. Calculations coupling six states become flat, and exhibit degeneracy for + and - parities, for α > 2 fm.

5.3 The State Vector in the Coherent State Representation

In order to implement the GCM, it is necessary to represent the state vector $|\alpha\rangle$ as a fully quantal state. For this we use the coherent state representation of σ rather than mean field interpretation of the σ_0 obtained in solving (5.1). We need boson operators which are independent α. Corresponding in (3.1) we set

$$\sigma = \sigma_v + \sum \left(\frac{1}{2\omega_k V}\right)^{\frac{1}{2}} (a_k^\dagger e^{-i\vec{k}\cdot\vec{r}} + a_k e^{i\vec{k}\cdot\vec{r}})$$

$$\pi = i \sum \left(\frac{\omega_k}{2V}\right)^{\frac{1}{2}} (a_k^\dagger e^{-i\vec{k}\cdot\vec{r}} - a_k e^{i\vec{k}\cdot\vec{r}}) \tag{5.18}$$

where V is the box-normalization volume (which will go out). Then

$$|\alpha\rangle = \prod_n c_n^\dagger(\alpha)\; e^{\sum \left(\frac{\omega_k}{2}\right)^{\frac{1}{2}} f_k(\alpha) a_k^\dagger} |0\rangle\; N_\alpha^{-\frac{1}{2}} \tag{5.19}$$

where

$$N_\alpha = e^{\frac{1}{2}\sum\omega_k |f_k(\alpha)|^2} \tag{5.20}$$

and the f_k are the Fourier transforms of the field $\sigma_0(\vec{r},\alpha)$:

$$f_k(\alpha) = V^{-\frac{1}{2}} \int (\sigma_0(\vec{r},\alpha) - \sigma_v) e^{-i\vec{k}\cdot\vec{r}} d^3r \tag{5.21}$$

Although it is not essential to do so, a great **conceptual** simplification occurs if we **choose** $\omega_k = \Omega$, independent of k. In section 6.3 we face this question again and there consider the "natural" choice $\omega_k = (k^2 + U''(\sigma_v))^{\frac{1}{2}}$, and work with the f_k. However, with the choice $\omega_k = \Omega$, all required expressions can be evaluated in configuration space. Recall that because we normal order our Hamiltonian, a_k brings down an $f_k(\alpha)(\Omega/2)^{\frac{1}{2}}$ and $a_k^\dagger$ an $f_k^*(\alpha')(\Omega/2)^{\frac{1}{2}}$. We obtain

$$\langle\alpha'|\alpha\rangle = \prod_n \left[\int \psi_n^\dagger(r,\alpha')\psi_n(\vec{r},\alpha)d^3r\right] e^{-\frac{\Omega}{4}\int[\sigma(\vec{r},\alpha') - \sigma(\vec{r},\alpha)]^2 d^3r} \tag{5.22}$$

N.B. In the quark overlap functions we have assumed the very special case that $\int \psi_n^\dagger(\vec{r},\alpha)\psi_{n'}(\vec{r},\alpha)d^3r = 0$ for $n \neq n'$ because of spin, parity, j_z and color symmetries. In more complex configurations, the quark overlap integral will contain exchange terms. The normalization factors have been used explicitly so that $\langle\alpha|\alpha\rangle = 1$.

Now one can calculate $\langle\alpha|:H:|\alpha'\rangle$, which enters in (5.2) in the following conceptually simple manner. Let the Hamiltonian **density** [see (2.1)] be written

$$\mathcal{H} = \mathcal{H}(\psi^\dagger, \psi, \sigma, \pi).$$ (5.23)

Then

$$\langle\alpha'|:H:|\alpha\rangle = \langle\alpha'|\alpha\rangle \int d^3r \; \mathcal{H}(\psi_n^\dagger(\alpha), \; \psi_n(\alpha'), \sigma_{\alpha\alpha'} \; \pi_{\alpha\alpha'})$$ (5.24)

with the arguments of $\mathcal{H}$ now functions rather than operators:

$$\sigma_{\alpha\alpha'}(\vec{r}) = \tfrac{1}{2}[\sigma_0(\vec{r},\alpha) + \sigma_0(\vec{r},\alpha')]$$

$$\pi_{\alpha\alpha'}(\vec{r}) = \tfrac{i}{2} \; \Omega[\sigma_0(\vec{r},\alpha) - \sigma_0(\vec{r},\alpha')]$$ (5.25)

It is worth noting that $\langle\alpha|:\pi^2:|\alpha'\rangle$ vanishes only for $\alpha' = \alpha$, which is also the case of the static bag.

The choice of Ω remains to be determined. Recall that it did not enter in the static, $\alpha' = \alpha$, calculations. There are at least two possibilities:

First, Ω is not a physical quantity, but the choice of value does affect our renormalization scheme. According to Stevenson's principle[14,15] of minimum sensitivity, such a parameter should be varied to find the value which extremizes or insensitizes the physical quantity of interest. In that case, we could even have $\Omega = \Omega(\bar{\alpha})$. One possible quantity of "interest" is $H_0(\bar{\alpha})$.

Second, one can argue that one should stick to $\omega_k = (k^2 + U''(\sigma_v))$ in order to nearly diagonalize the normal modes of the vacuum.

This matter is currently under study.

I conclude this section by emphasizing that in applying the GCM one need not go to the Schroedinger-like differential equation if the expansion in (5.4) is not valid. While the Schroedinger form has obvious conceptual advantages, the integral equation (5.2) is more generally valid.

6.0 RECOIL AND PROJECTION

A composite structure localized in a particular reference frame must be a wave packet containing a distribution of momentum components. This is the case for a bag described in the MFA. Denoting the total momentum operator by

$$\vec{P} = -\frac{i}{2} \int [\psi^{+} \overset{\leftrightarrow}{\nabla} \psi + \pi \overset{\leftrightarrow}{\nabla} \phi] \, d^3r \tag{6.1}$$

and by $|B\rangle$ a localized bag state, then $\langle B|\vec{P}|B\rangle = 0$ but $\langle B|\vec{P}^2|B\rangle > 0$. Since the full theory is Lorentz invariant, corrections beyond the MFA should tend to produce a state of good $\vec{P}$ and the lowest would correspond to $\vec{P} = 0$. Such a state would be spread over all space. Clearly $\langle r^2 \rangle$ is not an appropriate measure of hadronic size. We now consider two approaches to eliminate the effects of the spurious center-of-mass motion (momentum).

6.1 Center-of-energy Operator

A relativistic generalization of the center-of-mass operator is the center-of-energy operator

$$\vec{R} = \frac{\int d^3x \; \vec{x} \; \mathcal{H}(x)}{E} \quad , \quad E = \langle H \rangle \quad , \tag{6.2}$$

first proposed by Fokker[16] and discussed extensively by Pryce[17]. The numerator in (6.2) is the exact boost operator, which we will return to in section 7. The operator $\vec{R}$ satisfies the commutation relations

$$[R_i, P_j] = i\delta_{ij} \, H/E \quad , \tag{6.3}$$

$$\frac{d\vec{R}}{dt} = i[H, \vec{R}] = \vec{P}/E \quad , \tag{6.4}$$

and

$$[R_i, R_j] = -\epsilon_{ijk} \, M_k/E^2 \tag{6.5}$$

where $\vec{M}$ is the total angular momentum operator. (6.3) is the condition that $\vec{R}$ be canonically conjugate to P, which is satisfied in its expectation value or when operating on an eigenstate. (6.4) is a "very pleasant" and nontrivial result, which is valid for Bose fields,

Dirac fields and for interacting systems. Recall that for the Dirac operator $\vec{r}_{op} = \int \psi^\dagger \vec{r} \psi d^3 r$ we have

$$\frac{d\vec{r}_{op}}{dt} = \int \psi^\dagger \vec{\alpha} \psi d^3 r \tag{6.6}$$

which has eigenvalues equal, in magnitude, only to the speed of light. Wave packets satisfy (6.4), and so does the position operator in the Foldy-Wouthuysen Representation[18]. The operator $\vec{R}$ is closely related to the F-W position operator. Actually, the $\vec{Q}$ operator proposed by Pryce [17] is essentially the F-W position operator, but it is more cumbersome than the $\vec{R}$-operator. The $\vec{R}$-operator is not part of a Lorentz four-vector, and the failure of the components of $\vec{R}$ to commute are shortcomings in $\vec{R}$ as a center-of-energy operator. Indeed, there is no completely satisfactory candidate for a center-of-energy.

We consider the mean field approximation, where quarks move independently in a scalar field, and assume further that there are A quarks in the same spacial state of eigenvalue ε. Then

$$\langle (\vec{r} - \vec{R})^2 \rangle = [1 - \frac{2\varepsilon}{E} + \frac{A\varepsilon^2}{E^2}] \langle \vec{x}^2 \rangle + \frac{3A}{4E^2} , \tag{6.7}$$

where E is the total energy (mass) of the system. Here $\langle x^2 \rangle$ is the quark mean-square size measured with respect to the bag center. The last term in (6.7) arises from the zitterbewegung of each quark; it is the effective mean-square size of point quark. It is recognizeable as the factor in the Darwin term. Even for a single point Dirac particle we would find the term $3/4E^2$ when measured by (6.7). The experimental analysis does not include this term and we therefore identify $\langle (\vec{r}-\vec{R})^2 \rangle$ with $\langle r^2 \rangle + 3/4E^2$, where $\langle r^2 \rangle$ is the intrinsic size. Then

$$\langle r^2 \rangle \equiv \langle (\vec{r}-\vec{R})^2 \rangle - 3/4E^2$$

$$= [1 - \frac{2\varepsilon}{E} + \frac{A\varepsilon^2}{E^2}] \langle x^2 \rangle + \frac{3(A-1)}{4E^2} \tag{6.8}$$

In the MIT model, $\varepsilon = 2.04/R$ and $\langle x^2 \rangle = 0.532 \ R^2$. This leads to

$$\langle r^2 \rangle = [1 - \frac{15}{16A} + \frac{0.1907(A-1)}{A^2}] \langle x^2 \rangle , \ \text{MIT} \tag{6.9a}$$

$$= 0.7007 \ <x^2> \ , \quad A = 2 \ \text{(i.e. meson)} \tag{6.9b}$$

$$= 0.7298 \ <x^2> \ , \quad A = 3 \ \text{(i.e. nucleon)} \ . \tag{6.9c}$$

The recoil corrections are similar in the soliton model. For the standard parameters, we calculate

$$<r^2> \ = 0.737 \ <x^2> \ , \quad A = 3 \ . \tag{6.10}$$

Since we fit the proton size, the effect of these corrections is to increase the required static bag radius by about 15%. This also increases the calculated magnetic moment (which scales as a length) by the same fraction.

6.2 Energy Corrections[10]

The Einstein relation is

$$m^2 = H^2 - \vec{P}^2 \tag{6.11}$$

where $\vec{P}$ is the total momentum. A bound on the lowest mass of the system is

$$m_0^2 \ \leqslant \ <H^2> - <\vec{P}^2> \ . \tag{6.12}$$

We approximate

$$m_0 \ \simeq \sqrt{<H>^2 - <\vec{P}^2>} \ , \tag{6.13}$$

where the last term subtracts out the spurious square-momentum of the wave packet. In the mean field approximation

$$<\vec{P}^2> \ = \ <(\sum_i \vec{P}_i)^2> \ = \sum_i <\vec{P}_i^2> \ . \tag{6.14}$$

In the MIT bag model, because of the discontinuity in the wave function at the boundary, $<\vec{P}_i^2>$ is infinite for each quark; so also is $<H^2>$ infinite.

The corrections for the standard parameters of the soliton model is a decrease in the bag mass by a factor 0.84.

The results listed in Table 1 include the recoil corrections.

6.3 Projection

Given a state vector $|\vec{z}\rangle$ for a wave packet centered about the point $\vec{z}$, one can construct an unnormalized vector of good momentum $|\vec{P}\rangle$ by setting

$$|\vec{P}\rangle = \int e^{i\vec{P}\cdot\vec{z}} |\vec{z}\rangle d^3z. \tag{6.15}$$

It is well known that there are difficulties with this procedure, because the set of states so produced are not boosted states of one another. The states, in general, differ not only in total momentum but also in intrinsic excitation. To calculate the mass of the system, we consider the state $|\vec{P}=0\rangle$ and assume that of the set it has the lowest intrinsic excitation. Since the Hamiltonian is translationally invariant, we need only project on one side to evaluate

$$\frac{\int \langle \vec{z}=0|H|\vec{z}\rangle d^3z}{\int \langle \vec{z}=0|\vec{z}\rangle d^3z} = \frac{\int \langle -\frac{1}{2}\vec{z}|H|\frac{1}{2}\vec{z}\rangle d^3z}{\int \langle -\frac{1}{2}\vec{z}|\frac{1}{2}\vec{z}\rangle d^3z} \tag{6.16}$$

The mean field approach cannot be used directly to evaluate (6.16), because the representation of the σ-field is not the same for different $\vec{z}$. We use here the mode-expansion Sec. 3.1, and then use the one-mode coherent state approximation which is closely related to the mean field approximation. Thus we set

$$\sigma = \sigma_v + \sum \frac{1}{\sqrt{2\omega_k V}}(a_k e^{i\vec{k}\cdot\vec{r}} + a_k^\dagger e^{-i\vec{k}\cdot\vec{r}}) ,$$

$$\pi = -i\sum \sqrt{\frac{\omega_k}{2V}} (a_k e^{i\vec{k}\cdot\vec{r}} - a_k^\dagger e^{-i\vec{k}\cdot\vec{r}}) . \tag{6.17}$$

where V is normalizing box volume. The soliton part of the coherent state vector is

$$|f\rangle = \exp \left(\sum_k \sqrt{\frac{\omega_k}{2}} f_k a_k^\dagger\right)|0\rangle . \tag{6.18}$$

This is a one mode state of the form $\exp(\lambda A_0^\dagger)|0\rangle$ with $\lambda A_0^\dagger \equiv \sum \sqrt{\frac{\omega_k}{2}} f_k a_k^\dagger$, where in general $A_i^\dagger = \sum_j U_{ij}^* a_J^\dagger$. We have

$$a_{\vec{k}} | f> = \sqrt{\frac{\omega_k}{2}} \, f_{\vec{k}} | f> \tag{6.19}$$

and

$$<f | \sigma | f> = \sigma_v + V^{-\frac{1}{2}} \, \Sigma \, f_{\vec{k}} \, e^{i\vec{k}\cdot\vec{r}} \quad , \tag{6.20}$$

where we have used $f_{\vec{k}} = f^*_{-\vec{k}}$. We normal order the Hamiltonian in the operator a_k^+ and $a_{k'}$, which also means normal ordering with respect to the A_i and A_i^+. Everything goes through as in Sec. 3.1, including renormalized coefficients a, b and c. We have replaced a_0 of Sec. 3.1 with A_0). We can now identify

$$V^{-\frac{1}{2}} \, \Sigma f_{\vec{k}} \, e^{i\vec{k}\cdot\vec{r}} = \sigma_0(\vec{r}) - \sigma_v \quad , \tag{6.21}$$

where $\sigma_0(\vec{r})$ solves the mean field equations.

A bag state displaced from the origin by $\vec{z}$ (i.e., $s_0 \, (\vec{r}-\vec{z})$) is described by

$$f_{\vec{k}}(\vec{z}) = f_{\vec{k}}(0) \, e^{-i\vec{k}\cdot\vec{z}} \quad . \tag{6.22}$$

Our state vector

$$|\vec{z}> = \prod_{n=1}^{3} \psi_n(\vec{r}_n - \vec{z}) \, \exp \, (\Sigma \sqrt{\frac{\omega_k}{2}} \, f_{\vec{k}}(\vec{z}) a_{\vec{k}}^+) \, |0> \quad , \tag{6.23}$$

and we need to evaluate the following quantities:

$$N_\sigma(Z) = <f(-\tfrac{1}{2}\vec{z}) | f(\tfrac{1}{2}\vec{z})> = \exp \, (\Sigma \frac{\omega_k}{2} \, f_k^*(-\tfrac{1}{2}\vec{z}) f_k(\tfrac{1}{2}\vec{z})$$

$$= \exp \left(\frac{\omega}{2} \int [\sigma_0(\vec{r} - \tfrac{1}{2}\vec{z}) - \sigma_v][\sigma_0(\vec{r} + \tfrac{1}{2}\vec{z}) - \sigma_v] d^3 r \right); \tag{6.24}$$

$$<f(-\tfrac{1}{2}\vec{z}) | : \sigma^n : | f(+\tfrac{1}{2}\vec{z})>$$

$$= [\sigma_z(\vec{r})]^n \, N_\sigma(Z) \, N_q(Z) \; ; \tag{6.25}$$

$$\sigma_z(\vec{r}) = \frac{1}{2}[\sigma_0(\vec{r} - \tfrac{1}{2}\vec{z}) + \sigma_0(\vec{r} + \tfrac{1}{2}\vec{z})] \; ; \tag{6.26}$$

$$N_q(Z) = \prod_n \int \psi_n^\dagger (\vec{r} - \tfrac{1}{2}\vec{z}) \psi_n (\vec{r} + \tfrac{1}{2}\vec{z}) d^3r \ . \tag{6.27}$$

The expectation value of $H_q + H_{q\sigma}$ is given simply by

$$\frac{<P=0|H_q+H_{q\sigma}|P=0>}{<P=0|P=0>} =$$

$$\frac{\sum \int <-\tfrac{1}{2}\vec{z}|\psi_n^\dagger (\vec{r}-\tfrac{1}{2}\vec{z}) [\vec{\alpha}\cdot\vec{p}+\tfrac{1}{2}\beta g(\sigma_0(\vec{r}-\tfrac{1}{2}\vec{z}) + \sigma_0(\vec{r}+\tfrac{1}{2}\vec{z})]\psi(\vec{r}+\tfrac{1}{2}\vec{z})|\tfrac{1}{2}\vec{z}>d^3z}{\int <-\tfrac{1}{2}\vec{z}|\tfrac{1}{2}\vec{z}>d^3z} = \sum_n \epsilon_n \ .$$

$$\tag{6.28}$$

The final term is given by

$$\frac{<P=0|H_\sigma|P=0>}{<P=0|P=0>} = \frac{\int \mathcal{E}_\sigma(Z) N_\sigma(Z) N_q(Z) d^3z}{\int N_\sigma(Z) N_q(Z) d^3z} \tag{6.29}$$

where

$$\mathcal{E}_\sigma(Z) = \int [[\tfrac{1}{2}|\vec{\nabla}\sigma_z(\vec{r})|^2 + \tfrac{1}{2}\pi_z^2 + U(\sigma_z(\vec{r}))]d^3r \tag{6.30}$$

where

$$\pi_z = i \frac{\omega}{2} [\sigma_0(\vec{r} - \tfrac{1}{2}\vec{z}) - \sigma_0(\vec{r} + \tfrac{1}{2}\vec{z})] \ . \tag{6.31}$$

The program to project the wave function and evaluate the expectation value of the Hamiltonian has been completed by G. Lübeck. We do not quote numerical results here because it appears that further variation of the wave functions after projection is desirable (perhaps necessary) to obtain reliable results. We propose to introduce scale parameters into the soliton Fourier components and quark wave functions (not full functional variation) in order to extremize (not minimize) the energy.

6.4 Boosting the Bag

The exact boost operator, acting on a zero momentum state, for an interacting relativistic field theory leads to the Lorentz translated state

$$|\vec{v}\rangle = e^{i\vec{v}\cdot\vec{K}}|\vec{P} = 0\rangle, \tag{6.32}$$

where the boost operator is

$$\vec{K} = \int d^3x \; \mathcal{H}(\vec{x}) \; \vec{x} \; . \tag{6.33}$$

If

$$M_0 \simeq \langle\vec{P}=0|H|\vec{P}=0\rangle \tag{6.34}$$

is a good approximation, then we can construct the momentum state

$$|\vec{P}\rangle = e^{i\vec{P}\cdot\vec{K}/(M_0^2 + \vec{P}^2)^{\frac{1}{2}}}|P=0\rangle \tag{6.35}$$

where $\vec{K}/(M_0^2+\vec{P}^2)^{\frac{1}{2}} \simeq \vec{R}$ [c.f. Eq. (6.2)], the center-of-energy operator. This boost operator was exploited by Betz and Goldflam [12] to study corrections to the electric and magnetic form factors. For the electric rms size, they obtained results in agreement with Eq. (6.7). They found a small reduction in the magnetic moment.

The present program differs from that of Betz and Goldflam [12] in that they boosted the MFA localized bag (which is not an eigenstate of $\vec{P}$), while we boost the projected, P=0 eigenstate.

7.0 GLUONS AND COLOR

Color is the keystone of QCD. Although the full nonlinearity of QCD could be incorporated in the effective Lagrangian, the objective in modelling QCD is to circumvent such extreme complexities by simulating the nonlinear effects through the scalar, color-singlet σ-field. At present, we restrict consideration to one-gluon exchange diagrams, which also implies linearization of the gluon field equations; the non-Abelian nature is preserved.

7.1 One Gluon Exchange in the MIT Bag

De Grand et al[8] in their classic 1975 paper fit a large body of hadronic spectra in the context of the MIT bag model, including a controversial gluon zero-point energy correction, and one-gluon exchange (OGE). The results involving only u and d (zero mass) quarks are presented in table 2. There are three adjustable parameters in their model. Four masses are well fitted; they had modest success with other properties. I count the degenerate ρ-ω pair a single success. The η-meson was not calculated.

Table 2.

One gluon exchange corrections in the MIT model, from De Grand, Jaffe, Johnson and Kiskis[8]. All energies are in GeV; the bag radius is in fm. There are no recoil corrections in these calculations.

	M_{exp}	M_{Bag}	R_0 (fm)	E_0	E_V	E_Q	E_M
Δ	1.236	1.233	1.081	-.336	.308	1.119	.141
p	0.938	0.938	.987	-.367	.234	1.226	-.151
ω	0.783	0.783	.929	-.390	.196	.868	.110
ρ	0.77	0.783	.929	-.390	.196	.868	.110
π	0.139	0.280	.659	-.549	.070	1.222	-.462

$$E_0 = \frac{-Z_0}{R} \qquad Z_0 = 1.84 \; \hbar C = 0.363 \text{ GeV-fm}$$

$$E_V = \frac{4}{3}\pi R^3 B \qquad B = 57.5 \text{ MeV/fm}^3$$

$$E_Q = \frac{A \; 2.04}{R} \hbar C$$

$$E_M \propto \frac{\alpha_s}{R} \qquad \alpha_s = 4\alpha_c = 2.2$$

The π splits off and is driven down by the OGE, but still appears at roughly twice the physical value. This we regard as a success, not a failure. We expect the OGE contributions to be qualitatively similar to the soliton model as in the MIT model.

De Grand et al[8] did not include recoil corrections in their original works, although there has been subsequent work on this problem in the context of the MIT model. Our own estimates, based on the considerations of Sec. 6.1 and ref. [11] lead to a significant reduction in the pion mass. Within a range of "acceptable" model parameters we can fit the pion mass, obtain zero, or even obtain a negative m^2. We await evaluation of the energy obtained with projected wave functions and the consistent inclusion of the OGE energies in the soliton model.

7.2 The Linear Gluon Propagator in Media

The linearized gluon field equations are essentially identical to Maxwell's equations. The chromo-electric and magnetic fields, in the Coulomb gauge, can be written

$$\vec{E}^c = -\vec{\nabla}A_0^c \; , \quad \vec{B} = \vec{\nabla} \times \vec{A}^c \; ,$$

$$\vec{D}^c = \kappa \vec{E}^c \; , \quad \vec{H} = \mu^{-1}\vec{B} = \kappa\vec{B} \; , \tag{7.1}$$

since $\kappa\mu = 1$. Then

$$-\vec{\nabla}\cdot\kappa\vec{\nabla}A_0^c = j_0^c$$

$$(\partial^2/\partial t^2 + \vec{\nabla}x\kappa\vec{\nabla}x)\vec{A}^c = \vec{j}_t^c \; . \tag{7.2}$$

where subscript "t" means the transverse component.

I will only outline the solution for the scalar part of the gluon propagator (Green function) here. The mathematical formulation and computer program for the general scalar and vector frequency-dependent propagators in an inhomogeneous frequency-dependent medium has been completed by M. Bickeboeller and is being written up as a Diplom thesis for the University of Bonn.

$\kappa(\sigma(\vec{r}))$ is an operator, but is treated in the MFA as a c-number. Present calculations are restricted to axially and reflectionally symmetric functions. This is not a fundamental restriction to the method. The scalar propagator is defined by

$$\vec{\nabla} \cdot \kappa \vec{\nabla} \, G_{00}(\vec{r},\vec{r}') = \delta^3(\vec{r}-\vec{r}') \ . \tag{7.3}$$

Note that in the Coulomb gauge, G_{00} is time (frequency) independent. Let

$$G_{00}(\vec{r},\vec{r}') = \kappa(r)^{-\frac{1}{2}} \mathcal{G}(\vec{r},\vec{r}') \kappa(r')^{-\frac{1}{2}} \tag{7.4}$$

Then

$$[-\nabla^2 + W(\vec{r})] \, \mathcal{G}(\vec{r},\vec{r}') = \delta^3(\vec{r},\vec{r}') \tag{7.5}$$

where

$$W(\vec{r}) = \frac{|\nabla\kappa|^2}{4\kappa^2} - \frac{1}{2}\frac{\nabla^2\kappa}{\kappa} \ . \tag{7.6}$$

We write

$$\mathcal{G}(\vec{r},\vec{r}') = \sum_{\alpha\alpha'} C_{\alpha\alpha'} \, J^\alpha(\vec{r}_<) \, N^{\alpha'}(\vec{r}_>) \ , \tag{7.7}$$

where $J^\alpha(\vec{r})$ and $N^\alpha(\vec{r})$ each satisfy the homogeneous differential equation (7.5) with regular boundary conditions at $r=0$ and $r=\infty$ respectively. For numerical reasons, we take $\kappa \to \varepsilon$ (a small constant) rather than zero as $r \to \infty$. Here $\vec{r}_< = \vec{r}$ and $\vec{r}_> = \vec{r}'$ if $|\vec{r}| < |\vec{r}'|$ and conversely. The set of constant coefficients $C^{\alpha\alpha'}$ are determined by the conditions on $\mathcal{G}(\vec{r},\vec{r}')$ near $|\vec{r}| = |\vec{r}'| = $ any arbitrary matching radius, r_m).

$$\mathcal{G}(r_m+\delta,\theta\phi;\, r_m,\theta',\phi') = \mathcal{G}(r_m-\delta,\theta,\phi;r_m,\theta',\phi') \tag{7.8}$$

and

$$-\int d\vec{s} \cdot \vec{\nabla} \mathcal{G} = 1 \tag{7.9}$$

for a small surface enclosing $\vec{r}'$.

$J^\alpha(\vec{r})$ and $N^\alpha(\vec{r})$ are expanded in terms of spherical harmonics, e.g.

$$J^\alpha(\vec{r}) = r^{-1} \sum_{\ell n} j^\alpha_{\ell m}(r) \, Y_{\ell m}(\theta,\phi) \ , \tag{7.10}$$

where the $j^\alpha(r)$ satisfy the coupled equations

$$\left(-\frac{d^2}{dr^2} + \frac{\ell(\ell+1)}{r^2}\right) j^\alpha_{\ell m} + \sum_{\ell'} <\ell m|W|\ell'm> j^\alpha_{\ell'm} = 0 \ . \tag{7.11}$$

There is no coupling among functions of different m or parity. The index α labels the various linearly independent solutions. For example, let α stand for the set (L, m). Then independent solutions can be generated by starting at the origin with

$$j^{\alpha}_{\ell m} = Cr^{\ell} \delta_{\ell, L} .$$

(7.12)

Similar conditions can be applied for large r.

The vector propagator is complicated by requiring an expansion in terms of vector spherical harmonics. The requirement of transverse vector functions leads to a third order differential equation (for the electric modes) and the inhomogeneous term is a transverse vector delta function. As noted, Bickeboeller has solved these problems.

Thus a tensor linear gluon propagator $G_{\mu\nu}(\{\kappa\}, \vec{r}, \vec{r}', \omega)$ is constructed which is a functional of $\kappa(\sigma(\vec{r}))$ and the frequency $G_{\mu\nu}$ is diagonal in the color, and is in fact color-independent since κ is a color singlet.

The OGE interaction between quarks is then written

$$\frac{q_s^2}{4} \int d^3r_1 d^3r_2 (\bar{\psi}\lambda^c\gamma^\mu\psi)_1 \, G_{\mu\nu}(\vec{r}_1, \vec{r}_2) \, (\bar{\psi}\lambda^c\gamma^\nu\psi)_2$$

(7.13)

where q_s is the color charge. The λ-matrices act on the 3-color components of the quark spinors. For use in the GCM, ω is the energy difference in the quark states and κ may be evaluated at the mean deformation $\bar{\alpha}$.

8.0 TRANSITION FROM NUCLEAR MATTER TO THE QUARK PLASMA

The study of the N-N interaction gives insight into the change in the structure of the quark wave functions during the collision process. An alternate, albeit simplistic approach, is to consider nuclear matter as a collection of bags like the holes in Swiss cheese, and to study the structure of the system as it is compressed. We anticipate that the interstices between the bags--the physical vacuum--should disappear as the density is increased leading ultimately to quark plasma. Is the transition continuous or discontinuous? What is the order of the "phase" transition?

Our initial studies make rather drastic assumptions, although such approximations have proved useful in solid and fluid state physics to calculate equations of state. The approximations certainly need to be improved.

We replace the moving fluctuation bags by a regular, period fcc lattice of bags, characterized by the lattice displacement vectors $\vec{a}_n$.

We take $\sigma_0(\vec{r})$ to be periodic:

$$\sigma_0(r) = \sigma_0(\vec{r} + \vec{a}_n) \tag{8.1}$$

and to contain the reflectional and discrete rotational symmetries of a regular fcc lattice.

In the absence of OGE interactions, the quark functions satisfy the Bloch theorem

$$\psi(\vec{r}) = e^{i\vec{k}\cdot\vec{r}} \, \psi_k(\vec{r}) \, , \tag{8.2}$$

where $\vec{k}$ is a continuous vector and $\psi_{\vec{k}}(\vec{r})$ is periodic,

$$\psi_{\vec{k}}(\vec{r}) = \psi_k(\vec{r} + \vec{a}_n) \, , \tag{8.3}$$

although it need not possess the other symmetries of σ_0. The ψ_k satisfy the Dirac equation

$$(\vec{\alpha}\cdot(\vec{p}+\vec{k}) + g\beta\sigma_0(\vec{r}))\psi_{\vec{k}} = \varepsilon_{\vec{k}} \, \psi_{\vec{k}} \, . \tag{8.4}$$

The ε_k have the characteristic band spectrum of a crystal.

At low density--well separated bags--the self-consistent solutions for σ and ψ are those of isolated bags; the energy spectrum is discrete. As the bags are moved closer together, the ε_k spread out into bands.

Although the lattice calculation is feasible (by choosing selected $\vec{k}$-vectors) we choose to use the Wigner-Seitz spherical cell approximation. A single bag is enclosed in a sphere of radius, r_0 such that its volume is the same as that ascribed to each bag in the crystal. Because of the assumed spherical symmetry, the lowest band assumes the form for s-states; s- can be represented by (2.8), so that

$$du_k/dr = (g\sigma_0 + \varepsilon_k)v_k \, , \tag{8.5a}$$

$$dv_k/dr + 2v_k/r = (-g\sigma_0 + \varepsilon_k)u_k \, , \tag{8.5b}$$

and

$$-\nabla^2\sigma_0 + U'(\sigma_0) + \frac{9g}{\bar{k}^3} \int_0^{\bar{k}} k^2 dk(u_k^2 - v_k^2). \tag{8.5c}$$

The quark functions are normalized to

$$\int_0^{r_0} r^2 dr (u_k^2 + v_k^2) = 1 \ . \tag{8.6}$$

The boundary conditions on σ are

$$\sigma'(0) = \sigma'(r_0) = 0 \ . \tag{8.7}$$

The lowest member of the quark band satisfies the boundary condition

$$v_{bot}(r_0) = 0 \Rightarrow u_{bot}(r_0) = 0 \ . \tag{8.8}$$

At the top of the band, we have

$$u_{top}(r_0) = 0 \ . \tag{8.9}$$

Using these boundary conditions and a given $\sigma_0(r)$, we can solve for the corresponding ε_{bot} and ε_{top}. The intermediate ε's lie in the continuum of the band and do not require the solution of an eigenvalue problem. Rather ε_k is to be specified and Eqs. (8.5a&b) integrated forthright. Let $t = k/k_{top}$ (k_{top} is inversely proportional to the lattice spacing); $t_{top} = 1$. We make the reasonable Ansatz that

$$\varepsilon(t) = \varepsilon_{bot} + (\varepsilon_{top} - \varepsilon_{bot})(1 - \cos\pi t)/2 \tag{8.10}$$

The inhomogeneous term in (5.6c) can be rewritten

$$\frac{9g}{\bar{t}^3} \int_0^{\bar{t}} t^2 dt (u_t^2 - v_t^2) \tag{8.11}$$

The factor $9/\bar{t}^3$ assures 3 quarks per bag irrespective of $\bar{t}$. The spin-flavor-color degeneracy of each state is 12-fold. Thus the band is only 1/4 filled.

We pack the band to a value $\bar{t} = (1/4)^{1/3}$. The self consistent solution of Eqs. (8.5) leads to a phase transition from bags to uniform quark ($v=0$) and σ fields at $r_0 \approx 1.7$ fm, well above normal nuclear density. Gluon exchange is needed to promote clustering into three-quark structures.

9.0   SUMMARY AND PROSPECTS

The soliton model represents an extension of the MIT bag model to allow for the dynamical degrees of freedom associated with the confinement mechanism. The soliton model has 5 parameters, MIT has 3, but the soliton model has the flexibility, by choice of the parameters, to reproduce either the MIT or the SLAC bags. With appropriate choice of parameters and inclusion of one gluon exchange, the resulting hadronic spectra is similar to the MIT model.

Because the model can be cast in Hamiltonian form, dynamical processes can be calculated using techniques developed for nuclear collective motion. This permits calculation of N-N collisions, recoil corrections and the construction of bag states of good momentum. The last is essential for the proper calculation of electromagnetic form factors.

In this paper, the pion has been alluded to frequently. It is currently being studied actively in the context of the soliton model. The pion appears here as an anomalously light particle, split off and pushed down from the meson multiplet by OGE. The nucleon bag should be soft to qq , virtual excitation with pion quantum numbers. In the soliton model, these virtual excitations are to be identified with the pion cloud. One can also calculate pi-nucleon coupling and the weak decay of the pion, $\pi \to \mu + \bar{\nu}_{\mu}$. Indeed, bags can be created and destroyed in the model.

This description of pion physics begins with a Lagrangian which does not respect chiral invariance and seeks to achieve PCAC from dynamics. The more fashionable approach is to begin with a chirally invariant Lagrangian from which the pion emerges as a massless Goldstone boson; somewhere, the pion must be given a mass and CAC broken. In all models, effective fields (σ or π or both [20] are introduced to describe degrees of freedom which are too difficult to handle explicitly. It seems prudent to pursue various models, since each has its advantage for particular physical phenomena.

10. ACKNOWLEDGMENTS

I wish to thank my many collaborators in this program, J. Achtzehnter, M. Bickeböller, M. Birse, J-L. Dethier, E. M. Henley, R. Horn, G. Lübeck, J. Rehr, and A. Schuh. A special acknowledgment is due to R. Goldflam who collaborated on many of these projects. During the CSIR Advanced Course, I benefitted greatly from discussions with the lecturers and participants, particularly W. Weise and J. R. Rafelski. I am especially indebted to S. K. Kauffmann who introduced me to the principle of minimum sensitivity [14,15], and for discussions in depth on its meaning and implementation.

This work was supported in part by the U.S. Department of Energy.

References

1. A. Chodos, R. L. Jaffe, K. Johnson and C. B. Thorn,
   Phys. Rev. D 10, 2599 (1974).

2. G. E. Brown and M. Rho, Phys. Lett. 82B, 177 (1979);
   G. E. Brown, M. Rho and V. Vento, ibid 84B, 383 (1979);
   A. Chodos and C. B. Thorn, Phys. Rev. D12, 2733 (1975);
   F. Myhrer, G. E. Brown and Z. Xu, Nucl. Phys. A362, 377 (1981).

3. G. A. Miller, A. W. Thomas and S. Theberge,
   Phys. Lett. 91B, 192 (1980);
   S. Theberge, A. W. Thomas and G. A. Miller,
   Phys. Rev. D 22, 2838 (1980);
   A. W. Thomas, S. Theberge and G. A. Miller, ibid 24, 216 (1981).

4. R. Friedberg and T. D. Lee, Phys. Rev. D 15, 1694 (1977);
   D 16, 1096 (1977); D 18, 2623 (1978);
   T. D. Lee, Particle Physics and Introduction to Field Theory
   (Harwood Academic, New York, 1981).

5. R. Goldflam and L. Wilets, Phys. Rev. D 25, 1951 (1982).

6. T. Celik, J. Engels and H. Satz,
   Physics Letters 129B, 323 (1983).

7. K. Ishikawa, G. Schierholtz and M. Teper,
   Phys. Lett. 116B, 429 (1982).

8. T. DeGrand, R. L. Jaffe, K. Johnson, and J. Kiskis,
   Phys. Rev. D 12, 2060 (1975).

9. E. Eichten et al., Phys. Rev. D 21, 203 (1980).

10. R. Berg and L. Wilets, Proc. Phys. Soc. London A68, 229 (1955);
    L. G. Henyey, L. Wilets, K. H. Böhm, R. Le Levier and R. K. Levee,
    Astrophys. J. 129, 628 (1959).

11. J-L. Dethier, R. Goldflam, E. M. Henley and L. Wilets,
    Phys. Rev. D 27, 2191 (1983).

12. M. Betz and R. Goldflam, Phys. Rev. D (accepted for publication)

13. U. Mosel, Particles and Nuclei, 3, 297 (1972).

14. P. M. Stevenson, Phys. Rev. D23, 2916 (1981).

15. S. K. Kauffmann and S. M. Perez, "Minimal Sensitivity Optimi-
    zation of Perturbative Wave Functions," Univ. of Capetown,
    Institute for Theoretical Physics Preprint (1983).

16. A. D. Fokker, "Relativiteitstheorie" (Groningen: P. Noordhoff,
    1929).

17. M. H. L. Pryce, Proc. R. Soc. London A195 (1948).

18. L. L. Foldy and S. A. Wouthuysen, Phys. Rev. 78, 29 (1949).

19. P. K. Haff and L. Wilets, Phys. Rev. C 10, 353 (1974).

20. M. Gell-Mann and M. Levy, Nuovo Cimento 16, 705 (1960);
    M. C. Birse and M. K. Banerjee, "A Chiral Soliton Model of
    the Nucleon and Delta," U. of Maryland Preprint No. 83-201.

NUCLEAR MATTER UNDER EXTREME CONDITIONS

J. Rafelski
Institute for Theoretical Physics and Astrophysics
University of Cape Town, Cape Town

M. Danos
National Bureau of Standards
Washington, D.C. 20234, U.S.A.

The aim of this work is to present an overview including recent advances in our understanding of the behavior of nuclear matter at extreme conditions such as high baryon density and/or high temperature. The following subjects are covered:

1. The World of Quarks and Gluons
2. From Quark Bag to Quark-Gluon Plasma
3. Strangeness in the Quark-Gluon Plasma
4. Thermodynamics of the Interacting Hadronic Gas
5. Formation and Cooling of a Baryon-Rich Quark-Gluon Plasma in Nuclear Collisions.

Until now most of the investigations on the fundamental properties of matter have been performed with two-body systems. While this is an essential first step, there exist important phenomena based on many-body effects which therefore are not observable in the simple systems. One such a phenomenon is the hypothetical new phase of matter, viz., the quark-gluon plasma, which is that state where owing to a suitable combination of baryon and energy densities the individual hadrons have melted together, or, said differently, have dissolved, freeing the hadron constituents to form a weakly interacting Fermi and Bose gas. The existence of such a state of matter is an almost inevitable consequence of quantum chromodynamics (QCD) and the study of its properties is clearly of utmost importance [1].

The rapid development of this new and exciting field of high energy nuclear physics is based largely on the hypothesis that the energy available in the collision of two relativistic heavy nuclei is equipartitioned among the accessible degrees of freedom. Of course, it is not needed that the whole system participate in this thermalization; in addition, it is likely that only a small fraction of nuclear collisions leads to this reaction channel. This means that occasionally a domain in space arises in a center-of-mass frame in which the energy of the longitudinal motion has been largely transferred to transverse degrees of freedom. We call this region the "fireball." The physical variables characterizing a fireball are: energy density, baryon number density, and total volume. The basic question concerns the internal structure of the fireball. It can consist either of individual hadrons, or, instead, of quarks and gluons in a new physical phase, the plasma, in which they are deconfined and can move freely over the volume of the fireball. It appears that the phase transition from the hadronic gas phase to the quark-gluon plasma is controlled mainly by the energy density of the fireball. Several estimates [2] lead to $0.6-1$ GeV/fm^3 for the critical energy density, to be compared with ca. 0.16 GeV/fm^3 in nuclear matter. Many fundamental questions about the nature of the strong interactions will be settled when the properties of the phase transition are determined.

An important aspect of the developments in this field concerns the observability of the plasma state. It seems that in order to observe the characteristics of the plasma one must either use electromagnetically interacting particles [3] (photons, lepton pairs) which can rather easily leave the plasma, or study the heavy-flavor abundance generated in the collision [4]. To understand the latter point imagine that strange quarks are very abundant in the plasma (and indeed they are!). Then, since the (sss)-state is bound and stable in the perturbative QCD-vacuum, it would be the most abundant baryon to emerge from the plasma. Surely the observation of such an "omegaization" of nuclear matter could not leave any doubts about the presence of the plasma. The observation of other exotic hadrons [5] such as, e.g., csq, cs̄, etc. would support this conclusion. But even the enhancement of the abundance of the more accessible Λ̄ may already be sufficient at least for demonstrating the existence of a plasma.

To continue to higher energy densities, one may speculate that the restoration of the perturbative QCD vacuum may be accompanied by the restoration of chiral symmetry, then followed by the restoration of the SU(2) symmetry, and finally of the SU(5) symmetry. This way one could trace back the evolution of the universe [6] in the laboratory. In figure 1 we display qualitatively the boundary of the different phases for T < 200 MeV.

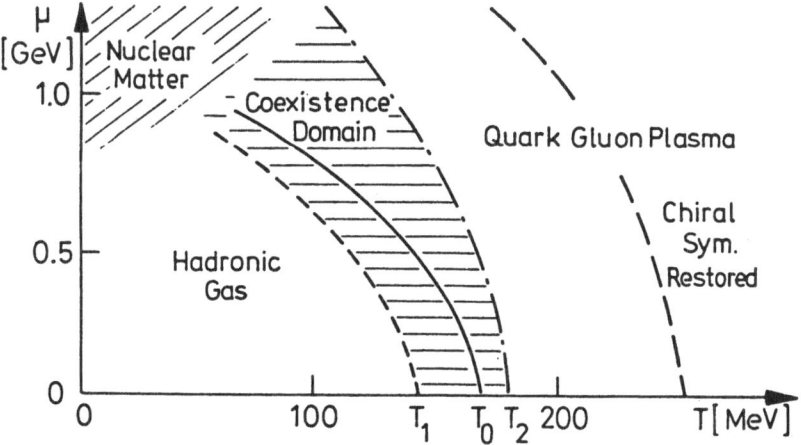

Fig. 1   Phase diagram of hadronic matter in the μ-T plane.

Another speculation concerns the fundamental aspect of the possible catalyzation of the baryon decay in the plasma [5]. A possible mechanism which has been discussed in the recent literature [7] involves the presence of magnetic monopoles. The quark-gluon plasma might just be the proper environment in which the catalyzer could continue to burn the baryon number at a rate sufficient to maintain the necessary particle densities and temperatures. However, in view of our ignorance of precisely how SU(5) tumbles down to SU(3) x SU(2) x U(1) we should be prepared for great surprises in this matter. Certainly, it would be most challenging to unlock the energy which had orginated in the Big Bang and has since remained frozen in the baryon number.

It appears that magnetic monopoles in centers of giant stars could generate sufficient amounts of energy in order to provide a quasar energy generation mechanism [7].

Coming back to earth we begin by recalling that in a statistical description of matter the un-handy microscopic variables, viz., energy, baryon number, etc. are replaced by thermodynamical quantities. To wit, the temperature T is a measure of energy per degree of freedom; the baryon chemical potential μ controls the mean baryon density (see fig. 1). The statistical quantities such as entropy (= measure of the number of available states), pressure, heat capacity, etc. also will be functions of T and μ and will have to be determined. The theoretical techniques required for the description of the two quite different phases, <u>viz</u>., the hadronic gas and the quark-gluon plasma, must allow

for the formation of numerous hadronic resonances [8], which then at
sufficiently high energy density dissolve into the state consisting of
their constituents. At this point we must appreciate the importance
and help in reaching the transition to the quark-gluon plasma provided
by a finite, i.e., non-zero temperature. To obtain a high particle
density, instead of compressing the matter (which as it turns out is
quite difficult), we may heat it up; many pions are generated easily,
allowing the transition to occur at moderate, even vanishing baryon
density [9].

We begin by considering a summary of the relevant postulates and
results that characterize the current understanding of strong interactions in QCD. The most important postulate is that the true physical
vacuum state in QCD is not the trivial perturbative state which is
changed little when the interactions between quarks and gluons are
turned off or on. In QCD the true vacuum state is believed to have a
complicated structure which originates in the gluon (gauge) sector of
the theory. It is supposed not to permit the presence of color fields.
The perturbative vacuum is an excited state with an energy density B
above the true vacuum. It is to be found inside hadrons where the perturbative quanta of the theory, in particular the quarks, therefore can
exist. The occurrence of the true vacuum state is intimately connected
with the gluon-gluon interaction; gluons also carry the color charge
that is responsible for the quark-quark interaction. The confinement
of quarks is a natural consequence of this hypothetical structure of
the true vacuum.

An important feature which arises as a consequence of the energy
density B of the perturbative vacuum is that the true vacuum exercises
a pressure on the surface of the region of the perturbative vacuum.
Indeed, this is just the idea behind the original MIT bag model [10].
The Fermi pressure of the confined almost massless light quarks is in
equilibrium with the vacuum pressure B. When many quarks are combined
to form a giant quark bag then their properties inside the bag can be
obtained using standard methods of many-body theory [2]. In particular, this also allows the inclusion of the effect of internal excitation through a finite temperature and through a change in the chemical
composition, a subject discussed in the subsequent lectures.

The essential role of strangeness as a characteristic observable
of the plasma state [4] with particular emphasis on strangeness generation in the plasma by elementary processes, and on expectations about

the normal hadronic gas phase is addressed next. From the comparison of the expectations for both phases of hadronic matter we are led to consider the strange-particle abundances as a possible approach to the observation of the properties and parameters of the quark-gluon plasma created in nuclear collisions. It suggests itself that strangeness is an excellent experimental trigger for the presence of plasma droplets in high energy nuclear, or slow p̄-nucleus, collisions.

A very important observation in this context is that strangness is a characteristic signal of the gluon abundance, which comprises the only essential difference in the structure of the plasma state and the hadronic gas. As we show [4], a large abundance of strangness is generated by the gluon fraction in the plasma state and can not be obtained from the hadronic gas phase owing to the fact that the then required numerous Bosonic degrees of freedom are available only when $T > 500$ MeV. But even then color is not free and the abundance of strangness would still be about 3-5 times smaller. The presence of gluons in the plasma state speeds the generation of strangeness allowing the equilibrium abundance to be reached so that in that case an up to 100 times larger strangeness abundance can be expected.

The state of hadronic matter formed by individual baryons and mesons, which we call the hadronic gas phase is described next. The present summary of the theoretical development of this field is based on the work of Hagedorn and Rafelski [9]. We content ourselves here with the presentation of the main results in so far as they influence our thinking about the phase transition to the quark-gluon plasma.

The attentive reader might question the validity of using simultaneously the bootstrap model and the bag model to describe hadronic states. We will indeed find that the description in terms of the statistical bootstrap for the hadronic gas on the one hand, and of hadrons as bound quark states on the other hand, have many properties in common and are quite complementary. Both the statistical bootstrap and the bag model of quarks are based on quite equivalent phenomenological observations. While it would be most interesting to derive the phenomenological models quantitatively from the accepted fundamental basis - the Lagrangian quantum field theory of a non-abelian SU(3) gluon gauge field coupled to colored quarks - in this report we will have to content ourselves with a qualitative understanding only. Already this will allow us to study the properties of hadronic matter in both aggregate states - with the emphasis in this report put in particular on the state in which individual hadrons have dissolved into the plasma consisting of quarks and of the gauge field quanta, the gluons.

Having described the properties of both hadronic phases, we present a discussion of the possible production and lifetime of the baryon-rich plasma in nuclear collisions in the central kinematic region and then describe the phase transition between the hadron gas and the plasma. In the final chapter we describe the formation mechanism for a baryon-rich plasma as possibly created at ∼ 5 GeV/Nuc c.m. energies. The conditions prevailing here are just opposite to those found in ultra-relativistic collisions in which the baryon density is expected [11] to be low in the central rapidity region. In our approach [11, 12] there is a substantial baryon density arising from pile-up of nuclear matter in the collision region. The description of ultra-relativistic collisions is based on extrapolations of pp and pA collisions, which in our view cannot lead to the pileup of matter, i.e., baryon number, which is needed in our description. In order to estimate the evolution of the plasma state we consider, contrary to popular belief [13] that hydrodynamical expansion dominates the plasma evolution, the losses arising from particle radiation through the plasma surface [14] and determine the corresponding time evolution of the baryon-rich plasma.

1. THE WORLD OF QUARKS AND GLUONS

From the study of the hadronic spectra as well as from hadron-hadron and hadron-lepton interactions there has emerged convincing evidence for the description of the hadronic structure in terms of quarks [15]. For many purposes it is entirely satisfactory to consider baryons as bound states of three fractionally charged particles, while mesons are quark-antiquark bound states. One of the central aims of this and the next section is to show how this picture of hadrons can be reconciled with the description of hot hadronic matter consisting of individual particles described in section 4.

We now recall some fundamental assumptions about the strong interactions, as needed here. The elementary quantum fields which appear in quantum chromodynamics are:

Spin 1: gauge bosons - gluons G_μ^i, $i = 1...8$

Spin 1/2: baryonic matter - quarks q_γ^α, $\alpha = R,G,B = $ color
$\gamma = d,u,s,c,b,(t) = $ flavor

The octet of gauge bosons G_μ mediates the quark-quark and quark-antiquark interactions between the color triplets {Red, Green, Blue} and antitriplets. The gauge vector fields are written as

$$G_\mu \equiv \sum_{i=1}^{8} G_\mu^i \frac{\lambda^i}{2} \tag{1.1}$$

where λ^i are the generators of the SU(3) algebra [16]

$$[\lambda^i, \lambda^j] = 2i\, f^{ijk}\, \lambda^k \quad . \tag{1.2}$$

Only quarks and antiquarks carry baryon number, i.e., $b_q \pm 1/3$. The flavor of the quarks represents all internal quantum numbers conserved in strong interactions - the up and down quarks carry $\pm 1/2$ units of I_z (isospin) and combine to form the lowest baryonic isospin doublet

$$\binom{p}{n} = \binom{uud}{ddu} \tag{1.3a}$$

and the mesonic isospin triplet

$$\begin{pmatrix} \pi^+ \\ \pi^0 \\ \pi^- \end{pmatrix} = \begin{pmatrix} u\bar{d} \\ \frac{1}{\sqrt{2}}(-u\bar{u} + d\bar{d}) \\ \bar{u}d \end{pmatrix} \quad . \tag{1.3b}$$

These are the input particles of the statistical bootstrap model discussed in section 4. The heavier flavors of quarks include the strange, charm, bottom and perhaps the as yet undiscovered top quark. The electric charge of u, c, t is + 2/3 and that of d,s,b is - 1/3.

It is the color-charge of the quarks that introduces the quark-quark interactions. The important empirical fact is that all known hadrons are color neutral (i.e., color singlets). Including color into the wave functions eq (1.3) and ignoring the space and spin degrees of freedom we have, e.g.,

$$p = \frac{1}{\sqrt{6}} (u^R u^G d^B - u^G u^R d^B + u^G u^B d^R - u^B u^G d^R + u^B u^R d^G - u^R u^B d^G) \qquad (1.4)$$

$$\pi^+ = \frac{1}{\sqrt{3}} (u^R \bar{d}^R + u^G \bar{d}^G + u^B \bar{d}^B) \qquad (1.5)$$

where the p, and baryons in general, are color-antisymmetric and π, and the mesons, are color symmetric. The antisymmetry of the baryonic wave functions in a hidden degree of freedom has been one of the original reasons for the introduction of color. Otherwise, e.g., $(\Delta^{++})^{I=3/2} = (uuu)^{I=3/2}$ could not have an antisymmetric quark wave function as required for Fermions. Further experimental evidence [17] of color includes the $\pi^\circ \to 2\gamma$ decay rate and the size of the $e^+e^- \to$ hadrons annihilation cross section. However, the evidence for color as a dynamical degree of freedom, in particular, as being responsible for quark-quark interactions, is derived from deep-inelastic lepton-nucleon scattering, from a detailed study of e^+e^- annihilation into hadrons, and in particular, from the flavor-independence of the charmonium and upsilonium potential which yields a quantitative agreement between the experimental and the theoretical excitation spectra.

The Lagrangian of quarks and gluons is very similar to that of electrons and photons, which is

$$L_{QED} = \bar{\phi}(\gamma \cdot (p-eA) - m) \phi - \frac{1}{4} F_{\mu\nu} F^{\mu\nu} , \qquad (1.6)$$

except for the required additional summations over flavor and color:

$$L_{QCD} = \sum_r^{\text{flavors}} (\sum_{\alpha=1}^{3} q_r^\alpha (\gamma \cdot p - m_r) q_r^\alpha + g \sum_{\alpha,\beta=1}^{3} \bar{q}_r^\alpha \gamma^\mu \sum_{i=1}^{8} (\frac{\lambda_{(\alpha\beta)}^i}{2} G_\mu^i) q_r^\beta)$$

$$- \frac{1}{4} \sum_{\alpha=1}^{8} F_{\mu\nu}^i F_i^{\mu\nu} + \text{herm. conj.} + \text{gauge fixing} \qquad (1.7)$$

The flavor-dependent masses m_r of the quarks are small. For u,d flavors one estimates $m_{u,d} \sim 5\text{-}20$ MeV when the strange quark mass is chosen in the range 150-280 MeV. In particular [18],

$$\frac{m_d - m_u}{m_d + m_u} \approx \frac{1}{3} \; ; \qquad\qquad (1.8a)$$

$$\frac{m_u}{m_d} = 0.38 \pm .13 \; ; \qquad\qquad (1.8b)$$

$$\frac{m_d}{m_s} = 0.045 \pm 0.011 \quad . \qquad\qquad (1.8c)$$

The heavy-quark mass differences can be obtained reliably from the detailed study of the quarkonium spectra [19],[20]

$$m_b - m_c = 3400 \text{ MeV}; \quad m_c - m_s = 1280 \text{ MeV} \quad . \qquad (1.9)$$

The color field strengths are now

$$F^i_{\mu\nu} = \partial_\mu G^i_\nu - \partial_\nu G^i_\mu + gf^{ijk} G^j_\mu G^k_\nu \; . \qquad\qquad (1.10)$$

We note the nonlinearity of F which is required to secure the invariance under local non-abelian gauge transformations. The presence of this gluon-gluon interaction leads to major differences between the properties of QED and QCD. As an example let us consider briefly the asymptotic freedom of gauge theories [21].

To introduce the subject we note that it is often convenient to define a q-dependent coupling constant by writing

$$\frac{e^2}{4\pi} D(q^2) = - \alpha(q^2) \frac{1}{q^2} \qquad\qquad (1.11)$$

where in the case of QED D is the QED longitudinal photon propagator. We ignore for the moment the transveral photon degrees of freedom. In terms of the polarization function $\Pi(q^2)$ we have

$$\alpha(q^2) = \frac{e^2/4\pi}{1+\Pi(q^2)} = \begin{cases} \dfrac{e^2}{4\pi} = \dfrac{1}{137} \; , \quad q^2 \to 0 \\[2ex] \dfrac{\alpha(0)}{1-\alpha(0)\dfrac{1}{6\pi}\,\ell n\left(-\dfrac{q^2}{m_e^2}\right)} \; , \quad |q^2| > m_e^2 \end{cases} \tag{1.12}$$

or, with the more complete form of the polarization function

$$\alpha^{-1}(q^2) = \alpha^{-1}(0) + \Pi(q^2)/\alpha(0)$$

$$= \alpha^{-1}(0) - \frac{q^2}{6\pi} \int_0^\infty \frac{dM^2}{M^2} \frac{\left(1 + \dfrac{2m^2}{M^2}\right)\sqrt{1 - \dfrac{4m^2}{M^2}}}{M^2 - q^2} \tag{1.13}$$

The electron-loop polarization function $\Pi(q^2)$ follows from the iteration of the standard lowest order diagram:

$$\tag{1.14}$$

As easily can be seen, $\alpha^{-1}(q^2)$ decreases with increasing $q^2 > 4m_e^2$. This means that for short distances the effective strength of the QED interaction increases. Only because of the magnitude of $\alpha^{-1}(0) = 137$ is this effect usually unimportant. However, it is part of the QED radiative corrections and has been quite precisely verified.

In QCD additional contributions originate in the gluon-gluon interaction

$$\tag{1.15}$$

Since gluons are massless we <u>cannot</u> select the point $q^2 = 0$ as a reference point. We have [21],

$$\alpha_s^{-1}(q^2) = \alpha_s^{-1}(-\mu^2) + \frac{1}{4\pi} \left[11 - \frac{2n_q}{3} \right] \ln \frac{-q^2}{\mu^2} \qquad (1.16)$$

with a certain space-like $q^2 = -\mu^2 \neq 0$ now serving as the reference.
n_q is the number of light quark flavors ($m_r^2 < |q^2|$). For large q^2,
absorbing the first term on the right hand side in eq (1.16) in the
definition of μ^2 we have the so called <u>asympototic freedom</u> formula:

$$\alpha_s(q^2) = \frac{12\pi}{33 - 2n_q} \frac{1}{\ln(-q^2/\Lambda^2)} \qquad (1.17)$$

which, unlike the case of QED leads to
falling α_s with rising $|q^2|$ for the
likely case $n_q < 16$. Hence, at
asymptotically short distances the
interaction diminishes and the theory
becomes free. We emphasize that
therefore the chain of approximations
leading to eq (1.17) here, <u>i.e.</u>, in
QCD, becomes more and more consistent
as $|q^2|$ increases. In figure 1.1 the
running coupling constant is shown for
space-like, $q^2 < 0$, and time-like,
$q^2 > 0$, momenta. In the latter case
we show Re α_s:

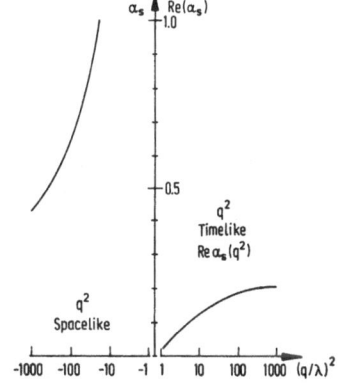

Fig. 1.1 $\alpha_s(q^2)$ for space-like
and Re $\alpha_s(q^2)$ for time-
like momenta.

$$\text{Re } \alpha_s(q^2 > 0) = \frac{12\pi}{33 - 2n_q} \frac{\ln|q^2/\Lambda^2|}{(\ln|q^2/\Lambda^2|)^2 + \pi^2} \qquad (1.18)$$

We notice that at the presently accessible momenta, <u>i.e.</u>, up to 100Λ
($\Lambda \sim 200\text{-}400$ MeV), α_s is considerably smaller for time-like q^2 than for
space-like q^2. For large $\alpha_s(q^2)$ other than order g^2 diagrams must be
included in the determination of $\alpha_s(q^2)$. This may change the value of
Λ which at this stage is a phenomenological parameter fitted to the
experiment and which reflects in its value the order of the expansion.
At present the actual value of Λ is rather uncertain since as can be

seen in eq (1.17) it manifests itself only in small logarithmic corrections. Quarkonium fits (space-like q^2) favor $\Lambda \sim 400$ MeV [19] while deep-inelastic experiments (time-like q^2) indicate $\Lambda = 100 \pm 100$ MeV [17].

As we have seen above, the strength of the gluon-gluon interaction influences significantly the gluon propagation in the (perturbative) vacuum. Little is known about the behavior of the gluon propagator at small q^2, i.e., at large distances. Attractive channels in the gluon-gluon interaction are expected to induce a gluonic structure onto the vacuum state [22],[23]. To appreciate this remark let us imagine a box of size R filled with a gas of N gluons. Including a 1/R kinetic energy and an attractive long range interaction we have for the energy density E/V:

$$\epsilon_{Box}(N) \sim N/R^4 - N^2 g^2/R^4 \qquad (1.19)$$

and hence for some $N = N_{cr}$ it would cost no energy to fill the box with gluons. Hence the empty box (perturbative vacuum) and the box with N_{cr} particles would be degenerate. We conclude that an improved gound state, i.e., the true vacuum, has to be constructed. Such a state would have a lower energy density than the value of the perturbative state.

The energy density of the perturbative state is defined with respect to the true vacuum state and hence is by definition a positive quantity, denoted by B. This notion has been introduced originally in the MIT bag model [10], but initially in a different context. The value of B is derived phenomenologically from a fit to the hadronic spectrum [10],[24] or from sum rule considerations [25] which give

$$B = [(140 - 210) \text{ MeV}]^4 = (50 - 250) \text{ MeV/fm}^3 \quad . \qquad (1.20)$$

The central assumption of the quark-bag approach is that inside a hadron where quarks are found the true vacuum structure is displaced or destroyed. One can turn this point around: quarks can only propagate in domains of space in which the true vacuum structure is absent. This statement is a resolution of the quark confinement problem. The remaining difficult problem is to show the incompatibility of quarks with the true vacuum structure. Examples of such behavior in ordinary physics are easily found: e.g., a light wave is reflected from a mirror surface; magnetic field lines are expelled from superconductors; etc.

In this spirit we may argue that all color-charged particles are reflected at the true vacuum surface (stationary waves) or alternatively, may under certain circumstances deform the surface. Whatever is the case, the presence of color electric fields in a volume element is incompatible with the presence of the true vacuum structure. It is interesting to note that the Lorentz covariance of the theory requires that a negative pressure $p = -B$ as seen from the perturbative vacuum acts on the surface between the true and the perturbative vacuum. Hence, in the absence of other forces the excited space domain containing the perturbative vacuum would quickly vanish.

In this picture of hadronic structure and quark confinement all colorless assemblies of quarks, antiquarks, and gluons can form stationary states, called a quark bag. In particular all higher combinations of the three-quark baryons (qqq) and quark-antiquark mesons ($q\bar{q}$) form a permitted state, _i.e._, a hadronic resonance, much in the spirit of the statistical bootstrap model of the hadronic gas, to be described later.

The energy of a hadronic bag of radius R including the particle and the volume bag terms is:

$$E(R) = (\sum_i X_i)/R + \frac{4}{3} \pi R^3 B \qquad (1.21)$$

where X_i/R are the appropriate eigenvalues, _i.e._, single particle energies of "confined" particles and the sum is over all quanta in the bag. Effects of interactions can be considered to be included in X_i, in which case the X_i become functions of the interaction strength and the number of particles present. For massive particles an additional dependence on mR is present. The radial pressure (force/area) on the surface is:

$$P_r = \frac{-\partial E/\partial R}{4\pi R^2} = -B + \frac{(\Sigma X_i)}{4\pi R^4} \qquad (1.22)$$

which, combined with eq (1.21) leads to the interesting relation

$$E(R) = (3P_r + 4B)V \quad . \qquad (1.23)$$

For a radially stable object P_r must vanish, or, in other words, E(R) must have a minimum. From eq (1.22) we have

$$R_{min} = B^{-1/4} \left(\frac{\Sigma X_i}{4\pi}\right)^{1/4} \tag{1.24}$$

and from eq (1.23)

$$E(R_{min}) = 4BV = B^{1/4}(\Sigma X_i)^{3/4} \frac{4}{3} (4\pi)^{1/4} \tag{1.25}$$

From eqs (1.24), (1.25) we learn that the radius of the bag grows with $(\Sigma X_i)^{1/4}$, while it decreases as $B^{-1/4}$, as could be expected from dimensional arguments. Similarly, the energy (mass) of the bag grows with $(\Sigma X_i)^{3/4}$, but also with $B^{1/4}$, as expected on dimensional grounds. The remarkable relation $E = 4BV$ is often called the virial relation as it follows alone from the dimensionality of space. We further notice that the dimensionless structure constant

$$R_{min} E(R_{min}) = \frac{4}{3} (\Sigma X_i) \tag{1.26}$$

can not be directly compared with the values known for example for protons:

$$R_{proton}^{charged} M_{proton} = 3.82 \tag{1.27}$$

since R_{min} is not the charge radius but the hadronic radius of the bag. Also note that eq (1.26) has been obtained without interactions.

To illustrate the conflict between both the quantities of (1.27) consider the true lowest eigenvalue X_0 for the quark wavefunctions; is found by solving the three-dimensional Dirac equation with the bag boundary condition [10], which leads to

$$R E = X_0 = 2.04 . \tag{1.28}$$

When inserted into eq (1.26), we would find with three quarks for the hadronic radius of a nucleon

$$R_{min} \sim 4 \times 2.04 \times 197 \text{ MeV fm}/940 \text{ MeV} \cong 1.7 \text{ fm} \tag{1.29}$$

which clearly is an unacceptable result.

Obviously, something is missing in eq (1.22), and it must be added in order for it to give the proper phenomenology of hadronic states. In the original MIT bag approach an additional zero-point energy

$$E_0 \approx \frac{-Z_0}{R} \tag{1.30}$$

was introduced. This can be taken care of by replacing (ΣX_i) in above formula by $(\Sigma X_i - Z_0)$. With this we find for the proton

$$R_{min} = \frac{4}{3} (\Sigma X_i - Z_0)/m_p \tag{1.31}$$

which requires $Z_0 \sim 2$ in order to make R_{min} sufficiently small, i.e., < 1 fm, as long as the noninteracting value $X_0 = 2.04$ is employed. The constraint arising from the fact that the sum of the bag energy and E_0 must not become negative has been so far little appreciated. Namely, a negative value is unacceptable, as it leads to stable empty bags; this would contradict the characteristics of the true vacuum. Using the virial relation eq (1.23), this constraint becomes:

$$0 < E_0 + BV = E_0 + \frac{1}{4} m_p = (\frac{1}{4} m_p R_{min} - Z_0) \frac{1}{R_{min}} \quad . \tag{1.32}$$

Recalling now eq (1.31) we find from eq (1.32) the constraint

$$0 < E_0 + BV \rightarrow Z_0 < \frac{1}{4} \Sigma X_i \quad . \tag{1.33}$$

This consideration is equally valid for mesons, but is less conclusive since other effects intervene, such as the restoration of translational invariance to the quark bound states. For nucleons, eq (1.33) implies

$$Z_0 < \frac{3}{4} X_0 \tag{1.34}$$

which is usually just barely satisfied once one includes the interactions. At any rate, for three quarks the introduction of Z_0 coupled with the constraint (1.33) reduces the numerical value eq (1.29) at most by a factor 3/4, which is not enough to yield the empirical value.

Clearly, this discussion shows that in order to resolve this apparent contradiction one must include the quark-quark interaction and eventually project on translationally invariant states. For our present discussion it is important to realize that the quark-bag picture can be made internally consistent only when the quarks are allowed to interact. Unfortunately, for "small" bags, i.e., for normal hadronic states, this opens the Pandora's box of all complicated self-energy, exchange and other contributions leading to the current confusion in the field of how such large corrections can mutually cancel; not to speak about such problems as the theoretically infinite values for Z_0, or the influence of pionic degrees of freedom when the bag radius is too small. However, we note that most of these problems disappear in "large" bags, i.e., those bags which contain many single-particle excitations. Also in this case the treatment of the quark-quark interactions by perturbative QCD becomes very simple, and hence we will introduce the interactions in this case below.

2. FROM QUARK BAG TO QUARK-GLUON PLASMA

A large quark-gluon bag, i.e., one which contains many particles, is characterized by the available modes X_i and their occupation numbers n_i. An important simplification of its description arises if it is possible to use a statistical treatment.

As the u and d quarks are almost massless inside a bag they can be produced in pairs, and at moderate internal excitations, i.e., temperatures, many $q\bar{q}$ pairs will be present. Similarly, $s\bar{s}$ pairs also will be produced. We will return to this point at length below. Furthermore, real transversal gluons can be excited and will be included here in our considerations. We now first convince ourselves that already a moderate number of quarks justifies the statistical approach. For a degenerate Fermi gas of quarks the number of light quarks (u and d) determines the quark Fermi energy μ_q. Omitting for the present the qq interactions we have

$$3 b = N_q = 2_s \times 2_f \times 3_c \times V \ \frac{4\pi}{(2\pi)^3} \ \frac{1}{3} \ \mu_q^3 = V \ \frac{2}{\pi^2} \ \mu_q^3 \qquad (2.1)$$

where the indices s,f,c refer to spin, flavor, and color degeneracies respectively. Equation (2.1) establishes a relation between a given baryon number b (quarks carry 1/3 unit of baryon number) and the variables V (volume) and μ_q. The energy of the quark bag is easily obtained noting that

$$E_{q,gas} = 2_s \times 2_f \times 3_c \times V \int \frac{d^3p}{(2\pi)^3} \sqrt{m_q^2 + p^2} \; \Theta(\mu^2 - (p^2 - m_q^2)) \quad (2.2)$$

Hence in the limit of small quark mass, i.e., $\mu_q \gg m_q$, we find, omitting here again for the sake of simplicity the qq-interaction term,

$$E(V, \mu(N_q, V)) = BV + V \frac{3}{2\pi^2} \mu_q^4 + O(m_q/\mu) \quad . \quad (2.3)$$

In order to determine the explicit dependence on a given quark number (baryon number) we use eq (2.1) to eliminate μ_q:

$$E(V, N_q) = BV + \frac{N_q^{4/3}}{V^{1/3}} \frac{3}{4} \left(\frac{\pi^2}{2}\right)^{1/3} \quad . \quad (2.4)$$

This expression has as before a minimum as function of the volume V, which corresponds to the equilibrium state:

$$P = -\left.\frac{\partial E}{\partial V}\right|_{s,b} \overset{!}{=} 0 = -B + \frac{1}{4} \left(\frac{\pi^2}{2}\right)^{1/3} \left(\frac{N_q}{V}\right)^{4/3}_{min} \quad . \quad (2.5)$$

Combining eqs (2.4) and (2.5), of course we find again

$$E_{min} = 4BV\big|_{min} \quad (2.6)$$

and hence we see that in agreement with the virial theorem the energy density is 4B also in the statistical bag. Combining eq (2.4) with (2.5), we find furthermore for the energy per quark the usual result:

$$(E/N_q)^{gas}_{min} = (\mu_q)_{min} = B^{1/4}(2\pi^2)^{1/4} = 2.11 \; B^{1/4} \quad . \quad (2.7)$$

Here $(\mu_q)_{min}$ is the chemical potential. It is found by inserting eq (2.5) into eq (2.1), upon which the N_q-dependence drops out. This result, eq (2.7) can be compared with a similar result for the smallest closed-shell bag which contains 12 quarks owing to $2_s \times 2_f \times 3_c = 12$. With $X_0 = 2.04$ we find from eq (1.25)

$$(E/12)_{bag} = \frac{(12 \times 2.04)^{3/4}}{12} \frac{3}{4} (4\pi)^{1/4} B^{1/4} = 2.3 \ B^{1/4} \quad . \quad (2.8)$$

Thus we conclude that the statistical result, eq (2.6) is in a remark-ably good agreement with a closed shell bag even when its baryon number is only 4. As the energy per quark in the statistical bag approach is slightly underestimated we conclude that the quark (i.e., baryon) density

$$N_q/V = (E/V) \ (N_q/E) = 4B/(E/N_q) \quad\quad\quad (2.9)$$

is somewhat overestimated.

As a final remark we note that eqs (2.3) and (2.6) imply that the energy per baryon in the bag is just μ, i.e., the baryon chemical potential

$$E/b = \frac{4}{3} E_q/(N_q/3) = 3\mu_q = \mu \quad\quad\quad (2.10a)$$

The factor 3 is necessary to account for the baryon number 1/3 of the quarks: three quarks form one baryon. We note that from eq (2.10) stems the conventional wisdom that $\mu_q = m_p/3$ at $T = 0$ where T is the temperature. Omitting the bag term in eq (2.2) one finds the well-known relativistic ideal-gas result

$$E_q/N = 3/4 \ \mu \quad . \quad\quad\quad (2.10b)$$

Thus we see that the bag term is a necessary ingredient for recovering the hadronic gas limit [9]

$$E/b|_{T=0} = \mu \quad . \quad\quad\quad (2.11)$$

Quarks will not always form a degenerate Fermi gas, especially inside a large bag. Depending on the creation history of the bag it is very likely that in an initial stage some of the quarks will be in excited states. In the statistical approach this situation easily can be described by introducing a quark temperature $T = 1/\beta$ which describes the internal excitations of each bag (= hadronic cluster) [26]. This does not imply an exact internal thermodynamic equilibrium of the quark gas in the bag. However, an assembly of excited bags in mutual thermal contact which is sharing to a certain extent the internal excitations, may be already similar in nature to the Gibb's grand canonical ensemble, i.e., an infinite number of interacting identical subsystems.

Hence, though the quarks in each individual bag may be far from thermodynamic equilibrium, in an assembly of bags which are able to scatter several times the average distribution may be much closer to the equilibrium. When making these remarks we have here particularly in mind highly excited nuclear matter as created in relativistic nuclear collisions, and, perhaps in antiproton annihilations in nuclei. Other circumstances prevail in $e^+e^- \rightarrow$ hadrons or even in pp reactions. But also in our case the word "kinetic equilibrium" has to be used with great care: the further in a particular bag the mean kinetic energy of the quarks is from $\sim T$, the less reliable becomes a priori the equilibrium assumption. We record here, however, that particle spectra from p-p collisions [8] behave as if a thermal equilibrium were always reached. Therefore the concept of "preformed" equilibrium has been introduced in thermodynamical models of hadron reactions.

With these remarks in mind we now turn to the description of excited quark bags with the help of quantum statistical methods. We will initially ignore the effect of quark-quark interactions and return to this problem further below. In principle, we could avoid the formal development and simply proceed by including the temperature through a Fermi distribution function in eq (2.2). However, as is well known, a complete description of the thermodynamical behavior of a many-particle system can be derived from the grand canonical partition function Z. Hence it is more useful for further developments to obtain right-away the grand partition functions for ideal Fermi and Bose gases. We follow here initially the standard textbooks [27] in calculating the grand canonical partition function which is defined as

$$Z(\beta, \mu_q, \ldots) = \text{Tr}\left(e^{-\beta(H - \mu Q)}\right) \ . \tag{2.12}$$

Here H is the Hamiltonian of the system and Q is the baryon charge operator. The chemical potential μ determines the average baryon number of the system. The trace is to be carried out over all allowed states of the many-body system. We note that

$$\langle Q \rangle \equiv \frac{\text{Tr } Q e^{-\beta(H-\mu Q)}}{\text{Tr } e^{-\beta(H-\mu Q)}} = \frac{1}{\beta} \frac{\partial}{\partial \mu} \ln Z(\beta, \mu, \dots) \qquad (2.13)$$

$$\langle H \rangle \equiv \frac{\text{Tr } H e^{-\beta(H-\mu Q)}}{\text{Tr } e^{-\beta(H-\mu Q)}} = - \frac{\partial}{\partial \beta} \ln Z(\beta, \mu, \dots) + \mu \langle Q \rangle \quad . \qquad (2.14)$$

The partition function may depend implicitly on other quantities such as the volume or even the shape of the considered quantum system.

In the particle-number representation the trace, eq (2.12), can be easily carried out for free quarks. Here

$$H = \sum_{i,\ell} \varepsilon_i^\ell n_i^\ell + \sum_{i,\ell} \varepsilon_i^\ell \bar{n}_i^\ell \qquad (2.15a)$$

$$Q = \sum_{i,\ell} b_\ell (n_i^\ell - \bar{n}_i^\ell) \qquad (2.15b)$$

where n_i^ℓ is the number operator of the ith single-particle state of a quark ($\bar{n}_i^\ell$ for antiquarks) with (discrete) quantum numbers "ℓ", such as flavor. b_ℓ is the baryon charge, i.e., $+ 1/3$ for quarks and $- 1/3$ for antiquarks as already introduced explicitly in eq (2.15b). A quantum state is characterized by the occupation numbers n_i^ℓ, $\bar{n}_i^\ell$ of the quarks and antiquarks. Hence the trace which sums only the diagonal matrix elements is

$$Z_q = \sum_{\{n_i^\ell, \bar{n}_i^\ell\}} e^{-[\beta \sum_{i,\ell} (\varepsilon_i^\ell - 1/3 \; \mu) n_i^\ell + \sum_{i,\ell} (\varepsilon_i^\ell + 1/3 \; \mu) \bar{n}_i^\ell]} \quad . \qquad (2.16)$$

Here the sum runs over all sets of numbers n_i^ℓ, $\bar{n}_i^\ell$. We factorize the partition function in terms of the discrete quantum numbers ℓ:

$$Z_q = \prod_\ell Z_\ell \qquad (2.17a)$$

$$Z_\ell = \sum_{\{n_i^\ell, \bar{n}_i^\ell\}} \prod_i e^{-\beta(\varepsilon_i^\ell - 1/3 \; \mu) n_i^\ell - \beta(\varepsilon_i^\ell + 1/3 \; \mu) \bar{n}_i^\ell} \quad . \qquad (2.17b)$$

The infinite product over all states can be interchanged with the infinite sum over all occupation numbers, leading to:

$$\ln Z_\ell = \sum_i \left[\left(\ln \sum_{n_i^\ell} e^{-\beta(\varepsilon_i^\ell - 1/3\,\mu)n_i^\ell}\right) + \left(\ln \sum_{\bar{n}_i^\ell} e^{-\beta(\varepsilon_i^\ell + 1/3\,\mu)\bar{n}_i^\ell}\right)\right] . \quad (2.18)$$

Only $n_i^\ell = 0,1$ is allowed for Fermions. Hence we find the well known result (f = flavor, s = spin, c = color)

$$\ln Z_q = \sum_{flavor} 2_s\, 3_c \left[\sum_i \ln\left(1 + e^{-\beta(\varepsilon_i^f - 1/3\,\mu)}\right) + \sum_i \ln\left(1 + e^{-\beta(\varepsilon_i^f + 1/3\,\mu)}\right)\right] \quad (2.19)$$

where the spin and color factors count the respective degeneracies. In the continuum limit

$$\sum_i \longrightarrow \int \frac{d^3x\, d^3p}{(2\pi)^3} = V \int \frac{d^3p}{(2\pi)^3} \quad (2.20a)$$

$$\varepsilon_i^f \longrightarrow \sqrt{p^2 + m_f^2} \quad (2.20b)$$

and we find

$$\ln Z_q = \sum_f 2_s\, 3_c\, V \int \frac{d^3p}{(2\pi)^3} \left[\ln\left(1+e^{-\beta(\sqrt{m_f^2+p^2}-1/3\,\mu)}\right)\right.$$

$$\left. + \ln\left(1+e^{-\beta(\sqrt{m_f^2+p^2}+1/3\,\mu)}\right)\right]. \quad (2.21)$$

For the light u and d quarks, for which usually $m_f \ll \mu$ is fulfilled, we can evaluate the momentum integrals analytically [2b], [28]:

$$(T\,\ln Z_q)_{light\ flavors} = \frac{2_s \times 2_f \times 3_c \times V}{24\pi^2}\left(\mu_q^4 + 2\mu_q^2(\pi T)^2 + \frac{7}{15}(\pi T)^4\right) . \quad (2.22)$$

The quark chemical potential

$$\mu_q = \frac{1}{3}\mu \quad (2.23)$$

controls the quark number $N_q - N_{\bar{q}}$.

In order to recover the limit $T \to 0$, eq (2.3), we must introduce a phenomenological bag, i.e., vacuum term

$$\ell nZ_{bag} \equiv -BV\beta \quad . \tag{2.24}$$

With eq (2.24) and

$$\ell nZ = \ell nZ_q + \ell nZ_{vac} \tag{2.25}$$

we indeed find

$$E = \langle H \rangle = -\frac{\partial}{\partial \beta} \ell nZ + \mu \frac{\partial}{\partial \mu} T \ell nZ$$

$$= BV + \frac{3}{2\pi^2} V \left[\mu_q^4 + 2\mu_q^2 (\pi T)^2 + \frac{7}{15} (\pi T)^4 \right] \tag{2.26}$$

and at the same time for the baryon number

$$b/3 = \langle \tfrac{1}{3} Q \rangle = N_q - N_{\bar{q}} = \frac{\partial}{\partial \mu_q} T \ell nZ$$

$$= \frac{2}{\pi^2} V(\mu_q^3 + \mu_q (\pi T)^2) \tag{2.27}$$

Equations (2.26) and (2.27) generalize the $T = 0$ results, eq (2.1) and (2.3) to finite temperatures. We note also that for finite T it is possible to eliminate μ_q analytically from eq (2.26) with the help of Cardan's formulae [28] for eq (2.27), and to obtain $E(V,b,T)$.

We have not yet considered one quite important aspect of the excited bag, viz., the possible presence of real transversal gluons. At present the evidence for the existence of perturbative gluons inside bags is not quite conclusive. Some theoretical calculations [29] indicate that gluons could be admixed to the quark wave functions in bags. However, gluonium, i.e., gluon-only bags have not yet been conclusively established experimentally [30]. None-the-less it is very likely that in a large quark bag at finite temperature the transversal gluons will be present with an abundance corresponding to that expected from the blackbody radiation law, i.e., as given by the Stefan-Boltzmann equation. To include these bosonic degrees of freedom we evaluate eq (2.16) taking into account the fact that the occupation number of the gluon modes can be $n_i = 1, 2, \ldots \infty$, and that gluons do not carry baryon charge. We thus find

$$Z_g = \prod_\ell \prod_i \sum_{n=0}^{\infty} e^{-\beta n \varepsilon_i^\ell} = \prod_\ell \prod_i \frac{1}{1 - e^{-\beta \varepsilon_i^\ell}} . \qquad (2.28)$$

Here ℓ counts the $N^2 - 1 = 8$ color degrees of freedom as well as the two transversal polarizations. All gluon single-particle energies are degenerate with respect to ℓ. Taking the continuum limit we find:

$$\ell n Z_g = -8_c \times 2_s \times V \int \frac{d^3 p}{(2\pi)^3} \ell n (1 - e^{-\beta |\vec{p}|}) . \qquad (2.29)$$

This expression is very well known and, except for the color factor, corresponds to the standard photon result. We find explicitly

$$T \, \ell n Z_g = 8 \frac{V}{45 \pi^2} (\pi T)^4 . \qquad (2.30)$$

As emphasized at the end of the last section the quark-quark inter-action still must be taken into account. We shall use here the contributions which are first order in the QCD running coupling constant $\alpha_s(q^2) = g^2/4\pi$. As $\alpha_s(q^2)$ increases when the average momentum exchanged between quarks decreases it would seem that this approach will have only limited validity at relatively low densities and/or temperatures. However, the collective screening and the phonon modes in the plasma are of comparable order of magnitude and cancel each other [31]. The influence of perturbative contributions are governed by the expansion factor $\delta = (4/3)(\alpha_s/\pi) \sim 0.15 - 0.3$. In other words, since $\delta^2 < 0.02 - 0.09$ the use of first order perturbation theory may be quite adequate [32]. For the case of the quark-gluon plasma in the perturbative vacuum, one finds for the partition function an analytic expression through first order in α_s when neglecting the quark masses. We obtain for the quark Fermi gas [2b], [2g], [28]:

$$\ell n Z_q(\beta, \mu) = \frac{gV}{6\pi^2} \beta^{-3} [(1 - \frac{2\alpha_s}{\pi})(\frac{1}{4} (\mu\beta)^4 + \frac{\pi^2}{2} (\mu\beta)^2) + (1 - \frac{50}{21} \frac{\alpha_s}{\pi}) \frac{7\pi^4}{60}]$$

$$(2.31)$$

where $g = (2s + 1)(2I + 1)N = 12$ counts the number of the components, i.e., the degeneracy of the quark gas. The gluon partition function is similarly reduced by the interactions:

$$\ell n Z_g(\beta, \lambda) = V \frac{8\pi^2}{45} \beta^{-3} \left(1 - \frac{15\alpha_s}{4\pi}\right) . \tag{2.32}$$

We notice above the second relevant difference from the photon gas: aside from the presence of the degeneracy factor 8 there is the term associated with the gluon-gluon interaction since gluons carry color charge.

Finally, let us recall the true-vacuum term $\ell n\, Z_{vac}$, eq (2.24), which describes the required positive energy density B within the volume occupied by the colored quarks and gluons and leads to a negative pressure on the surface of this region. As discussed above, at this stage this term is entirely phenomenological. The equations of state for the quark-gluon plasma are obtained by differentiating

$$\ell n Z = \ell n Z_q + \ell n Z_g + \ell n Z_{vac} \tag{2.33}$$

with respect to β, μ, and V. The energy density, baryon number density, pressure, and entropy density, of u,d quarks and gluons respectively, are written in terms of the baryonic chemical potential μ and the temperature T

$$\varepsilon = \frac{6}{\pi^2}\left[\left(1 - \frac{2\alpha_s}{\pi}\right)\left(\frac{1}{4}\left(\frac{\mu}{3}\right)^4 + \frac{1}{2}\left(\frac{\mu}{3}\right)^2(\pi T)^2\right) + \left(1 - \frac{50}{21}\frac{\alpha_s}{\pi}\right)\frac{7}{60}(\pi T)^4\right]$$

$$+ \frac{8}{15\pi^2}(\pi T)^4\left(1 - \frac{15}{4}\frac{\alpha_s}{\pi}\right) + B \tag{2.34}$$

$$\nu = \frac{2}{3\pi^2}\left[\left(1 - \frac{2\alpha_s}{\pi}\right)\left(\left(\frac{\mu}{3}\right)^3 + \frac{\mu}{3}(\pi T)^2\right)\right] \tag{2.35}$$

$$P = \frac{1}{3}(\varepsilon - 4B) \tag{2.36}$$

$$s = \frac{2}{\pi}\left(1 - \frac{2\alpha_s}{\pi}\right)\left(\frac{\mu}{3}\right)^2(\pi T) + \frac{14}{15\pi}\left(1 - \frac{50}{21}\frac{\alpha_s}{\pi}\right)(\pi T)^3 + \frac{32}{45\pi}\left(1 - \frac{15}{4}\frac{\alpha_s}{\pi}\right)(\pi T)^3 . \tag{2.37}$$

In eqs (2.34) and (2.37) the second T^4 and T^3 terms originate in the gluonic degrees of freedom. In eq (2.36) we have right away used the relativistic relation between the quark and gluon energy density and pressure

$$P_q = \frac{1}{3} \, \varepsilon_q, \quad p_g = \frac{1}{3} \, \varepsilon_g \qquad (2.38)$$

in order to derive this simple form of the equation-of-state of the quark-gluon plasma. This form is slightly modified when the finite quark masses are included, or when the dependence of the QCD coupling constant α_s on the dimensional parameter Λ is taken into account.

As we have already seen in the discussion of the hadronic bag structure in section 1, an assembly of quarks will assume a geometric configuration such as to make the total energy $E(V,b,S)$ as small as possible at given baryon number and fixed total entropy. As is apparent from the first law of thermodynamics

$$dE = - PdV + TdS + \mu db \qquad (2.39)$$

we have

$$P = - \frac{\partial E(V,b,S)}{\partial V} \qquad (2.40)$$

Hence, the geometrically stable configuration $\partial E / \partial V = 0$ corresponds also to the configuration with vanishing pressure P. Rather than to work in the microcanonical ensemble with fixed b and S, we exploit the advantages of the grand cononical ensemble and consider P as a function of μ and T:

$$P = \frac{\partial}{\partial V} (T \, \ell n Z(\mu, T, V)) \qquad (2.41)$$

with the result as given by eq (2.36). From eq (2.36) it follows that when the pressure vanishes in a static configuration the energy density is 4B, independent of the values of μ and T which fix the line P = 0. We recall that this has been precisely the kind of behavior found for the hadronic gas. For P > 0 we have ε > 4B. We recall that in the hadronic gas we always had ε < 4B. Thus, in this domain P > 0 of the μ - T plane the quark-gluon plasma must be exposed to an external force to achieve a stationary state.

In order to obtain an idea of the form and location of the P = 0 configuration in the μ - T plane for the quark-gluon plasma, we rewrite eq (2.36) for P = 0:

$$B = \frac{(1 - \frac{2\alpha_s}{\pi})}{162\pi^2} [\mu^2 + (3\pi T)^2]^2 - \frac{T^4\pi^2}{45} [12(1 - \frac{5\alpha_s}{3\pi}) - 8(1 - \frac{15\alpha_s}{4\pi})] .$$

(2.42)

Here, the last term is the contribution of the gluons to the pressure. We find that the greatest lower bound on the temperature T_q at μ = 0 is, for α_s = 1/2 about

$$T_q \approx .83B^{1/4} \sim 160 \text{ MeV} \approx T_0 .$$

(2.43)

This result shows the expected order of magnitude. The most remarkable point is that it leads to a numerically similar value as that which we will find below in the study of the hadronic gas. Another point worth mentioning is the influence of the strange quarks: they increase the quark pressure just by the amount needed to counter the effect of the interaction in eq (2.42). Hence we indeed have $T_0 \sim B^{1/4}$, including the strange quarks (see the discussion after eq (2.46)).

Let us here further note that for T << μ the baryon chemical potential tends to

$$\mu = 3\mu_q ==> 3B^{1/4} [\frac{2\pi^2}{(1 - \frac{2\alpha_s}{\pi})}]^{1/4} = 1320 \text{ MeV} [\alpha_s = 1/2, B^{1/4} = 190 \text{ MeV}].$$

(2.44)

In concluding this discussion of the P = 0 line of the quark-gluon plasma, let us note that the choice $\alpha_s \sim 1/2$ is motivated by the fits to the charmonium and upsilonium spectra as well as to deep inelastic scattering data. In both these cases space-like domains of momentum transfer are explored. The much smaller value of $\alpha_s \sim 0.2$ is found in time-like regions of momentum transfer in $e^+e^- \rightarrow$ hadron experiments. We recall that this was the behavior derived from eq (1.17) (see fig. 1.1). In the quark-gluon plasma, described up to first order perturbation theory, both positive and negative q^2 occur: the perturbative corrections to the radiative T^4 contribution is dominated by time-like momentum transfers, while the correction to the μ^4 term originates from space-like quark-quark scattering. Hence it is preferable that two different values of α_s be used in the above expressions.

Consider now the energy density at $\mu = 0$. We find the simple result, restating again some factors

$$\varepsilon(\mu = 0) = B + \frac{\pi^2}{30} T^4 [2_S \times 8_C \times (1 - \frac{15}{4} \frac{\alpha_S}{\pi}) + 2_I \times 2_S \times 3_C \times \frac{7}{4} (1 - \frac{50}{21} \frac{\alpha_S}{\pi})] .$$
$$(2.45)$$

We note that in both quarks and gluons the interaction conspires to reduce the effective available number of degrees of freedom. At $\alpha_S = 0$ we find the handy relation

$$\varepsilon_q + \varepsilon_g = (\frac{T}{160 \text{ MeV}})^4 [\frac{GeV}{fm^3}] .$$
$$(2.46)$$

At $\alpha_S = 1/2$ we seem to be left with only $\sim 50\%$ of the degrees of freedom, and the temperature "unit" in the above formula drops to 135 MeV. However, as mentioned above, we rather should use $\alpha_S \sim .2$ in eq (2.45) in which case the contribution of strange quarks, which is about 30% of the last term in (2.45) just compensates these interaction effects. Hence (2.46) is the proper rough estimate to be kept in mind.

We now discuss briefly the influence of the heavy flavors. For a charm quark with a mass of about 1500 MeV the thermodynamic abundance is sufficiently low to ignore its influence on the properties of the plasma. While its production is exceedingly slow, even the influence of its equilibrium abundance on the thermodynamic properties of the plasma would be quite negligible. To wit, evaluating the phase-space integrals we find that the ratio of charm to light antiflavor (either $\bar{u}$ or $\bar{d}$) is

$$c/\bar{q} = \bar{c}/\bar{q} = e^{-(m_c - \mu/3)/T} (\frac{m_c}{T})^{3/2} \frac{1}{2} \sqrt{\frac{\pi}{2}} .$$
$$(2.47)$$

Taking as a numerical example $m_c = 1500$ MeV, $T = 200$ MeV, $\mu = 0$ one finds $c/\bar{q} = 7.10^{-3}$. Thus the energy fraction carried by the plasma charm here would be $\sim 0.2\%$ and unimportant for the thermodynamic properties of the plasma, but quite significant in direct charm detection experiments. However, the approach to chemical equilibrium (see below) is too slow to saturate in nuclear collisions the phase space even within the most optimistic scenarios, except in circumstances in which $T \sim \frac{1}{2} m_c$ were reached.

Clearly, we must turn our attention to strangness. With a current quark mass of about 150-180 MeV we are actually at the threshold $T = m_s$ and indeed one finds that there is quite an appreciable s-abundance. An explicit calculation [4b] has shown that chemical equilibrium will be reached during the short time of a heavy ion reaction. The motion of the particles being already semi-relativistic, the s̄s productiuon results in a significant increase in the number of available degrees of freedom of quarks in eq (2.45). Thus for $T > m_s$ we have to increase the number of flavors to 3 while at $T \sim m_s$ the effective flavor number is 2.8. The appearance of strangeness is a very important qualitative feature and we will return to its discussion in section 3.

As a final aspect of the perturbative quark-gluon plasma we consider now the role of the color charge in the statistical description. We note that for finite-size bags it is essential to ensure the color neutrality of the considered states: much of the hadronic structure is a consequence of the requirement of color neutrality and of the symmetries of the quark wave functions in the bags. However, we have not yet included this effect of color into our considerations. As long as only very few particles are present, color neutral states can be constructed explicitly. But how can we treat an excited, relativistic many-body system? The answer is quite simple in principle: in eq (2.12) the trace has to include only color neutral states. That is, we should consider

$$Z_{c=o} = \mathrm{Tr}_{c=o} \; e^{-\beta(H - \mu_q Q)} \quad . \tag{2.48}$$

However, in order to arrive at a manageable result we had to allow <u>all</u> states in the trace. In order to solve this problem [33] we borrow the main technical idea from the work of Redlich and Turko [34].

Actually, for simple cases, an answer to this problem can be directly written down [35]. Each state of the Hamiltonian H can be classified within the irreducible subspaces according to its transformation properties under the representations of the SU(3) color group. In order to compute the singlet contribution eq (2.48) we first introduce the generating functional

$$\mathcal{Z} = \sum_c \frac{1}{d_c} \; \chi_c(\phi_i) \; Z_c \tag{2.49}$$

where the sum is carried out over all irreducible representations characterized by the index c. The variables ϕ_i of the coefficient functions, i.e., group characters χ_c will permit the inversion of eq (2.49) through a relation of the type found when solving for a set of complete orthogonal functions, i.e.,

$$\int d^n\phi\; M(\phi_i)\; \chi_{c'}^*(\phi_i)\; \chi_c(\phi_i) = \delta_{cc'} \qquad (2.50a)$$

$$\sum_c \chi_c^*(\phi_i')\; \chi_c(\phi_i) = \delta^n(\phi' - \phi) \qquad (2.50b)$$

Hence d_c in eq (2.49) is a suitable normalization constant (dimension of the representation), while $M(\chi_i)$ is a function defining the Haar measure [35]. With these relations we have

$$Z_c = d_c \int d^n\phi\; M(\phi_i)\; \chi_c(\phi_i)\; Z(\beta,V\ldots;\phi_i) \quad . \qquad (2.51)$$

The problem is to obtain a suitable set of functions $\phi_c(\chi_i)$ such that

    i) Z eq (2.49) can be explicitly computed, e.g., in the particle number representation, and

    ii) eqs (2.50) are satisfied.

A hint of how to proceed is contained in eq (2.12): since the baryon number operator commutes with the Hamiltonian we could use Q in the exponent in order to divide the Hilbert space into sectors of given baryon number. We proceed now in this fashion with the non-abelian group SU(3). There are two mutually commuting charges; in the standard representation of SU(3) they are the 3 and 8 directions of the color space. We therefore consider the following Ansatz for the generating function

$$Z = \mathrm{Tr}(e^{-\beta(H-\mu_q Q)}\tilde{u}(\phi_3,\phi_8)) \qquad (2.52a)$$

here

$$\tilde{u}(\phi_3,\phi_8) = e^{-i\phi_3 Q_3 - i\phi_8 Q_8} \qquad (2.52b)$$

where we have introduced the new factor $\tilde{u}$. Since the norm of $\tilde{u}$ is bounded by unity we have no trouble to establish the existence of the generating function (2.52): its absolute value must always be smaller than Z, which is obtained replacing $\tilde{u}$ by unity.

It is our first aim now to show that eq (2.52) yields the desired form (2.49). The Hilbert space spanned by H is a direct product of orthogonal subspaces characterized by the color c of states belonging to each sector (i.e., transformation properties under the $SU(3)_c$ group). Hence we can write:

$$Z = \sum_c \text{Tr}[e^{-\beta(H-\mu_q Q)} e^{-i\phi_3 Q_3 - i\phi_8 Q_8}] \tag{2.53}$$

where the sum c is a symbolic sum over all irreducible representations of SU(3), usually characterized by two positive integers $(p,q) \in (0,\infty)$. We can now constrain our discussion to the "good color" subsectors of the entire Hilbert space of the color space. Within each sector, a complete orthonormal set of states is generated in the particle number representation of H:

$$1_c = \sum_{\xi_c=1}^{d_c} \sum_{\nu_c} |\nu_c, \xi_c\rangle \langle \xi_c, \nu_c| \tag{2.54}$$

where d_c is the degeneracy, i.e., dimensionality of each color multiplet, while ξ_c denotes all quantum numbers within a given irreducible representation which are related to the internal symmetry. ν_c is a short-hand notation for the set of states determined by the occupation numbers n_i of the ith momentum state. Inserting eq (2.54) into (2.53) we find

$$Z = \sum_c \{\sum_{\xi_c} \sum_{\nu_c} \sum_{\xi_c'} \sum_{\nu_c'} \langle \xi_c, \nu_c| e^{-\beta(H-\mu_q Q)}$$

$$\times \langle \xi_c', \nu_c'| e^{-i\beta\phi_3 Q_3 - i\beta\phi_8 Q_8} |\nu_c, \xi_c\rangle\} \quad . \tag{2.55}$$

Owing to the assumed exact color symmetry the first factor in eq (2.55) is diagonal in ξ. Similarly, the second factor is diagonal in ν. Thus we have both $\delta_{\xi\xi'}$ and $\delta_{\nu\nu'}$ as factors and two of the sums collapse. Dropping the irrelevant indices we find

$$\mathcal{Z} = \sum_c \{\sum_{\nu_c} \langle \nu_c | e^{-\beta(H-\mu_q Q)} | \nu_c \rangle \sum_{\xi_c=1}^{d_c} \langle \xi_c | e^{-i\phi_3 Q_3 - i\phi_8 Q_8} | \xi_c \rangle \} \ . \quad (2.56)$$

Recalling that by definition

$$Z_c = Tr_c \ e^{-\beta(H-\mu_q Q)} = d_c \sum_{\nu_c} \langle \nu_c | e^{-\beta(H-\mu_q Q)} | \nu_c \rangle \quad (2.57)$$

we find the desired decomposition (2.49) with

$$\chi_c(\phi_3,\phi_8) = \sum_{\xi_c=1}^{d_c} \langle \xi_c | e^{-i\phi_3 Q_3 - i\phi_8 Q_8} | \xi_c \rangle \quad (2.58)$$

now being recognized as the character of the irreducible representation. The relations (2.50) are automatically satisfied for all compact semi-simple Lie groups. Even without the use of methods of group theory one easily can verify in particular cases that eq (2.58) indeed defines a suitable set of functions. Returning for a moment to the Lie group SU(2) we find

$$\chi_J^{SU(2)}(\phi_3) = \sum_{m_z=-J}^{+J} e^{-i\phi_3 m_z} = \frac{\sin(2J+1)\phi_3}{\sin 1/2 \ \phi_3} \ . \quad (2.59)$$

Hence, with the Haar measure [36]

$$M^{SU(2)}(\phi_3) = \frac{1}{\pi} \sin^2\phi_3/2 \ , \quad \phi_3 \ c- \ (0,2\pi) \quad (2.60)$$

equations (2.50) easily are verified. Note that the rank of the irreducible representation enters in eq (2.59) only via the summation over the eigenvalues of I_3. An equivalent remark holds for SU(3), eq (2.58).

The crucial point of this approach is the fact that the generating function $\tilde{Z}$, eq (2.53) can be explicitly determined! Actually the steps are identical to those used above, c.f., eqs (2.16)-(2.21). The reader is invited to repeat the derivation now with complex quantities $i\phi$ instead of the chemical potential. One simply finds for quarks the analogue of eq (2.21):

$$\ln Z_q = \sum_{flavor} 2_s \times V \int \frac{d^3p}{(2\pi)^3} \sum_{c,b} \ln[1 + \exp(-\beta(\sqrt{p^2+m_f^2} - \mu b) - i\alpha_c)] \quad (2.61)$$

for colored quarks. Instead of a color factor 3_c we now find the sum over the eigenvalues α_i of the charges Q_3, Q_8; in the triplet $(c \frac{1}{2} (1,0))$ and antitriplet $(c \frac{1}{2} (0,1))$ representation. In the triplet representation for which $b = + 1/3$ we have

$$\alpha_R = \phi_1/2 + \phi_2/3 \qquad (2.62a)$$

$$\alpha_G = -\phi_1/2 + \phi_2/3 \qquad (2.62b)$$

$$\alpha_B = -2/3 \phi_2 \qquad (2.62c)$$

where R,B,G, refers to the usual red, blue, green colors. In the antitriplet representation in which $b = - 1/3$ the sign of all the three angles reverses. We note that except when $\mu = 0$, the generating function Z, eq (2.61) will not be real. Of course, the integration over the group with the proper Haar measure [36] leads to a real result for the partition function Z_c. This measure is for SU(3)

$$d^2\phi M(\phi_1,\phi_2) = \frac{8}{3\pi^2} d\left(\frac{\phi_1}{2}\right) d\left(\frac{\phi_2}{3}\right)[\sin\frac{1}{2}\left(\frac{\phi_1}{2} + \phi_2\right) \sin\frac{\phi_1}{2} \sin\frac{1}{2}\left(-\frac{\phi_1}{2} + \phi_2\right)]^2$$

$$\phi_1, \phi_2 \in [-\pi,\pi]$$
$$(2.63)$$

We note that for massless quarks, i.e., neglecting the strange quark fraction eq (2.61) can be evaluated analytically following the results of eq (2.22) and replacing $\mu \rightarrow \mu \pm i\alpha$. In order to judge the influence of the color conservation it is sufficient to consider computationally the particular case of a baryon-less, $\mu = 0$, plasma droplet. In this case we find, substituting $\mu/T \rightarrow i\alpha$ in eq (2.22)

$$\ln Z_q = \ln Z_q^{(0)} \ln Z_q^{(1)} \qquad (2.64a)$$

$$\ln Z_q^{(0)} = VT^3 \pi^2 \frac{7}{30} \qquad (2.64b)$$

$$\ln Z_q^{(1)} = VT^3 \frac{\pi^2}{3} \sum_{i=R,G,B} [\frac{1}{2} \left(\frac{\alpha_i}{\pi}\right)^4 - \left(\frac{\alpha_i}{\pi}\right)^2] \qquad (2.64c)$$

Here $\ln Z_q^{(0)}$ is the partition function without the color constraint while $\ln Z_q^{(1)}$ vanishes at $\alpha_i = 0$.

It is important to appreciate that in the quark-gluon plasma the projection, eq (2.51) on a good color sector has to be carried out for both quarks and gluons simultaneously. Hence one has to carry out the integral

$$Z_c = \frac{1}{d_c} \int d^2\phi \, M(\phi_1, \phi_2) \, \chi_c \, (\phi) \, Z_q \, (\beta, V, \mu; \phi) \, Z_g \, (\beta, V; \phi) \quad (2.65)$$

where for the singlet, $c = 0$, the character χ_0 is just unity. In order to obtain Z_g we have to evaluate the analogue of eq (2.29)

$$\ln Z_g = 2_s \sum_{i=1}^{8} V \int \frac{d^3p}{(2\pi)^3} \, \ln(1 - e^{-\beta|p| - i\alpha_i}) \quad (2.66)$$

where the color sum i runs over the octet of gluons with the eigenvalues of the Q_3, Q_8 charges in the octet representation $(1,1)$.

As the octet results from the product of the triplet and antitriplet representations of SU(3), omitting the singlet, we have simply

$$\alpha_i^g = \alpha_j - \alpha_k \quad , \quad j,k \, \epsilon \, (R,B,G) \quad . \quad (2.67a)$$

In detail,

$$\alpha_i^g = (\alpha_R - \alpha_G, \, \alpha_G - \alpha_B, \, \alpha_B - \alpha_R,$$

$$- (\alpha_R - \alpha_G), - (\alpha_G - \alpha_B), - (\alpha_B - \alpha_R), 0,0) \quad . \quad (2.67b)$$

We therefore find

$$\ln Z_g = \ln Z_g^{(0)} + \ln Z_g^{(1)} \quad (2.68a)$$

$$\ln Z_g^{(0)} = VT^3 \, \pi^2 \, \frac{8}{45} \quad (2.68b)$$

$$\ln Z_g^{(1)} = VT^3 \, \frac{\pi^2}{6} \, \sum_{i=1}^{3} \, [-\frac{1}{2} + (\frac{|\alpha_i^g|}{\pi} - 1)^2 - \frac{1}{2} (\frac{|\alpha_i^g|}{\pi} - 1)^4] \quad (2.68c)$$

We note that the color sum in eq (2.68) was simplified and now includes only the first three terms of eq (2.67b). Again, $Z_g^{(0)}$ is the partition function without the color constraint eq (2.30), while $\ln Z_q^{(1)}$ vanishes at $\alpha_i^g = 0$.

From the preceeding derivations the seemingly enormous influence of color neutrality is apparent. Practically <u>all</u> states have to be rejected when evaluating the trace (2.48) constrained to color singlets. A global quark-gluon color correlation is introduced in view of eq (2.65). Both quark and gluon fractions of the plasma separately can have color as long as overall neutrality is assured. However, in the limit $V \to \infty$ one can see that color neutrality actually should not be that important. To wit, color fluctuations should go only as the square root of n, hence, the influence of the requirement of exact color conservation should not be felt. While this argument easily can be proven analytically it turns out [33] that the limit $V \to \infty$ actually means

$$VT^3 > 10 \qquad (2.69)$$

To illustrate this point we show in figure 2.1 the energy density as derived from eq (2.65) and divided by σT^4:

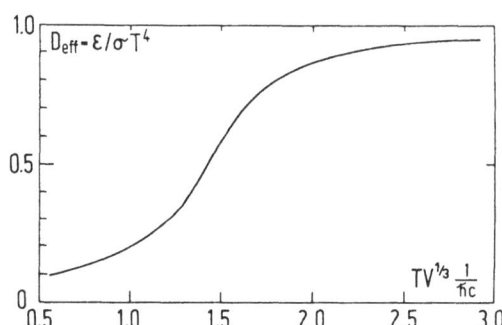

Fig. 2.1   Relative degeneracy D_{eff} of the quark-gluon plasma, as a function of $TV^{1/3}$.

$$D_{eff} \equiv \varepsilon / \sigma T^4 \quad , \qquad\qquad (2.70a)$$

$$\sigma = (2 \times 8 + 2 \times 2 \times 2 \times 3 \times \tfrac{7}{8}) \; \pi^2/30 \qquad (2.70b)$$

for a quark-gluon plasma, c.f., eq (2.45). On the other hand we observe that at $TV^{1/3} \sim 1.5$ only half of the expected number of excitations is available.

We also can evaluate D_{eff} at finite μ, in which instance we expect to obtain a less stringent constraint: The larger number of the available particles per unit volume reduces the importance of the constraint. The constraint can only become relevant when in the available volume the number of charge carrying quanta is of the order of one per degree of freedom. In figure 2.2 we display D_{eff} for $\mu = 0$ and $\mu = \pi T$ as function of the volume. The dot-dashed line shows the influence of the perturbative QCD corrections.

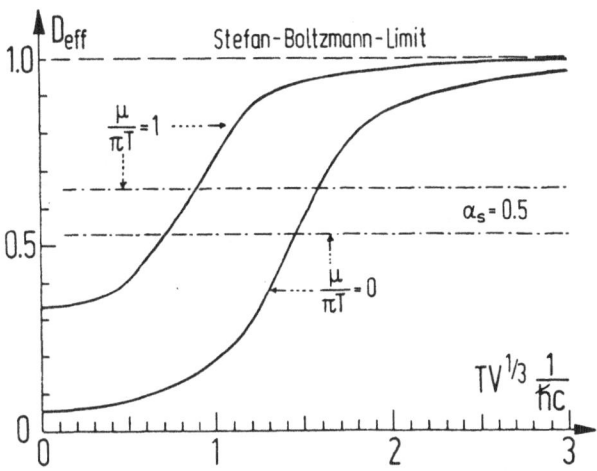

Fig. 2.2   The effective relative degeneracy D_{eff} of single particle energy levels.

We can thus conclude that the color constraints will alter the properties of quark-gluon plasma droplets, whenever they have $b < 4$. In particular this observation concerns the study of the phase transition from the plasma state to the hadronic gas; color configurations and correlations here will be of great relevance. However, in such an approach to the phase transition the inclusion of the effects of the interaction together with the color constraints is of essence. Therefore, no analytical understanding of the properties of the plasma near to the phase transition limit is available as of now. Though lattice gauge calculations have recently been carried out for such cases, no conclusions have been reached in view of the formidable numerical difficulties. The understanding of the phase transition region from the quark-gluon plasma to the hadronic gas phase is at present an open physical question.

3.   STRANGENESS IN THE QUARK-GLUON PLASMA

We now show in some detail why the strange particle abundances are so helpful [4] in observing the formation and the properties of the quark-gluon plasma.   First we note that at a given temperature the quark-gluon plasma will contain an equal number of strange (s) quarks and antistrange ($\bar{s}$) quarks.   Thus, assuming chemical and thermal equilibrium in the quark plasma we find the density of the strange quarks to be (two spins and three colors):

$$s/V = \bar{s}/V = 6 \int \frac{d^3p}{(2\pi)^3} \frac{1}{(e^{\sqrt{p^2+m_s^2}/T} + 1)} \approx 3 \frac{T m_s^2}{\pi^2} K_2(m_s/T) \quad (3.1)$$

neglecting, for the time being, the QCD perturbative corrections.   We recall that the mass of the strange quarks, m_s, in the perturbative vacuum is believed to be of the order of 140-200 MeV [18].   In eq (3.1) we were able to use the Boltzmann limit since the phase space density of strangeness is not too high.   Similarly, the light antiquark density ($\bar{q}$ stands for either $\bar{u}$ or $\bar{d}$) is:

$$\bar{q}/V = 6 \int \frac{d^3p}{(2\pi)^3} \frac{1}{(e^{|p|/T+\mu_q/T} + 1)} \approx e^{-\mu_q/T} T^3 \frac{6}{\pi^2} \quad . \quad (3.2)$$

The chemical potential of the quarks surpresses the $\bar{q}$ density.   This phenomenon reflects on the chemical equilibrium between q-$\bar{q}$ and the presence of a light-quark density associated with the net baryon number: the $\bar{q}$ are easily destroyed by the abundant q's when the q-density is large.

We now intend to show that often more $\bar{s}$ quarks are present than antiquarks of either light flavor.   Indeed:

$$\bar{s}/\bar{q} = \frac{1}{2} \left(\frac{m_s}{T}\right)^2 K_2\left(\frac{m_s}{T}\right) e^{\mu/3T} \quad . \quad (3.3)$$

This ratio is shown in figure 3.1.   We notice that in our case of interest, i.e., $\mu_q \sim T$ the abundances of $\bar{s}$ and $\bar{q}$ quarks are comparable and, in many cases of interest, $\bar{s}/\bar{q} \sim 5$.   For $\mu \to 0$ at $T \gtrsim m_s$ there are about as many $\bar{u}$ and $\bar{d}$ quarks as there are $\bar{s}$ quarks.

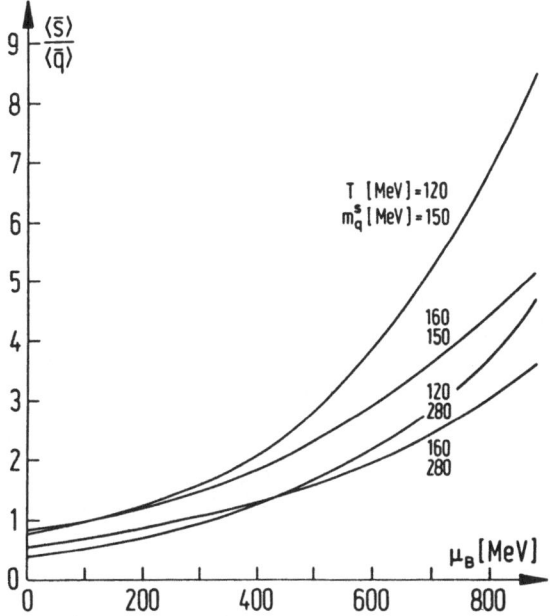

Fig. 3.1 Abundance of strange (or antistrange) quarks relative to the light quark abundance as function of μ for several choices of the temperature T and of the strange quark mass m_s.

When the quark matter hadronizes some of the numerous s and $\bar{s}$ may form strangness clusters such as Ξ, Ω, and their antiparticles, and also exotic strange objects instead of being bound in kaons. The probability for this process seems to be of similar magnitude as the production by the quarks of the $\bar{s}$ carrying antibaryons. It is particularly noteworthy that conventionally, i.e., in pp collisions, they can be produced only in direct pair production reactions. This process is suppressed by energy-momentum conservation and phase space considerations up to high energies since the final state has to contain four particles. This leads to the argument that a study of the $\bar{\Lambda}$, $\bar{\Sigma}$, $\bar{\Xi}$, $\bar{\Omega}$ in high-energy nuclear collisions could shed light on the early stages of the reaction in which a quark-gluon plasma may have been present. However, these antibaryons have a large transformation probability into kaons. To wit, the abundance of $\bar{\Omega}$ would be equal to Ω if the plasma state would freeze out directly into a low density hadron gas. In contrast, in a long collision process the $\bar{\Omega}$ abundance may be depleted by a number of transfer, i.e., exchange, processes, in particular by

$$\Omega + p \rightarrow KKK + X$$

owing to the strongly exothermic character of this reaction. As these remarks demonstrate, strangness is not only a tag of the plasma state but also a diagnostic tool for the transition, plasma → hadronic gas, and for the evolution of the hadronic gas.

The crucial aspects of the proposal to use strangeness as a signature for the quark-gluon plasma involve:

i)   assumption of thermal and chemical equilibrium
ii) comparison between results anticipated in both hadronic phases at given T and μ, where the chemical potential must be determined by other considerations.

We now turn to the discussion of both these points and begin by calculating the abundance of strangeness as function of the lifetime and excitation of the plasma state [4b].

In lowest order in perturbative QCD, $s\bar{s}$-quark pairs can be created by annihilation of light quark-antiquark pairs (fig. 3.2a) and in collisions of two gluons (fig. 3.2b). The averaged total cross sections for these processes were calculated by Combridge [37]. For fixed invariant mass-squared $s = (k_1 + k_2)^2$, where k_i are the four-momenta of the incoming particles $(w(s) = (1 - \frac{4M^2}{s})^{1/2})$ we have:

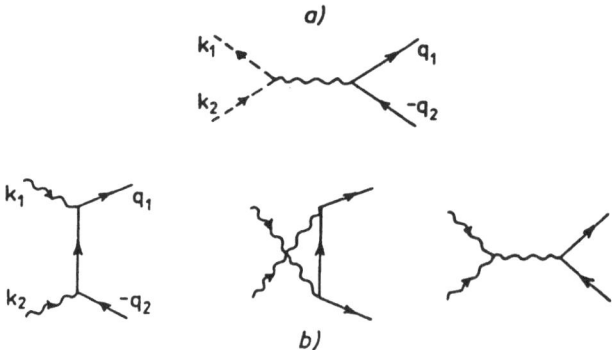

Fig. 3.2 Lowest order QCD diagrams for ss production: a) qq → $s\bar{s}$; b) gg → $s\bar{s}$.

$$\bar{\sigma}_{q\bar{q}\to s\bar{s}} = \frac{8\pi\alpha_s^2}{27\,s}\left(1 + \frac{2M^2}{s}\right) w(s) \qquad (3.4)$$

$$\bar{\sigma}_{gg \to s\bar{s}} = \frac{2\pi\alpha_s^2}{3s} \left[\left(1 + \frac{4M^2}{s} + \frac{M^4}{s^2}\right) \tanh^{-1} w(s) - \left(\frac{7}{8} + \frac{31}{8}\frac{M^2}{s}\right) w(s) \right] . \quad (3.5)$$

For the mass of the strange quark we will assume, a) the value [10] fitted within the MIT bag model: m_s = 280 MeV and, b) the typical value [18] found in the study of quark currents: m_s = 150 MeV. When discussing light quark production below we will use m_q = 15 MeV. The effective QCD coupling constant $\alpha_s = g^2/4\pi$ is an average over space-like and time-like domains of momentum transfers in the reactions shown in figure 3.2 as discussed in section 1. We use: (a) α_s = 2.2, the value consistent with m_s = 280 MeV in the MIT bag model, and (b) the value α_s = 0.6, expected at the momentum transfers in this process. We believe the choice (b) of the parameters to be realistic and to be consistent with the spirit of this work. The choice (a) is used as a reference; even when m_s = 280 MeV we will see that the chemical equilibrium will be reached.

Given the averaged cross sections it is easy to calculate the rate of events per unit time, summed over all final and initial states:

$$\frac{dN}{dt} = \int d^3x \int \frac{d^3k_1}{(2\pi)^3 |k_1|} \sum_i \rho_i(k_1,x) \int \frac{d^3k_2}{(2\pi)^3 |k_2|}$$

$$\times \sum_i \rho_i(k_2,x) \int_{4M^2}^{\infty} ds\, \delta(s - (k_1 + k_2)^2)\, k_1^\mu k_{2\mu}\, \bar{\sigma}(s) \quad (3.6)$$

The sum over initial states involves the discrete quantum numbers i (color, spin, etc.) over which eq (3.5) was averaged. The factor $\frac{k_1 \cdot k_2}{|k_1||k_2|}$ is the relative velocity for massless particles, and we have introduced a dummy integration over s in order to facilitate the calculations. We now replace the phase space densities $\rho_i(k,x)$ by momentum distributions $f_g(k)$, $f_q(k)$, $f_{\bar{q}}(k)$ of gluons, quarks, and antiquarks that can still have a parametric x-dependence through a space-dependence of the temperature $T = T(x)$ and the chemical potential $\mu = \mu(x)$. The invariant rate per unit time and volume for the elementary processes shown in figure 3.2 is then:

$$A = \frac{dN}{dt \, d^3x} = \frac{1}{2} \int_{4M^2}^{\infty} s \, ds \, \delta(s - (k_1 + k_2)^2) \int \frac{d^3k_1}{(2\pi)^3 |k_1|} \int \frac{d^3k_2}{(2\pi)^3 |k_2|}$$

$$\times \{(2 \times 8)^2 \, f_g(k_1) \, f_g(k_2) \, \bar{\sigma}_{gg \to s\bar{s}}(s)$$

$$+ 2 \times (2 \times 3)^2 \, f_q(k_1) \, f_{\bar{q}}(k_2) \, \bar{\sigma}_{q\bar{q} \to s\bar{s}}(s)\} \quad , \tag{3.7}$$

where the numerical factors count the spin, color and isospin degrees of freedom.

Assuming that in the rest frame of the plasma the distribution functions f depend only on the absolute value of the momentum, $|k| = k_0 \equiv k$, we can evaluate the angular integrals in eq (3.7):

$$A = \frac{8}{\pi^4} \int_{4M^2}^{\infty} s \, ds \, \bar{\sigma}_{gg \to s\bar{s}} [\int_0^{\infty} dk_1 \int_0^{\infty} dk_2 \Theta(4k_1 k_2 - s) \, f_g(k_1) \, f_g(k_2)]$$

$$+ \frac{9}{4\pi^4} \int_{4M^2}^{\infty} s \, ds \, \bar{\sigma}_{q\bar{q} \to s\bar{s}} [\int_0^{\infty} dk_1 \int_0^{\infty} dk_2 \Theta(4k_1 k_2 - s) \, f_q(k_1) \, f_{\bar{q}}(k_2)], \tag{3.8}$$

where the step function Θ requires that $k_1 k_2 > \frac{s}{4} > M^2$. We now turn to the discussion of the momentum distribution and related questions. The anticipated lifetime of the plasma created in nuclear collisions, as discussed below in section 5, is about 6 fm/c $= 2 \times 10^{-23}$ sec. After this time the high internal excitation will most likely have dissipated to below the energy density required for the quark-gluon plasma. We recall again that the transition between the hadronic and the quark-gluon phase is expected at an energy density of approximately 0.6 - 1 GeV/fm^3. Under these conditions each perturbative quantum (light quark, gluon) in the plasma state will rescatter several times during the lifetime of the plasma. Hence the momentum distribution functions f(p) can be approximated by the statistical Bose or Fermi distribution functions, regardless of the shortness of time:

$$f_g(p) \approx (e^{\beta \cdot p} - 1)^{-1}, \quad \text{(gluons)} \tag{3.9a}$$

$$f_{q/\bar{q}}(p) \approx (e^{\beta \cdot p} \lambda^{\pm} + 1)^{-1}, \quad \text{(quarks-antiquarks)} \tag{3.9b}$$

where β is the covariant temperature, $\beta \cdot p = \beta_0 |\vec{p}| - \vec{\beta} \cdot \vec{p}$ for massless particles, $(\beta \cdot \beta)^{-1/2} = T$ is the temperature and $\lambda^{\pm}$ is the baryon number (antibaryon number) fugacity. In the rest frame of the plasma, $\beta \cdot p = |\vec{p}|/T$. The distributions (3.9) can be taken seriously only for $|\vec{p}|$ not very much larger than T; to populate the high energy tail of the distributions too many collisions are required for which there may not be enough time during the lifetime of the plasma. Furthermore, we note that while in each individual nuclear collision the momentum distribution may vary, the ensemble of many collisions may have a better statistical distribution.

Finally, consider the values of the fugacities $\lambda^{\pm}$ in eq (3.9b). As we will show the gg → q$\bar{q}$ reaction time is much shorter than that for q$\bar{q}$ → s$\bar{s}$ production since the light quark masses are only of the order of ~ 15 MeV. Consequently we may assume chemical equilibrium between q and $\bar{q}$, i.e.,

$$\lambda^+ = \frac{1}{\lambda^-} = e^{-\mu_q/T} \quad , \quad \mu = 3\mu_q \tag{3.10}$$

and the baryon density is given by eq (2.35) omitting for the present the $0(\alpha)$ corrections, i.e.;

$$\nu(T, \mu_q) = \frac{2}{3\pi^2} \left(\mu_q^3 + \mu_q (\pi T)^2 \right) \quad .$$

We note that since gluons dominate the s$\bar{s}$ production in the plasma state, the conditions at the phase transition, such as the abundances of q and $\bar{q}$, will not matter for the s$\bar{s}$ abundances at times comparable to the lifetime of the plasma.

We now return to the evaluation of the rate integrals, eq (3.8). In the gluon part of the rate A, eq (3.8), the k_1, k_2 integral can be carried out exactly by expanding the Bose function in a power series in $\exp(-k/T)$:

$$A_g = \frac{8}{\pi^4} T \int_{4M^2}^{\infty} ds \, s^{3/2} \, \bar{\sigma}_{gg \to s\bar{s}}(s) \sum_{n,n'=1}^{\infty} (nn')^{-1/2} K_1 \left(\frac{(nn's)^{1/2}}{T} \right) \quad . \tag{3.11}$$

In the quark contribution an analytic treatment of the Fermi function is not feasible and the integrals must be evaluated numerically. It is found that the gluon contribution, eq (7.11), dominates the rate A. For $T/M \gtrsim 1$ we find:

$$A \approx A_g = \frac{7}{3\pi^2} \alpha_s^2 MT^3 e^{-2M/T} (1 + \frac{51}{14} \frac{T}{M} + \ldots) \qquad (3.12)$$

The abundance of $s\bar{s}$-pairs cannot grow forever; at some point the $s\bar{s}$-annihilation reaction will deplete the strange quark population. The loss term of the strangeness population is proportional to the square of the density n_s of strange and antistrange quarks. With $n_s(\infty)$ being the saturation density at large times, the following differential equation determines n_s as function of time:

$$\frac{dn_s}{dt} \approx A[1 - (n_s(t)/n_s(\infty))^2] \qquad (3.13)$$

We note that eq (3.13) in principle should also include a term linear in $n_s(t)$. Namely, when the plasma density is sufficiently high the produced strange quarks have difficulty to quickly get away from each other. With a scattering length of the order of 1/3 fm in extreme cases one has to consider diffusion rather than free motion. Hence in this limiting case we have always a $s\bar{s}$ pair in a given unit volume, leading to

$$\frac{dn_s}{dt} \approx A(1 - n_s(t)/n_s(\infty)) . \qquad (3.13b)$$

The solutions of eq (3.13) are, respectively

$$n_s(t) = n_s(\infty) \tanh(t/\tau) \qquad (3.14a)$$

$$n_s(t) = n_s(\infty) (1 - e^{-t/\tau}) \qquad (3.14b)$$

with

$$\tau = n_s(\infty)/A . \qquad (3.14c)$$

Both solutions are monotonically rising saturating functions with similar behavior, controlled by the characteristic time constant τ. In a thermally equilibrated plasma the asymptotic strangeness density, $n_s(\infty)$, is that of a chemically unconstrained relativistic Fermi gas ($\lambda = 1$):

$$n_s(\infty) = \frac{2 \times 3}{2\pi^2} T M^2 \sum_{n=1}^{\infty} \frac{(-)^{n-1}}{n} K_2(nM/T) \quad , \qquad (3.15)$$

We find for the relaxation time (3.14c) from eq (3.12), (3.15)

$$\tau \approx \tau_g = \frac{9}{7} \left(\frac{\pi}{2}\right)^{1/2} \alpha_s^{-2} M^{1/2} T^{-3/2} e^{M/T} \left(1 + \frac{99}{56} \frac{T}{M} + \ldots\right)^{-1} \qquad (3.16)$$

which falls rapidly with increasing temperature.

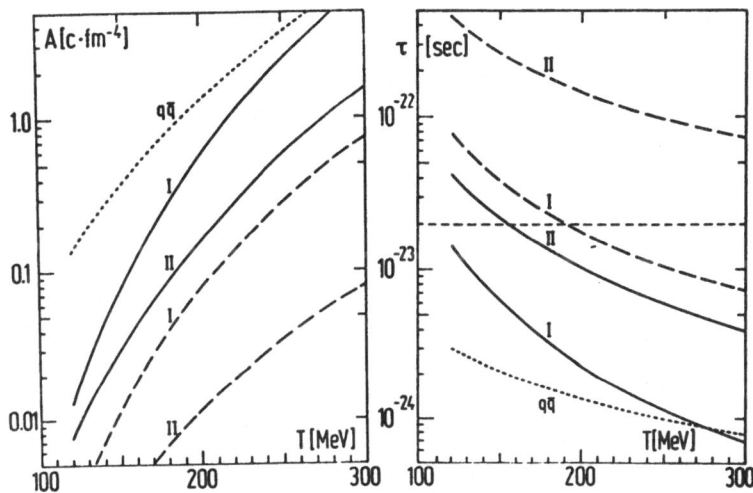

Fig. 3.3   Chemical relaxation times as functions of the temperature T. Full lines: $q\bar{q} \to s\bar{s}$ and $gg \to s\bar{s}$; dashed lines $q\bar{q} \to s\bar{s}$; dotted lines $gg \to q\bar{q}$ ($m_q = 15$ MeV). Curves marked I are for $\alpha_s = 2.2$ and $m_s = 280$ MeV, those marked II are for $\alpha_s = 0.6$ and $m_s = 150$ MeV: a) rates A;   b) time constants τ.

We now discuss the numerical results for the rates, time constants, and the expected strangeness abundance.   In figure 3.3a we compare the rates for strangeness production by the processes depicted in figure 3.2 for the two different choices of parameters discussed above, after eq (3.5).   The rate for $q\bar{q} \to s\bar{s}$ alone (shown separately) contributes less than 10 percent to the total rate. In figure 3.3b we show the corresponding characteristic relaxation times toward chemical equilibrium, τ, defined in eq (3.14). While our results for strangeness production by light quarks agree only in order of magnitude with those of Biró and Zimányi [38] owing to the difference in the chosen values of the parameters, it is obvious from our results that gluonic strangeness production, which was not discussed initially by these authors [39], is the dominant process. If we compare the time constant τ with the estimated lifetime of the plasma state we find that the strangeness abundance will be chemically saturated for temperatures of 160 MeV and above, i.e., for an energy density above 1 GeV/fm^3. We note that τ is quite sensitive to the choice of the strange-quark mass parameter and the coupling constant α_s which must, however, be chosen consistently. A measure of the uncertainty associated with the choice of parameters is illustrated by the difference between our results for the two parameter sets taken here.

Also included in figures 3.3a, and 3.3b are our results for gluon conversion into light quark-antiquark pairs. The shortness of τ for this process indicates that gluons and light quarks reach chemical equilibrium during the beginning stage of the plasma state, even if the quark/antiquark, i.e., baryon/meson ratio was quite different in the prior hadronic compression phase.

The evolution of the density of strange quarks, eq (3.14), relative to the baryon number content of the plasma state, is shown in figure 3.4 for various temperatures. The saturation of the abundance is clearly visible for T $>$ 160 MeV. To obtain the experimentally accessible abundance of strange quarks, the corresponding values reached after the typical lifetime of the plasma state, 2 x 10^{-23} sec, can be read off in figure 3.4 as a function of the temperature. The strangeness abundance shows a pronounced threshold behavior at T $\sim$ 120 -160 MeV.

We thus conclude that the strangeness abundance saturates in a sufficiently excited quark-gluon plasma with T $>$ 160 MeV, ϵ $>$ 1 GeV/fm^3 owing to the high gluon density. This allows strangeness to be an important observable indicating the presence of gluons in the reaction. We hence turn to the study of the strangeness in normal nuclear matter in order to gain insight into the relevance of strangeness as a characteristic signature of the quark-gluon plasma.

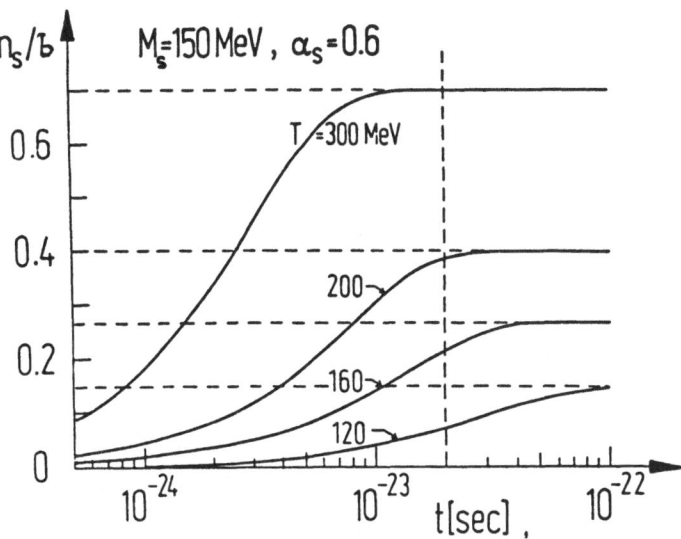

Fig. 3.4   Time evolution of the relative strange quark to baryon number
abundance in the plasma for various temperatures T.
m_s = 150 MeV, α_s = 0.6.

To this end we must first establish [4c, 40] the relevant relative
strange particle rates originating from highly excited matter consisting of individual hadrons, the hadronic gas phase.   The main hypothesis which allows us to simplify the situation is to postulate the
resonance-dominance of hadron-hadron interactions (see section 4).   In
this case the hadronic gas phase is a superposition of different
hadronic gases and all information about the interaction is hidden in
the mass spectrum $\tau(m^2,b)$ which describes the number of hadrons of
baryon number b in a mass interval dm^2.   When considering strangeness
carrying particles, all we then need to include is the influence of the
non-strange hadrons in the baryon chemical potential established by the
non-strange particles.   The total partition function is approximately
additive in these degrees of freedom:

$$\ln Z = \ln Z^{\text{non-strange}} + \ln Z^{\text{strange}} \quad . \qquad (3.17)$$

In order to determine the particle abundances it is sufficient to list
the strange particles separately and we find

$$\ell n\ Z^{strange}\ (T,V,\lambda_s,\lambda_q)\ =\ C\ \{2W(x_k)\ [\lambda_s\lambda_q^{-1} + \lambda_s^{-1}\lambda_q]$$

$$+\ 2[W(x_\Lambda) + 3W(x_\Sigma)]\ [\lambda_s\lambda_q^2 + \lambda_s^{-1}\lambda_q^{-2}]\} \qquad (3.18a)$$

$$W(x_i)\ =\ (\frac{m_i}{T})^2\ K_2(\frac{m_i}{T})\ . \qquad (3.18b)$$

We have $C = VT^3/2\pi^2$ for a fully equilibrated state. The case of chemical non-equilibrium can be effectively taken care of by using smaller values of C. Since the strangeness-exchange cross sections are very large, strangeness always will be distributed among all particles in (3.18a) according to the values of the fugacities $\lambda_q = \lambda_B^{1/3}$ and λ_s. Hence we can speak of relative strangeness chemical equilibrium, see below. We have neglected to write down quantum statistics corrections as well as the multi-strange particles, Ξ and Ω^-, as our considerations remain valid in this simple approximation [40]. Interactions are effectively included through explicit reference to the baryon number content of the strange particles as just discussed. Non-strange hadrons influence the strange fraction by establishing the value of λ_q at the given temperature and baryon density.

As introduced here, λ_s controls the strange quark content while the up- and down-quark content is controlled by $\lambda_q = \lambda_B^{1/3}$.

Using the partition function eq (3.18a) and (3.18b) we calculate for given μ_B, T, and V the mean strangeness by evaluating

$$<n_s - n_{\bar{s}}>\ =\ \lambda_s\ \frac{\partial}{\partial\lambda_s}\ \ell n\ Z^{strange}(T,V,\lambda_s,\lambda_q)\ , \qquad (3.19)$$

which is the difference between strange and anti-strange components. This expression must be equal to zero since strangeness is a conserved quantum number with respect to the strong interactions. From this condition we get:

$$\lambda_s\ =\ \lambda_q\ \left[\frac{W(x_k) + \lambda_B^{-1}[W(x_\Lambda) + 3W(x_\Sigma)]}{W(x_k) + \lambda_B[W(x_\Lambda) + 3W(x_\Sigma)]}\right]^{1/2}\ \equiv\ \lambda_q\gamma\ . \qquad (3.20)$$

We notice a strong dependence of γ on the baryon number. For large μ_B the term with $\lambda_B{}^{-1}$ will tend to zero and the term with λ_B will dominate the expression for λ_s and γ. As a consequence the particles with fugacity λ_s and strangeness $S = -1$ (note that by convention strange quarks carry $S = -1$, while strange antiquarks carry $S = 1$) are suppressed by a factor γ which is always smaller than unity. Conversely, the production of particles which carry the strangeness $S = +1$ will be favored by γ^{-1}. This is the consequence of the presence of nuclear matter; for $\mu = 0$ we find $\gamma = 1$.

In order to calculate the mean abundance of strange particles we must introduce for each species its own fugacity which subsequently must be set equal to unity since all different strange particles are in mutual chemical equilibrium by assumption. This assumption is made as a consequence of the large strangeness exchange cross sections, in reactions such as

$$N + K \leftrightarrow Y + \pi \quad ; \qquad (3.21a)$$

here Y stands for a hyperon Λ, Σ. These are much larger then the strangeness production cross sections, such as

$$N + N \rightarrow N + \Lambda + K \qquad (3.21b)$$

or even

$$\pi + N \rightarrow \Sigma + K \qquad (3.21c)$$

when considered at moderate temperatures (energy threshold > 500 MeV). Hence in nuclear collisions the mutual chemical equilibrium, that is, a proper distribution of strangness among the strange hadrons, most likely will be achieved. By studying the relative yields we can exploit this fact and eliminate the absolute normalization, C, eq (3.18a) from our considerations. We recall that the value of C is uncertain for several reasons: (i) V is unknown, (ii) T^3 is strongly $T(t,r)$-dependent, and (iii) most importantly, the absolute normalization assumes chemical saturation which is not achieved owing to the shortness of the collision. Indeed we have (cf., eq (3.3))

$$\frac{dC}{dt} = A_H \left(1 - C(t)^2/C^2(\infty)\right) \quad , \qquad (3.22)$$

and the time constant $\tau_H = C(\infty)/A_H$ for strangness production in nuclear matter can be estimated to be 10^{-21} sec. [41]. The generation of strangeness is most likely driven by reaction (3.21c). Thus C does not reach $C(\infty)$ in plasma-less nuclear collisions. If the plasma state is formed, then the relevant $C > C(\infty)$. Details of the time dependence of the chemical composition of the hadronic gas are being studied [42].

We now compute the relative strangness abundances. Using eq (3.20) we find from eq (3.18) the grand canonical partition sum for zero average strangeness:

$$\ln Z_0^{strange} = C \{2W(x_K) [\gamma \lambda_K + \gamma^{-1} \lambda_{\bar{K}}] + 2W(x_\Lambda) [\gamma \lambda_B \lambda_\Lambda + \gamma^{-1}\lambda_B^{-1}\lambda_{\bar{\Lambda}}]$$

$$+ 6W(x_\Sigma) [\gamma \lambda_B \lambda_\Sigma + \gamma^{-1}\lambda_B^{-1}\lambda_{\bar{\Sigma}}]\} . \tag{3.23}$$

The strange particle multiplicities follow from ($i = K,\bar{K},\Lambda,\bar{\Lambda},\Sigma,\bar{\Sigma}$):

$$\langle n_i \rangle = \lambda_i \frac{\partial}{\partial \lambda_i} \ln Z_0^{strange}\Big|_{\lambda_i = 1} . \tag{3.24}$$

Explicitly we find

$$\langle n_{K^\pm} \rangle = C \gamma^{\mp 1} W(x_K) \tag{3.25a}$$

$$\langle n_{\Lambda/\bar{\Lambda}} \rangle = C \gamma^{\pm 1} W(x_\Lambda) e^{\pm\mu_B/T} \tag{3.25b}$$

and hence the ratio $\langle n_{K^+} \rangle/\langle n_{K^-} \rangle = \gamma^{-2}$. This is shown in figure 3.5 as function of the baryo-chemical potential μ_B for several temperatures.

We note that this particular particle ratio is a good measure of the baryon chemical potential in the hadronic gas phase, provided that the temperatures are approximately known. The mechanism for this process is: the strangeness exchange reaction (eq (3.21a)) tilts to the left (K^-) or the right (abundance $\gamma \sim K^+$) depending on the value of the baryo-chemical potential.

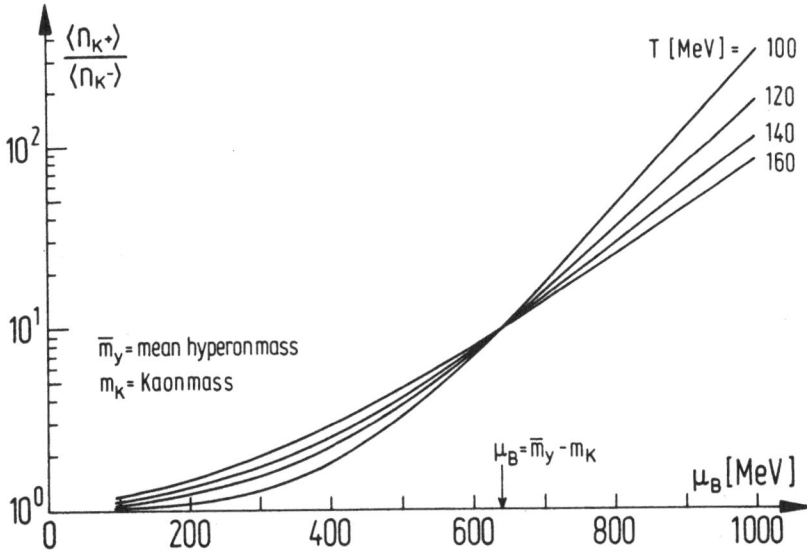

Fig. 3.5 The ratio $\langle n_{K+}\rangle\langle n_{K-}\rangle = \gamma^{-2}$ as a function of the baryo-chemical potential for several temperatures.

We turn our further interest to the rarest of all singly strange particles, and show in figure 3.6 the ratio $\langle n_{\bar{\Lambda}}\rangle/\langle n_{\Lambda}\rangle$. We notice an expected suppression of $\bar{\Lambda}$ due to the baryo-chemical potential as well as the strangeness chemistry. This ratio exhibits both a strong temperature and μ_B dependence. The remarkably small abundance of $\bar{\Lambda}$, e.g., $10^{-4}\Lambda$, under conditions likely to be reached in an experiment at the end of the hadronization phase (T ~ 120 - 180 MeV, μ_B ~ (4-6)T) is characteristic of the nuclear nature of the hot hadronic matter phase. Our estimates for the quark-gluon plasma based on flavor content are two to three orders of magnitude higher. One may observe that the formation of $\bar{\Lambda}$ in nuclear matter will probably be even much less than shown here since $\bar{\Lambda}$ will be much further away from the equilibrium abundance than Λ's. Hence the ratio of figure 3.6 may be viewed as an upper limit for the case of hot hadronic matter.

We have already shown that the strangeness abundance is chemically equilibrated in the quark-gluon plasma phase and indicated that this is not the case in the hadronic gas phase. We now further note that even assuming, probably much too optimistically, absolute chemical equilibrium in the gas phase, we find 3 to 5 times more strangeness in the plasma at comparable thermodynamic parameters, i.e., equal μ,T. This

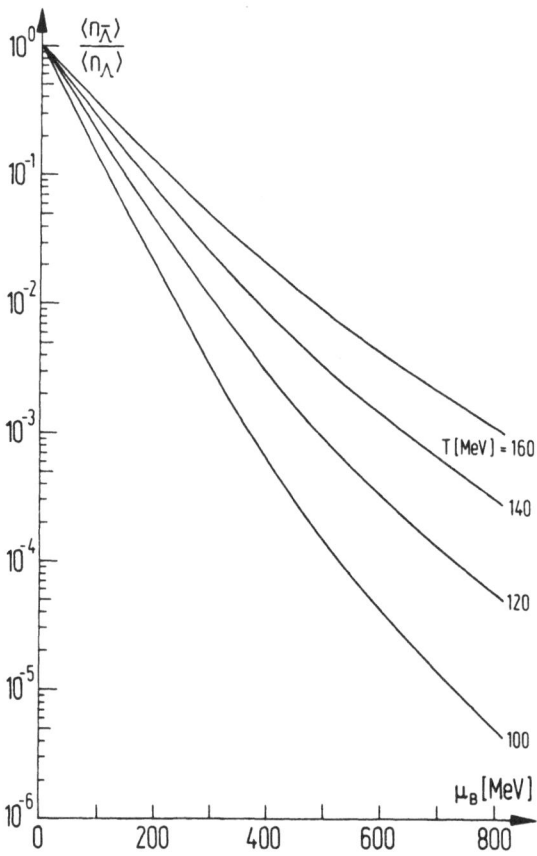

Fig. 3.6 The ratio $<n_{\bar{\Lambda}}>/<n_{\Lambda}>$ as a function of μ_B for several temperatures as an upper limit for $\bar{\Lambda}$ abundance in the hadronic gas
phase.

is shown in figure 3.7 as function of μ at some selected values of T
and m_s, where the conversion from μ as a variable to baryon density has
been done using perturbative QCD. Thus the simplest of all observations pointing to the quark-gluon plasma is the measurement of an
anomolously high strange particle abundance as function of CM energy in
the colliding nuclei, i.e., preferably at high $p_\perp$. Furthermore we have
argued that $\bar{\Lambda}$ are strongly suppressed in the nuclear gas. Thus an
anomalous yield of $\bar{\Lambda}$'s is an even more characteristic observable of the
plasma.

More speculative is the observation that strangeness may cluster
in the dilute plasma to form strangeness impurities such as ss, sss,
etc., owing to the attractive QCD-Coulomb interaction. Although the $s\bar{s}$
state would be more bound than $\bar{s}\bar{s}$ or ss state, its statistical weight

associated with the color selection factors is so much smaller
that we believe the latter state
to be the dominant strangeness
cluster. This coagulation of
strangeness can proceed even
further with $\bar{s}\bar{s}\bar{s}$ or sss states
(free Ω's) being formed. It is
very hard to make a quantitative
prediction of this process, but
clearly one must look out to
measure the abundances of such
rare baryons as $\bar{\Omega}$, possibly on
event by event basis. However
we observe that hadrochemical
calculations, i,e., those in hot
nuclear matter eq (3.18) also
lead to anomalous abundances of
multistrange baryons [42, 43]
quite similar to the results of
a recent hyperon beam experiment
[44].

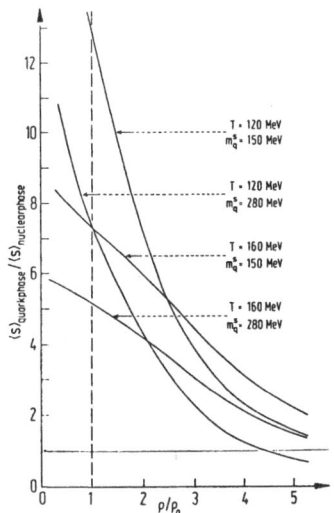

Fig. 3.7   Ratio of strangeness
along the transition line
between the plasma and
the hadronic gas phase as
a function of assumed
baryon density on the
plasma side.

In the above discussion, the rapid production of strangeness in
the plasma phase and its higher statistical abundance as compared with
the hadronic gas phase, are the central features of strangeness as a
characteristic observable for the quark-gluon plasma. We have
discussed the large plasma domains. Now we turn our attention to
"small" plasma droplets which may be either created in collisions of
light nuclei [45] or perhaps in antiproton annihilations on nuclear
targets [2h], [46]. Before turning to these phenomenological details
we first derive another effect that further enhances the role of
strangeness as a signature for plasma droplets.

As we have described at length in section 2, exact conservation of
quantum numbers, here of the total strangeness, greatly influences the
actual partition function. Rewriting eq (2.61) for strange quarks we
have the generating partition function

$$\ln Z_{q_s} = g_s V \int \frac{d^3p}{(2\pi)^3} \{\ln(1 + \exp[-\beta\sqrt{p^2+m^2} - \frac{\beta\mu}{3} + i\phi])$$

$$+ \ln(1 + \exp[-\beta\sqrt{p^2+m^2} + \frac{\beta\mu}{3} - i\phi])\} \tag{3.26}$$

where the statistical degeneracy of strange quarks is $g_s = 2_s \times 3_c = 6$. In eq (3.26) we have included the baryochemical potential associated with strange quarks. The angle ϕ is associated with the U(1) group and will allow to ensure exact strangeness conservation.

From eq (3.26) we can extract the partition function of given strangeness s by a simple integration

$$Z_{n_s}(T,V,\mu) = \int_0^{2\pi} \frac{d\phi}{(2\pi)} e^{-i\phi n_s} Z_{q_s}(T,V,\mu;\phi) \tag{3.27}$$

which in the Boltzmann limit can be carried out analytically. We have in this case

$$\ln Z_{q_s} \approx g_s V \int \frac{d^3p}{(2\pi)^3} \exp[-\beta\sqrt{p^2+m^2} - \frac{\beta\mu}{3}] e^{i\phi}$$

$$+ g_s V \int \frac{d^3p}{(2\pi)^3} \exp[-\beta\sqrt{p^2+m^2} + \frac{\beta\mu}{3}] e^{-i\phi}$$

$$= g_s \frac{VT^3}{2\pi^2} W(m/T) [e^{-\beta\mu/3} e^{i\phi} + e^{\beta\mu/3} e^{-i\phi}] \tag{3.28}$$

where as usual $W(x) = x^2 K_2(x)$.

We now recall the generating series of the modified Bessel functions $I_k(z)$

$$e^{z(t+1/t)/2} = \sum_{k=-\infty}^{+\infty} t^k I_k(z), \quad t \neq 0 . \tag{3.29}$$

We introduce the definitions

$$t = e^{-\beta\mu/3+i\phi} , \tag{3.30a}$$

$$z = 2Z^{(1)} = 2g_s \frac{VT^3}{2\pi^2} \left(\frac{m}{T}\right)^2 K_2\left(\frac{m}{T}\right) \quad , \tag{3.30b}$$

where $Z^{(1)}$ is the one-particle partition function. For the generating partition function in the Boltzman approximation we have hence

$$Z_{q_s}^s = \sum_{k=-\infty}^{+\infty} e^{-\beta\mu k/3} e^{ik\phi} I_k(2Z^{(1)}) \quad . \tag{3.31}$$

The integral (3.27) can now be carried out:

$$Z_{n_s}^s = e^{-\beta\mu n_s/3} I_{n_s}(2Z^{(1)}) \quad . \tag{3.32}$$

For finite n_s the chemical potential regulates the particle abundances in the conventional fashion. For $n_s = 0$ we have a very interesting special case [35]

$$Z_0(T,V) = I_0(2Z^{(1)}(T,V)) \tag{3.33}$$

$Z_0(T,V)$ is now μ-independent and describes an arbitrary number of $s\bar{s}$ pairs in the volume V and at temperature T, but with the number of s-quarks being $\underline{exactly}$ equal to that of $\bar{s}$ quarks. However, having used the Boltzman limit, we have constrained the validity of eq (3.33) to temperatures not much larger than the strange quark mass; only then is the expression (3.28) valid, $\underline{i.e.}$, the phase space sufficiently thinly populated allowing the neglect of quantum symmetry effects.

Next we notice that the argument in the I_0 function is the number of strange and antistrange quarks, computed as if we had neglected the influence of exact strangeness conservation. In the limit that this number is large we can employ the asymptotic expansion

$$I_0(z) \sim \frac{e^z}{\sqrt{2\pi z}} \left(1 + \frac{1}{8z} + \dots\right) \tag{3.34}$$

to recover the usual result

$$\ln Z_0(T,V) \xrightarrow[V\to\infty]{} 2Z^{(1)}(T,V) = Z_s^{(1)}(T,V) + Z_{\bar{s}}^{(1)}(T,V) \quad . \tag{3.35}$$

In the above equation we have the volume V as the only quantity that controls the argument of the partition function at fixed temperature.

The actual number of s-pairs present in the plasma is

$$\langle n_s \rangle = \lambda \frac{\partial}{\partial \lambda} \ln I_0(\lambda^{1/2} 2Z^{(1)})\Big|_{\lambda=1} = \frac{I_1(2Z^{(1)})}{I_0(2Z^{(1)})} Z^{(1)}(T,V) \quad (3.36)$$

where we note the appearence of a suppression factor

$$\eta(T,V) = \frac{I_1(2Z^{(1)})}{I_0(2Z^{(1)})} \ . \quad (3.37)$$

This factor is shown in figure 3.8 for T = 160 MeV and strange quark mass m_s = 160 MeV as a function of the volume. The volume is measured in terms of the elementary hadronic volume $V_h = \frac{4\pi}{3}$ (1 fm)^3. The important aspect of this result is that as the volume goes from 1/2 V_h to 3 V_h, η more than triples. Hence, we expect that in the plasma droplet strangeness would be enhanced by the relaxation of phase space constraints arising for small volumes from the fact that strangeness is generated in $s\bar{s}$ production which is the physical fact expressed by the quenching factor (3.37).

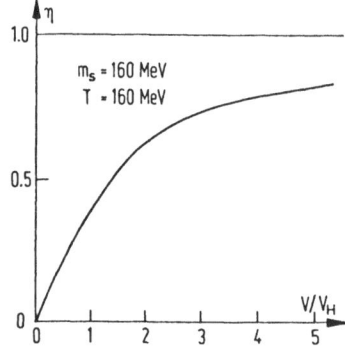

Fig. 3.8 Suppression factor for strangeness production as a function of the reaction volume .

This way we have two effects leading to a significant increase in the strangeness abundance even in small plasma droplets:

(a) The nonlinear volume effect: the abundance of strangeness is not only proportional to V through $Z^{(1)}$ eq (3.35), but in addition there is the disappearance of the phase space quenching factor η eq (3.37) for volumes exceeding V_H. Even for small droplets with V $\approx$ 2V_H this effect leads to an enhancement by a factor $\sim (V/V_h)^2$ = 4.

(b) The word "plasma" implies the equidistribution of the energy into the available degrees of freedom and a lifetime of the droplet of more than 10^{-23} sec. As the QCD cross sections indicate (see preceeding sections) this time almost allows the chemical equilibrium state to be reached. As discussed in section 7 this means that the strangeness abundance found would be larger by a factor of 2 to 4 than in the equilibrated hadronic gas phase. As the equilibration is also unlikely in the gas phase, we must also expect at least a factor 5 or so more strangeness from the plasma droplet than from the gas phase.

Taken together, both (a) and (b) imply that the strangeness originating from plasma droplets is more than ten times as abundant than that expected from the hadronic gas phase, making the appearance of a plasma easily visible as the plasma production threshold is passed in suitable experiments. Actually, this factor very likely will be more than 100 rather than 10, since in the hadronic gas phase the phase space of strange hadrons is not going to be saturated.

While no systematic experimental information is available as of now, we have found one piece of data which seems to confirm these considerations. Namely, instead of using high-energy nuclear collisions, one can employ antiproton annihilation in nuclei in order to produce a local plasma droplet [46]. We would like to argue that when slow, i.e., LEAR antiprotons penetrate into a nucleus the first step in the annihilation process will be the formation of a baryon-number zero fireball, filled with colored gluons and quark-antiquark pairs. As it turns out, this picture allows us to describe satisfactorily the π-multiplicities in annihilations where it is important to consider the conservation of isospin [47]. In $\bar{p}p$ reactions such a state would then break up into several mesons, a process that may last sufficiently long to allow the fireball to sometimes collide with one or more of the nearby nucleons in a nucleus. Very likely, this will lead to the dissolving of some as yet unspecified number of nucleons into the fireball. What do equations of state, cf. section 2, tell us about the physical properties of such a state? We have the energy density of a (relaxed) droplet, eq (2.6):

$$4B = E/V = (E/b)(b/V) \quad , \quad b = A - 1 \qquad (3.38)$$

where b is the baryon number of the droplet, i.e., one less than the total number of nucleons A that have reacted with the incoming antiproton. We also know the total energy E of the fireball,

$$E = (A + 1)m_N + E_{kin}^{\bar{p}} \tag{3.39}$$

and hence can solve eq (3.38) for the baryon density ν of the droplet:

$$\nu = b/V = 4B \frac{b}{E} = \frac{4B}{m_N} \frac{A-1}{A+1} \left(1 + \frac{1}{A+1} E_{kin}^{\bar{p}}/m_N\right)^{-1} . \tag{3.40}$$

At low energy the last factor containing the kinetic energy of $\bar{p}$ may be dropped. Figure 3.9 shows the "compression" ν/ν_0 as function of A, where ν_0 as usual is the normal nuclear density in heavy nuclei, $\nu_0 = .145/fm^3$.

We see that the baryon density of the droplet remains similar to that of the normal nuclear matter, i.e., we find no appreciable compression [46b].

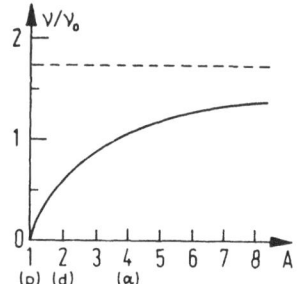

Fig. 3.9 Nuclear matter compression as a function of A.

Certainly, the quark droplet will disintegrate into A-1 baryons and several mesons. Our understanding of this process is still unsatisfactory but some properties of the emerging particles nonetheless can be estimated. First we observe that inside the low density droplet a temperature of the order of about 160 MeV (A = 1) prevails. This value is actually slightly A-dependent but will be about 140 MeV even for large A, owing to the low baryon density. Particles originating from the disintegration of the droplet will in the event-ensemble show a momentum distribution with characteristic slope ($\sim 1/T$). Parallel to this effect we should expect a significant enhancement of the strange particle abundance, as just described.

Apparently, an experiment to observe a plasma droplet would consist of using a strangeness trigger and measuring the momentum distribution of high-momentum so called spectator protons. Such an experiment has actually been carried out for the lightest nuclei. In figure 3.10 we show the results of the $\bar{p}$-d annihilation taken from reference [48]. Here the event rate as a function of the recoiling proton momentum is shown if the annihilation is accompanied by $K\bar{K}$ production.

Indeed we see a strong enhancement at proton momenta p > .3 GeV which nicely follows a T = 160 MeV slope. It is very interesting to note that this clear signal seems to disappear in the background when the $K\bar{K}$ trigger is not used (see fig. 2 of reference [49]). Another confirmation of this interpretation is obtained by considering the reaction

$$\bar{p}d \rightarrow \Lambda + X \ . \qquad (3.41)$$

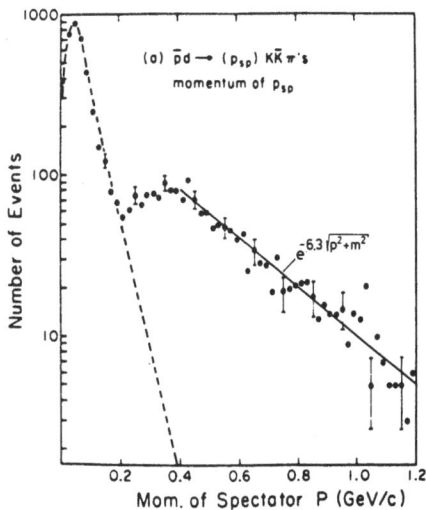

Fig. 3.10 Momentum distribution of spectator protons in coincidence with strangeness production.

The strangeness is now attached to the nucleon and the reaction is self-analyzing in the sense that the recoiling particle has the trigger quantum number. Indeed, in reference [50] Oh and Smith record that the Λ $p_\perp$ spectrum is identical to their p spectrum in the bump above $p_\perp$ > .3 GeV. Recent measurements of the reaction (3.41) [51] attempt also an alternative interpretation in terms of on-the-mass-shell $\bar{K}$-exchange which, however, seems to fall short of the data. Another experimental evidence against the $\bar{K}$-exchange mechanism is the anomalous enhancement of "K-d reactions" when the spectator momentum exceeds 200 MeV/c [52]. In favor of our present interpretation is the recent result that a hadronic gas with conserved charge and strangeness leads quantitatively to the observed momentum distribution and absolute normalization for the $\bar{p}$-d spectators at p = 200 MeV/c [53]. This indicates a very high degree of thermalization which would require an intermediate plasma state.

We can thus conclude that a first signal for the annihilation on two nucleons leading to a b = 1 plasma droplet may have been seen in the $\bar{p}$ - d reaction. It would be of great interest to see if a similar signal can be obtained, e.g., in $\bar{p}$ - α annihilations. Here in particular a 4π geometry would be of great help in order to select events in which all three remaining nucleons share the annihilation energy. A simultaneous enhancement of the s-yield would give a confirmation of the presented arguments.

4. THERMODYNAMICS OF THE INTERACTING HADRONIC GAS

The description of the hadronic gas can be very much simplified by the hypothesis that the hadron-hadron interaction is dominated by the hadron resonances [8]. In this case the hadronic gas phase is essentially a superposition of different hadronic gases and all the information about the interactions is hidden in the mass spectrum $\tau(m^2, b)$ which describes the number of hadrons of baryon number b in a mass interval dm^2 [9].

Let us first assume that the mass spectrum $\tau(m^2, b)$ is already known. Then, following the developments of Refs. [8], [9], the grand canonical level density σ is given by an invariant phase space integral. The extreme richness of the spectrum, $\tau(m^2, b) \sim \exp(m/T_0)$, enables us to neglect Fermi and Bose statistics above $T \approx 50$ MeV and to treat all particles as "Boltzmannions" in the external volume V_{ex}. We find for given $p_\mu = (E, \vec{p})$ and baryon number b ($\delta_K =$ Kronecker δ-function)

$$\sigma(p, V_{ex}, b) = \delta^4(p) \delta_K(b) + \sum_{N=1}^{\infty} \frac{1}{N!} \delta^4(p - \sum_{i=1}^{N} p_i)$$

$$\times \sum_{\{b_i\}} \delta_K(b - \sum_{i=1}^{N} b_i) \prod_{i=1}^{N} \frac{2\Delta_\mu p_i^\mu}{(2\pi)^3} \tau(p_i^2, b_i) d^4 p_i \quad . \quad (4.1)$$

Here, the first term corresponds to the vacuum state. The Nth term is the sum over all possible partitions of the total baryon number and the total momentum p among N Boltzmannions, each having an internal number of quantum states given by $\tau(p_i^2, b_i)$. These Boltzmannions are hadronic resonances of baryon number b_i ($-\infty < b_i < \infty$). Every resonance can move freely in the remaining volume Δ left over after subtracting the proper volumes V_c of all hadrons from the external volume V_{ex}:

$$\Delta^\mu = V_{ex} - \sum_{i=1}^{N} V_{c,i}^\mu \quad . \quad (4.2)$$

V^μ is a covariant generalization of V_i; in the rest frame $V_\mu = (V, 0)$.

The generalization (4.1) of the familiar phase space formula includes the following three essential features of the hadronic inter-actions:

a) The dominance of the particle scattering by the dense set of hadronic resonances via $\tau(m_i^2, b_i)$.

b) The proper natural volumes of hadronic resonances via Δ^μ.

c) The conservation of baryon number and the clustering of hadrons into lumps of matter with $|b| > 1$.

The thermodynamic properties of a hot hadronic gas follow from the study of the grand partition function $Z(\beta, V, \lambda)$, as obtained from the level density $\sigma(p, V, b)$:

$$Z(\beta, V, \lambda) = \sum_{b=-\infty}^{\infty} \lambda^b \int e^{-\beta \cdot p} \, \sigma(p, V, b) d^4 p \qquad (4.3)$$

Here the covariant generalization of thermodynamics with the inverse temperature four-vector β_μ has been used. In the rest frame of the relativistic baryon the chemical potential μ is defined by

$$\lambda = \exp(\mu/T) \quad ; \qquad (4.4)$$

it is introduced in order to conserve the baryon number in the statis-tical ensemble. All quantities of physical interest can be derived as usual by differentiating $\ell n Z$ with respect to its variables.

Equations (4.1)-(4.3) leave us with the task of finding the mass spectrum τ. Experimental knowledge of τ is limited to low excitations and/or low baryon number. Hagedorn [8] has introduced a theoretical model, "the statistical bootstrap," in order to obtain a mass spectrum consistent with direct and indirect experimental evidence. The quali-tative arguments leading to an integral equation for $\tau(m^2, b)$ are the following. When V_{ex} in eq (4.1) is just the proper volume V_c of a hadronic cluster then, up to a normalization factor σ in eq (4.1) is essentially the mass spectrum τ. Indeed, one cannot distinguish between a composite system as described by eq (4.1) compressed to the natural volume of a hadronic cluster and an "elementary" cluster having the same quantum numbers. Thus we demand

$$\sigma(p,V,b)\,|_{V=V_C} \equiv H\tau(p^2,b) \qquad (4.5)$$

where the "bootstrap constant" H is to be determined below. It is not sufficient simply to insert eq (4.5) into eq (4.1) to obtain the bootstrap equation for τ; more involved arguments are necessary [9b] in order to obtain the following "bootstrap equation" for the mass spectrum:

$$H\tau(p^2,b) = Hz_b\delta_0(p^2 - M_b^2) + \sum_{N=2}^{\infty} \frac{1}{N!} \int \delta^4(p - \sum_{i=1}^{N} p_i)$$

$$\times \sum_{\{b_i\}} \delta_K(b - \sum_{i=1}^{N} b_i) \prod_{i=1}^{N} H\tau(p_i^2,b_i)d^4p_i \quad . \qquad (4.6)$$

The first term is the lowest one-particle contribution to the mass spectrum, z_b is its statistical weight $(2I + 1)(2J + 1)$. The index "o" restricts the δ function to the positive root. Only terms with $b = 0, \pm 1$, corresponding to lowest energy $q\bar{q}$ (pion) and qqq (nucleon) states contribute in the first term of eq (4.6). All excitations are contained in the second term since arbitrary quark configurations can be achieved by combining $[(q\bar{q})^n (qqq)^m]$. The small influence of heavy flavors is ignored at this point but easily can be introduced.

In the course of deriving the bootstrap equation (4.6) it turns out that the cluster volume V_C grows proportionally to the invariant cluster mass [9],

$$V_C(p^2) = \sqrt{p^2}/(4B) \quad . \qquad (4.7)$$

The proportionality constant has been called 4B in order to establish a close relationship with the quark bag model [10]. The value of B can be derived from different considerations involving the true and perturbative QCD states. While the original MIT-bag fit has been $B^{1/4} = 145$ MeV, the (unweighted) average of different fits is today

$$B^{1/4} = 190 \text{ MeV}$$

$$B = 170 \text{ MeV/fm}^3 \qquad (4.8)$$

As far as the bootstrap is concerned the constant H and the bag constant B are free parameters. However, as just pointed out, B is determined from other considerations, while H turns out to be inversly proportional to B [9b]. Hence, if one wishes to believe the statistical bootstrap approach to the last detail there remains no free parameter in this approach. The implications of this for the transition, gas to plasma, will now be discussed.

Instead of solving eq (4.6), which leads to the exponential mass spectrum [8],

$$\tau(m^2,b) \sim e^{m/T_0} \tag{4.9}$$

we wish to concentrate here on the double integral, i.e., the Laplace transform of eq (4.6) which will be all we need to establish the physical properties of the hadronic gas phase. Introducing the transforms of the one-particle term, eq (4.6)

$$\phi(\beta,\lambda) = \sum_{b=-\infty}^{\infty} \lambda^b \, Hz_b \delta_0(p^2 - M_b^2)e^{-\beta \cdot d} \, d^4p \tag{4.10}$$

with pions and nucleons only

$$\phi(\beta,\lambda) = 2\pi HT \, [3m_\pi K_1(\frac{m_\pi}{T}) + 4(\lambda + \frac{1}{\lambda})m_N K_1(\frac{m_N}{T})] \tag{4.11}$$

(K_n is the modified Bessel function), and the mass spectrum:

$$\phi(\beta,\lambda) = \sum_{b=-\infty}^{\infty} \lambda^b \int H\tau(p^2,b)e^{-\beta \cdot d} \, d^4p \, , \tag{4.12}$$

we find for the entire eq (4.6) the simple relation

$$\phi(\beta,\lambda) = \varphi(\beta,\lambda) + \exp[\phi(\beta,\lambda)] - \phi(b,\lambda) - 1 \, . \tag{4.13}$$

To study the behavior of $\phi(\beta,\lambda)$ we make use of the apparent implicit dependence:

$$\phi(\beta,\lambda) = G(\varphi(\beta,\lambda)) \tag{4.14a}$$

with the function G being defined by eq (4.13)

$$\varphi = 2G + 1 - \exp G \quad . \tag{4.14b}$$

This function $G(\varphi)$ is shown in figure 2.1. As is apparent there is a maximal value φ_0

$$\varphi_0 = \ln(4/e) = 0.3863 \tag{4.14c}$$

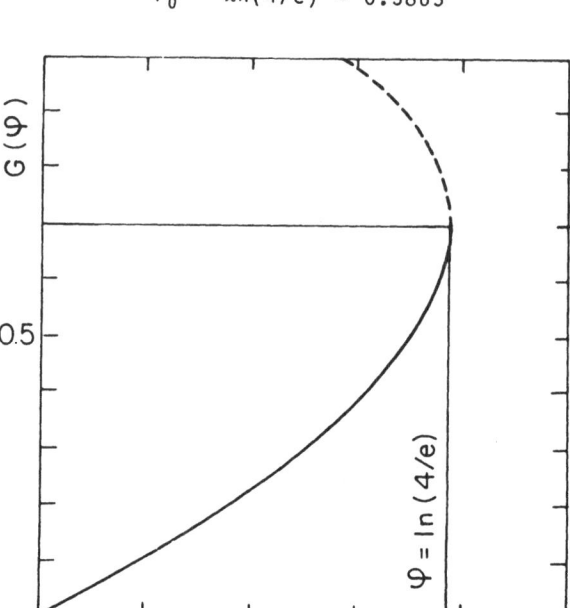

Fig. 4.1 Bootstrap function $G(\varphi)$. The dashed line represents the unphysical branch.

beyond which the function G has no real solutions. Recalling the physical meaning of G, eq (4.12, 4.14a) we conclude that eq (4.14c) establishes a boundary for the values of λ, i.e., μ, and T beyond which the hadronic gas phase cannot exist. This boundary is implicitly given by the relation (4.11):

$$\ln(4/e) = 2\pi HT_{cr}[3m_\pi K_1(\frac{m_\pi}{T_{cr}}) + 8m_N K_1(\frac{m_N}{T_{cr}}) \cosh(\frac{\mu_{cr}}{T_{cr}})] \tag{4.15}$$

shown in figure 4.2. The region denoted "Hadronic Gas Phase" is described by our current approach. With H correlated to B as given by eq (4.8) we find that

$$T_{cr}(\mu_{cr} = 0) = T_0 \sim 160\text{-}170 \text{ MeV} \quad . \tag{4.16}$$

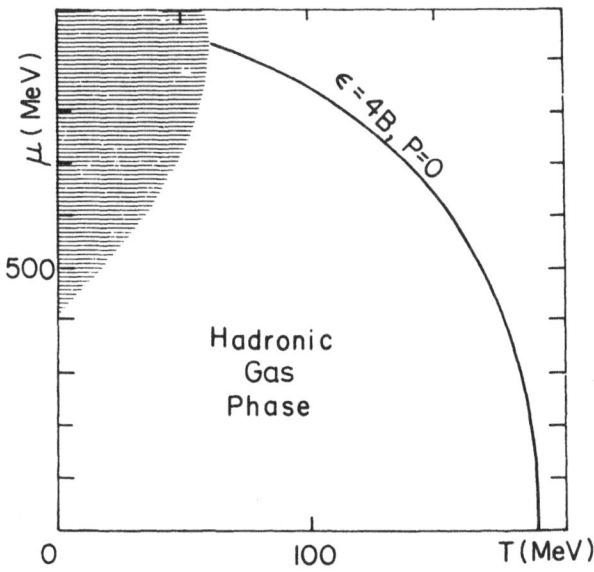

Fig. 4.2 Boundary of the "hadronic gas phase" in the bootstrap model.
In the shaded region quantum statistics cannot be neglected.

However, in view of the uncertainties involved it is more prudent to argue that the value $T_{cr} \sim 160\text{-}170$ MeV which is required in the description of hadronic reactions determines the value of the parameter H. Note that $\mu = 0$ implies zero baryon number of the plasma state. For $\mu_0 = \mu_{cr}(T_{cr} = 0)$ the solution of eq (4.15) is simply $\mu_{cr} \sim m_N$ since no quantum statistics effects have been included. Thus the dashed region in figure 4.2 "nuclear matter" must be excluded from our considerations. As we shall see shortly, the boundary to the hadronic gas phase is also characterized by a constant energy density $\varepsilon = 4B$.

Given the function $G(\varphi) = \phi(\beta,\lambda)$ we can in principle study the form of the hadronic mass spectrum. As it turns out we can obtain the partition function directly from ϕ. Namely, the formal similarity between eq (4.3) and eq (4.12) can be exploited to derive a relation between their integral transforms [9] (from here on: $\beta = \sqrt{\beta_\mu \beta^\mu}$)

$$\ln Z(\beta, V_{ex}, \lambda) = -\frac{2\Delta(V_{ex})}{H(2\pi)^3} \frac{\partial}{\partial\beta} \phi(\beta,\lambda) \tag{4.17}$$

where V_{ex} is the external volume, i.e., the volume not occupied by the hadrons. Equation (4.17) can also be written in a form which makes more explicit the different physical inputs:

$$\ell n Z(\beta, V_{ex}, \lambda) = \frac{\Delta(V_{ex})}{V_{ex}} \frac{\partial G(\varphi)}{\partial \varphi} Z_1(\beta, \lambda, V) \qquad (4.18)$$

In the absence of a finite hadronic volume and of the interactions described by the first two terms respectively, we would simply have an ideal Boltzman gas described by the one-particle partition function Z_1:

$$Z_1 = Z_{q\bar{q}} + 2\cosh(\mu/T) Z_{qqq} \qquad (4.19)$$

where

$$Z_{q\bar{q}/qqq} \sim (2I + 1)(2S + 1) \frac{VT^3}{2\pi^2} \left(\frac{m_{\pi/N}}{T}\right)^2 K_2\left(\frac{m_{\pi/N}}{T}\right) . \qquad (4.20)$$

The remainder of the discussion of the hadronic gas is an application of the rules of statistical thermodynamics. However, when working out the relevant physical consequences we must always remember that the fireball is an isolated physical system for which the statistical approach has been taken in view of the internal disorder (high number of available states) rather than because of a coupling to a heat bath. Let us first discuss the role of the available volume. As we have explicitly assumed, all hadrons have an internal energy density 4B (actually at finite pressure there is a small correction, see Ref. [4a] for details). Hence the total energy of the fireball E_F can be written as

$$E_F \equiv \varepsilon V_{ex} = 4B(V_{ex} - \Delta) \qquad (4.21)$$

where $V_{ex} - \Delta$ is the volume occupied by the hadrons. We thus find

$$\Delta = V_{ex} - E_F/4B = V_{ex}(1 - \varepsilon/(4B)) . \qquad (4.22)$$

By investigating the meaning of the thermodynamic averages it turns out that the apparent (β, λ) dependence of the available volume Δ in eq (4.22) must be disregarded when differentiating $\ell n Z$ with respect to β and λ. As eq (4.1) shows explicitly, the density of states of extended particles in V_{ex} is the same as that of point particles in Δ. Therefore also

$$\ell nZ(\beta, V_{ex}, \lambda) \equiv \ell nZ_{pt}(\beta, \Delta, \lambda) \quad . \tag{4.23}$$

We thus first calculate the point particle energy, and baryon number densities, pressure, and entropy density

$$\varepsilon_{pt} = -\frac{1}{\Delta} \frac{\partial}{\partial\beta} \ell nZ_{pt} = \frac{2}{H(2\pi)^3} \frac{\partial^2}{\partial\beta^2} \phi(\beta, \lambda) \tag{4.24}$$

$$\nu_{pt} = \frac{1}{\Delta} \lambda \frac{\partial}{\partial\lambda} \ell nZ_{pt} = -\frac{2}{H(2\pi)^3} \lambda \frac{\partial^2}{\partial\lambda\partial\beta} \phi(\beta, \lambda) \tag{4.25}$$

$$P_{pt} = \frac{T}{\Delta} \ell nZ_{pt} = -\frac{2T}{H(2\pi)^3} \frac{\partial}{\partial\beta} \phi(\beta, \lambda) \tag{4.26}$$

$$s_{pt} = \frac{1}{\Delta} \frac{\partial}{\partial T} (T\ell nZ_{pt}) = \frac{P_{pt}}{T} + \frac{\varepsilon_{pt} - \mu\nu_{pt}}{T} \quad . \tag{4.27}$$

From this, we easily find the energy density as

$$\varepsilon = \frac{\langle E \rangle}{V_{ex}} = -\frac{1}{V_{ex}} \frac{\partial}{\partial\beta} \ell nZ(\beta, V_{ex}, \lambda) = \frac{\Delta}{V_{ex}} \varepsilon_{pt} \quad . \tag{4.28}$$

Inserting eq (4.22) into eq (4.28) and solving for ε we find:

$$\varepsilon(\beta, \lambda) = \frac{\varepsilon_{pt}(\beta, \lambda)}{1 + \varepsilon_{pt}(\beta, \lambda)/4B} \quad . \tag{4.29}$$

Hence we can write eq (4.22) also in another form:

$$V_{ex} = \Delta (1 + \varepsilon_{pt}(\beta, \lambda)/4B) \quad . \tag{4.30}$$

Using eq (4.30) we find for the baryon density, pressure, and entropy density:

$$\nu = \frac{\nu_{pt}}{1 + \varepsilon_{pt}/4B} \tag{4.31}$$

$$p = \frac{P_{pt}}{1 + \epsilon_{pt}/4B} \tag{4.32}$$

$$s = \frac{s_{pt}}{1 + \epsilon_{pt}/4B} \; . \tag{4.33}$$

We now have a complete set of equations of state for the observable quantities as functions of the chemical potential μ, the temperature T, and the external volume V_{ex}. While these equations are semi-analytic, one has to evaluate the different quantities numerically owing to the implicit definition of $\phi(\beta,\lambda)$ that determines $\ell n Z$. However, when β,λ approach the critical curve, figure 2.2, we easily find from the singularity of ϕ that ϵ_{pt} diverges, and therefore

$$\epsilon \longrightarrow 4B$$
$$p \longrightarrow 0 \tag{4.34}$$
$$\Delta \longrightarrow 0$$

These limits indicate that at the critical line matter has lumped into one large cluster with the energy density 4B. No free volume is left and as only one cluster is present the pressure has vanished. However, the baryon density varies along the critical curve; it falls with increasing temperature. This is easily understood: as the temperature is increased more mesons are produced that take up some of the available space. Therefore hadronic matter then can saturate at lower baryon density. We further note here that in order to properly understand the apoproach to the phase boundary one has to incorporate and understand the properties of the hadronic world beyond the critical curve. Therefore we now turn to the study of the world of quarks and gluons and ultimately of the phase of matter consisting of these quanta.

5. FORMATION AND COOLING OF A BARYON RICH QUARK-GLUON PLASMA IN NUCLEAR COLLISIONS

Two extreme pictures of a high energy collision between two heavy nuclei suggest themselves:

(a) collision between two rather transparent bodies where the reaction products remain essentially in the projectile and the target reference frames respectively,

(b) collision between two rather absorbent bodies in which matter
piles up in the collision and where therefore the reaction
products appear in the central rapidity region.

Off hand picture (a) would seem to be the more reasonable one
considering the rather small high energy hadron-hadron cross sections.
This is the basis of a number of models purporting to describe high
energy nuclear collisions [11]. However, recent experimental evidence
from p-nucleus collisions and cosmic ray data indicate that case (b) is
a more frequent reaction channel for the formation of a quark-gluon
plasma. In particular, according to the analysis of Busza and
Goldhaber [54], the recent 100 GeV p-nucleus experiment of Barton
et al. [54b] indicate that the pp-data seriously underestimate the
extent to which heavy nuclei slow one another down. Instead of losing
one unit of rapidity in traversing the other nucleus, they find that a
heavy nucleus would lose perhaps 2.8 units of rapidity. Thus there
would be nothing left of the central baryon-free region. While this
substantial collective slowing effect is verified experimentally only
in 100 GeV lab energy collisions, the cosmic ray data indicate a
similar phenomenon at ultra-high energies [55]. We further recall the
recently observed rather narrow rapidity distributions at $\sqrt{s}$ = 540 GeV
from the CERN p$\bar{\text{p}}$ collider [56] which indicate hadronic non-transparency
at a level not anticipated before.

We conjecture here, that in order to create a large size high-
density region a quark-gluon plasma seed [12] must have been formed by
a statistical fluctuation. Thereafter the plasma can begin to grow by
capture of the trailing nucleons of the colliding nuclei. In such a
scenario, the densest plasma will result when the seed is formed early
in the collision, i.e., in the central rapidity region for symmetric
collisions (A_p = A_t). However, plasma production will occur according
to this mechanism with a non-negligible distribution towards projectile
and target rapidity limits. In events with an early plasma seed the
baryon number content of the plasma would be appreciable for large
nuclei, peaking in the central rapidity region.

In order to fulfill its role the above introduced seed must indeed
be a high particle density region similar to the quark-gluon plasma,
albeit small in size, with sufficiently thermalized momentum distributions and with some color deconfinement; however, chemical equilibrium
between different particle species, i.e., quark flavors, is not

required. In such a case the quark mean free path, λ, can become comparable to the seed size, R, and we can have $R/\lambda \gtrsim 1$. Occasional formation of such seeds is assured by inspection of actual numerical results obtained with relativistic cascade calculations [57]. We have good reason to believe not only in occasional, but perhaps even in relatively frequent, creation of such a seed, through an accidental local large fluctuation of particle density in a region of the size of a hadronic volume.

The energy influx to the plasma seed is controlled by the nuclear matter inflow. We consider here a) the kinematic conditions for the occurrence of the instability, seed → plasma; and b) the maximum achievable temperature in the most favorable case. For this purpose we do not need to consider the influence of the likely increase of the energy and particle density of the projectile or target in their rest frames arising from the entrance channel interactions. In order to err on the conservative side we compute as if all of the interacting region would instantly turn into the plasma state without compressions of nuclear degrees of freedom. Namely, if the formation of the seed is delayed, the increase of the densities would make the environment even more suitable for the occurrence of the plasma seed. However, the crucial condition to be respected follows from the observation that once the seed is there it can lead to a large-scale plasma state only if the energy loss of the seed is exceeded by its energy gain. Even below this "sharp" boundary defined as the instability without nuclear compression in target or projectile, occasional formation of plasma drops in the dense regions of compressed nuclear matter will occur. These precursor phenomena will smear out the kinematic limit, otherwise already spread out by fluctuations of the seed location, range of the impact parameters, etc. We believe that a detailed discussion of these effects is premature. Therefore we now determine the conditions which must be fulfilled for the ignition of a large-scale central plasma state.

While the plasma receives energy and baryon number by the nucleons impacting on it, it also inevitably loses energy by thermal radiation. Thus, in order to grow there must hold for the total plasma energy E,

$$\frac{dE}{dt} = \frac{dE^A}{dt} - \frac{dE^R}{dt} > 0 \qquad (5.1)$$

where dE^A/dt is the heating by the incoming nucleons absorbed in the seed, and dE^R/dt is the energy loss by thermal radiation. If dE/dt is negative the plasma will fizzle rather than grow. We now discuss the two terms, beginning with the gain term.

The energy influx into the plasma is controlled by the nuclear four-velocity, $u^\nu = \gamma(1,\vec{v})$; the plasma surface normal vector as seen from the CM-frame, $n^\mu = (0,\vec{n})$; the nuclear energy-momentum tensor, $T_{\mu\nu}$; and the probability for the absorption of an incoming nucleon by the plasma, a. Thus we have, with d^2A the surface element,

$$\frac{dE^A}{dt} = \int d^2A(- T_{\mu\nu} u^\mu n^\nu a) \ . \tag{5.2}$$

As is well known

$$T_{\mu\nu} = \varepsilon_0 u_\mu u_\nu \tag{5.3}$$

where ε_0 is the energy density in the rest frame of the projectile or target nucleus, respectively. Hence we have

$$T_{\mu\nu} u^\mu n^\nu = \rho_0 m \gamma \vec{n} \cdot \vec{v} \tag{5.4}$$

where ρ_0 is the equilibrium nuclear density, i.e., $\rho_0 = 1/6$ fm^{-3}. Furthermore, seen from the CM frame and expressed in terms of the projectile laboratory energy per nucleon, E_p, we have

$$v = \left(\frac{E_p - m}{E_p + m}\right)^{1/2} \tag{5.5a}$$

$$\gamma = \frac{(2E_p m + 2m^2)^{1/2}}{2m} \tag{5.5b}$$

The absorption coefficient a is assumed, as usual, to be

$$a(z) = 1 - e^{-z/\lambda} \tag{5.6}$$

where z is the thickness of the plasma region and λ is the absorption length of a hadron in the plasma. When weighted with $\vec{n} \cdot \vec{v}$ over the plasma surface this leads to

$$\bar{a}(R) = \frac{1}{2} \{1 + 2e^{-2R/\lambda} [\frac{\lambda}{2R} + (\frac{\lambda}{2R})^2] - 2(\frac{\lambda}{2R})^2\} . \qquad (5.7)$$

The overall factor 1/2 reflects the ratio between the surface of a circle with radius R and a half sphere, for $\lambda/R \to 0$. The absorption coefficient $\bar{a}(R)$ is indeed the average <u>absorption probability</u>. Through λ it depends on the particle density in the plasma, <u>i.e.</u>, temperature and baryon density. The final expression is, in detail,

$$\frac{d^3E^A}{d^2Adt} = \frac{1}{2} \rho_0 \left(\frac{E_p - m}{E_p + m}\right)^{1/2} (2E_p m + 2m^2)^{1/2}$$

$$\times \frac{1}{2} [1 + 2e^{-2R/\lambda} (\frac{\lambda}{2R} + (\frac{\lambda}{2R})^2) - 2(\frac{\lambda}{2R})^2] . \qquad (5.8)$$

We now turn to the description of the energy loss term of eq (5.1). In general, two mechanisms for the cooling of a plasma are possible, viz., adiabatic expansion and thermal radiation. At least in the beginning, <u>i.e.</u>, at the time of decision between ignition and fizzle, the expansion should play no role as the impacting nucleons provide an inertial confinement for the plasma. However, pion evaporation from the plasma is still possible, and the cooling associated with this process provides the energy loss of eq (5.1). Of course, some of the emitted pions will be returned to the plasma by the incoming nucleons. However, this return will be too late to have an impact on the question fizzle or grow: once the process has fizzled, <u>i.e.</u>, the plasma seed has hadronized, the collision is back to the hadron cascade regime. On the other hand, if plasma growth has taken place the returning pions will of course return their evaporation energy to the plasma and contribute to the ultimate energy density of the plasma. Also, the influence of the plasma expansion has to be reconsidered then.

We now develop a quantitative model [14] suitable for surface temperatures of 150 - 220 MeV and moderate baryon densities, such that the particle density is less than $\sim$ 10 particles/fm^3. Under these circumstances surface collisions involving more than one particle per fm^2 are rare. Hence we can limit ourselves to consider sequential one-particle events. In such instance, the emission of pions from a large and highly excited quark bag is described by bag models incorporating the chiral symmetry [58]. In such a model the pions are supposed to interact linearly with the pseudoscalar quark density at the bag surface. This is described by the Lagrangian

$$L_{q\pi} = \frac{i}{2f} \, \bar{q} \, \gamma_s \, \tau \cdot \varphi_\pi \, q \, \Delta_s \qquad (5.9)$$

where Δ_s is the surface δ-function and f is the pion decay constant (f = 93 MeV). Equation (5.9) describes the following processes:

    (a)   a quark or antiquark hits the plasma surface and emits a bremsstrahlung pion while being reflected back

    (b)   a quark-antiquark pair hits the surface and converts into a pion.

As the pion emission by plasma surface is a direct process the resulting pion spoectrum intensity is non-thermal, while the spectral form is determined by the thermal quark spectra [59]. Consequently, the surface pion radiance can substantially exceed the black body limit.

This treatment, however, must be viewed as being semiphenomenological, as the true approach would have to be based on QCD and would at least require the proper understanding of the pion formation. An attempt at this is terms of color flux tubes was made by Glendening et al. [60]. However, the strong quark binding in the pion in the bag model is difficult to account for, and, also, the role of the pion as the Goldstone meson of chiral symmetry is ignored in such a treatment. In contrast, these features are emphazied in the form (5.9) of the interaction.

For the purpose of estimating the importance of the pion radiation process we consider a model based on kinematics only: in order for the surface collision to lead to pion emission the particle momentum normal to the surface must exceed a certain threshold. In particular, this momemtum has to be larger than the normal momentum of the emitted pion. We take this threshold momentum to be of the order of 1/4 GeV/c for quarks leading to pions; our results are quite insensitive to the precise choice, as well as to the actual shape, of the threshold function θ describing the probability of pion emission. Hence we will use:

$$\theta(p) = \begin{cases} 1, & p_\perp \gtrsim p_M \sim 1/4 \text{ GeV} \\[2mm] 0, & 0 < p_\perp \lesssim p_M \end{cases} \qquad (5.10)$$

We note that the average energy of the practically massless quarks is about 3T $\sim$ 500-650 MeV and that the particle densities peak at $\sim$ 2T.

Hence almost half of all quarks and antiquarks can participate in the radiation cooling. We also include the pion bremsstrahlung by gluons impinging on the surface.

The energy per unit surface and unit time that leaves the quark-gluon plasma is now simply given by

$$\frac{d^3E}{d^2Adt} = g\int \frac{d^3p}{(2\pi)^3} \rho(p)f(E) \; E(p)\theta(p) \; \frac{d^3V}{d^2Adt} \tag{5.11}$$

where g are the degeneracies. As only light quarks lead to the dominant pion channel we have $g_q = 3_c \times 2_s \times 2_f = 12$, and $g_G = 2_s \times 8_c = 16$. Here $\rho(p)$ is the phase space density of colored particles,

$$\rho(p) = g_q \{[\exp((p - \mu_q)/T) + 1]^{-1} + [\exp((p + \mu_q)/T) + 1]^{-1}\}$$

$$+ g_G[\exp(p/T) - 1]^{-1} \quad . \tag{5.12}$$

The differential in (5.11) is simply the normal velocity of particles impinging on the plasma surface

$$\frac{d^3V}{d^2Adt} = \frac{d^2Adz}{d^2Adt} = \frac{dz}{dt} = v_\perp = \frac{p_\perp}{E(p)} = \frac{p_\perp}{(p_\perp^2 + p_\parallel^2)^{1/2}} \tag{5.13}$$

Since the energy leaving the plasma region is not the total energy contained in the leading particle we have in (5.11) included the efficiency factor f. In the present case only one pair is created to form the emitted pion. A naive degree-of-freedom counting leads to $f \approx 2/3$. f probably approaches unity for very high energy leading particles. We disregard the energy dependence of f; choosing the value $\bar{f} = 2/3$ we obtain a lower limit on the energy transfer.

In view of the qualitative nature of our model it is sufficient to expand in eq (5.12) the quantum distributions and to retain only the Boltzmann term for the $q,\bar{q},G$ distributions:

$$\rho(p) \approx (g_q \; \eta(3) \; 2\cosh(\mu/T) + g_G \; \xi(3)) \; e^{-\sqrt{p_\parallel^2 + p_\perp^2}/T}$$

$$\equiv \frac{8}{3} \; g \; e^{-\sqrt{p_\parallel^2 + p_\perp^2}/T} \quad , \tag{5.14}$$

where we have corrected the counting of the Bose and Fermi degrees of freedom by indlucing the phase space integral weights $\eta(3) \approx 0.9$ and $\zeta(3) \approx 1.2$ in the above. Finally, we must still account for the requirement that the color and spin degrees of freedom of the emitting particles, i.e., the quarks or the gluons, must be coupled to the quantum number of the emitted pions. This introduces a factor which is 3/8 for both cases. We already have included this factor in the definition of g; hence the factor 8/3 in (5.14). Collecting all factors we see that the effective number of Boltzmann degrees of freedom of quarks and antiquarks at $\mu_q = T$ is 12.5 while that of gluons is 7.5. At $\mu_q = 0$ the number of quark degrees of freedom (22) is about that of gluons. Thus g varies between 16 and 21 as function of μ_q.

Combining eqs (5.11) and (5.13) with eq (5.14) we obtain the generalized Stefan-Boltzmann law:

$$\frac{d^3E}{d^2Adt} = \bar{f}g \int_{P_M}^{\infty} \frac{dp_\perp}{(2\pi)} p_\perp \int_0^{\infty} \frac{p_\parallel dp_\parallel}{(2\pi)} e^{-\sqrt{p_\parallel^2 + p_\perp^2}/T}$$

$$= \bar{f} \frac{g}{2\pi^2} T^4 \, 3e^{-P_M/T} \left(\frac{1}{3} \left(\frac{P_M}{T}\right)^2 + \left(\frac{P_M}{T}\right) + 1 \right) . \qquad (5.15)$$

In figure 5.1 we show the cooling rate calculated from eq (5.15) as a function of the surface temperature T, choosing $\mu_q/T = 1$. For $\mu_q \approx 0$ the values are lower by about 20%. Our current values for the radiance of the plasma about half of those given by us earlier in Ref. [12] where the pion radiation by gluons and the coupling to the pion quantum numbers had not yet been included. From figure 5.1 we further see that the precise value of P_M, or, said differently, the precise form of the threshold function θ, eq (5.10), does not matter. However, we note here that our estimate may be uncertain by

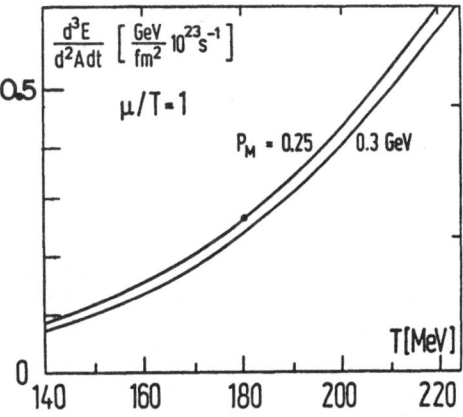

Fig. 5.1 Pion radiation surface brightness as function of temperature.

perhaps a factor 2 considering the qualitative nature of our considerations. We see that indeed the precise value of P_M, or, said differently, the precise form of the threshold function θ, eq (5.10), does not matter. Concerning the choice of μ_q we note that even though equilibrium baryon density does not very likely prevail, the best choice for μ_q in the initial stages of the plasma formation would be $\mu_q \lesssim M_N/3 \approx 2T$, consistent with the non-degeneracy assumption for $T \sim \frac{1}{6} M_N$. As local thermalization occurs, μ_q diminishes and approaches T.

Before returning to the ignition condition given by the inequality (5.1) we discuss our result in terms of a numerical example chosen to represent a typical case of a quark-gluon plasma. Our example is a spherical plasma droplet of $R = 4$ fm, a surface temperature of $T = 180$ MeV, and $\mu/T = 1$. The energy density then is 2.1 GeV/fm^3 according to eq (2.34) and recalling that strange quarks compensate for a large part of the interaction which is of order $0(\alpha_s)$. The baryon density is according to eq (2.34), $\sim .5/$fm^3, i.e., about 3 ρ_0. The baryon number exceeds 150 if T is larger in the interior. Since 0.7 GeV/fm^3 is needed for creation of the final baryons implied by the assumed value of μ, the available energy density is about 1.4 GeV/fm^3 and the total available energy is ca. 400 GeV. For this example we find for the rate of energy loss through the surface A

$$\frac{dE}{dt} = A \bar{f} \, 0.25 \, \frac{GeV}{fm^2} \frac{c}{fm} = A \, 0.5 \, \frac{GeV}{fm^2} \, 10^{+23} \, sec^{-1} \; .$$

We note that this confirms the assumption of a sequential individual-particle process: when one particle of 0.430 MeV impinges on a surface area of 1 fm^2 the next particle following it with light velocity would be behind by a distance of about 1 fm (i.e., several mean free paths). On the other hand, this indeed is a very large energy loss rate. In our example, the energy loss in the first 10^{-23} sec is ($A = 200$ fm^2)

$$\Delta t \, \frac{dE}{dt} = 120 \, GeV \quad ,$$

which represents a substantial fraction of the total available energy of about 400 GeV. Clearly the smaller the plasma droplet, the more relevant becomes the radiation loss for the lifetime of the plasma. As the available excitation energy scales with R^3 and the radiation loss with R, a small plasma droplet of $b \sim 18$, $R \sim 2$ fm and available energy

40 GeV radiates 30 GeV in the first 10^{-23} sec. Hence we are led to urge that experiments involving very heavy nuclei be performed to allow for the creation of sufficiently longlived (_i.e._, large) plasma regions.

We now return to the discussion of the ignition conditions: we set $\frac{dE}{dt} = 0$ in eq (5.1). In figure 5.2 we show the minimum size a plasma seed must have in order for it to grow, _i.e._, the minimum seed size for plasma ignition, as a function of projectile energy for a selection of plasma ignition temperatures, T_I, computed taking $\mu/T = 2$. In the initial stages of the nuclear colli- sion this is the more likely nature of the parameters. q dominate $\bar{q}$ and we err on the conservative side by enhancing the radiation losses by that choice. While at density of 2 GeV/fm^3 in the plasma the particle density is about 4/fm^3 leading to $\lambda \sim \frac{1}{3} - \frac{1}{2}$ fm, we anticipate that in the initial stages of the collision we have a particle density of about 1/fm^3 and hence $\lambda \sim 1 - 1.5$ fm.

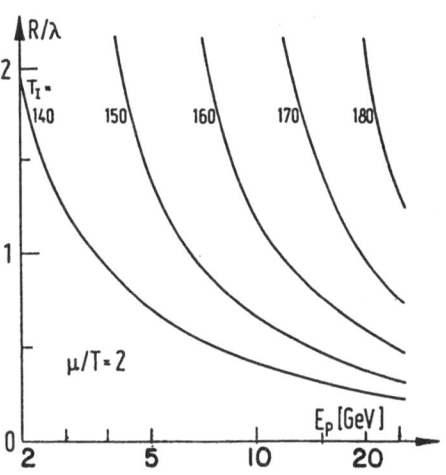

Fig. 5.2   Minimum size of a plasma seed as function of beam laboratory energy for different radiation temperatures.

For $R/\lambda \sim 1$ we notice that at $T_I \sim 150 - 160$ MeV beam energies of 10 to 20 GeV/nucleon should suffice to lead to plasma ignition with $R \sim \lambda$. We note that the seed size considered is of the order of the nucleon size. We note that the obtained lower limits for the heavy ion kinetic energy is above the kinematic limit obtained neglecting the loss term in eq (5.1) and requiring an ignition termperature of 160 MeV. On the other hand, it seems rather unlikely that ignition can be achieved at much lower beam energies if the phase transition is of first order. Thus below our limit the collision will fizzle and we just achieve a superheated nucleon gas.

Once the plasma has ignited the temperature of the plasma will grow until the nuclear collision terminates or until the temperature has risen to a level at which the pion radiation overwhelms the energy influx. At this point one must re-examine the question of the cooling mechanism, i.e., first, whether the evaporation of pions will lead to a cooling off of the surface and hence to a shut-off of the evaporation process, or whether the plasma heat conductivity is sufficiently large to maintain a surface temperature high enough for pion radiation to continue; second, whether other processes, principally expansion, con-tribute substantially to the cooling process.

We begin by considering the heat conductivity. Since the plasma consists of rather free particles the naive expectation is that a sufficently high conductivity obtains. Indeed, the basic relation between the heat flow $\vec{Q}$ and the energy density ε is

$$\vec{Q} = \ell \, \vec{\nabla} \, \varepsilon(T; \mu_q/T) \qquad (5.16)$$

where ℓ is the mean free path. Assuming that only a radial gradient of T exists, with $\mu_q/T \sim$ const over the volume, the radiation equilibrium requires

$$\frac{d^3 E}{d^2 A dt} = Q_r = \ell \, \frac{\partial T}{\partial r} \frac{\partial \varepsilon}{\partial T} = \ell \, \frac{1}{T} \frac{\partial T}{\partial r} 4\varepsilon \; . \qquad (5.17)$$

In our numerical example the required temperature gradient at the surface is, with ℓ in the range 1/2 - 1/3 fm:

$$\frac{\partial T}{\partial r} = \frac{T}{\ell} \, \frac{0.215 \text{ GeV/fm}^3}{4 \times 2.1 \text{ GeV/fm}^3} = (5 - 8) \, \frac{\text{MeV}}{\text{fm}} \quad .$$

It appears that this temperature gradient is just within sensible bounds, leading for a plasma radius of 4 fm to a temperature differen-tial between the origin and the surface of $\sim$ 15-20 MeV. We further note that unlike in non-relativistic gases, the mean free path ℓ here is inversely proportional to $\partial\varepsilon/\partial T$ since it is inversely proportional to the particle density. For $\mu_q/T < 2$ the energy per particle in the plasma is just 3T and hence the particle density $\rho = \varepsilon/3T$. Therefore the necessary temperature gradient, eq (5.17), turns out to be

$$\frac{\partial T}{\partial r} = \frac{d^3 E}{d^2 A dt} \frac{1}{20} \, \bar{\sigma} \qquad (5.18)$$

where $\bar{\sigma}$ is the average particle-particle cross section. The range of values given above for $\frac{\partial T}{\partial r}$ corresponds to $\bar{\sigma} \sim \frac{1}{3}$ to $\frac{1}{2}$ fm^2.

We now turn to the discussion of the adiabatic expansion of the plasma. To begin with one must recognize that in contrast to the above discussed pion radiation process the expansion requires a collective flow, i.e., a flow in which a hydrodynamic velocity is superimposed over the random thermal motion of all the quarks and gluons. Therefore the relevant time constant is given by the speed of sound and thus is about three times larger than the radiation time constant. Furthermore, the expansion is driven by the excess of the internal pressure over that exerted on the surface by the physical vacuum. Now, the effect of the internal pressure on the surface is reduced by the pion radiation. The point is that those particles which penetrate the surface do not exert their full force on the surface. We now demonstrate that they are responsible for a substantial fraction of the internal surface pressure. Balancing the momenta at the surface we find that instead of $2p_\perp$ the momentum transferred to the surface is

$$\Delta p = \begin{cases} 2p_\perp & : \; p_\perp < p_M \\ 2p_\perp(1 - f) & : \; p_\perp > p_M \end{cases} \tag{5.19}$$

where f is the fraction of the normal momentum carried away by the emitted pion. We now recompute the effective pressure on the plasma surface:

$$\bar{P}_q = \bar{g}_q \left[\int_0^{p_M} \frac{dp_\perp}{(2\pi)} \, 2p_\perp v_\perp \frac{p_\parallel dp_\parallel}{(2\pi)^2} \, \rho(p) \right.$$

$$\left. + (1 - f) \int_{p_M}^{\infty} \frac{dp_\perp}{(2\pi)} \, 2p_\perp v_\perp \int_0^{\infty} \frac{p_\parallel dp_\parallel}{(2\pi)^2} \, \rho(p) \right] \tag{5.20}$$

where we have used eq (5.13). Also, $\bar{g}_q$ is the effective number of degrees of freedom for the quarks as devined in eq (5.14). We notice that the effective quark pressure $\bar{P}_q$ is equal to the expected quark pressure $P_q = 1/3 \; \varepsilon_q$, reduced by the contribution of high normal momentum particles, weighted by the factor f:

$$\bar{P}_q = P_q - fg \int_{p_M}^{\infty} \frac{dp_\perp}{(2\pi)} \, 2p_\perp v_\perp \int_0^{\infty} \frac{p_\parallel dp_\parallel}{(2\pi)^2} \, \rho(p) \quad . \tag{5.21}$$

The important point to realize is that the contributions of particles with $p_\perp > p_M$ to the particle pressure P_q are dominant. To see this we evaluate, in obvious notation,

$$\frac{P_q(p_\perp > p_M)}{P_q} \equiv \frac{\int_{p_M}^{\infty} dp_\perp p_\perp^2 \int_0^{\infty} \frac{p_\parallel dp_\parallel}{\sqrt{p_\perp^2 + p_\perp^2}} \rho(p)}{\int_0^{\infty} dp_\perp p_\perp^2 \int_0^{\infty} \frac{p_\parallel dp_\parallel}{\sqrt{p_\perp^2 + p_\parallel^2}} \rho(p)} = \frac{\int_{p_M}^{\infty} dp_\perp p_\perp^2 e^{-p_\perp/T}}{\int_0^{\infty} dp_\perp p_\perp^2 e^{-p_\perp/T}}$$

$$= e^{-p_M/T} \left(\frac{1}{2} \left(\frac{p_M}{T}\right)^2 + \left(\frac{p_M}{T}\right) + 1\right) \tag{5.22}$$

This is a monotonically falling function of p_M/T; for $p_M/T \sim 1 - 1.5$ we find that the ratio eq (5.21) varies between .92 and .81. Hence, inserting eq (5.22) into eq (5.21) we find for $f \sim 2/3$

$$\bar{P}_q = P_q \left(1 - f \frac{P_q(p_\perp > p_M)}{P_q}\right) \cong 0.4 \, P_q \quad . \tag{5.23}$$

A similar calculation can be carried out for the gluons with a similar outcome for the reduction at the pressure. The overall result is that only about half of the internal pressure acts on the surface. Thus, in effect, the time constant relevant for the cooling process through expansions is extended by a factor of almost two. Thus we are led to the conclusion that the expansion contributes only about 10-20% to the cooling of the plasma. Even though this effect is somewhat reduced for a baryonless plasma, i.e., $\mu = 0$, it still relieves 1/3 of the total pressure.

The physical distinction between the cooling of the plasma by pion radiation vs. by expansion resides in that the former leads to a reduction of the plasma temperature without a significant increase of the plasma volume. This, of course, has important consequences in the dynamics of the plasma development, and, in particular, eventually on the observable quantities. In particular, cooling by radiation seems to convert the internal energy more efficiently into pions than the expansion mechanism. In an expansion this energy is converted into collective motion and is manifested in the form of additional kinetic energy of the produced particles. Hence in the radiation cooling the available entropy is used to create more new particles, i.e., pions,

while in the adiabatic expansion it is essentially contained in the
kinetic motion. In both instances cooling is approximately adiabatic.

We next discuss the maximally obtainable plasma temperature,
neglecting the effect of the cooling by expansion. As already
remarked, once the plasma has ignited a fraction of the radiated pions
will be swept along by the incoming nucleons and re-enter the plasma.
This process introduces a dependence of the loss term on the beam
characteristics. Even though this turn-around of the pions does not
change the ignition conditions it influences the maximal achievable
plasma energy density. Since the thermal radiation is isotropic the
returned fraction, η, will be of the order $\eta \lesssim 1/2$. To obtain an estimate of this maximum plasma energy density one has to multiply the
energy radiation term, eq (5.15), with $(1 - \eta)$ and balance it with the
unmodified gain term, eq (5.8). We recall that in the derivation of
eq (5.15) a non-degenerate quark-gas has been assumed, and μ_q/T is
expected to be less than 2. As the collision process continues the
temperature of the plasma will grow until the nuclear collision terminates or until the temperature has risen to a level at which the pion
radiation overwhelms the energy influx. This maximum achievable
temperature is shown in figure 5.3 for a few choices of the pion turn-
around coefficient η, as a function of projectile beam energy. In view
of high plasma density here we have used $R/\lambda = 5$, $\mu/T = 1$. As one can
see the maximal temperature achievable in the collision does not depend

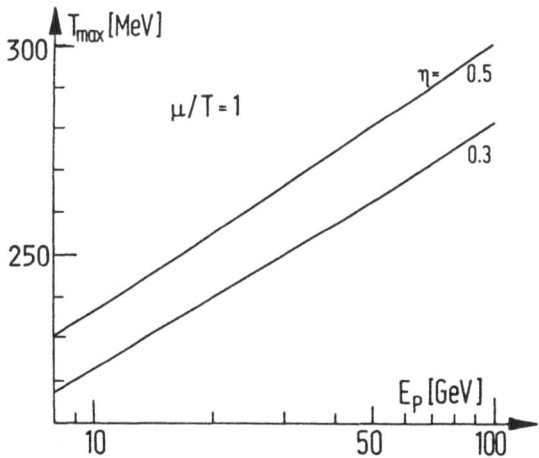

Fig. 5.3 Maximum achievable plasma temperature as function of beam
laboratory energy for two values of the pion turn-around
coefficient.

too sensitively on the choice of the parameters and reaches for 50 GeV
a value around 230 MeV. Hence, once a plasma has ignited one can
expect that a full-fledged quark-gluon plasma event will take place,
with energy density reaching 4-5 GeV/fm^3. However, we note that underlying this scenario is the requirement that the collisions take place
between two quite heavy nuclei.

After the end of the build-up phase, i.e., at the termination of
the nuclear collision, the dynamics is governed by a collaboration of
pion radiation and hydrodynamic expansion of the plasma. At this point
one must ask whether the density of the radiated pions is large enough
for them to undergo multiple scattering, so that a pion gas cloud could
be formed which would exert a back-pressure on the radiated pions, and
thus could slow down the radiative energy loss of the plasma, and also
the expansion.

Considering that here we deal with hadronic (rather than QCD)
cross sections and moderate particle densities one should think that
the effect of the surrounding pion gas on the radiation should not be
too large.

In order to illuminate this question, consider the case when the
emitted pions would form a density ρ surrounding the plasma droplet of
the form

$$\rho = \rho_0 \left(\frac{R}{r}\right)^2 . \tag{5.24}$$

Let us consider that a given pion travels through a gas having the
density distribution (5.24). In that case the scattering probability
is given by (j is the radial current of the considered pion)

$$\frac{1}{r^2} \frac{d}{dr} (jr^2) = -j\sigma\rho = -j\sigma\rho_0 \left(\frac{R}{r}\right)^2 , \tag{5.25}$$

and hence we have

$$j = \frac{j_0 R^2}{r^2} e^{(\sigma\rho_0 R)((R/r) - 1)} . \tag{5.26}$$

For $\sigma\rho_0 R \ll 1$ we find the unperturbed pion current which behaves like r^{-2}. The exponential describes the scattering in the gas. Taking for a numerical example $R \approx 4$ fm, $\rho_0 \simeq 1$ fm^{-3}, and, considering that the pion-pion scattering peaks at the ρ-meson mass, which is several linewidths above a typical c.m. pion-pion energy, assuming $\sigma \simeq 0.2 - 0.5$ fm^2, we find considering the value of $j(r = \infty)$ which represents the unscattered part of the beam that a pion will scatter one or two times on the way out to infinity.

We now turn our attention to the nature of the transformation of the quark-gluon plasma into individual hadrons. We recall here that we have developed two inherently different descriptions which neverthe-less leads to the prediction of a qualitatively similar thermodynamic region for the transition between both phases of hadronic matter. As we shall see in a moment the physics which went into these theoretical approaches requires that this is a first order phase transition. How-ever, of course, we cannot actually deduce the order of the transition in the presented considerations. We record here that recent Monte-Carlo simulations on a lattice show phase coexistence in SU(3) gauge theories which is characteristic of first order phase transitions [6a], [41]. This is contrary to results found in SU(2) simulations [42].

Consider the P-V diagram shown in figure 5.4. Here we distinguish three domains. The hadronic gas region is simply a Boltzmann gas where the pressure raises with reduction of the volume. When the internal excitation rises, the individual hadrons begin to cluster. This reduces the increase in the Boltzmann pressure since a smaller number of particles exercises a smaller pressure. In a complete description of the different phases we have to allow for a coexistence of hadrons with the plasma state in the sense that the internal degrees of freedom of each cluster, i.e., quarks and gluons contribute to the total pres-sure even before the dissolution of individual hadrons. This indeed becomes necessary when the clustering overtakes the compressive effects and the hadronic gas pressure falls to zero as V reaches the proper volume of hadronic matter. At this point the pressure rises again very quickly, since in absence of individual hadrons we now compress only the hadronic constituents. By performing the Maxwell construction between volumes V_1 and V_2 as indicated in figure 5.4 we can in part account for the complex process of hadronic compressibility alluded to above. We find this way the most likely path taken by the compressed hadronic gas in a nuclear collision. This discussion shows that in our approach we are straightforwardly led to a first order phase transi-tion, as first conjectured in ref. [2g,h].

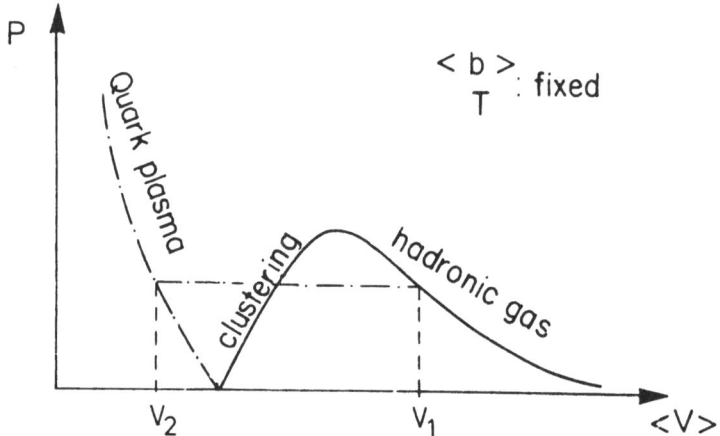

Fig. 5.4 P-V diagram for the gas-plasma first order transition.

It is interesting to follow the path taken by an isolated quark-gluon plasma fireball in the μ-T plane, or equivalently in the ν-T plane. Several cases are depicted in figure 5.5. In the Big Bang expansion the cooling shown by the dashed line occurs in a universe in which most of the energy is in the radiation. Hence, we have for the chemical potential μ << T. Similarly, the baryon density ν is quite small. In normal stellar collapse leading to cold neutron stars we follow the dash-dotted line parallel to the μ-resp. ν-axis. The compression is accompanied by little heating.

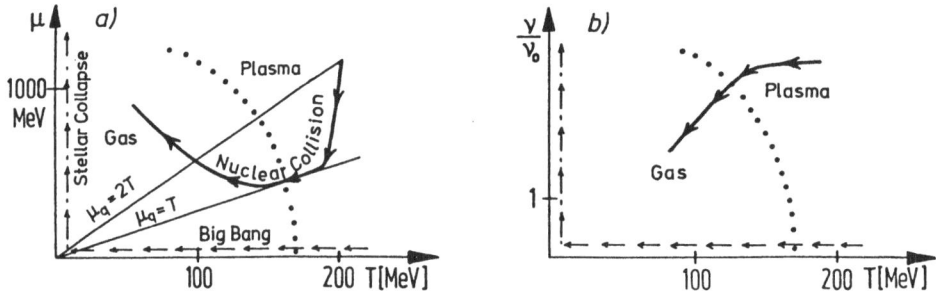

Fig. 5.5 Paths taken in the (a) μ-T plane and (b) ν-T plane by different physical events.

In contrast, in nuclear collisions almost the entire μ-T and ν-T can be explored by varying the parameters of the colliding nuclei. As we have already argued the most easily accessible region corresponds to $\mu_q/T \lesssim 2$. To appreciate this further consider the baryon density, eq (2.35). Call $\mu_q/\pi T = \delta < 1$. Then the baryon density is

$$\nu = \frac{2\pi}{3} \left(1 - \frac{2\alpha_s}{\pi}\right) (\delta^2 + 1) T^3 \delta \quad . \tag{5.27}$$

Since $\delta < 1$ by assumption, we neglect δ^2 against 1, that is

$$\nu \approx 1.4 \, T^3 \delta \quad , \quad \frac{\mu_q}{\pi T} = \delta < 1 \quad , \quad \alpha_s = .55 \quad . \tag{5.28}$$

At $T = 160$ MeV we verify that (6.2) leads to $\nu(160) = 3/4\delta/\text{fm}^3$. Hence $\nu = 2\nu_0$ implies $\delta = 2/9$ in agreement with our prior assumption of a small δ (ν_0 is the normal baryon density in nuclei, $\nu_0 = 1/6 \, 1/\text{fm}^3$). Thus, as long as we are interested in the domain $T \gtrsim 160$ MeV, $\nu/\nu_0 \lesssim 2.6$ we are allowed to use eq (5.28) to replace μ (i.e., δ) by ν in the expressions for the energy density, eq (2.34) and entropy density eq (2.37). We find

$$s \approx 6.55 \, \frac{\nu^2}{T^3} + 8.12 \, T \tag{5.29}$$

$$\varepsilon \approx 9.82 \, \frac{\nu^2}{T^2} + 6.1 \, T^4 + B = \frac{3}{2} \, (6.55) \, \frac{\nu^2}{T^2} + \frac{3}{4} \, (8.12) \, T^4 + B \quad , \tag{5.30}$$

where in the last equality we emphasise the relation to eq (5.29). It is interesting to note that at constant ν both s and ε have a minimum at the same value of T, which is

$$0 = \left.\frac{\partial s}{\partial T}\right|_\nu \to T_\nu = 0.965 \, \nu^{1/3} = 105 \text{ MeV } (\nu/\nu_0)^{1/3} \quad . \tag{5.31}$$

Thus we find that both entropy density __and__ energy density decrease as the temperature decreases from its initial value around 200 MeV. This supports the proposition of pion radiation from the plasma at constant baryon density, at least until the minimum value T_ν of the temperature is approached. For $\nu/\nu_0 = 2.5$ we have $T_{2.5} = 142$ MeV while apparently $T_1 = 105$ MeV.

We can further evaluate how much entropy each radiated pion removes from the plasma and how much must be generated in the radiation process. Here we recall that each pion carries away about 3T MeV of energy and as a Boltzman particle carries four units of entropy . From eqs (5.29) and (5.30) we find

$$s/\epsilon = \frac{1}{T}\frac{3}{2}\frac{6.55\ \nu^2/T^3 + 8.12\ T^3}{6.55\ \nu^2/T^3 + \frac{3}{4}\ 8.12\ T^3 + B/T}$$

$$\approx 4\ \frac{1}{3T}\ [1 - \frac{6.55\ \nu^2}{8.12\ T^6} + \cdots - 0.094\ B/T^4]$$

$$\approx 4\ \frac{1}{3T}\ [1 - 1.6\ \delta^2 + \cdots - 0.094\ (B^{1/4}/T)^4]\ , \qquad (5.32)$$

Since $B^{1/4} < T$ and $\delta^2 \ll 1$, to the precision of our approximation ($\delta < 1$) we find that when lowering the energy of the plasma by 3T (i.e., by about the energy of the radiated pion) we must lower its entropy content by about four units. As this is exactly the same as the entropy content of an emitted pion we conclude that the pion radiation is not a strongly entropy generating process, as it should be, in order for it to proceed without impediment.

Some information about the evolution of the plasma volume can be derived from the first law of thermodynamics (b is the baryon number)

$$dE = -PdV + TdS + \mu db\ \ .$$

Let us now finally, but briefly consider the question: is the transition hadronic-gas ↔ quark-gluon plasma in principle a phase transition or is it only a transformation that is a change in the nature of hadronic matter, not associated with any kind of singularity of the partition function in the limit of infinite volume? The spirit of the theoretical approaches taken here requires a first order transition. However, this conclusion is only preliminary. Contrary arguments can be found [61] by arguing that only a finite number of incompressible hadrons can be fitted into a given volume. Here it turns out that one must very carefully study the meaning of thermo-dynamical limits (see references [9d] and [62]). Even worse: for compressible individual hadrons we might find a second order phase transition. We see that the phenomenological hadronic gas theory and the quark bag picture have little predictive power about the initial behavior. It is likely that only the experiment will help us under-stand this important aspect of strong interactions. Numerical digital experiments concluded recently allow one to believe that we have been on the right track with our description of the hadronic world. However

even these elaborate numerical experiments seem to lose their predictive power when fermions are included on the lattice. Some numerical
experimental groups continue to see a phase transition of first order,
while others claim only to observe a transformation. It is unfortunate
that we probably will have to wait for some years for the needed actual
physical experiments - and we hope that as soon as the just mentioned
required experimental advances occur, we will review and re-analyze its
content hoping to unravel this important question.

6. SUMMARY

These lectures aimed to provide an overview of the theory of
highly excited hadronic matter. By considering matter in kinetic and
chemical equilibrium we have been able to develop a thermodynamic
description valid for high temperatures. In the present work we have
described two physically different domains: the hadronic gas phase, in
which individual hadrons can exist as separate entities, but are sometimes combined to larger hadronic clusters; and the quark-gluon plasma,
where individual hadrons dissolve into one large cluster consisting of
the hadron constitutents. Our emphasis has been on the world of quarks,
which is the more fundamental approach.

In order to obtain a theoretical description of both phases we
have used some "common" knowledge and plausible interpretations of the
currently available experimental observations. The obtained equations
of state of hadronic matter, of course, reflect in certain ways on what
we have included in our considerations. It is the quantitative nature
of our work that allows a detailed comparison with experiment. It is
important to observe that the predicted temperatures and mean transverse momenta of particles agree with the experimental results available at $E_{k,lab}/A = 2$ GeV [BEVELAC] and at 100 GeV [ISR] as far as a
comparison is permitted [4a].

The internal theoretical consistency of our description of the gas
phase leads us in a straightforward fashion to the postulate of a first
order phase transition to the quark-gluon plasma. In order to describe
this phase in addition to the standard Lagrangian quantum field theory
of "weakly" interacting particles at finite temperature and density, we

also introduce the phenomenological vacuum pressure and energy density B. This term is required in a consistent theory of hadronic structure. It turns out that $B^{1/4} \sim 150$ MeV is just, to within 20%, the temperature of the plasma phase before condensation into hadrons. This is similar to the well known Hagedorn temperature $T_0 \cong 160$ MeV.

An interesting aspect of our studies is the realization that the transition to the quark-gluon plasma will occur at a lower baryon density for highly excited hadronic matter than for matter in the ground state (T = 0). Using the currently accepted value for B we find that at $\nu \sim 2 - 3 \nu_0$, T = 150 MeV, the plasma phase may indeed already be formed.

One of the possible plasma signatures is the enormous strangeness abundance in the plasma. We show that gluons have a sufficiently large reaction rate for the strangeness abundance in the plasma to reach chemical equilibrium during the lifetime of the plasma formed in high-energy nuclear collisions. The subsequent depletion of the strangeness during the plasma hadronization as well as its preferred hadronization channels are currently being studied in detail. However, only if the plasma hadronization is an extremely slow process, lasting on the order of 10^{-22} sec., a significant depletion of the high s-abundance created at the maximal temperature reached in the collision can be anticipated. As shown in figure 3.3 the reaction rates drop quite rapidly with decreasing temperature, leading to a rapid increase of the equilibrium time constant τ. Hence the strangeness abundance decouples from the equilibrium and remains a witness of the hot collision period. We have further shown that strangness may be a useful experimental trigger on plasma formation.

It is apparent from our results that the measurement of production cross sections of anti-strange baryons already could be quite helpful in the observation of the phase transition. The high suppression of these degrees of freedom in the hadronic gas phase is not maintained in the plasma phase where the $\bar{s}$ abundance is larger than the $\bar{u}, \bar{d}$ abundances. A measurement of the relative K^+/K^- yield, while indicative of the value of the chemical potential in a hot nuclear gas may carry less specific information about the plasma. The K/π ratio may also contain relevant information. However, since the π originates from diverse sources its abundance is controlled by the total entropy created in N-N collisions. Hence, it will be much more difficult to decipher the message. Perhaps a steep rise of K/π ratio at high $p_\perp$ could be helpful here.

On the other hand it appears that the abundances of otherwise quite rare strange hadrons will be enhanced, on the one hand by the relatively high phase space density of strangeness in the plasma, on the other hand in view of the attractive ss-QCD interaction in the $\bar{3}_c$ and $\bar{s}s$ in 1_c channels. Hence we should search for the strangeness abundance in the yields of particles like $\Xi, \bar{\Xi}, \Omega, \bar{\Omega}, \phi$, rather than in the K-channels. It may be that such experiments would uniquely determine the existence, and eventually the characteristics, of the phase transition to the quark-gluon plasma.

It is important to appreciate that the experiments discussed above would certainly be quite complementary to the measurements utilizing electromagnetically interacting probes, e.g., di-leptons or direct photons. Strangeness-based measurements have the advantage over the measurement of the electromagnetic particles in that they involve the observation of a <u>strongly interacting particle</u> ($s, \bar{s}$ quark) which happened to be a direct constituent of the hot plasma phase.

Finally, we have described the radiation cooling of the plasma by the emission of pions, and the conditions under which a plasma seed will grow in a high energy nuclear collision. Using the value T = 150-160 MeV as the phase transition temperature we find that 15 to 20 GeV/nucleon on a fixed target should suffice to lead to a quark-qluon plasma. The frequency of such events is controlled by the probability that a plasma seed of adequate size be formed early in the collision. A plasma seed is needed since the hadron cross sections are too small to lead to energy confinement for the required length of time. The cross sections in a plasma can be much larger since the phase space there is much larger than in the hadronic gas phase owing to the absence of the color constraint; for example, a 3 quark system in the plasma has 27 color states while only the color singlet is permitted in the hadronic phase. We also have briefly discussed the possibility of a phase transition plasma → hadronic gas.

ACKNOWLEDGEMENTS

We would like to thank all who have, through collaboration or stimulating discussions helped us in our study of this research field: H.-Th. Elze, W. Greiner, R. Hagedorn, P. Koch, B. Müller, H. Rafelski, and G. Staadt. Several research periods at CERN-Theory Division by one of us (J.R.) have contributed essentially to the theoretical developments presented here.

REFERENCES

1. a. J. Rafelski and M. Danos, "Perspectives in High Energy Nuclear Collisons," NBSIR 83-2725, Washington, D.C. 1983.
    b. Workshop on Future Relativistic Heavy Ion Experiments, Proceedings edited by R. Stock and R. Bock, GSI 81-6, Orange Report 1981.
    c. Workshop on Quark Matter Formation and Heavy Ion Collisions, Proceedings edited by M. Jacob and H. Satz, World Scientific Publ. Co., Singapore 1982.

2. An incomplete list of quark-gluon plasma papers includes:
    a. B. A. Freedman and L. D. McLerran, Phys. Rev. D16 (1977) 1169;
    b. S. A. Chin, Phys. Lett. 78B (1978) 552;
    c. P. D. Morley and M. B. Kislinger, Phys. Rep. 51 (1979) 63;
    d. J. I. Kapusta, Nucl. Phys. B148 (1979) 461;
    e. O. K. Kalashnikov and V. V. Kilmov, Phys. Lett. 88B (1979) 328;
    f. E. V. Shuryak, Phys. Lett. 81B (1979) 65; also Phys. Rep. 61 (1980) 71;
    g. J. Rafelski and R. Hagedorn "From Hadron Gas to Quark Matter II," in Thermodynamics of Quarks and hadrons, Ed. H. Satz, North Holland, Amsterdam 1981.
    h. J. Rafelski, H.-Th. Elze, and R. Hagedorn, "Hot Hadronic and Quark Matter in p̄-Annihilation on Nuclei," CERN Preprint TH2912 (1980), in Proceedings of 5th European Symposium on Nucleon-Antinucleon Interactions, Bressanone 1980, CLEUP, Padua, 1980

3. a. G. Domokos and J. I. Goldman, Phys. Rev. D23 (1981) 203;
    b. K. Kajantie and H. I. Mietinnen, Z. Phys. C9 (1981) 341.
    c. K. Kajantie and H. I. Mietinnen, Z. Phys. C14 (1982) 357.

4. a. J. Rafelski, "Extreme States of Nuclear Matter" in reference [1], p. 282; also Universität Frankfurt Preprint UFTP 52/1981;
    b. J. Rafelski and B. Müller, Phys. Ref. Lett. 48 (1982) 1066;
    c. P. Koch, J. Rafelski, and W. Greiner, Phys. Lett. 123B (1983), 151.
    d. J. Rafelski, "Strangeness in Quark-Gluon Plasma," Universität Frankfurt preprint UFTP86/1982.
    e. J. Rafelski, Nucl. Phys. A in print 1984.

5. J. Rafelski, "Hot Hadronic Matter" in New Flavours and Hadron Spectroscopy, Editions Frontières 1981, J. Tran Thanh Van, editor, page 619.

6. a. J. Kogut, M. Stone, H. Wyld, J. Shigemitsu, S. Shenker and D. Sinclair, Phys. Rev. Lett $\underline{48}$ (1982) 114.
    b. J. Ellis, "Phenomenology of Unified Gauge Theories" CERN-preprint TH 3174.
    c. L. Van Hove, "Very Dense States of Matter in Particle Physics and Early Cosmology," lecture at the Institut-Lorentz, 1981/1982, Leiden, Netherlands.

7. a. V. A. Rubakov, JETP Lett $\underline{33}$ (1981) 699, and Nucl. Phys. $\underline{B203}$ (1982) 311.
    b. C. G. Callan, Jr., Nucl. Phys. $\underline{B212}$ (1983) 391.
    c. C. G. Callan, "Catalysis of Baryon Decay," lecture at the Magnetic Monopoles Conference, Racine, Wisconsin, October 1982.
    d. G. Schmidt, W. Greiner, and J. Rafelski, "Quasars: Convective giant stars with quark gluon plasma core," University of Cape Town preprint UCT-TP 5/84.

8. These ideas originate in Hagedorn's statistical bootstrap theory, see:
    R. Hagedorn, Suppl. Nuovo Cimento $\underline{3}$ (1964) 147; and Nuovo Cimento $\underline{6}$ (1968) 311, also
    R. Hagedorn, "How to Deal with Relativistic Heavy Ion Collisions," p. 236 in ref. [1].

9. a. R. Hagedorn and J. Rafelski, Phys. Lett. $\underline{97B}$ (1980) 136.
    b. The extension of statistical bootstrap to finite baryon number and volume has been introduced in:
    R. Hagedorn, I. Montvay, and J. Rafelski, Lecture at Erice Workshop "Hadronic Matter at Extreme Energy Density," edited by N. Cabibbo, Plenum Press, New York (1980) p. 49.
    c. R. Hagedorn and J. Rafelski, Manuscript in preparation for Physics Reports.
    d. R. Hagedorn, Z. Phys. $\underline{C17}$ (1983) 265 (CERN Preprint TH-3392).

10. a. A. Chodos, R. L. Jaffe, K. Johnson, C. B. Thorn, V. F. Weisskopf, Phys. Rev. $\underline{D9}$ (1974) 3471.
    b. K. Johnson, Acta Phys. Polon. $\underline{B6}$ (1975) 865; and
    c. T. de Grand, R. L. Jaffe, K. Johnson, and J. Kiskis, Phys. Rev. $\underline{D12}$ (1975) 2060.

11. a. R. Anishetty, P. Koehler, and L. McLerran, Phys. Rev. D22 (1980) 2793.
    b. J. D. Bjorken, Phys. Rev. D27 (1983) 140.

12. M. Danos and J. Rafelski, "Formation of Quark-Gluon Plasma at Central Rapidity," University of Frankfurt preprint UFTP 94/1982.

13. a. S. A. Chin, Phys. Lett 119B (1982) 51.
    b. K. Kajantie and L. McLerran, "Energy Densities, Initial Conditions and Hydrodynamic Equations for Ultrarelativistic Nucleus-Nucleus Collisions," University of Helsinki Report HU-TPT-82-30.

14. M. Danos and J. Rafelski, Phys. Rev. D27 (1983) 671. See also, M. Danos and J. Rafelski, "Pion Radiation by Hot Quark-Gluon Plasmas" CERN preprint TH-3607.

15. S. Gasiorowicz and J. L. Rosner, "Hadron Spectra and Quarks," Am. J. Phys. 49 (1981) 954.

16. E. S. Abers and B. W. Lee, "Gauge Theories," Phys. Rep. 9C (1973) 1.

17. P. Soding and G. Wolf, "Experimental Evidence on QCD," DESY-Report 81-013; (1981); also, Ann. Rev. Nuc. Part. Sci. (1982).

18. a. P. Langacker and H. Pagels, Phys. Rev D19 (1979) 2070, and references therein.
    b. S. Narison, N. Paver, E. de Rafael, and D. Treleani, Nucl. Phys. B212 (1983) 365.

19. R. D. Viollier and J. Rafelski, "Quarkonium Spectra in the Framework of Quantum Chromodynamics," Helv. Phys. Acta 53 (1980) 352.

20. A. Martin, "Heavy Quark Systems," CERN-TH 3162, appeared in Proceedings of the Int. Conf. on High Energy Physics, Lisbon, July 1981.

21. a. H. D. Politzer, Phys. Rev. Lett. 30 (1973) 1346.
    b. D. Gross and F. Wilczek, Phys. Rev. Lett. 30 (1973) 1343.
    c. W. Marciano and H. Pagels, "Quantum Chromodynamics," Phys. Rep. 36C (1978) 137.

22. a. K. Johnson, "A Simple Model of the Ground State of Quantum
         Chromodynamics," SLAC-Publ 2436 (1979), in AIP Conf. Proc.
         No. 59, Am. Inst. of Phys., N.Y. 1979.
     b. M. Danos, D. Gogny, and D. Iracani, "Simple model of the QCD
         Vacuum," NBSIR 83-2759.

23. J. Rafelski, "Particle Condensates in Strongly Coupled Quantum
     Field Theory," UFTP Preprint 67/1981 in Quantum Electrodynamics of
     Strong Fields, edited by W. Greiner, D. Reidel Publ. Co. (1981).

24. P. Hasenfratz, R. R. Horgan, J. Kuti, and J. M. Richard, Phys.
     Lett. 95B (1980) 299.

25. See, e.g., M. A. Shifman, Z. Phys. 69 (1981) 347 and references
therein.

26. G. Peressutti and B.-S. Skagerstam, "Hydrodynamics and the Bag
     Model," Phys. Ref D18 (1978) 4304.

27. See, e.g., R. R. Feynman, "Statistical Mechanics," W. A. Benjamin,
     Inc. 1972.

28. H.-Th. Elze, W. Greiner, and J. Rafelski, "The Relativistic Ideal
     Fermi Gas Revisited," J. Phys. G6 (1980) L149.

29. a. See, T. Barnes, F. E. Close, and F. deViron, "QQg
         Hermaphrodite mesons in the MIT bag model," Rutherford
         preprint RL-82-088 T.311 and references therein.
     b. D. Robson, "Toroidal Bags," Z. Physik C3 (1980) 199.

30. a. S. J. Lindenbaum, in Proceedings XVIth Rencounte de Moridae -
         "New Flavours and Hadron Sopectroscopy," edited by J. Tran
         Thanh Van, Frontieres (1981) p. 187.
     b. See, J. Donoghue, and H. Gomm, Phys. Lett. 112B (1982) 409,
         and references therein.
     c. D. G. Aschman, (Crystal Ball collaboration) "Possible Gluonium
         States in Radiative ψ Decay," in Proc. of XVIIth Recontre de
         Moriond, Workshop on New Flavors, Les Arcs, France, 1982.

31. P. Carruthers, "Role of the Phonon in the QCD Plasma," Los Alamos
     preprint LA-UR-83-130.

32. C.-G. Kallman and C. Montonen, Phys. Lett. 115B (1982) 473.

33. H.-Th. Elze, W. Greiner, and J. Rafelski, Phys. Lett. 124B (1983) 515; and H.-Th. Elze, W. Greiner, and J. Rafelski, "Frozen Color Degrees of Freedom," UFTP preprint 126/1984 and UCT-TP 6/84.

34. a. K. Redlich and L. Turko, Z. Phys. C5 (1980) 201,
    b. L. Turko, Phys. Lett. 104B (1981) 153.

35. J. Rafelski an M. Danos, Phys. Lett 97B (1980) 279.

36. For a discussion of the Haar measure we recommend G. Rosen, "Formulations of Classical and Quantum Dynamical Theory," Academic Press, New York 1969, Appendix C; or J. P. Elliot and P. G. Dawber, "Symmetry in Physics," Vols. 1/2, McMillan, London 1979, Appendix A.4.3.

37. B. L. Combridge, Nucl. Phys. B151 (1979) 429.

38. T. S. Biro, J. Zimanyi, Phys. Lett. 113B (1982) 6.

39. From the study of the process $q\bar{q} \rightarrow s\bar{s}$ it was initially concluded [48] that strangness would not saturate. The authors of ref. [48] have since corrected their calculations: T. S. Biro, J. Zimanyi, Nucl. Phys. A395 (1983) 525 and agree now in their results with those given here, as originally presented in ref. [4b].

40. P. Koch, Diploma Thesis Frankfurt 1983 (unpublished).

41. A. Z. Mekjian, Nucl. Phys A384 (1982) 492.

42. P. Koch and J. Rafelski, in preparation.

43. W. Greiner, P. Koch, and J. Rafelski, "Strange Particle Production in pp and pN reactions," CERN preprint TH-3781.

44. M. Bourquin et al, Nucl. Phys B153 (1970) 13.
    M. Bourquin et al, Z. Phys. C5 (1980) 275.
    S. F. Biagi et al., Phys. Lett 122B (1983) 455.

45. Oxygen impinging at PS-energies (9-13GeV) on heavy targets is actively considered as a next step in the study of hot nuclear matter. See the GSI proposal reproduced in ref. (1b) page 557.

46. a) J. Rafelski, Phys. Lett. 91B (1980) 281.
    b) J. Rafelski, "Quark-Gluon Plasma in $\bar{p}$ - Annihilation on Nuclei," UFTP preprint 76/1982; in Proceedings of Workshop on Physics at LEAR, Erice, May 1982.

47. B. Müller and J. Rafelski, Phys. Lett 116B (1982) 274.

48. B. Y. Oh, P. S. Eastman, Z. Ming Ma, D. L. Parker, G. A. Smith and
    R. J. Sprafka, Nucl. Phys. B51 (1973) 57.

49. P. S. Estman, Z. Ming Ma, B. Y. Oh, D. L. Parker, G. A. Smith nd
    R. J. Sprafka, Nucl. Phys B51 (1973) 29.

50. B. Y. Oh and G. A. Smith, Nucl. Phys. B40 (1972) 151.

51. M. A. Mandelkern, L. R. Price, J. Schulz and D. W. Smith, Phys.
    Rev. D27 (1983) 19.

52. O. Braun, V. Hepp, H. Strobele, and W. Wittek, Nucl. Phys. B160
    (1979) 467.

53. Ch. Derret, H.-Th. Elze, P. Koch, W. Greiner, and J. Rafelski, in
    preparation.

54. a. W. Busza and A. S. Goldhaber, MIT Reporty No. MIT/LNS/CSC 82-2
       (unpublished) and Phys. Lett. (in print).
    b. D. S. Barton et al. Phys. Rev. D27, (1983) 2580.

55. G. B. Yodh, "Review of Recent Developments in Cosmic Ray
    Experiments at Very High Energies," p. 213-236 in reference [1c]
    and references therein.

56. UA1-Collaboration-CERN, G. Arnison, et al., Phys. Lett. 123B
    (1983) 108.

57. M. Danos and R. K. Smith, unpublished; see also, A. Ausden,
    J. N. Ginochio, F. H. Harlow, J. R. Nix, M. Danos, E. E. Halbert,
    and R. K. Smith, Phys. Rev. Lett. 38 (1977) 1055.

58. A. W. Thomas, J. Phys. G7 L283 (1981).

59. A. Schnabel and J. Rafelski, in preparation.

60. B. Banerjee, N. K. Glendening and T. Matsui "Fission of a
    Chromoelectric Flux Tube and Meson Radiation from a Quark-Gluon
    Plasma," Prepriunt LBL 15349.

61. F. Karsch and H. Satz, Phys. Rev. 21D (1980) 1168.

62. R. Hagedorn, "On a Possible Phase Transition Between Hadron Matter
    and Quark-Gluon Matter" CERN Preprint TH-3207.

PARTICIPANTS

BEDFORD, D	University of Natal, Durban.
BLELOCH, A.L.	University of the Witwatersrand, Johannesburg.
CARTER, J.	University of the Witwatersrand, Johannesburg.
COLE, B.J.	University of the Witwatersrand, Johannesburg.
COWLEY, A.A.	National Accelerator Centre, CSIR, Faure.
DAVIS, E.D.	University of the Witwatersrand, Johannesburg.
DELIC, G.	University of the Witwatersrand, Johannesburg.
DE WET, J.A.	Institute for Basic Research, "Mount Marlow" Witmos.
DONOVAN, S.J.	University of the Witwatersrand, Johannesburg.
EGGERS, H.C.	University of Pretoria.
ENGELBREGHT, C.A.	University of Stellenbosch.
EYRE, D.	NRIMS - CSIR, Pretoria.
FOLSCHER, G.C.K.	University of the Witwatersrand, Johannesburg.
FRIEDLAND, E.	University of Pretoria.
GADINABOKAO, W.L.	University of Bophuthatswana, Mafikeng.
GAVIN, MS E.J.O.	University of Cape Town, Rondebosch.
GERING, M.	University of the Witwatersrand, Johannesburg.
GEYER, H.B.	NUCOR, Pretoria.
GREBEN, J.M.	University of Alberta, Alberta, Canada.
HAHNE, F.J.W.	University of Stellenbosch.
HEISS, W.D.	University of the Witwatersrand, Johannesburg.
HENNING, J.J.	NUCOR, Pretoria.
HEYMANN, G.	CSIR, Pretoria.
HNIZDO, V.	University of the Witwatersrand, Johannesburg.
HOFMEYR, C.	NUCOR, Pretoria.
KAUFFMANN, S.K.	University of Cape Town, Rondebosch.
KOEN, J.W.	University of Stellenbosch.
KRANOLD, H.U.	NUCOR, Pretoria.

LITTLEWORT, Miss G.C.   University of Cape Town, Rondebosch.

LOCKETT, R.   University of Natal, Durban.

McMURRAY, W.R.   National Accelerator Centre, CSIR, Faure.

MILLER, H.G.   NRIMS - CSIR, Pretoria.

MILLS, S.J.   National Accelerator Centre, CSIR, Faure.

NAUDE, W.J.   University of Stellenbosch.

PROZESKY, V.M.   University of Pretoria.

QUICK, Miss R.M.   NRIMS - CSIR, Pretoria.

RAUTENBACH, W.L.   University of Stellenbosch.

RICHTER, W.A.   University of Stellenbosch.

SCHOLTZ, F.G.   NUCOR, Pretoria.

SELLSCHOP, J.P.F.   University of the Witwatersrand, Johannesburg.

VAN DER MERWE, P du T.   NUCOR, Pretoria.

VAN RENSBURG,
    E. J. Janse   University of the Witwatersrand, Johannesburg.

VAN RENSBURG,
    M. P. Janse   University of Stellenbosch.

VIOLLIER, R.D.   University of Cape Town, Rondebosch.

WHITE, H.   University of the Witwatersrand, Johannesburg.

WHITTAL, D.M.   National Accelerator Centre, CSIR, Faure.

Lecture Notes in Physics

Vol. 195: Trends and Applications of Pure Mathematics to Mechanics. Proceedings, 1983. Edited by P. G. Ciarlet and M. Roseau. V, 422 pages. 1984.

Vol. 196: WOPPLOT 83. Parallel Processing: Logic, Organization and Technology. Proceedings, 1983. Edited by J. Becker and I. Eisele. V, 189 pages. 1984.

Vol. 197: Quarks and Nuclear Structure. Proceedings, 1983. Edited by K. Bleuler. VIII, 414 pages. 1984.

Vol. 198: Recent Progress in Many-Body Theories. Proceedings, 1983. Edited by H. Kümmel and M. L. Ristig. IX, 422 pages. 1984.

Vol. 199: Recent Developments in Nonequilibrium Thermodynamics. Proceedings, 1983. Edited by J. Casas-Vázquez, D. Jou and G. Lebon. XIII, 485 pages. 1984.

Vol. 200: H. D. Zeh, Die Physik der Zeitrichtung. V, 86 Seiten. 1984.

Vol. 201: Group Theoretical Methods in Physics. Proceedings, 1983. Edited by G. Denardo, G. Ghirardi and T. Weber. XXXVII, 518 pages. 1984.

Vol. 202: Asymptotic Behavior of Mass and Spacetime Geometry. Proceedings, 1983. Edited by F. J. Flaherty. VI, 213 pages. 1984.

Vol. 203: C. Marchioro, M. Pulvirenti, Vortex Methods in Two-Dimensional Fluid Dynamics. III, 137 pages. 1984.

Vol. 204: Y. Waseda, Novel Application of Anomalous (Resonance) X-Ray Scattering for Structural Characterization of Disordered Materials. VI, 183 pages. 1984.

Vol. 205: Solutions of Einstein's Equations: Techniques and Results. Proceedings, 1983. Edited by C. Hoenselaers and W. Dietz. VI, 439 pages. 1984.

Vol. 206: Static Critical Phenomena in Inhomogeneous Systems. Edited by A. Pękalski and J. Sznajd. Proceedings, 1984. VIII, 358 pages. 1984.

Vol. 207: S. W. Koch, Dynamics of First-Order Phase Transitions in Equilibrium and Nonequilibrium Systems. III, 148 pages. 1984.

Vol. 208: Supersymmetry and Supergravity/Nonperturbative QCD. Proceedings, 1984. Edited by P. Roy and V. Singh. V, 389 pages. 1984.

Vol. 209: Mathematical and Computational Methods in Nuclear Physics. Proceedings, 1983. Edited by J. S. Dehesa, J. M. G. Gomez and A. Polls. V, 276 pages. 1984.

Vol. 210: Cellular Structures in Instabilities. Proceedings, 1983. Edited by J. E. Wesfreid and S. Zaleski. VI, 389 pages. 1984.

Vol. 211: Resonances – Models and Phenomena. Proceedings, 1984. Edited by S. Albeverio, L. S. Ferreira and L. Streit. VI, 369 pages. 1984.

Vol. 212: Gravitation, Geometry and Relativistic Physics. Proceedings, 1984. Edited by Laboratoire "Gravitation et Cosmologie Relativistes", Université Pierre et Marie Curie et C.N.R.S., Institut Henri Poincaré, Paris. VI, 336 pages. 1984.

Vol. 213: Forward Electron Ejection in Ion Collisions. Proceedings, 1984. Edited by K. O. Groeneveld, W. Meckbach and I. A. Sellin. VII, 165 pages. 1984.

Vol. 214: H. Moraal, Classical, Discrete Spin Models. VII, 251 pages. 1984.

Vol. 215: Computing in Accelerator Design and Operation. Proceedings, 1983. Edited by W. Busse and R. Zelazny. XII, 574 pages. 1984.

Vol. 216: Applications of Field Theory to Statistical Mechanics. Proceedings, 1984. Edited by L. Garrido. VIII, 352 pages. 1985.

Vol. 217: Charge Density Waves in Solids. Proceedings, 1984. Edited by Gy. Hutiray and J. Sólyom. XIV, 541 pages. 1985.

Vol. 218: Ninth International Conference on Numerical Methods in Fluid Dynamics. Edited by Soubbaramayer and J. P. Boujot. X, 612 pages. 1985.

Vol. 219: Fusion Reactions' Below the Coulomb Barrier. Proceedings, 1984. Edited by S. G. Steadman. VII, 351 pages. 1985.

Vol. 220: W. Dittrich, M. Reuter, Effective Lagrangians in Quantum Electrodynamics. V, 244 pages. 1985.

Vol. 221: Quark Matter '84. Proceedings, 1984. Edited by K. Kajantie. VI, 305 pages. 1985.

Vol. 222: A. García, P. Kielanowski, The Beta Decay of Hyperons. Edited by A. Bohm. VIII, 173 pages. 1985.

Vol. 223: H. Saller, Vereinheitlichte Feldtheorien der Elementarteilchen. IX, 157 Seiten. 1985.

Vol. 224: Supernovae as Distance Indicators. Proceedings, 1984. Edited by N. Bartel. VI, 226 pages. 1985.

Vol. 225: B. Müller, The Physics of the Quark-Gluon Plasma. VII, 142 pages. 1985.

Vol. 226: Non-Linear Equations in Classical and Quantum Field Theory. Proceedings, 1983/84. Edited by N. Sanchez. VII, 400 pages. 1985.

Vol. 227: J.-P. Eckmann, P. Wittwer, Computer Methods and Borel Summability Applied to Feigenbaum's Equation. XIV, 297 pages. 1985.

Vol. 228: Thermodynamics and Constitutive Equations. Proceedings, 1982. Edited by G. Grioli. V, 257 pages. 1985.

Vol. 229: Fundamentals of Laser Interactions. Proceedings, 1985. Edited by F. Ehlotzky. IX, 314 pages. 1985.

Vol. 230: Macroscopic Modelling of Turbulent FLows. Proceedings, 1984. Edited by U. Frisch, J. B. Keller, G. Papanicolaou and O. Pironneau. X, 360 pages. 1985.

Vol. 231: Hadrons and Heavy Ions. Proceedings, 1984. Edited by W. D. Heiss. VII, 458 pages. 1985.

Selected Issues from

Lecture Notes in Mathematics

Vol. 932: Analytic Theory of Continued Fractions. Proceedings, 1981. Edited by W.B. Jones, W.J. Thron, and H. Waadeland. VI, 240 pages. 1982.

Vol. 934: M. Sakai, Quadrature Domains. IV, 133 pages. 1982.

Vol. 935: R. Sot, Simple Morphisms in Algebraic Geometry. IV, 146 pages. 1982.

Vol. 936: S.M. Khaleelulla, Counterexamples in Topological Vector Spaces. XXI, 179 pages. 1982.

Vol. 937: E. Combet, Integrales Exponentielles. VIII, 114 pages. 1982.

Vol. 938: Number Theory. Proceedings, 1981. Edited by K. Alladi. IX, 177 pages. 1982.

Vol. 942: Theory and Applications of Singular Perturbations. Proceedings, 1981. Edited by W. Eckhaus and E.M. de Jager. V, 363 pages. 1982.

Vol. 953: Iterative Solution of Nonlinear Systems of Equations. Proceedings, 1982. Edited by R. Ansorge, Th. Meis, and W. Törnig. VII, 202 pages. 1982.

Vol. 956: Group Actions and Vector Fields. Proceedings, 1981. Edited by J.B. Carrell. V, 144 pages. 1982.

Vol. 957: Differential Equations. Proceedings, 1981. Edited by D.G. de Figueiredo. VIII, 301 pages. 1982.

Vol. 963: R. Nottrot, Optimal Processes on Manifolds. VI, 124 pages. 1982.

Vol. 964: Ordinary and Partial Differential Equations. Proceedings, 1982. Edited by W.N. Everitt and B.D. Sleeman. XVIII, 726 pages. 1982.

Vol. 968: Numerical Integration of Differential Equations and Large Linear Systems. Proceedings, 1980. Edited by J. Hinze. VI, 412 pages. 1982.

Vol. 970: Twistor Geometry and Non-Linear Systems. Proceedings, 1980. Edited by H.-D. Doebner and T.D. Palev. V, 216 pages. 1982.

Vol. 972: Nonlinear Filtering and Stochastic Control. Proceedings, 1981. Edited by S.K. Mitter and A. Moro. VIII, 297 pages. 1983.

Vol. 978: J. Ławrynowicz, J. Krzyż, Quasiconformal Mappings in the Plane. VI, 177 pages. 1983.

Vol. 979: Mathematical Theories of Optimization. Proceedings, 1981. Edited by J.P. Cecconi and T. Zolezzi. V, 268 pages. 1983.

Vol. 982: Stability Problems for Stochastic Models. Proceedings, 1982. Edited by V.V. Kalashnikov and V.M. Zolotarev. XVII, 295 pages. 1983.

Vol. 989: A.B. Mingarelli, Volterra-Stieltjes Integral Equations and Generalized Ordinary Differential Expressions. XIV, 318 pages. 1983.

Vol. 994: J.-L. Journé, Calderón-Zygmund Operators, Pseudo-Differential Operators and the Cauchy Integral of Calderón. VI, 129 pages. 1983.

Vol. 999: C. Preston, Iterates of Maps on an Interval. VII, 205 pages. 1983.

Vol. 1000: H. Hopf, Differential Geometry in the Large, VII, 184 pages. 1983.

Vol. 1003: J. Schmets, Spaces of Vector-Valued Continuous Functions. VI, 117 pages. 1983.

Vol. 1005: Numerical Methods. Proceedings, 1982. Edited by V. Pereyra and A. Reinoza. V, 296 pages. 1983.

Vol. 1007: Geometric Dynamics. Proceedings, 1981. Edited by J. Palis Jr. IX, 827 pages. 1983.

Vol. 1015: Equations différentielles et systèmes de Pfaff dans le champ complexe – II. Seminar. Edited by R. Gérard et J.P. Ramis. V, 411 pages. 1983.

Vol. 1021: Probability Theory and Mathematical Statistics. Proceedings, 1982. Edited by K. Itô and J.V. Prokhorov. VIII, 747 pages. 1983.

Vol. 1031: Dynamics and Processes. Proceedings, 1981. Edited by Ph. Blanchard and L. Streit. IX, 213 pages. 1983.

Vol. 1032: Ordinary Differential Equations and Operators. Proceedings, 1982. Edited by W.N. Everitt and R.T. Lewis. XV, 521 pages. 1983.

Vol. 1035: The Mathematics and Physics of Disordered Media. Proceedings, 1983. Edited by B.D. Hughes and B.W. Ninham. VII, 432 pages. 1983.

Vol. 1037: Non-linear Partial Differential Operators and Quantization Procedures. Proceedings, 1981. Edited by S.I. Andersson and H.-D. Doebner. VII, 334 pages. 1983.

Vol. 1041: Lie Group Representations II. Proceedings 1982–1983. Edited by R. Herb, S. Kudla, R. Lipsman and J. Rosenberg. IX, 340 pages. 1984.

Vol. 1045: Differential Geometry. Proceedings, 1982. Edited by A.M. Naveira. VIII, 194 pages. 1984.

Vol. 1047: Fluid Dynamics. Seminar, 1982. Edited by H. Beirão da Veiga. VII, 193 pages. 1984.

Vol. 1048: Kinetic Theories and the Boltzmann Equation. Seminar, 1981. Edited by C. Cercignani. VII, 248 pages. 1984.

Vol. 1049: B. Iochum, Cônes autopolaires et algèbres de Jordan. VI, 247 pages. 1984.

Vol. 1054: V. Thomée, Galerkin Finite Element Methods for Parabolic Problems. VII, 237 pages. 1984.

Vol. 1055: Quantum Probability and Applications to the Quantum Theory of Irreversible Processes. Proceedings, 1982. Edited by L. Accardi, A. Frigerio and V. Gorini. VI, 411 pages. 1984.

Vol. 1057: Bifurcation Theory and Applications. Seminar, 1983. Edited by L. Salvadori. VII, 233 pages. 1984.

Vol. 1058: B. Aulbach, Continuous and Discrete Dynamics near Manifolds of Equilibria. IX, 142 pages. 1984.

Vol. 1059: Séminaire de Probabilités XVIII, 1982/83. Proceedings. Edité par J. Azéma et M. Yor. IV, 518 pages. 1984.

Vol. 1063: Orienting Polymers. Proceedings, 1983. Edited by J.L. Ericksen. VII, 166 pages. 1984.

Vol. 1065: A. Cuyt, Padé Approximants for Operators: Theory and Applications. IX, 138 pages. 1984.

Vol. 1066: Numerical Analysis. Proceedings, 1983. Edited by D.F. Griffiths. XI, 275 pages. 1984.

Vol. 1071: Padé Approximation and its Applications, Bad Honnef 1983. Prodeedings Edited by H. Werner and H.J. Bünger. VI, 264 pages. 1984.

Vol. 1072: F. Rothe, Global Solutions of Reaction-Diffusion Systems. V, 216 pages. 1984.

Vol. 1085: G.K. Immink, Asymptotics of Analytic Difference Equations. V, 134 pages. 1984.

Vol. 1086: Sensitivity of Functionals with Applications to Engineering Sciences. Proceedings, 1983. Edited by V. Komkov. V, 130 pages. 1984

Vol. 1100: V. Ivrii, The Precise Spectral Asymptotics for Elliptic Operators Acting in Fiberings over Manifolds with Boundary. V, 237 pages. 1984.